D1226717

Small Area Estimation

WILEY SERIES IN SURVEY METHODOLOGY
Established in part by Walter A. Shewhart and Samuel S. Wilks

Editors: *Robert M. Groves, Graham Kalton, J. N. K. Rao, Norbert Schwarz, Christopher Skinner*

A complete list of the titles in this series appears at the end of this volume.

Small Area Estimation

J. N. K. RAO
Carleton University

WILEY-
INTERSCIENCE

A JOHN WILEY & SONS, INC., PUBLICATION

Copyright © 2003 by John Wiley & Sons, Inc. All rights reserved.

Published by John Wiley & Sons, Inc., Hoboken, New Jersey.
Published simultaneously in Canada.

No part of this publication may be reproduced, stored in a retrieval system or transmitted in any form or by any means, electronic, mechanical, photocopying, recording, scanning or otherwise, except as permitted under Section 107 or 108 of the 1976 United States Copyright Act, without either the prior written permission of the Publisher, or authorization through payment of the appropriate per-copy fee to the Copyright Clearance Center, Inc., 222 Rosewood Drive, Danvers, MA 01923, (978) 750-8400, fax (978) 750-4744, or on the web at www.copyright.com. Requests to the Publisher for permission should be addressed to the Permissions Department, John Wiley & Sons, Inc., 111 River Street, Hoboken, NJ 07030, (201) 748-6011, fax (201) 748-6008, e-mail: permreq@wiley.com.

Limit of Liability/Disclaimer of Warranty: While the publisher and author have used their best efforts in preparing this book, they make no representation or warranties with respect to the accuracy or completeness of the contents of this book and specifically disclaim any implied warranties of merchantability or fitness for a particular purpose. No warranty may be created or extended by sales representatives or written sales materials. The advice and strategies contained herein may not be suitable for your situation. You should consult with a professional where appropriate. Neither the publisher nor author shall be liable for any loss of profit or any other commercial damages, including but not limited to special, incidental, consequential, or other damages.

For general information on our other products and services please contact our Customer Care Department within the U.S. at 877-762-2974, outside the U.S. at 317-572-3993 or fax 317-572-4002.

Wiley also publishes its books in a variety of electronic formats. Some content that appears in print, however, may not be available in electronic format.

Library of Congress Cataloging-in-Publication Data:

Rao, J. N. K., 1937–
Small area estimation / J.N.K. Rao.
p. cm. — (Wiley series in survey methodology)
Includes bibliographical references and index.
ISBN 0-471-41374-7 (cloth)
1. Sampling (Statistics) 2. Estimation theory. I. Title. II. Series.

QA276.6 .R344 2003
519.5'2—dc21 2002033197

Printed in the United States of America.

10 9 8 7 6 5 4 3 2 1

To my mother, Sakuntalamma

and

to my wife, Neela

Contents

List of Figures

List of Tables

Foreword

The history of modern sample surveys dates back to the nineteenth century, but the field did not fully emerge until the 1930. It grew considerably during the World War II, and has been expanding at a tremendous rate ever since. Over time, the range of topics investigated using survey methods has broadened enormously as policy makers and researchers have learned to appreciate the value of quantitative data and as survey researchers – in response to policy makers' demands – have tackled topics previously considered unsuitable for study using survey methods. The range of analyses of survey data has also expanded, as users of survey data have become more sophisticated and as major developments in computing power and software have simplified the computations involved. In the early days, users were mostly satisfied with national estimates and estimates for major geographic regions and other large domains. The situation is very different today: more and more, policy makers are demanding estimates for small domains for use in making policy decisions. For example, population surveys are often required to provide estimates of adequate precision for domains defined in terms of some combination of such factors as age, sex, race/ethnicity, and poverty status. A particularly widespread demand from policy makers is for estimates at a finer level of geographic detail than the broad regions that were commonly used in the past. Thus estimates are frequently needed for such entities as states, provinces, counties, school districts, and health service areas.

The need to provide estimates for small domains has led to developments in two directions. One direction is toward the use of sample designs that can produce domain estimates of adequate precision within the standard design-based mode of inference used in survey analysis (i.e., "direct estimates"). Many sample surveys are now designed to yield sufficient sample sizes for key domains to satisfy the precision requirements for those domains. This approach is generally used for socio-economic domains and for some larger geographic domains. However, the increase in overall sample size that this approach entails may well exceed the survey's funding resources and capabilities, particularly so when estimates are required for many geographic areas. In the United States, for example, few surveys are large enough to be capable of providing reliable subpopulation estimates for all 50 states, even if the sample is optimally allocated across states for this purpose. For very small geographic areas like school districts, either a complete census or a sample of at least the size of the census long-form sample (on average about 1 in 6 households nationwide) is required. Even censuses, however, although valuable, cannot be the complete solution for the production of small area estimates. In most countries censuses are conducted only once a decade. They cannot, therefore, provide satisfactory small area estimates for intermediate time points during a decade for population characteristics that change markedly over time. Furthermore, census content is inherently severely restricted, so a census cannot provide small area estimates for all the characteristics that are of interest. Hence another approach is needed.

The other direction for producing small area estimates is to turn away from conventional direct estimates toward the use of indirect model-dependent estimates. The model-dependent approach employs a statistical model that "borrows strength" in making an estimate for one small area from sample survey data collected in other small areas or at other time periods. This approach moves away from the design-based estimation of conventional direct estimates to indirect model-dependent estimates. Naturally, concerns are raised about the reliance on models for the production of such small area estimates. However, the demand for small area estimates is strong and increasing, and models are needed to satisfy that demand in many cases. As a result, many survey statisticians have come to accept the model-dependent approach in the right circumstances, and the approach is being used in a number of important cases. Examples of major small area estimation programs in the United States include: the Census Bureau's Small Area Income and Poverty Estimates program, which regularly produces estimates of income and poverty measures for various population subgroups for states, counties and school districts; the Bureau of Labor Statistics' Local Area Unemployment Statistics program, which produces monthly estimates of employment and unemployment for states, metropolitan areas, counties and certain subcounty areas; the National Agricultural Statistics Service's County Estimates Program, which produces county estimates of crop yield; and the estimates of substance abuse in states and metropolitan areas, which are produced by the Substance Abuse and Mental Health Services Administration (see Chapter 1).

The essence of all small area methods is the use of auxiliary data available at the small area level, such as administrative data or data from the last census. These data are used to construct predictor variables for use in a statistical model that can be used to predict the estimate of interest for all small areas. The effectiveness of small area estimation depends initially on the availability of good predictor variables that are uniformly measured over the total area. It next depends on the choice of a good prediction model. Effective use of small area estimation methods further depends on a careful, thorough evaluation of the quality of the model. Finally, when small area estimates are produced, they should be accompanied by valid measures of their precision.

Early applications of small area estimation methods employed only simple methods. At that time the choice of the method for use in particular case was relatively simple, being limited by the computable methods then in existence. However, the situation has changed enormously in recent years, and particularly in the last decade. There now exists a wide range of different, often complex models that can be used, depending on the nature of the measurement of the small area estimate (e.g., a binary or continuous variable) and on the auxiliary data available. One key distinction in model construction is between situations where the auxiliary data are available for the individual units in the population and those where they are available only at the aggregate level for each small area. In the former case, the data can be used in unit level models, whereas in the latter they can be used only in area level models. Another feature involved in the choice of model is whether the model borrows strength cross-sectionally, over time, or both. There are also now a number of different approaches, such as empirical best linear prediction (EBLUP), empirical Bayes (EB) and hierarchical Bayes (HB), that can be used to estimate the models and the variability of the model-dependent small area estimates. Moreover, complex procedures that would have been extremely difficult to apply a few years ago can now be implemented fairly straightforwardly, taking advantage of the continuing increases in computing power and the latest developments in software.

The wide range of possible models and approaches now available for use can be confusing to those working in this area. J.N.K. Rao's book is therefore a timely contribution, coming at a point in the subject's development when an integrated, systematic, treatment is needed. Rao has done a great service in producing this authoritative and comprehensive account of the subject. This book will help to advance the subject and be a valuable resource for practitioners and theorists alike.

GRAHAM KALTON

Preface

Sample surveys are widely used to provide estimates of totals, means and other parameters not only for the total population of interest but also for subpopulations (or domains) such as geographic areas and socio-demographic groups. Direct estimates of a domain parameter are based only on the domain-specific sample data. In particular, direct estimates are generally "design-based" in the sense that they make use of "survey weights" and the associated inferences (standard errors, confidence intervals, etc.) are based on the probability distribution induced by the sample design, with the population values held fixed. Standard sampling texts (e.g., the 1977 Wiley book *Sampling Techniques* by W.G. Cochran) provide extensive accounts of design-based direct estimation. Models that treat the population values as random may also be used to obtain model dependent direct estimates. Such estimates in general do not depend on survey weights and the associated inferences are based on the probability distribution induced by the assumed model (e.g., the 2001 Wiley book *Finite Population Sampling and Inference: A Prediction Approach* by R. Valliant, A.H. Dorfman and R.M. Royall).

We regard a domain as large if the domain sample size is large enough to yield direct estimates of adequate precision; otherwise, the domain is regarded as small. In this text, we generally use the term "small area" to denote any subpopulation for which direct estimates of adequate precision cannot be produced. Typically, domain sample sizes tend to increase with the population size of the domains, but this is not always the case. For example, due to oversampling of certain domains in the U.S. Third Health and Nutrition Examination Survey, sample sizes in many states were small (or even zero).

It is seldom possible to have a large enough overall sample size to support reliable direct estimates for all the domains of interest. Therefore, it is often necessary to use indirect estimates that "borrow strength" by using values of the variables of interest from related areas, thus increasing the "effective" sample size. These values are brought into the estimation process through a model (either implicit or explicit) that provides a link to related areas (domains) through the use of supplementary information related to the variables of interest, such as recent census counts and current administrative records. Availability of good auxiliary data and determination of suitable linking models are crucial to the formation of indirect estimates.

In recent years, the demand for reliable small area estimates has greatly increased worldwide. This is due, among other things, to their growing use in formulating policies and programs, the allocation of government funds and in regional planning. Demand from the private sector has also increased because business decisions, particularly those related to small businesses, rely heavily on the local conditions. Small area estimation is particularly important for studying the economies in transition in central and eastern European countries and the former Soviet Union countries because these countries are moving away from centralized decision making.

The main aim of this text is to provide a comprehensive account of the methods and theory of small area estimation, particularly indirect estimation based on explicit small area linking models. The model-based approach to small area estimation offers several advantages, most importantly, increased precision. Other advantages include the derivation of "optimal" estimates and associated measures of variability under an assumed model, and the validation of models from the sample data.

Chapter 1 introduces some basic terminology related to small area estimation and presents some important applications as motivating examples. Chapter 2 contains a brief account of direct estimation, which provides a background for later chapters. It also addresses survey design issues that have a bearing on small area estimation. Traditional demographic methods that employ indirect estimates based on implicit linking models are studied in Chapter 3. Typically, demographic methods only use administrative and census data and sampling is not involved, whereas indirect estimation methods studied in later chapters are largely based on sample survey data in conjunction with auxiliary population information. Chapter 4 gives a detailed account of traditional indirect estimation based on implicit linking models. The well-known James-Stein method of composite estimation is also studied in the context of sample surveys.

Explicit small area models that account for between area variation are presented in Chapter 5, including linear mixed models and generalized linear mixed models, such as logistic models with random area effects. The models are classified into two broad groups: (i) Area level models that relate the small area means to area-specific auxiliary variables, (ii) Unit level models that relate the unit values of study variables to unit-specific auxiliary variables. Several extensions to handle complex data structures, such as spatial dependence and time series structures, are also presented. Chapters 6–8 study in more detail linear mixed models involving fixed and random effects. General results on empirical best linear unbiased prediction (EBLUP) under the frequentist approach are presented in Chapter 6. The more difficult problem of estimating the mean squared error (MSE) of EBLUP estimators is also considered. A basic area level model and a basic unit level model are studied thoroughly in Chapter 7 by applying the EBLUP results developed in Chapter 6. Several important applications are also presented in this chapter. Various extensions of the basic models are considered in Chapter 8.

Chapter 9 presents empirical Bayes (EB) estimation. This method is more generally applicable than the EBLUP method. Various approaches to measuring the variability of EB estimators are presented. Finally, Chapter 10 presents a self-contained account of hierarchical Bayes (HB) estimation, by assuming prior distributions on model parameters. Both chapters include actual applications with real data sets.

Throughout the text, we discuss the advantages and limitations of the different methods for small area estimation. We also emphasize the need for both internal and external evaluations for model selection. To this end, we provide various methods of model validation, including comparisons of estimates derived from a model with reliable values obtained from external sources, such as previous census values.

Proofs of basic results are given in Sections 2.7, 3.5, 4.4, 6.4, 9.9 and 10.14, but proofs of results that are technically involved or lengthy are omitted. The reader is referred to relevant papers for details of omitted proofs. We provide self-contained accounts of direct estimation (Chapter 2), linear mixed models (Chapter 6), EB estimation (Chapter 9) and HB estimation (Chapter 10). But prior exposure to a standard course in mathematical statistics, such as the 1990 Wadsworth & Brooks/Cole book *Statistical Inference* by G. Casella and R.L. Berger, is essential. Also, a course in linear mixed models, such as the 1992 Wiley book *Variance Components* by S.R. Searle, G. Casella and C.E. McCulloch, would be helpful in understanding model based small area estimation. A basic course in survey sampling methods, such as the 1977 Wiley book *Sampling Techniques* by W.G. Cochran, is also useful but not essential.

This book is intended primarily as a research monograph, but it is also suitable for a graduate level course on small area estimation. Practitioners interested in learning small area estimation methods may also find portions of this text useful; in particular, Chapters 4, 7, 9, and Sections 10.1–10.3 and 10.5 as well as the applications presented throughout the book.

Special thanks are due to Gauri Datta, Sharon Lohr, Danny Pfeffermann, Graham Kalton, M.P. Singh, Jack Gambino and Fred Smith for providing many helpful comments and constructive suggestions. I am also thankful to Ming Yu and Yong You for typing portions of this text, to Gill Murray for the final typesetting and preparation of the text, and to Roberto Guido of Statistics Canada for designing the logo on the cover page. Finally, I am grateful to my wife Neela for her long enduring patience and encouragement and to my children, Sunil and Supriya, for their understanding and support.

J.N.K. Rao

Ottawa, Canada
January, 2003

Chapter 1

Introduction

1.1 What is a Small Area?

Sample surveys have long been recognized as cost-effective means of obtaining information on wide-ranging topics of interest at frequent intervals over time. They are widely used in practice to provide estimates not only for the total population of interest but also for a variety of subpopulations (domains). Domains may be defined by geographic areas or socio-demographic groups or other subpopulations. Examples of a geographic domain (area) include a state/province, county, municipality, school district, unemployment insurance (UI) region, metropolitan area and health service area. On the other hand, a socio-demographic domain may refer to a specific age-sex-race group within a large geographic area. An example of "other domains" is the set of business firms belonging to a census division by industry group.

In the context of sample surveys, we refer to a domain estimator as "direct" if it is based only on the domain-specific sample data. A direct estimator may also use known auxiliary information, such as the total of an auxiliary variable, x, related to the variable of interest, y. A direct estimator is typically "design based" but it can also be motivated by and justified under models (see Section 2.1 of Chapter 2). Design based estimators make use of survey weights, and the associated inferences are based on the probability distribution induced by the sampling design with the population values held fixed (see Chapter 2). "Model assisted" direct estimators that make use of "working" models are also design based, aiming at making the inferences "robust" to possible model misspecification (see Chapter 2).

A domain (area) is regarded as large (or major) if the domain-specific sample is large enough to yield "direct estimates" of adequate precision. A domain is regarded as "small" if the domain-specific sample is not large enough to support direct estimates of adequate precision. Some other terms used to denote a domain with small sample size include "local area", "subdomain", "small subgroup", "subprovince" and "minor domain". In some applications, many domains of interest (such as counties) may have zero sample size.

In this text, we generally use the term "small area" to denote any domain for which direct estimates of adequate precision cannot be produced. Typically domain sample size tends to increase with the population size of the domain, but this is not always the case. Sometimes the sampling fraction is made larger than the average fraction in small domains in order to increase the domain sample sizes and thereby increase the precision of domain estimates. Such oversampling was, for example, used in the U.S. Third Health and Nutrition Examination Survey (NHANES III) for certain domains in the cross-classification of sex, race/ethnicity, and age, in order that direct estimates of acceptable precision could be produced for those domains. This oversampling led to a greater concentration of the sample in certain states (e.g., California and Texas) than normal, and thereby exacerbated the common problem in national surveys that sample sizes in many states are small (or even zero). Thus, while direct estimates may be used to estimate characteristics of demographic domains with NHANES III, they cannot be used to estimate characteristics of many states. States may therefore be regarded as small areas in this survey. Even when a survey has large enough state sample sizes to support the production of direct estimates for the total state populations, these sample sizes may well not be large enough to support direct estimates for subgroups of the state populations, such as school-age children or persons in poverty. Clearly, it is seldom possible to have a large enough overall sample size to support reliable direct estimates for all domains. Furthermore, in practice it is not possible to anticipate all uses of the survey data, and "the client will always require more than is specified at the design stage" (Fuller (1999), p. 344).

In making estimates for small areas with adequate level of precision, it is often necessary to use "indirect" estimators that "borrow strength" by using values of the variable of interest, y, from related areas and/or time periods and thus increase the "effective" sample size. These values are brought into the estimation process through a model (either implicit or explicit) that provides a link to related areas and/or time periods through the use of supplementary information related to y, such as recent census counts and current administrative records. Three types of indirect estimators can be identified (Schaible (1996), Chapter 1): "domain indirect', "time indirect" and "domain and time indirect". A domain indirect estimator makes use of y-values from another domain but not from another time period. A time indirect estimator uses y-values from another time period for the domain of interest but not from another domain. On the other hand, a domain and time indirect estimator uses y-values from another domain as well as another time period. Some other terms used to denote an indirect estimator include "non-traditional", "small area", "model dependent" and "synthetic".

Availability of good auxiliary data and determination of suitable linking models are crucial to the formation of indirect estimators. As noted by Schaible (1996, Chapter 10), expanded access to auxiliary information through coordination and cooperation among different agencies is needed.

1.2 Demand for Small Area Statistics

Historically, small area statistics have long been used. For example, such statistics existed in eleventh century England and seventeenth century Canada, based on either census or on administrative records (Brackstone (1987)). Demographers have long been using a variety of indirect methods for small area estimation of population and other characteristics of interest in postcensal years. Typically, sampling is not involved in the traditional demographic methods (Chapter 3).

In recent years, the demand for small area statistics has greatly increased worldwide. This is due, among other things, to their growing use in formulating policies and programs, in the allocation of government funds and in regional planning. Legislative acts by national governments have increasingly created a need for small area statistics, and this trend will likely continue. Demand from the private sector has also increased because business decisions, particularly those related to small businesses, rely heavily on the local socio-economic, environmental and other conditions. Schaible (1996) provides an excellent account of the use of traditional and model-based indirect estimators in U.S. Federal Statistical Programs.

Small area estimation is of particular interest for the economies in transition in central and eastern European countries and the former Soviet Union countries. In the 1990s, these countries have moved away from centralized decision making. As a result, sample surveys are now used to produce estimates for large areas as well as small areas. Prompted by the demand for small area statistics, an International Scientific Conference on Small Area Statistics and Survey Designs was held in Warsaw, Poland, in 1992 and an International Satellite Conference on Small Area Estimation was held in Riga, Latvia, in 1999 to disseminate knowledge on small area estimation. See Kalton, Kordos and Platek (1993) and IASS Satellite Conference (1999) for the published conference proceedings.

Some other proceedings of conferences on small area estimation include National Institute on Drug Abuse (1979), Platek and Singh (1986) and Platek, Rao, Särndal and Singh (1987). Review papers on small area estimation include Rao (1986), Chaudhuri (1992), Ghosh and Rao (1994), Rao (1999), Marker (1999), Rao (2001) and Pfeffermann (2002). A text on the theory of small area estimation has also appeared (Mukhopadhyay (1998)).

1.3 Traditional Indirect Estimators

Traditional indirect estimators, based on implicit linking models, include synthetic and composite estimators (Chapter 4). These estimators are generally design based and their design variances (i.e., variances with respect to the probability distribution induced by the sampling design) are usually small relative to the design variances of direct estimators. However, the indirect estimators will be generally design biased and the design bias will not decrease as the overall sample size increases. If the implicit linking model is approximately true, then the design bias will be small, leading to significantly smaller design mean squared error (MSE) compared to the MSE of a direct estimator. Reduction in MSE is the main reason for using indirect estimators.

1.4 Small Area Models

Explicit linking models based on random area-specific effects that account for between area variation beyond that is explained by auxiliary variables included in the model will be called "small area models" (Chapter 5). Indirect estimators based on small area models will be called "model-based estimators". We classify small area models into two broad types: (i) Aggregate (or area) level models that relate small area direct estimators to area-specific covariates. Such models are necessary if unit (or element) level data are not available. (ii) Unit level models that relate the unit values of a study variable to unit-specific covariates. A basic area level and a basic unit level are introduced in Sections 5.2 and 5.3, respectively. Various extensions of the basic small area models are outlined in Section 5.4. Models given in Sections 5.2–5.4 are relevant for continuous responses y, and may be regarded as special cases of a general linear mixed model (Section 6.3). However, for binary or count variables y, generalized linear mixed models are used (Section 5.6); in particular, logistic linear mixed models for the binary case and loglinear mixed models for the count case.

A critical assumption for the unit level models is that the sample values obey the assumed population model, that is, sample selection bias is absent (see Section 5.3). For area level models, we assume the absence of informative sampling of the areas in situations where only some of the areas are selected to the sample, that is, the sample area values (the direct estimates) obey the assumed population model.

Inferences from model-based estimators refer to the distribution implied by the assumed model. Model selection and validation, therefore, play a vital role in model-based estimation. If the assumed models do not provide a good fit to the data, the model-based estimators will be model biased which, in turn, can lead to erroneous inferences. Several methods of model selection and validation are presented in Chapters 6, 7 and 10. It is also useful to conduct external evaluations by comparing indirect estimates (both traditional and model-based) to more reliable estimates or census values based on past data; see Examples 7.1.1 and 7.1.2 for both internal and external evaluations.

1.5 Model-Based Estimation

It is now generally accepted that when indirect estimators are to be used they should be based on explicit small area models. Such models define the way that the related data are incorporated in the estimation process. The model-based approach to small area estimation offers several advantages: (1) "Optimal" estimators can be derived under the assumed model. (2) Area-specific measures of variability can be associated with each estimator unlike global measures (averaged over small areas) often used with traditional indirect estimators. (3) Models can be validated from the sample data. (4) A variety of models can be entertained depending on the nature of the response variables and the complexity of data structures (such as spatial dependence and time series structures).

In this text, we focus on empirical best linear unbiased prediction (EBLUP) estimators (Chapters 6–8), parametric empirical Bayes (EB) estimators (Chapter 9) and parametric hierarchical Bayes (HB) estimators (Chapter 10) derived from small area models. For the HB method, a further assumption on the prior distribution of model parameters is also needed. EBLUP is applicable for linear mixed models, whereas EB and HB are more generally valid.

The EBLUP method for general linear mixed models has been extensively used in animal breeding and other applications to estimate realized values of linear combinations of fixed and random effects. An EBLUP estimator is obtained in two steps: (i) The best linear unbiased predictor (BLUP), which minimizes the model MSE in the class of linear model unbiased estimators of the quantity of interest is first obtained. It depends on the variances (and covariances) of random effects in the model. (ii) An EBLUP estimator is obtained from the BLUP by substituting suitable estimators of variance parameters. Chapter 6 presents some unified theory of the EBLUP method for the general linear mixed model, which covers many small area models considered in the literature (Chapters 7 and 8). Estimation of model MSE of EBLUP estimators is studied in detail in Chapters 6–8. PROC MIXED in SAS software can be used to implement the EBLUP method.

Under squared error loss, the best predictor (BP) of a (random) small area quantity of interest (such as mean or proportion) is the conditional expectation of the quantity given the data and the model parameters. Distributional assumptions are needed for calculating the BP. The empirical BP (or EB) estimator is obtained from BP by substituting suitable estimators of model parameters. On the other hand, the HB estimator under squared error loss is obtained by integrating the BP with respect to the (Bayes) posterior distribution of model parameters derived from an assumed prior distribution of model parameters. The HB estimator under squared error loss is the posterior mean of the estimand, where the expectation is with respect to the posterior distribution of the quantity of interest given the data. The HB method uses the posterior variance as a measure of uncertainty associated with the HB estimator. Posterior (or credible) intervals for the quantity of interest can also be constructed from the posterior distribution of the quantity of interest. Currently, the HB method is being extensively used for small area estimation because it is straightforward, inferences are "exact" and complex problems using recently developed Markov chain Monte Carlo (MCMC) methods can be handled. Software for implementing the HB method is also available (subsection 10.2.5). Chapter 10 gives a self-contained account of the HB method and its applications to small area estimation.

"Optimal" model-based estimates of small area totals or means may not be suitable if the objective is to produce an ensemble of estimates whose distribution is in some sense close enough to the distribution of the corresponding estimands. We are also often interested in the ranks (e.g., ranks of schools, hospitals or geographical areas) or in identifying domains (areas) with extreme values. Ideally, it is desirable to construct a set of "triple-goal" estimates that can produce good ranks, a good histogram and good area-specific estimates. However, simultaneous optimization is not feasible, and it is necessary to seek a compromise set that can strike an effective balance between the three goals. Triple-goal EB estimation and constrained EB estimation that preserves the ensemble variance are studied in Section 9.6. A HB version of constrained HB estimation is studied in Section 10.13.

1.6 Some Examples

We conclude the introduction by presenting some important applications of small area estimation as motivating examples. Details of some of these applications, including auxiliary information used, are given in Chapters 7–10.

Health

Small area estimation of health related characteristics has attracted a lot of attention in the U.S. because of a continuing need to assess health status, health practices and health resources at both the national and subnational levels. Reliable estimates of health-related characteristics help in evaluating the demand for health care and the access individuals have to it. Health care planning often takes place at the state and sub-state levels because health characteristics are known to vary geographically. Health System Agencies in the United States, mandated by the National Health Planning Resource Development Act of 1994, are required to collect and analyze data related to the health status of the residents and to the health delivery systems in their health service areas (Nandram (1999)).

(i) The U.S. National Center for Health Statistics pioneered the use of synthetic estimation, based on implicit linking models, developing state estimates of disability and other health characteristics for different groups from the National Health Interview Survey(NHIS). Examples 4.2.2 and 10.11.3 give health applications from national surveys. Malec, Davis, and Cao (1999) studied HB estimation of overweight prevalence for adults by states, using data from NHANES III. Folsom, Shah and Vaish (1999) produced survey-weighted HB estimates of small area prevalence rates for states and age groups, for up to 20 binary variables related to drug use, using data from pooled National Household Surveys on Drug Abuse. Chattopadhyay, Lahiri, Larsen and Reimnitz (1999) studied EB estimates of state-wide prevalences of the use of alcohol and drugs (e.g., marijuana) among civilian non-institutionalized adults and adolescents in the United States. These estimates are used for planning and resource allocation, and to project the treatment needs of dependent users.

(ii) *Mapping* of small area mortality (or incidence) rates of diseases, such as cancer, is a widely used tool in public health research. Such maps permit the analysis of geographical variation which may be useful for formulating and assessing aetiological hypotheses, resource allocation, and the identification of areas of unusually high risk warranting intervention; see Section 9.5. Direct (or crude) estimates of rates, called standardized mortality ratios (SMRs) can be very unreliable, and a map of crude rates can badly distort the geographical distribution of disease incidence or mortality because the map tends to be dominated by areas of low population. Disease mapping, using model-based estimators, has received increased attention in recent years. The October 2000 issue of *Statistics in Medicine* is devoted to the new developments in disease mapping. We give several examples of disease mapping in this text: see Examples 9.5.1, 9.7.1, 10.10.11 and 10.10.3. Typically, sampling is not involved in disease mapping applications.

Agriculture

The U.S. National Agricultural Service (NASS) publishes model-based county estimates of crop acreage using remote sensing satellite data as auxiliary information; see Example 7.2.1 for an application. County estimates assist the agricultural authorities in local agricultural decision making. Also, county crop yield estimates are used to administer federal programs involving payments to farmers if crop yields fall below certain levels. Example 4.2.2 gives an application of synthetic estimation to produce county estimates of wheat production in the state of Kansas based on a non-probability sample of farms. Chapters 6 and 7 of Schaible (1996) provide details of traditional and model-based indirect estimation methods used by NASS for county crop acreage and production.

Remote sensing satellite data and crop surveys are currently being used in India to produce crop yield estimates at the district level (Singh and Goel (2000)). Small area estimation techniques to obtain estimates of crop production at lower administrative units like "tehsil" or block level are also under study.

Income for small places

Example 7.1.1 gives details of an application of the EB (EBLUP) method of estimation of small area incomes, based on a basic area level linking model (see Section 7.1). This method, proposed originally by Fay and Herriot (1979), was adopted by the U.S. Bureau of the Census to form updated per capita income (PCI) for small places. This was the largest application (prior to 1990) of model-based estimators in a U.S. Federal Statistical Program. The PCI estimates are used to determine fund allocations to local government units (places) under the General Revenue Sharing Program.

Poverty counts

The Fay-Herriot method has been recently used to produce model-based current county estimates of poor school-age children in U.S.A. (National Research Council (2000)). Using these estimates, the U.S. Department of Education allocates annually over $7 billion of funds, (called Title I funds), to counties, and then states distribute the funds among school districts. The allocated funds support compensatory education programs to meet the needs of educationally disadvantaged children. In the past, funds were allocated on the basis of estimated counts from the previous census, but this allocation system had to be changed since the poverty counts vary significantly over time. EBLUP estimates in this application obtained from the Current Population Survey (CPS) using administrative data as auxiliary information. Example 7.1.2 presents details of this application. The reader is referred to National Research Council (2000) for further details.

Median income of four-person families

Estimates of the current median income of four-person families in each of the U.S. states are used to determine the eligibility for a program of energy assistance to low-income families administered by the U.S. Department of Health and Human Services. CPS and administrative information are used to produce model-based estimates, using extensions of the basic area level linking model; see Examples 8.1.1 and 8.3.3.

Chapter 2

Direct Domain Estimation

2.1 Introduction

Sample survey data are extensively used to provide reliable direct estimates of totals and means for the whole population and large areas or domains. As noted in Chapter 1, a direct estimator for a domain uses values of the variable of interest, y, only from the sample units in the domain. Sections 2.2–2.5 provide a brief account of direct estimation under the design-based or repeated sampling framework. We refer the reader to standard text books on sampling theory (e.g., Cochran (1977), Hedayat and Sinha (1991), Särndal, Swensson and Wretman (1992), Thompson (1997), Lohr (1999)) for a more extensive treatment of direct estimation.

Model-based methods have also been used to develop direct estimators and associated inferences. Such methods provide valid conditional inferences referring to the particular sample that has been drawn, regardless of the sampling design (see Brewer (1963), Royall (1970) and Valliant, Dorfman and Royall (2001)). But unfortunately, model-based strategies can perform poorly under model misspecification as the sample size in the domain increases. For instance, Hansen, Madow and Tepping (1983) introduced a model misspecification that is not detectable through tests of significance from sample sizes as large as 400, and then showed that the repeated sampling coverage probabilities of model-based confidence intervals on the population mean, $\bar{Y}$, are substantially less than the desired level and that the understatement becomes worse as the sample size decreases. This poor performance is largely due to asymptotic design-inconsistency of the model-based estimator with respect to the stratified random sampling design employed by Hansen et al. (1983). We do not consider model-based direct estimators in this monograph, but model-based methods will be extensively used in the context of indirect estimators and small sample sizes in the domains of interest. As noted in Chapter 1, an indirect estimator for a domain "borrows strength" by using the values of the study variable, y, from sample units outside the domain of interest.

The main intention of Chapter 2 is to provide some background material for later chapters and to indicate that direct estimation methods may sometimes suffice, particularly after addressing survey design issues that have a bearing on small area estimation (see Section 2.6). Effective use of auxiliary information through ratio and regression estimation is also useful in reducing the need for indirect small area estimators (Sections 2.3–2.5).

2.2 Design-based Approach

We assume the following somewhat idealized set-up and focus on the estimation of a population total or mean in Section 2.3. Direct estimators for domains are obtained in Section 2.4, using the results for population totals. A survey population U consists of N distinct elements (or ultimate units) identified through the labels $j = 1, \ldots, N$. We assume that a characteristic of interest, y, associated with element j can be measured exactly by observing element j. Thus measurement errors are assumed to be absent. The parameter of interest is the population total $Y = \sum_U y_j$ or the population mean $\bar{Y} = Y/N$, where $\sum_U$ denotes summation over the population elements j.

A sampling design is used to select a sample s from U with probability $p(s)$. The sample selection probability $p(s)$ can depend on known design variables such as stratum indicator variables and size measures of clusters. In practice, a sampling scheme is used to implement a sampling design. For example, a simple random sample of size n can be obtained by drawing n random numbers from 1 to N without replacement. Commonly used sampling designs include stratified simple random sampling (e.g., establishment surveys) and stratified multistage sampling (e.g., large-scale socio-economic surveys such as the Canadian Labour Force Survey and the Current Population Survey of the United States).

To make inferences on the total Y, we observe the y-values associated with the selected sample s. For simplicity, we assume that all the elements $j \in s$ can be observed, that is, complete response. In the design-based approach, an estimator $\hat{Y}$ of Y is said to be design-unbiased (or p–unbiased) if the design expectation of $\hat{Y}$ equals Y; that is,

$$E_p(\hat{Y}) = \sum p(s)\hat{Y}_s = Y, \tag{2.2.1}$$

where the summation is over all possible samples s under the specified design and $\hat{Y}_s$ is the value of $\hat{Y}$ for the sample s. The design variance of $\hat{Y}$ is denoted as $V_p(\hat{Y}) = E_p\left[\hat{Y} - E_p(\hat{Y})\right]^2$. An estimator of $V_p(\hat{Y})$ is denoted as $v(\hat{Y}) = s^2(\hat{Y})$, and the variance estimator $v(\hat{Y})$ is p-unbiased for $V(\hat{Y})$ if $E_p[v(\hat{Y})] \equiv V_p(\hat{Y})$. An estimator of $\hat{Y}$ is design-consistent (or p-consistent) if $\hat{Y}$ is p-unbiased (or its p-bias tends to zero as the sample size increases) and $V_p(\hat{Y})$ tends to zero as the sample size increases. Strictly speaking, we need to consider p-consistency in the context of a sequence of populations U_ν such that both the sample size n_ν and the population size N_ν tend to ∞ as $\nu \to \infty$. p-consistency of the variance estimator $v(\hat{Y})$ is similarly defined. If the estimator $\hat{Y}$ and variance estimator $v(\hat{Y})$ are both p-consistent, then the design-based approach provides valid inferences on Y regardless of the population values in the sense that the pivotal $t = (\hat{Y} - Y)/s(\hat{Y})$ converges in distribution ($\to_d$) to a $N(0,1)$ variable as the sample size increases. Thus in repeated sampling about $100(1-\alpha)\%$ of the confidence intervals $[\hat{Y} - z_{\alpha/2}s(\hat{Y}), \hat{Y} + z_{\alpha/2}s(\hat{Y})]$ contain the true value Y as the sample size increases, where $z_{\alpha/2}$ is the upper $\alpha/2$- point of a $N(0,1)$ variable. In practice, one often reports only the estimate (realized value of $\hat{Y}$) and associated standard error (realized value of $s(\hat{Y})$) or coefficient of variation (realized value of $c(\hat{Y}) = s(\hat{Y})/\hat{Y}$). Coefficient of variation or standard error is used as a measure of variability associated with the estimate.

The design-based (or probability sampling) approach has been criticized on the grounds that the associated inferences, although assumption-free, refer to repeated sampling instead of just the particular sample s that has been drawn. A conditional design-based approach that allows us to restrict the set of samples used for inference to a "relevant" subset has also been proposed. This approach leads to conditionally valid inferences. For example, in the context of poststratification (i.e., stratification after selection of the sample) it makes sense to make design-based inferences conditional on the realized poststrata sample sizes (Holt and Smith (1979); Rao (1985)). Similarly, when the population total X of an auxiliary variable x is known, conditioning on the estimator $\hat{X}$ of X is justified because the distance $|\hat{X} - X|/X$ provides a measure of imbalance in the realized sample (Robinson (1987); Rao (1992); Casady and Valliant (1993)).

2.3 Estimation of Totals

2.3.1 Design-unbiased Estimator

Design weights $w_j(s)$ play an important role in constructing design-based estimators $\hat{Y}$ of Y. These basic weights may depend both on s and the element j ($j \in s$). An important choice is $w_j(s) = 1/\pi_j$, where $\pi_j = \sum_{\{s:j\in s\}} p(s)$, $j = 1, 2, \ldots, N$ are the inclusion probabilities and $\{s : j \in s\}$ denotes summation over all samples s containing the element j. To simplify the notation, we write $w_j(s) = w_j$ except when the full notation $w_j(s)$ is needed. The weight w_j may be interpreted as the number of elements in the population represented by the sample element j.

In the absence of auxiliary population information, we use the expansion estimator

$$\hat{Y} = \Sigma_s w_j y_j, \tag{2.3.1}$$

where Σ_s denotes summation over $j \in s$. In this case, the design-unbiasedness condition (2.2.1) reduces to

$$\sum_{\{s:j\in s\}} p(s)w_j(s) = 1;\; j = 1,\ldots,N. \tag{2.3.2}$$

The choice $w_j(s) = 1/\pi_j$ satisfies the unbiasedness condition (2.3.2) and leads to the well-known Horvitz-Thompson (H-T) estimator (see Cochran (1977), p. 259).

It is convenient to denote $\hat{Y} = \Sigma_s w_j y_j$ in an operator notation as $\hat{Y} = \hat{Y}(y)$. Using this notation, we have $\hat{Y}(x) = \Sigma_s w_j x_j$ for another variable x, whereas the traditional notation is to denote $\Sigma_s w_j x_j$ as $\hat{X}$. Similarly, we denote a variance estimator of $\hat{Y}$ as $v(\hat{Y}) = v(y)$. Using this notation, we have $v(\hat{X}) = v[\hat{Y}(x)] = v(x)$. Note that $\hat{Y}(x)$ and $v(x)$ are obtained by attaching the subscript j to the character, x, in the brackets and then replacing y_j by x_j in the formulas for $\hat{Y}(y)$ and $v(y)$. Hartley (1959) introduced the operator notation. We refer the reader to Cochran (1977), Särndal et al. (1992) and Wolter (1985) for details on variance estimation. Rao (1979) has shown that a nonnegative unbiased quadratic estimator of variance of $\hat{Y}$ is necessarily of the form

$$v(\hat{Y}) = v(y) = -\sum_{j<k,\; j,k\in s}\sum w_{jk}(s)b_j b_k\left(\frac{y_j}{b_j} - \frac{y_k}{b_k}\right)^2, \tag{2.3.3}$$

where the weights $w_{jk}(s)$ satisfy the unbiasedness condition and the nonzero constants b_j are such that the variance of $\hat{Y}$ becomes zero when $y_j \propto b_j$ for all j. For example, in the special case of $w_j = 1/\pi_j$ and a fixed sample size design, we have $b_j = \pi_j$ and $w_{jk}(s) = (\pi_{jk} - \pi_j\pi_k)/(\pi_{jk}\pi_j\pi_k)$ where $\pi_{jk} = \sum_{\{s:(j,k)\in s\}} p(s)\; j < k = 1,\ldots,N$ are the joint inclusion probabilities assumed to be positive. The variance estimator (2.3.3) in this case reduces to the well-known Sen-Yates-Grundy (S-Y-G) variance estimator (see Cochran (1977), p. 261).

Under stratified multistage sampling, the expansion estimator $\hat{Y}$ may be written as

$$\hat{Y} = \hat{Y}(y) = \Sigma_s w_{hlk} y_{hlk}, \tag{2.3.4}$$

where w_{hlk} is the design weight associated with the kth element in the l-th primary sampling unit (cluster) belonging to the hth stratum, y_{hlk} is the associated y-value and Σ_s is the summation over all elements $j = (hlk) \in s$ $(h = 1,\ldots,L; l = 1,\ldots,n_h)$. It is common practice to treat the sample as if the clusters are sampled with replacement and subsampling is done independently each time a cluster is selected. This leads to overestimation of the variance but the variance estimator is greatly simplified. We have

$$v(\hat{Y}) = v(y) = \sum_h \frac{1}{n_h(n_h-1)}\sum_l (y_{hl} - \bar{y}_h)^2, \tag{2.3.5}$$

where $y_{hl} = \sum_k (n_h w_{hlk}) y_{hlk}$ are weighted sample cluster totals and $\bar{y}_h = \sum_l y_{hl}/n_h$. Note that $v(\hat{Y})$ depends on the y_{hlk}'s only through the totals y_{hl}. The relative bias of $v(\hat{Y})$ is small if the first-stage sampling fraction is small in each stratum. Note that $\hat{Y}(x)$ and $v(x)$ are obtained by attaching the subscripts hlk to the character, x, in the brackets and then replacing y_{hlk} by x_{hlk} in the formulas (2.3.4) and (2.3.5) for $\hat{Y}(x)$ and $v(y)$, respectively.

2.3.2 Generalized Regression Estimator

Suppose now that auxiliary information in the form of known population totals $\mathbf{X} = (X_1, \ldots, X_p)^T$ is available and that the auxiliary vector $\mathbf{x}_j$ for $j \in s$ is also observed, that is, the data $(y_j, \mathbf{x}_j)$ for each element $j \in s$ are observed. An estimator that makes efficient use of this auxiliary information is the generalized regression (GREG) estimator which may be written as

$$\hat{Y}_{\mathrm{GR}} = \hat{Y} + (\mathbf{X} - \hat{\mathbf{X}})^T \hat{\mathbf{B}}, \qquad (2.3.6)$$

where $\hat{\mathbf{X}} = \Sigma_s w_j \mathbf{x}_j = \hat{Y}(\mathbf{x})$ and $\hat{\mathbf{B}} = (\hat{B}_1, \ldots, \hat{B}_p)^T = \hat{\mathbf{B}}(y)$ is the solution of the sample weighted least squares equations:

$$\left(\Sigma_s w_j \mathbf{x}_j \mathbf{x}_j^T / c_j\right) \hat{\mathbf{B}} = \Sigma_s w_j \mathbf{x}_j y_j / c_j \qquad (2.3.7)$$

with specified constants $c_j (> 0)$. It is also useful to write $\hat{Y}_{\mathrm{GR}}$ in the expansion form with design weights w_j changed to "revised" weights w_j^*. We have

$$\hat{Y}_{\mathrm{GR}} = \Sigma_s w_j^* y_j = \hat{Y}_{\mathrm{GR}}(y) \qquad (2.3.8)$$

in operator notation, where $w_j^* = w_j^*(s) = w_j(s) g_j(s)$ with

$$g_j(s) = 1 + (\mathbf{X} - \hat{\mathbf{X}})^T \left(\Sigma_s w_j \mathbf{x}_j \mathbf{x}_j^T / c_j\right)^{-1} \mathbf{x}_j / c_j. \qquad (2.3.9)$$

The revised weight $w_j^*(s)$ is the product of the design weight $w_j(s)$ and the estimation weight $g_j(s)$. The form (2.3.8) shows that the same weight w_j^* is applied to all variables of interest as in the case of the expansion estimator. This ensures consistency of results when aggregated over different variables, that is,

$$\hat{Y}_{\mathrm{GR}}(y_1) + \ldots + \hat{Y}_{\mathrm{GR}}(y_r) = \hat{Y}_{\mathrm{GR}}(y_1 + \ldots + y_r)$$

for different variables $y_1, \ldots, y_r$ attached to each element.

An important property of the GREG estimator is that it ensures consistency with the known auxiliary totals $\mathbf{X}$ in the sense

$$\hat{Y}_{\mathrm{GR}}(\mathbf{x}) = \Sigma_s w_j^* \mathbf{x}_j = \mathbf{X}. \qquad (2.3.10)$$

A proof of (2.3.10) is given in subsection 2.7.1. This property does not hold for the basic expansion estimator $\hat{Y}$. Many agencies regard this property as desirable from the user's viewpoint. Because of the property (2.3.10), the GREG estimator is also called a calibration estimator (Deville and Särndal (1992)). In fact, among all calibration estimators of the form $\Sigma_s b_j y_j$ with weights b_j satisfying the calibration constraints $\Sigma_s b_j \mathbf{x}_j = \mathbf{X}$, the GREG weights w_j^* minimize a chi-squared distance, $\Sigma_s c_j(w_j - b_j)^2/w_j$, between the basic weights w_j and the calibration weights b_j (see subsection 2.7.2). Thus the GREG weights w_j^* modify the design weights as little as possible subject to the calibration constraints.

The GREG estimator takes a simpler form when $c_j = \boldsymbol{\nu}^T \mathbf{x}_j$ for all $j \in U$ and some constant column vector $\boldsymbol{\nu}$. In this case we have

$$\hat{Y}_{\text{GR}} = \mathbf{X}^T \hat{\mathbf{B}} = \Sigma_s \tilde{w}_j y_j \tag{2.3.11}$$

because $\Sigma_s w_j e_j(s) = \hat{Y} - \hat{\mathbf{X}}^T \hat{\mathbf{B}} = 0$ (see subsection 2.7.3), where $e_j(s) = e_j = y_j - \mathbf{x}_j^T \hat{\mathbf{B}}$ are the sample residuals and $\tilde{w}_j = \tilde{w}_j(s) = w_j(s)\tilde{g}_j(s)$ with

$$\tilde{g}_j(s) = \mathbf{X}^T \left(\Sigma_s w_j \mathbf{x}_j \mathbf{x}_j^T / c_j\right)^{-1} \mathbf{x}_j / c_j. \tag{2.3.12}$$

The GREG estimator covers many practically useful estimators as special cases. For example, in the case of a single auxiliary variable x we get the well-known ratio estimator

$$\hat{Y}_R = \frac{\hat{Y}}{\hat{X}} X, \tag{2.3.13}$$

by setting $c_j = x_j$ in (2.3.12) and noting that $\tilde{g}_j(s) = X/\hat{X}$. The ratio estimator $\hat{Y}_R$ uses the weights $\tilde{w}_j = w_j(X/\hat{X})$. If only the population size N is known, we set $x_j = 1$ in (2.3.13) so that $X = N$ and $\hat{X} = \hat{N} = \Sigma_s w_j$. If we set $\mathbf{x}_j = (1, x_j)^T$ and $c_j = 1$, then $\boldsymbol{\nu} = (1, 0)^T$ and (2.3.11) reduces to the familiar linear regression estimator

$$\hat{Y}_{\text{LR}} = \hat{Y} + \hat{B}_{\text{LR}}(X - \hat{X}), \tag{2.3.14}$$

where

$$\hat{B}_{\text{LR}} = \hat{B}_2 = \Sigma_s w_j (x_j - \hat{\bar{X}})(y_j - \hat{\bar{Y}}) / \Sigma_s w_j (x_j - \hat{\bar{X}})^2$$

with $\hat{\bar{Y}} = \hat{Y}/\hat{N}$ and $\hat{\bar{X}} = \hat{X}/\hat{N}$. The GREG estimator also covers the familiar poststratified estimator as a special case. Suppose we partition U into G poststrata U_g (e.g., age/sex groups) with known population counts N_g ($g = 1, \ldots, G$). Then we set $\mathbf{x}_j = (x_{1j}, \ldots, x_{Gj})^T$ with $x_{gj} = 1$ if $j \in U_g$ and $x_{gj} = 0$ otherwise so that $\mathbf{X} = (N_1, \ldots, N_G)^T$. Noting that $\mathbf{1}^T \mathbf{x}_j = 1$ for all j we take $c_j = 1$ and $\boldsymbol{\nu} = \mathbf{1}$ and the GREG estimator (2.3.11) reduces to the poststratified estimator

$$\hat{Y}_{\text{PS}} = \sum_g \frac{N_{\cdot g}}{\hat{N}_{\cdot g}} \hat{Y}_{\cdot g}, \tag{2.3.15}$$

where $\hat{N}_{.g} = \sum_{s_{.g}} w_j$ and $\hat{Y}_{.g} = \sum_{s_{.g}} w_j y_j$, with $s_{.g}$ denoting the sample of elements belonging to poststratum g.

Turning to variance estimation, the traditional Taylor linearization method gives

$$v_L(\hat{Y}_{\mathrm{GR}}) = v(e), \tag{2.3.16}$$

which is obtained by substituting the residuals e_j for y_j in $v(y)$. Simulation studies have indicated that $v_L(\hat{Y}_{\mathrm{GR}})$ may lead to slight underestimation, whereas an alternative estimator

$$v(\hat{Y}_{\mathrm{GR}}) = v(ge), \tag{2.3.17}$$

obtained by substituting $g_j e_j$ for y_j in $v(y)$, reduces this underestimation, where $g_j = g_j(s)$ (Estevao, Hidiroglou and Särndal (1995)). The alternative variance estimator $v(\hat{Y}_{\mathrm{GR}})$ also performs better for conditional inference in the sense that it is approximately unbiased for the model variance of $\hat{Y}_{\mathrm{GR}}$ conditionally on s for several designs, under the following linear regression model (or GREG model):

$$y_j = \mathbf{x}_j^T \boldsymbol{\beta} + \epsilon_j, \ j \in U \tag{2.3.18}$$

where $E_m(\epsilon_j) = 0, V_m(\epsilon_j) = c_j\sigma^2, \mathrm{Cov}_m(e_j, e_k) = 0$ for $j \neq k$, and E_m, V_m and Cov_m, respectively, denote the model expectation, variance and covariance (Särndal, Swensson and Wretman (1989); Rao (1994)). In the model-based framework, y_j is a random variable and s is fixed. The GREG estimator is also model-unbiased under (2.3.18) in the sense $E_m(\hat{Y}_{\mathrm{GR}}) = E_m(Y)$ for every s. In the design-based framework, $\hat{Y}_{\mathrm{GR}}, v_L(\hat{Y}_{\mathrm{GR}})$ and $v(\hat{Y}_{\mathrm{GR}})$ are p-consistent.

We refer the reader to Särndal et al. (1992) for a detailed account of GREG estimation and to Estevao et al. (1995) for the highlights of a Generalized Estimation System at Statistics Canada based on GREG estimation theory.

2.4 Domain Estimation

2.4.1 Case of no Auxiliary Information

Suppose U_i denotes a domain (or subpopulation) of interest and that we are required to estimate the domain total $Y_i = \sum_{U_i} y_j$ or the domain mean $\bar{Y}_i = Y_i/N_i$, where N_i, the number of elements in U_i, may or may not be known. If y_i is binary (1 or 0), then $\bar{Y}_i$ reduces to the domain proportion P_i; for example, the proportion in poverty in the ith domain. Much of the theory in Section 2.3 for a total can be adapted to domain estimation by using the following relationships. Writing Y in the operator notation as $Y(y)$ and defining

$$y_{ij} = \begin{cases} y_j & \text{if } j \in U_i \\ 0 & \text{otherwise,} \end{cases}$$

$$a_{ij} = \begin{cases} 1 & \text{if } j \in U_i \\ 0 & \text{otherwise,} \end{cases}$$

we have

$$Y(y_i) = \sum_{j \in U} y_{ij} = \sum_{j \in U_i} y_j = Y_i \tag{2.4.1}$$

and

$$Y(a_i) = \sum_{j \in U} a_{ij} = \sum_{j \in U_i} 1 = N_i. \tag{2.4.2}$$

Note that y_{ij} may also be written as $a_{ij}y_j$. If the domains of interest, say, $U_1, \ldots, U_m$ form a partition of U (or of a larger domain), it is desirable from a user's viewpoint to ensure that the estimates of domain totals add up to the estimate of population total.

In the absence of auxiliary population information, we use the expansion estimator

$$\hat{Y}_i = \hat{Y}(y_i) = \sum_{j \in s} w_j y_{ij} = \sum_{j \in s_i} w_j y_j, \tag{2.4.3}$$

where s_i denotes the sample of elements belonging to domain U_i. It readily follows from (2.4.1) that $\hat{Y}_i$ is p-unbiased for Y_i if $\hat{Y}$ is p-unbiased for Y. It is also p-consistent if the expected domain sample size is large. Similarly, $\hat{N}_i = \hat{Y}(a_i)$ is p-unbiased for N_i, using (2.4.2). We note from (2.4.3) that the additive property is satisfied: $\hat{Y}_1 + \ldots + \hat{Y}_m = \hat{Y}$.

Noting that $v(\hat{Y}) = v(y)$, an estimator of the variance of $\hat{Y}_i$ is simply obtained from $v(y)$ by changing y_j to y_{ij}:

$$v(\hat{Y}_i) = v(y_i). \tag{2.4.4}$$

It follows from (2.4.3) and (2.4.4) that no new theory is required for domain estimation.

The domain mean $\bar{Y}_i = Y(y_i)/Y(a_i)$ is estimated by

$$\hat{\bar{Y}}_i = \frac{\hat{Y}(y_i)}{\hat{Y}(a_i)} = \frac{\hat{Y}_i}{\hat{N}_i}. \tag{2.4.5}$$

If $y_j = 1$ or 0, then $\hat{\bar{Y}}_i$ reduces to $\hat{P}_i$, an estimator of the domain proportion P_i. If the expected domain sample size is large, the ratio estimator (2.4.5) is p-consistent, and a Taylor linearization variance estimator is given by

$$v_L(\hat{\bar{Y}}_i) = v(\tilde{e}_i)/\hat{N}_i^2, \tag{2.4.6}$$

where $\tilde{e}_{ij} = y_{ij} - \hat{\bar{Y}}_i a_{ij}$. Note that $v(\tilde{e}_i)$ is obtained from $v(y)$ by changing y_j to $\tilde{e}_{ij}$. It follows from (2.4.5) and (2.4.6) that no new theory is required for domain means as well. Note that $\tilde{e}_{ij} = 0$ if $j \in s$ and $j \notin U_i$. We refer the reader to Hartley (1959) for domain estimation.

2.4.2 GREG Estimation

GREG estimation of Y is also easily adapted to estimation of a domain total Y_i. It follows from (2.3.8) that the GREG estimator of Y_i is

$$\hat{Y}_{i\text{GR}} = \hat{Y}_{\text{GR}}(y_i) = \sum_{j \in s_i} w_j^* y_j, \tag{2.4.7}$$

when the population total of the auxiliary vector $\mathbf{x}$ is known. It follows from (2.4.7) that the GREG estimator also satisfies the additive property: $\hat{Y}_{1\text{GR}} + \ldots + \hat{Y}_{m\text{GR}} = \hat{Y}_{\text{GR}}$. The estimator $\hat{Y}_{i\text{GR}}$ is approximately p-unbiased if the overall sample size is large, but p-consistency requires a large expected domain sample size as well. The special case of a ratio estimator (2.3.13) gives

$$\hat{Y}_{iR} = \frac{\hat{Y}_i}{\hat{X}} X$$

by changing y_j to $a_{ij} y_j$. Similarly, a poststratified estimator is obtained from (2.3.15) by changing y_{gj} to $a_{ij} y_{gj}$:

$$\hat{Y}_{i\text{PS}} = \sum_g \frac{N_{\cdot g}}{\hat{N}_{\cdot g}} \sum_{s_{ig}} w_j y_j,$$

where s_{ig} is the sample falling in the (ig)th cell of the cross-classification of domains and poststrata.

A Taylor linearization variance estimator of $\hat{Y}_{i\text{GR}}$ is simply obtained from $v(y)$ by changing y_j to $e_{ij} = y_{ij} - \mathbf{x}_{ij}^T \hat{\mathbf{B}}(y_i)$, where $\hat{\mathbf{B}}(y_i)$ is obtained from $\hat{\mathbf{B}}(y)$ by changing y_j to y_{ij}. Note that $e_{ij} = -\mathbf{x}_{ij}^T \hat{\mathbf{B}}(y_i)$ if $j \in s$ and $j \notin U_i$. The large negative residuals for all sampled elements not in U_i lead to inefficiency, unlike in the case of $\hat{Y}_{\text{GR}}$ where the variability of the e_j's will be small relative to the variability of the y_j's. This inefficiency of the GREG estimator $\hat{Y}_{i\text{GR}}$ is due to the fact that the auxiliary population information used here is not domain-specific. But $\hat{Y}_{i\text{GR}}$ has the advantage that it is approximately p-unbiased even if the expected domain sample size is small, whereas the GREG estimator based on domain specific auxiliary population information is p-biased unless the expected domain sample size is also large.

2.4.3 Domain-specific Auxiliary Information

We now turn to GREG estimation of a domain total Y_i under domain-specific auxiliary information. We assume that the domain totals $\mathbf{X}_i = (X_{i1}, \ldots, X_{ip})^T = Y(\mathbf{x}_i)$ are known, where $\mathbf{x}_{ij} = \mathbf{x}_j$ if $j \in U_i$ and $\mathbf{x}_{ij} = 0$ otherwise. In this case, a GREG estimator of Y_i is given by

$$Y_{i\text{GR}}^* = \hat{Y}_i + (\mathbf{X}_i - \hat{\mathbf{X}}_i)^T \hat{\mathbf{B}}_i, \tag{2.4.8}$$

where $\hat{\mathbf{X}}_i = \hat{Y}(\mathbf{x}_i)$ and

$$\left(\sum_{j \in s} w_j \mathbf{x}_{ij} \mathbf{x}_{ij}^T / c_j \right) \hat{\mathbf{B}}_i = \sum_{j \in s} w_j \mathbf{x}_{ij} y_{ij} / c_j.$$

We may also write (2.4.8) as

$$Y^*_{i\mathrm{GR}} = \sum_{j\in s} w^*_{ij} y_{ij}, \tag{2.4.9}$$

where $w^*_{ij} = w_j g^*_{ij}$ with

$$g^*_{ij} = 1 + (\mathbf{X}_i - \hat{\mathbf{X}}_i)^T \left(\sum_{j\in s} w_j \mathbf{x}_{ij}\mathbf{x}^T_{ij}/c_j \right)^{-1} \mathbf{x}_{ij}/c_j.$$

Note that the weights w^*_{ij} now depend on i unlike the weights w^*_j. Therefore, the estimators $Y^*_{i\mathrm{GR}}$ do not add up to $\hat{Y}_{\mathrm{GR}}$. Also, $Y^*_{i\mathrm{GR}}$ is not approximately p-unbiased unless the domain sample size is large.

In the special case of a single auxiliary variable x with known domain total X_i, we set $c_j = x_j$ in (2.4.9) to get the ratio estimator

$$Y^*_{iR} = \frac{\hat{Y}_i}{\hat{X}_i} X_i. \tag{2.4.10}$$

If domain-specific poststrata counts N_{ig} are known, then a poststratified count (PS/C) estimator is obtained from the GREG estimator (2.4.9) as

$$Y^*_{i\mathrm{PS/C}} = \sum_g \frac{N_{ig}}{\hat{N}_{ig}} \hat{Y}_{ig}, \tag{2.4.11}$$

where $\hat{N}_{ig} = \Sigma_{s_{ig}} w_j$ and $\hat{Y}_{ig} = \Sigma_{s_{ig}} w_j y_j$. If the cell totals X_{ig} of an auxiliary variable x are known, then we can use a poststratified-ratio (PS/R) estimator

$$Y^*_{i\mathrm{PS/R}} = \sum_g \frac{X_{ig}}{\hat{X}_{ig}} \hat{Y}_{ig}, \tag{2.4.12}$$

where $\hat{X}_{ig} = \Sigma_{s_{ig}} w_j x_j$.

If the expected domain size is large, then a Taylor linearization variance estimator of $Y^*_{i\mathrm{GR}}$ is obtained from $v(y)$ by changing y_j to $e^*_{ij} = y_{ij} - \mathbf{x}^T_{ij}\hat{\mathbf{B}}_i$. Note that $e^*_{ij} = 0$ if $j \in s$ and $j \notin U_i$ unlike the large negative residuals e_{ij} in the case of $\hat{Y}_{i\mathrm{GR}}$. Thus the domain-specific GREG estimator $Y^*_{i\mathrm{GR}}$ will be more efficient than $\hat{Y}_{i\mathrm{GR}}$, provided the expected domain-specific sample size is large.

Example 2.4.1 Wages and Salaries. Särndal and Hidiroglou (1989) and Rao and Choudhry (1995) considered a population U of $N = 1,678$ unincorporated tax filers (units) from the province of Nova Scotia, Canada, divided into 18 census divisions. This population is actually a simple random sample but it was treated as a population for simulation purposes. The population is also classified into four mutually exclusive industry groups: retail (515 units), construction (496 units) accommodation (114 units) and others (553 units).

Domains (small areas) are formed by a cross-classification of the four industry types with 18 census divisions. This leads to 70 nonempty domains out of 72 possible domains. The objective is to estimate the domain totals Y_i of the y-variable (wages and salaries), utilizing the only auxiliary variable x (gross business income) assumed to be known for all the N units in the population. A simple random sample, s, of size n is drawn from U and y-values observed. The sample s_i consist of n_i (≥ 0) units. The data consist of (y_j, x_j) for $j \in s$ and the auxiliary population information.

Under the above set-up, we have $\pi_j = n/N$, $w_j = N/n$ and the expansion estimator (2.4.3) reduces to

$$\hat{Y}_i = \begin{cases} (N/n)\sum_{s_i} y_j & \text{if } n_i \geq 1 \\ 0 & \text{if } n_i = 0. \end{cases} \qquad (2.4.13)$$

The estimator $\hat{Y}_i$ is p-unbiased for Y_i unconditionally, but it is p-biased conditional on the realized domain sample size n_i. In fact, conditional on n_i, the sample s_i is a simple random sample of size n_i from U_i, and the conditional p-bias of $\hat{Y}_i$ is

$$B_2(\hat{Y}_i) = E_2(\hat{Y}_i) - Y_i = N\left(\frac{n_i}{n} - \frac{N_i}{N}\right)\bar{Y}_i, \; n_i \geq 1$$

where E_2 denotes conditional expectation. Thus the conditional bias is zero only if the sample proportion n_i/n equals the population proportion N_i/N.

Suppose we form G poststrata based on the x-variable known for all the N units. Then (2.4.11) and (2.4.12) reduce to

$$Y^*_{i\text{PS/C}} = \sum_g N_{ig}\bar{y}_{ig} \qquad (2.4.14)$$

and

$$Y^*_{i\text{PS/R}} = \sum_g X_{ig}\frac{\bar{y}_{ig}}{\bar{x}_{ig}}, \qquad (2.4.15)$$

where $\bar{y}_{ig}$ and $\bar{x}_{ig}$ are the sample means for the n_{ig} units falling in the cell (ig), and the cell counts N_{ig} and the cell totals X_{ig} are known. If $n_{ig} = 0$, we set $\bar{y}_{ig} = 0$ and $\bar{y}_{ig}/\bar{x}_{ig} = 0$. The estimator $Y^*_{i\text{PS/C}}$ is p-unbiased conditional on the realized sample sizes n_{ig} (≥ 1 for all g), whereas $Y^*_{i\text{PS/R}}$ is only approximately p-unbiased conditional on the n_{ig}'s, provided all the expected sample sizes $E(n_{ig})$ are large.

If poststratification is not used, we can use the ratio estimator (2.4.10):

$$Y^*_{iR} = X_i\frac{\bar{y}_i}{\bar{x}_i}, \qquad (2.4.16)$$

where $\bar{y}_i$ and $\bar{x}_i$ are the sample means for the n_i units falling in domain i. If an x-variable is not observed but N_i is known, we can use an alternative estimator

$$Y^*_{iC} = N_i\bar{y}_i. \qquad (2.4.17)$$

This estimator is p-unbiased conditional on the realized sample size n_i (≥ 1).

It is desirable to make inferences conditional on the realized sample sizes, but this may not be possible under designs more complex than simple random sampling. Even under simple random sampling, we require $n_{ig} \geq 1$ for all g which limits the use of poststratification when n_i is small.

2.5 Modified Direct Estimators

We now consider modified direct estimators that use y-values from outside the domain but remain p-unbiased or approximately p-unbiased as the overall sample size increases. In particular, we replace $\hat{\mathbf{B}}_i$ in (2.4.8) by the overall regression coefficient $\hat{\mathbf{B}}$, given by (2.3.7), to get

$$\tilde{Y}_{i\text{GR}} = \hat{Y}_i + \left(\mathbf{X}_i - \hat{\mathbf{X}}_i\right)^T \hat{\mathbf{B}} = \sum_{j\in s} \tilde{w}_{ij} y_j \tag{2.5.1}$$

with

$$\tilde{w}_{ij} = w_j a_{ij} + \left(\mathbf{X}_i - \hat{\mathbf{X}}_i\right) \left(\Sigma_s w_j \mathbf{x}_j \mathbf{x}_j^T / c_j\right)^{-1} (w_j \mathbf{x}_j / c_j),$$

where a_{ij} is the domain indicator variable. The estimator $\tilde{Y}_{i\text{GR}}$ is approximately p-unbiased as the overall sample size increases, even if the domain sample size is small. This estimator is also called the "survey regression" estimator (Battese, Harter and Fuller (1988), Woodruff (1966)). The modified estimator (2.5.1) may also be viewed as a calibration estimator $\Sigma_s \tilde{w}_{ij} y_j$ with weights $b_{ij} = \tilde{w}_{ij}$ minimizing a chi-squared distance $\Sigma_s c_j (w_j a_{ij} - b_{ij})^2 / w_j$ subject to the constraints $\Sigma_s b_{ij} \mathbf{x}_j = \mathbf{X}_i$ (Singh and Mian (1995)).

A ratio form of (2.5.1) in the case of a single auxiliary variable x with known domain total X_i is given by

$$\tilde{Y}_{iR} = \hat{Y}_i + \frac{\hat{Y}}{\hat{X}}(X_i - \hat{X}_i). \tag{2.5.2}$$

Although the modified direct estimator borrows strength for estimating the regression coefficient, it does not increase the effective sample size, unlike indirect estimators. To illustrate this, consider the simple random sample of Example 2.4.1. In this case, $\tilde{Y}_{iR}$ reduces to

$$\tilde{Y}_{iR} = N_i \left[\bar{y}_i + \frac{\bar{y}}{\bar{x}}(\bar{X}_i - \bar{x}_i)\right], \tag{2.5.3}$$

where $\bar{y}$ and $\bar{x}$ are the overall sample means and $\bar{X}_i = X_i/N_i$. For large n, we can replace $\bar{y}/\bar{x}$ by $R = \bar{Y}/\bar{X}$ and the conditional variance is

$$V_2(\tilde{Y}_{iR}) \approx N_i^2 \left(\frac{1}{n_i} - \frac{1}{N_i}\right) S_{Ei}^2, \tag{2.5.4}$$

where $S_{Ei}^2 = \sum_{j\in U_i}(E_j - \bar{E}_i)^2/(N_i - 1)$ with $E_j = y_j - Rx_j$ and $\bar{E}_i$ is the domain mean of the E_j's. It follows from (2.5.4) than $V_2(\tilde{Y}_{iR})/N_i^2$ is of order n_i^{-1} so that the effective sample is not increased, although the variability of the E_j's may be smaller than the variability of the y_j's for $j \in U_i$. Note that the variability of the E_j's will be larger than the variability of the domain specific residuals $y_j - R_i x_j$ for $j \in U_i$ unless $R_i = \bar{Y}_i/\bar{X}_i \approx R$.

The modified GREG estimator (2.5.1) may be expressed as

$$\tilde{Y}_{i\text{GR}} = \mathbf{X}_i^T\hat{\mathbf{B}} + \sum_{j\in s_i} w_j e_j. \tag{2.5.5}$$

The first term $\mathbf{X}_i^T\hat{\mathbf{B}}$ is the synthetic-regression estimator (see Chapter 4) and the second term $\sum_{s_i} w_j e_j$ approximately corrects the p-bias of the synthetic estimator. We can improve on $\tilde{Y}_{i\text{GR}}$ by replacing the expansion estimator $\sum_{s_i} w_j e_j$ in (2.5.5) with a ratio estimator (Särndal and Hidiroglou (1989)):

$$\hat{E}_{iR} = N_i(\sum_{s_i} w_j e_j)/(\sum_{s_i} w_j); \tag{2.5.6}$$

note that $\hat{N}_i = \sum_{s_i} w_j$. The resulting estimator

$$\tilde{Y}_{i\text{GR}}(m) = \mathbf{X}_i^T\hat{\mathbf{B}} + \hat{E}_{iR}, \tag{2.5.7}$$

however, suffers from the ratio bias when the domain sample size is small, unlike $\tilde{Y}_{i\text{GR}}$.

A Taylor linearization variance estimator of $\tilde{Y}_{i\text{GR}}$ is obtained from $v(y)$ by changing y_j to $a_{ij}e_j$:

$$v_L(\tilde{Y}_{i\text{GR}}) = v(a_i e). \tag{2.5.8}$$

This variance estimator is valid even when the small area sample size is small, provided the overall sample size is large.

2.6 Design Issues

"Optimal" design of samples for use with direct estimators of large area totals or means has received a lot of attention over the past 60 years or so. In particular, design issues, such as number of strata, construction of strata, sample allocation and selection probabilities, have been addressed (see, e.g., Cochran (1977)). The ideal goal here is to find an "optimal" design that minimizes the MSE of a direct estimator subject to a given cost. This goal is seldom achieved in practice due to operational constraints and other factors. As a result, a "compromise" design that is "close" to the optimal design is adopted.

In practice, it is not possible to anticipate and plan for all possible areas (or domains) and uses of survey data as "the client will always require more than is specified at the design stage" (Fuller (1999), p. 344). As a result, indirect estimators will always be needed in practice, given the growing demand for reliable small area statistics. However, it is important to consider design issues that have an impact on small area estimation, particularly in the context of planning and designing large-scale surveys. In this section, we present a brief discussion on some of the design issues. A proper resolution of these issues could lead to enhancement in the reliability of direct (and also indirect) estimates for both planned and unplanned domains. For a more detailed discussion, we refer the reader to Singh, Gambino and Mantel (1994) and Marker (2001).

(i) *Minimization of clustering*

Most large-scale surveys use clustering to a varying degree in order to reduce the survey costs. Clustering, however, results in a decrease in the "effective" sample size. It can also adversely affect the estimation for unplanned domains because it can lead to situations where some domains become sample rich while others may have no sample at all. It is therefore useful to minimize the clustering in the sample. The choice of sampling frame plays an important role in this respect; for example, the use of a list frame, replacing clusters wherever possible, such as Business Registers for business surveys and Address Registers for household surveys. Also, the choice of sampling units, their sizes and the number of sampling stages have significant impact on the effective sample size.

(ii) *Stratification*

One method of providing better sample size distribution at the small area level is to replace large strata by many small strata from which samples are drawn. By this approach, it may be possible to make an unplanned small domain contain mostly complete strata. For example, each Canadian province is partitioned into both Economic Regions (ERs) and Unemployment Insurance Regions (UIRs), and there are 71 ERs and 61 UIRs in Canada. In this case, the number of strata may be increased by treating all the areas created by the intersections of the two partitions as strata. This strategy will lead to 133 intersections (strata). As another example of stratification, the United States National Health Interview Survey (NHIS) used stratification by region, metropolitan area status, labor force data, income and racial composition until 1994. The resulting sample sizes for individual states did not support state-level direct estimates for several states; in fact, two of the states did not have NHIS sampled units. The NHIS stratification scheme was replaced by state and metropolitan area status in 1995, thus enabling state-level direct estimation for all states; see Marker (2001) for details.

(iii) *Sample allocation*

By adopting compromise sample allocations, it may be possible to satisfy reliability requirements at a small area level as well as large area level, using only direct estimates. Singh et al. (1994) presented an excellent illustration of compromise sample allocation in the Canadian LFS to satisfy reliability requirements at the provincial level as well as sub-provincial level. For the LFS with a monthly sample of 59,000 households, "optimizing" at the provincial level yields a coefficient of variation (CV) of the direct estimate for "unemployed" as high as 17% for some UIR's (small areas). On the other hand, a two-step compromise allocation with 42,000 households allocated at the first step to get reliable provincial estimates and the remaining 17,000 households allocated at the second step to produce best possible UIR estimates reduced the worst case of 17% CV for UIR to 9.4% at the expense of a small increase at the provincial and national levels: CV for Ontario increased from 2.8% to 3.4% and for Canada from 1.36% to 1.51%. Thus, by oversampling small areas it is possible to decrease the CV of direct estimates for these areas significantly at the expense of a small increase in CV at the national level. The U.S. National Household Survey on Drug Abuse used stratification and oversampling to produce direct estimates for every state. The 2000 Danish Health and Morbidity Survey used two national samples, each of 6,000 respondents, and distributed an additional 8,000 respondents to guarantee at least 1,000 respondents in each county (small area). Similarly, the Canadian Community Health Survey (CCHS) conducted by Statistics Canada accomplishes its sample allocation in two steps. First, it allocates 500 households to each of its 133 Health Regions and then the remaining sample (about one half of 130,000 households) is allocated to maximize the efficiency of provincial estimates; see Béland, Bailie, Catlin and Singh (2000) for details.

(iv) *Integration of surveys*

Harmonizing questions across surveys of the same population leads to increased effective sample sizes for the harmonized items. The increased sample sizes, in turn, lead to improved direct estimates for small areas. However, caution should be exercised because the data may not be comparable across surveys even if the questionnaire wording is consistent. As noted by Groves (1989), different modes of data collection and the placement of questions can cause differences.

A number of current surveys in Europe are harmonized both within countries and between countries. For example, the European Community Household Panel Survey (ECHP) collects consistent data across member countries. Statistics Netherlands uses a common procedure to collect basic information across social surveys.

(v) *Dual frame surveys*

Dual frame surveys can be used to increase the effective sample size in a small area. In a dual frame survey, samples are drawn independently from two overlapping frames that together cover the population of interest. For example, suppose frame A is a complete area frame and data collected by personal interviewing, while frame B is an incomplete list frame and data collected by telephone interviewing. In this case, the dual frame design augments the expensive frame A information with inexpensive additional information from B. There are many surveys using dual frame designs. For example, the Dutch Housing Demand Survey collects data by personal interviewing, but uses telephone supplementation in over 100 municipalities to produce dual frame estimates for those municipalities (small areas). Statistics Canada's CCHS is another recent example, where an area sample is augmented with a telephone list sample in selected Health Regions.

Hartley (1974) discussed dual frame designs, and developed a unified theory for dual frame estimation of totals by combining information from the two samples. Skinner and Rao (1996) developed dual frame estimators that use the same survey weights for all the variables. Lohr and Rao (2000) applied the jackknife method to obtain variance estimators for dual frame estimators of totals.

(vi) *Repeated surveys*

Many surveys are repeated over time and effective sample size can be increased by combining data from two or more consecutive surveys. For example, the United States National Health Interview Survey (NHIS) is an annual survey that uses nonoverlapping samples across years. Combining consecutive annual NHIS samples leads to improved estimates, although the correlation between years, due to the use of the same psu's, reduces the effective sample size. Such estimates, however, can lead to significant bias if the characteristic of interest is not stable over the time period.

Marker (2001) studied the level of accuracy for state estimates by combining the 1995 NHIS sample with the previous year sample or the previous two years samples. He showed that aggregation helps achieve CV's of 30% and 20% for four selected variables, but 10% CV cannot be achieved for many states even after aggregation across 3 years.

Kish (1999) recommended "rolling samples" as a method of cumulating data over time. Unlike the customary periodic surveys, such as the NHIS with the same psu's over time or the Canadian LFS and the United States Current Population Survey (CPS) with the same psu's and partial overlap of sample elements, rolling samples (RS) aim at a much greater spread to facilitate maximal spatial range for cumulation over time. This, in turn, will lead to improved small area estimates when the periodic samples are cumulated. The American Community Survey (ACS), scheduled to begin in 2003, is an excellent example of RS design. It aims to provide monthly samples of 250,000 households and detailed annual statistics based on 3 million households spread across all counties in the United States. It will also provide quinquennial and decennial census samples later.

2.7 Proofs

2.7.1 Proof of $\hat{Y}_{\mathrm{GR}}(\mathbf{x}) = \mathbf{X}$

We have

$$\begin{aligned}\hat{Y}_{\mathrm{GR}}(\mathbf{x}^T) &= \Sigma_s w_j g_j \mathbf{x}_j^T \\ &= \Sigma_s w_j \left[\mathbf{x}_j^T + (\mathbf{X}-\hat{\mathbf{X}})^T \left(\Sigma_s w_j \mathbf{x}_j \mathbf{x}_j^T / c_j\right)^{-1} \mathbf{x}_j \mathbf{x}_j^T / c_j\right] \\ &= \hat{\mathbf{X}}^T + (\mathbf{X}-\hat{\mathbf{X}})^T = \mathbf{X}^T.\end{aligned}$$

2.7.2 Derivation of Calibration Weights w_j^*

We minimize the chi-squared distance $\Sigma_s c_j (w_j - b_j)^2 / w_j$ with respect to the b_j's subject to the calibration constraints $\Sigma_s b_j \mathbf{x}_j = \mathbf{X}$, that is, minimize $\phi = \Sigma_s c_j (w_j - b_j)^2 / w_j - 2\boldsymbol{\lambda}^T (\Sigma_s b_j \mathbf{x}_j - \mathbf{X})$, where $\boldsymbol{\lambda}$ is the vector of Lagrange multipliers. We get

$$b_j = w_j (1 + \mathbf{x}_j^T \boldsymbol{\lambda} / c_j),$$

where

$$\boldsymbol{\lambda} = \Sigma_s (w_j \mathbf{x}_j \mathbf{x}_j^T / c_j)^{-1} (\mathbf{X} - \hat{\mathbf{X}}).$$

Thus $b_j = w_j^*$, where $w_j^* = w_j g_j(s)$ and $g_j(s)$ is given by (2.3.9).

2.7.3 Proof of $\hat{Y} = \hat{\mathbf{X}}^T \hat{\mathbf{B}}$ when $c_j = \boldsymbol{\nu}^T \mathbf{x}_j$

If $c_j = \boldsymbol{\nu}^T \mathbf{x}_j$, then

$$\begin{aligned}\hat{\mathbf{X}}^T \hat{\mathbf{B}} &= \left(\Sigma_s w_j \mathbf{x}_j^T\right) \hat{\mathbf{B}} \\ &= \boldsymbol{\nu}^T \left(\Sigma_s w_j \mathbf{x}_j \mathbf{x}_j^T / c_j\right) \hat{\mathbf{B}} \\ &= \boldsymbol{\nu}^T \left(\Sigma_s w_j \mathbf{x}_j y_j / c_j\right) \\ &= \Sigma_s w_j y_j = \hat{Y}.\end{aligned}$$

This establishes the result $\Sigma_s w_j e_j(s) = 0$ noted below (2.3.11).

Chapter 3

Traditional Demographic Methods

3.1 Introduction

Population censuses are usually conducted at 10-year or 5-year intervals to provide population counts for detailed geographical areas of a country as well as for domains (or subpopulations) defined by age, sex, marital status and other demographic variables. Such counts serve a variety of purposes including the calculation of revenue transfers and grants from federal governments to state and local governments. For example, the U.S. census counts are used in the allocation of federal funds to the 50 states and the 39,000 general purpose governmental units.

Information from a census becomes outdated due to changes in the size and composition of the resident population over time. In the absence of population registers maintained over time (as in some Scandinavian countries), it becomes necessary to develop suitable methods of population estimation in the noncensal years, exploiting administrative files that contain valuable demographic information related to population changes. Such postcensal estimates of population are used for a variety of purposes both in the public and private sectors. Some specific uses include the determination of fund allocations, the calculation of social and economic indicators such as vital rates and unemployment rates in which the population count serves as the denominator, and the calculation of survey weights for use in ongoing large-scale sample surveys. The current population trends derived from postcensal population estimates are also used for planning purposes, market research, and making decisions about site locations, advertising and so on. For example, more than 200 health system agencies in the United States use the postcensal estimates to develop health plans and review proposed health programs (National Research Council (1980)).

Traditional demographic methods employ indirect estimators based on implicit linking models. Typically, sampling is not involved in these methods, excepting the sample regression method (subsection 3.3.2).

Sections 3.2–3.4 give a brief account of traditional demographic methods used in developing postcensal estimates for local areas. These methods may be categorized as (a) symptomatic accounting techniques (SAT) and (b) regression symptomatic procedures. The SAT methods include the vital rates (VR) method (Bogue (1950)), the composite method (Bogue and Duncan (1959)), the composite method-II (CM-II) (U.S. Bureau of the Census, 1966), the administrative records (AR) method (Starsinic (1974)), and the housing unit (HU) method (Smith and Lewis (1980)). The regression symptomatic procedures include the ratio correlation method (Schmitt and Crosetti (1954)), the difference correlation method (O'Hare (1976)), and the sample regression method utilizing current survey estimates (Ericksen (1974)). We refer the reader to the following reports/monographs for detailed accounts of the traditional demographic methods: Purcell and Kish (1979), National Research Council (1980), Rives, Serow, Lee and Goldsmith (1989), Statistics Canada (1987) and Zidek (1982). In recent years, demographers have also been using sophisticated model-based methods, as noted in Section 1.3.

Demographic methods make use of census counts in conjunction with demographic information derived from administrative files, but censuses are often subject to omissions, duplications and misclassification. In fact, the issue of adjusting a census for undercount, utilizing estimates of net undercount from a postcensal survey, has received considerable attention in recent years because of the impact of undercount on the allocation of funds. Section 3.4 gives a brief account of the dual-system method of estimating the total population, using the census counts in conjunction with undercount data from a post-enumeration survey.

3.2 Symptomatic Accounting Techniques

Administrative registers contain current data on various demographic variables, changes in which are strongly related to changes in local population. Such variables are called "symptomatic" indicators. For example, the number of births and deaths and net migration during the period since the last census are obvious components of population change. The diverse registration data used in the U.S. include births and deaths as well as school enrollments and number of existing and new housing units. Data on the number of children receiving family allowance and the number of health care recipients are also used in some countries (e.g., Canada).

3.2.1 Vital Rates Method

The vital rates (VR) method uses only birth and death data for the current year t but assumes that an independent estimate of the current population total, P_t, for a larger area (say a state) containing the local area of interest is available from official sources.

Let (b_t, d_t) and (b_0, d_0) respectively denote the annual number of births and deaths for the local area for the current year t and the last census year $(t = 0)$; b_t and d_t are determined from administrative registers. Further, let (r_{1t}, r_{2t}) and (r_{10}, r_{20}) respectively denote the crude birth and death rates for the local area for the current year and the last census. Note that $r_{1t} = b_t/p_t$ and $r_{2t} = d_t/p_t$, where p_t is the current population for the local area; r_{10} and r_{20} are similarly defined.

If r_{1t} and r_{2t} were known, then both b_t/r_{1t} and d_t/r_{2t} would yield the current population of the local area, p_t. This suggests the estimation of r_{1t} and r_{2t}, utilizing the census values r_{10}, r_{20} and the independent estimate, $\hat{P}_t$, of P_t for the larger area. The VR method essentially finds updating factors ϕ_1 and ϕ_2 such that $r_{1t} = \phi_1 r_{10}$ and $r_{2t} = \phi_2 r_{20}$. Estimates of ϕ_1 and ϕ_2 are obtained by assuming that the same factors also apply for the larger area, that is $R_{1t} = \phi_1 R_{10}$ and $R_{2t} = \phi_2 R_{20}$, where $R_{1a} = B_a/P_a, R_{2a} = D_a/P_a$ are the rates for the larger area and (B_a, D_a) denote the number of births and deaths in the larger area for time $a(= 0, t)$. Under this assumption, using the known census rates R_{10} and R_{20} we obtain estimates of ϕ_1 and ϕ_2 as

$$\hat{\phi_1} = \hat{R_{1t}}/R_{10} \qquad \text{and} \qquad \hat{\phi_2} = \hat{R_{2t}}/R_{20},$$

where $\hat{R_{1t}} = B_t/\hat{P}_t$ and $R_{2t} = D_t/\hat{P}_t$. The estimate of current population, p_t, is then obtained as

$$\hat{p}_t = \frac{1}{2}\left(\frac{b_t}{\hat{r_{1t}}} + \frac{d_t}{\hat{r_{2t}}}\right), \tag{3.2.1}$$

where $\hat{r_{1t}} = \hat{\phi_1} r_{10}$ and $\hat{r_{2t}} = \hat{\phi_2} r_{20}$.

The success of the VR method depends heavily on the validity of the implicit model that the updating factors, ϕ_1 and ϕ_2, for the local area remain valid for the larger area containing the local area. Such a strong assumption is often questionable in practice.

Example 3.2.1. County Population. Govindarajulu (1999, Chapter 17) considered the estimation of the population of a small county in Kentucky. Here $b_t = 400, d_t = 350$ for the current year t and from the 1990 census $R_{10} = 2\%$ and $R_{20} = 1.8\%$. It was assumed that $R_{10} = r_{10}$ and $R_{20} = r_{20}$. The current state rates are $\hat{R}_{1t} = 2.1\%$ and $\hat{R}_{2t} = 1.9\%$ so that $\hat{\phi}_1 = 2.1/2$ and $\hat{\phi}_2 = 1.9/1.8$. Also,

$$\hat{r}_{1t} = (2.1/2)2 = 2.1\%, \hat{r}_{2t} = (1.9/1.8)(1.8) = 1.9\%.$$

It now follows from (3.2.1) that

$$\hat{p}_t = \frac{1}{2}\left(\frac{400}{0.021} + \frac{350}{0.019}\right) \approx 18{,}735.$$

3.2.2 Composite Method

The composite method is a refinement on the VR method. It uses the VR method to compute group-specific population estimates separately and then sums these estimates across the groups to obtain a "composite" estimate of the current population, p_t. This method requires group-specific birth and death counts for the local area as well as current population in each group for the larger area. Alternative data sources may be used for specific groups. For example, the school age population, 5–14, in the United States is estimated by using school enrollment and school enrollment rates (see Zidek (1982)).

3.2.3 Component Methods

Component methods derive current population estimates by taking census values, adding births, subtracting deaths, and adding an estimate of net migration (which can be negative). Let $b_{0,t}$, $d_{0,t}$ and $m_{0,t}$ respectively denote the numbers of births and deaths and net migration in the local area during the period $[0, t]$. Net migration $m_{0,t}$ is the sum of immigration, $i_{0,t}$, and net interarea migration, $n_{0,t}$, minus emigration $e_{0,t}$. The current population, p_t, may be expressed as

$$p_t = p_0 + b_{0,t} - d_{0,t} + m_{0,t},$$

where p_0 is the baseline census population. Registration of births and deaths are usually complete in the United States and Canada, but net migration figures are not directly available. In the United States, net migration is divided into military and civilian migration since the former is obtainable from administrative records. For estimating civilian migration, the component method-II (CM-II) uses school enrollments, whereas the Administrative Records (AR) method employs income tax returns (see Zidek (1982)). An earlier version of CM-II is called CM-I. In Canada, emigration and interarea migration are inferred from personal income tax files for census divisions (or areas). Immigration statistics at the province level are obtained from official sources, but such official statistics are not available for all census divisions. As a result, immigration figures at the census division level are also estimated from income tax returns but benchmarked to the official figures for the provinces.

Post-censal population estimates by age, sex and marital status are produced in Canada using the cohort component method which is similar to the component method but uses certain modifications because of the nature of disaggregation (Statistics Canada (1987), Chapter II).

3.2.4 Housing Unit Method

Housing units and group quarters (e.g., college dormitories, prisons, nursing homes) may be distinguished as places of residence. Accordingly, the current population, p_t, may be expressed as

$$p_t = (h_t)(pph_t) + gq_t,$$

where h_t is the number of occupied housing units at time t, pph_t is the average number of persons per housing unit at time t and gq_t is the number of persons in group quarters at time t. The quantities h_t, pph_t, and gq_t all need to be estimated. In particular, h_t is estimated from the change in the number of housing units, $hu_t - hu_0$, which, in turn, is given by

$$hu_t - hu_0 = bp_t - dl_t.$$

Here bp_t is the number of building permits issued during $[0, t]$ adjusted for completion times and dl_t is the number of demolitions during $[0, t]$. It now follows that h_t may be obtained as

$$h_t = h_0 + (bp_t - dl_t)(ocr),$$

where ocr denotes the occupancy rate which may be estimated by, ocr_0, the value at the time of the last census. The quantity pph_t is estimated by the value pph_0 at the time of the last census or estimated by extrapolating pph values from the previous two censuses. Finally, gq_t is estimated by the value gq_0 at the time of the last census or determined by extrapolation.

Various refinements and modifications of the housing unit (HU) method have been proposed. For example, h_t may be obtained as $h_0 + rec_t - rec_0$, where rec_α is the number of active residential electrical units at time α, or as $(h_0/rec_0)(rec_t)$. A more accurate estimate of h_t may be obtained by stratifying the housing units according to type (single family, mobile homes, etc.) and then applying the HU method separately in each stratum. The resulting strata estimators are summed to get an estimate of h_t.

The HU method has been used in the United States to estimate county and sub-county level populations. It is an attractive method but getting the relevant data is not easy.

3.3 Regression Symptomatic Procedures

Regression symptomatic procedures use multiple linear regression to estimate local area populations, utilizing symptomatic variables as independent variables in the regression equation. Two such procedures are the ratio correlation and the difference correlation methods.

3.3.1 Ratio Correlation and Difference Correlation Methods

Let $0, 1$ and $t(> 1)$ denote two consecutive census years and the current year, respectively. Also, let p_{ia} and s_{ija} be the population size and the value of the jth symptomatic variable ($j = 1, \ldots, p$) for the ith local area ($i = 1, \ldots, m$) in the year a ($= 0, 1, t$). Also, let p_{ia}/P_a and s_{ija}/S_{ja} be the corresponding proportions, where $P_a = \sum_i p_{ia}$ and $S_{ja} = \sum_i s_{ija}$ are the values for the larger area (state or province).

The change in proportional values of the dependent variable between census years 0 and 1, say U_i for the ith area, are related to the corresponding changes in proportional values of the symptomatic variables, say z_{ij} for the jth symptomatic variable and the ith area, through multiple linear regression:

$$U_i = \gamma_0 + \gamma_1 z_{i1} + \ldots + \gamma_p z_{ip} + u_i, \tag{3.3.1}$$

where the u_i's are random errors assumed to be uncorrelated with zero means and constant variance σ_u^2. In the ratio correlation method ratios are used to measure the changes:

$$U_i = \frac{p_{i1}/P_1}{p_{i0}/P_0}, \qquad z_{ij} = \frac{s_{ij1}/S_{j1}}{s_{ij0}/S_{j0}},$$

whereas the difference correlation method uses differences:

$$U_i = p_{i1}/P_1 - p_{i0}/P_0$$

and

$$z_{ij} = s_{ij1}/S_{j1} - s_{ij0}/S_{j0}.$$

Equation (3.3.1) is fitted to the data $(U_i, z_{i1}, \ldots, z_{ip}; i = 1, \ldots, m)$ by the ordinary least squares method, to get the least squares estimates $\hat{\gamma}_0, \ldots, \hat{\gamma}_p$. The change in the postcensal period, Y_i, is then predicted from the fitted regression equation:

$$Y_i^* = \hat{\gamma}_0 + \hat{\gamma}_1 x_{i1} + \ldots + \hat{\gamma}_p x_{ip}$$

using the known changes, x_{ij}, in the symptomatic proportions in the postcensal period. The ratio correlation method uses

$$Y_i = \frac{p_{it}/P_t}{p_{i1}/P_1}, \qquad x_{ij} = \frac{s_{ijt}/S_{jt}}{s_{ij1}/S_{j1}},$$

whereas the difference correlation method uses

$$Y_i = p_{it}/P_t - p_{i1}/P_1$$

and

$$x_{ij} = s_{ijt}/S_{jt} - s_{ij1}/S_{j1}.$$

The current population counts, p_{it}, are finally estimated as

$$\hat{p}_{it} = Y_i^*(p_{i1}/P_1)\hat{P}_t$$

in the ratio correlation method, and estimated as

$$\hat{p}_{it} = [Y_i^* - (p_{i1}/P_1)]\,\hat{P}_t$$

in the difference correlation method, where an independent estimate, $\hat{P}_t$, of the count for the larger area is ascertained from other sources. A choice between the two methods is customarily based on empirical evaluations.

Multicollinearity in fitting the regression equation (3.3.1), unequal error variances and other problems are handled by the customary methods used in linear regression analysis before estimating the regression parameters γ_i. For example, Statistics Canada (1987) uses a weighted regression method to handle heteroscedasticity and ridge regression to control multicollinearity.

In Canada, the regression method was used to produce preliminary estimates for local areas (census divisions and census metropolitan areas), while the final estimates were based on the component method. A "regression nested" method was also used to improve the preliminary estimates. If t and $t+1$ denote two consecutive postcensal years, then a preliminary estimate for time $t+1$ based on the regression nested method is given by

$$\hat{p}_{i,t+1}(RN) = \hat{p}_{it}(C) + [\hat{p}_{i,t+1}(R) - \hat{p}_{it}(R)] .$$

Here $\hat{p}_{it}(C)$ and $\hat{p}_{it}(R)$ denote the tth year estimates for the ith local area based on the component method and the regression method, respectively. A comparison with the 1981 census counts, taking 1981 as the current year t and 1971 and 1976 as the census years 0 and 1, showed that the regression nested method performed better than the regression and component methods for Canada as a whole in terms of absolute relative error (ARE). The ARE criterion is defined as

$$ARE = \frac{1}{m}\sum_{i=1}^{m} |\hat{p}_{it} - p_{it}|/p_{it},$$

where p_{it} is the census count for the ith local area, $\hat{p}_{it}$ is the corresponding estimated count for a particular method and m is the number of census divisions in Canada. The difference between the regression nested and component methods, however, was not statistically significant. For individual provinces, the component method was slightly better than the regression nested method in 5 out of 10 provinces, whereas the converse was true in the remaining 5 provinces. On the other hand, the regression method was inferior to the component and regression nested methods except in the province of Saskatchewan. In Saskatchewan, information from health insurance files was used as the symptomatic variable in the regression method whereas in the remaining provinces, information on family allowance recipients was used as the symptomatic variable.

3.3.2 Sample Regression Method

The ratio-correlation and difference-correlation methods use the regression coefficients, $\hat{\gamma}_j$, based on the data from the two previous censuses, to predict the change in the postcensal period, Y_i. But changes in the statistical relationship can lead to biases in the predictions Y_i^*, that is, the current relationship

$Y_i = \beta_0 + \beta_1 x_{i1} + \ldots + \beta_p x_{ip} + v_i$ might involve regression parameters, β_j, significantly different from γ_j. The sample regression method (Ericksen (1974)) avoids this problem by using survey estimates, $\hat{Y}_i$ to establish the current regression equation.

Suppose survey estimates $\hat{Y}_i$ are available for k out of m local areas ($i = 1, \ldots, k$). Then the regression equation is fitted to the data $(\hat{Y}_i, x_{i1}, \ldots, x_{ip})$ from the k sampled areas and the Y_i for all the areas are then predicted from the fitted regression equations:

$$\tilde{Y}_i = \hat{\beta}_0 + \hat{\beta}_1 x_{i1} + \ldots + \hat{\beta}_p x_{ip}, \qquad i = 1, \ldots, m.$$

The current counts, p_{it}, are estimated as before by changing Y_i^* to $\tilde{Y}_i$.

Ericksen (1974) compared the mean percentage errors of the sample regression and ratio-correlation methods, using 1970 U.S. census data and sample data from the Current Population Survey (CPS). His results have shown that the reduction of average error is slight compared to the ratio-correlation method but the reduction of larger errors (10 % or greater) is more substantial.

To study the overall performances of the survey estimates $\hat{\mathbf{Y}} = (\hat{Y}_1, \ldots, \hat{Y}_k)^T$, the sample regression estimates $\tilde{\mathbf{Y}} = (\tilde{Y}_1, \ldots, \tilde{Y}_k)^T$ and the ratio or difference correlation estimates $\mathbf{Y}^* = (Y_1^*, \ldots, Y_k^*)^T$ for the sampled areas, we compute the average mean squared errors under the assumed model

$$\mathbf{Y} = \mathbf{X}\boldsymbol{\beta} + \mathbf{v}, \tag{3.3.2}$$

where $\mathbf{Y} = (Y_1, \ldots, Y_k)^T$, $\mathbf{X}$ is a $k \times (p+1)$ matrix of symptomatic variables with ith row $\mathbf{x}_i^T = (1, x_{i1}, \ldots, x_{ip})^T$, $\boldsymbol{\beta} = (\beta_0, \ldots, \beta_p)^T$. The vector of errors $\mathbf{v}^T = (v_1, \ldots, v_k)^T$ has mean $\mathbf{0}$ and covariance matrix $\sigma_v^2\mathbf{I}$, where $\mathbf{0}$ is the k-vector of 0's and $\mathbf{I}$ is the $k \times k$ identity matrix. We also assume that the vector of survey estimators $\hat{\mathbf{Y}}$ is design-unbiased for $\mathbf{Y}$ with uncorrelated sampling errors:

$$\hat{\mathbf{Y}} = \mathbf{Y} + \mathbf{e}, \tag{3.3.3}$$

where $\mathbf{e}$ is the vector of sampling errors with mean $\mathbf{0}$ and covariance matrix $\sigma_e^2\mathbf{I}$, and $\mathbf{e}$ is uncorrelated with the model errors $\mathbf{v}$. Combining (3.3.2) and (3.3.3) leads to

$$\hat{\mathbf{Y}} = \mathbf{X}\boldsymbol{\beta} + \mathbf{v} + \mathbf{e}, \tag{3.3.4}$$

where the composite error $\boldsymbol{\varepsilon} = \mathbf{v} + \mathbf{e}$ has mean $\mathbf{0}$ and covariance matrix $(\sigma_v^2 + \sigma_e^2)\mathbf{I}$.

The average (model) MSE's, derived in Section 3.5, are given by

$$\frac{1}{k} E_m(\hat{\mathbf{Y}} - \mathbf{Y})^T(\hat{\mathbf{Y}} - \mathbf{Y}) = \sigma_e^2 \tag{3.3.5}$$

$$\frac{1}{k} E_m(\tilde{\mathbf{Y}} - \mathbf{Y})^T(\tilde{\mathbf{Y}} - \mathbf{Y}) = \sigma_v^2 + \frac{p+1}{k}(\sigma_e^2 - \sigma_v^2) \tag{3.3.6}$$

and

$$\frac{1}{k}E_m(\mathbf{Y}^* - \mathbf{Y})^T(\mathbf{Y}^* - \mathbf{Y})$$
$$= \sigma_v^2 + \frac{1}{k}\left[\mathrm{tr}\left(\mathbf{X}^T\mathbf{X}\right)\left(\mathbf{Z}^T\mathbf{Z}\right)^{-1}\sigma_u^2 + (\boldsymbol{\gamma} - \boldsymbol{\beta})^T\left(\mathbf{X}^T\mathbf{X}\right)(\boldsymbol{\gamma} - \boldsymbol{\beta})\right], \quad (3.3.7)$$

where E_m denotes expectation with respect to the combined model (3.3.4), tr denotes the trace operator, $\mathbf{Z}$ is a $k \times (p+1)$ matrix with ith row $\mathbf{z}_i^T = (1, z_{i1}, \ldots, z_{ip})^T$, $\boldsymbol{\gamma} = (\gamma_0, \ldots, \gamma_p)^T$ and the model errors u_i in (3.3.1) are assumed to be uncorrelated with the current model errors v_i in deriving (3.3.7).

It now follows from (3.3.5) and (3.3.6) that the sample regression estimator $\tilde{\mathbf{Y}}$ has smaller average MSE than the survey estimator $\hat{\mathbf{Y}}$ if $\sigma_v^2 < \sigma_e^2$, noting that

$$\sigma_e^2 - \left\{\sigma_v^2 + \frac{p+1}{k}(\sigma_e^2 - \sigma_v^2)\right\} = \frac{k-p-1}{k}(\sigma_e^2 - \sigma_v^2).$$

In practice, the sampling variance σ_e^2 is likely to be significantly larger than the model variance σ_v^2, especially when the regression fit is good. The comparison between the average MSE's of $\tilde{\mathbf{Y}}$ and $\mathbf{Y}^*$ depends on the size of structural change as measured by the distance $(\boldsymbol{\gamma}-\boldsymbol{\beta})^T\left(\mathbf{X}^T\mathbf{X}\right)(\boldsymbol{\gamma}-\boldsymbol{\beta})$. In the absence of structural change and assuming $\mathrm{tr}\left(\mathbf{X}^T\mathbf{X}\right)\left(\mathbf{Z}^T\mathbf{Z}\right)^{-1} \approx \mathrm{tr}\,\mathbf{I} = p+1$, the sample regression estimator $\tilde{\mathbf{Y}}$ has larger average MSE than $\mathbf{Y}^*$ if $\sigma_e^2 > \sigma_u^2 + \sigma_v^2$ which is likely to be satisfied. Thus the success of the sample regression method depends on the dynamics of the underlying regression relationship and the precision of the survey estimates $\hat{\mathbf{Y}}$.

It is possible to get a more efficient estimator than $\tilde{\mathbf{Y}}$ and $\hat{\mathbf{Y}}$, by taking an optimal linear combination of $\tilde{\mathbf{Y}}$ and $\hat{\mathbf{Y}}$, assuming a known proportion of model variance to total variance, $\sigma_v^2/(\sigma_v^2 + \sigma_e^2) = \delta_0$. The combined (or composite) estimator is given by

$$\hat{\mathbf{Y}}(\delta_0) = \delta_0\hat{\mathbf{Y}} + (1-\delta_0)\tilde{\mathbf{Y}}, \quad (3.3.8)$$

with average MSE (see Section 3.5)

$$\frac{1}{k}E_m\left[(\hat{\mathbf{Y}}(\delta_0) - \mathbf{Y})^T(\hat{\mathbf{Y}}(\delta_0) - \mathbf{Y})\right]$$
$$= \left(1 - \frac{p+1}{k}\right)\sigma_v^2(1-\delta_0) + \frac{p+1}{k}\sigma_e^2 \quad (3.3.9)$$
$$< \left(1 - \frac{p+1}{k}\right)\sigma_v^2 + \frac{p+1}{k}\sigma_e^2$$
$$= \text{average MSE of } \tilde{\mathbf{Y}}.$$

The quality of the estimator $\tilde{\mathbf{Y}}$ may be measured by its estimated average MSE which is obtained from (3.3.6) by using estimators of σ_v^2 and σ_e^2. Ericksen

(1974) used a random groups estimator of σ_e^2 by splitting the sample into half samples within each sampled area and then averaging the resulting variance estimators over the areas. Denote the half-sample estimators in the ith area by $\hat{Y}_{i1}$ and $\hat{Y}_{i2}$ and assume that $\hat{Y}_{i1}$ and $\hat{Y}_{i2}$ are uncorrelated and unbiased for Y_i. The estimator of σ_e^2 is then given by

$$\hat{\sigma}_e^2 = \frac{1}{4k}\sum_{i=1}^{k}(\hat{Y}_{i1} - \hat{Y}_{i2})^2, \tag{3.3.10}$$

noting that $\hat{Y}_i = (\hat{Y}_{i1} + \hat{Y}_{i2})/2$ and $V_p(\hat{Y}_{i1}) = V_p(\hat{Y}_{i2}) = 2\sigma_e^2$, where V_p denotes the variance with respect to the sampling design. The estimator (3.3.10) will be stable if the number of sampled areas, k, is not small. To estimate the model variance σ_v^2, Ericksen (1974) used the moment estimator

$$\hat{\sigma}_v^2 = \frac{k}{k-p-1}\left\{\frac{1}{k}\sum_i\left(\hat{Y}_i - \tilde{Y}_i\right)^2\right\} - \hat{\sigma}_e^2. \tag{3.3.11}$$

Substituting (3.3.10) and (3.3.11) in (3.3.6), we get an unbiased estimator of the average MSE of $\tilde{\mathbf{Y}}$ under the model:

$$\overline{\text{mse}}(\tilde{\mathbf{Y}}) = \frac{1}{k}\sum_i\left(\hat{Y}_i - \tilde{Y}_i\right)^2 - \frac{k-2p-2}{k}\hat{\sigma}_e^2. \tag{3.3.12}$$

To use the optimal estimator $\hat{\mathbf{Y}}(\delta_0)$ in practice we substitute $\hat{\delta} = \hat{\sigma}_v^2/(\hat{\sigma}_v^2 + \hat{\sigma}_e^2)$ in (3.3.8). The average MSE of the resulting estimator $\hat{\mathbf{Y}}(\hat{\delta})$ will be larger than (3.3.9) due to uncertainty associated with $\hat{\delta}$. As a result, estimating (3.3.9) and using it as an estimator of average MSE of $\hat{\mathbf{Y}}(\hat{\lambda})$ can lead to significant underestimation. Chapter 7 discusses methods that take account of the variability in $\hat{\delta}$ in estimating the MSE of $\hat{Y}_i(\hat{\delta})$.

Example 3.3.1. Radio Listening. Hansen, Hurwitz and Madow (1953, pp. 483–486) described the first application of sample regression. The objective of a radio listening survey was to estimate the median number of stations heard during the day in each of the more than 500 county areas in the United States. A mail survey was first conducted by sampling 1,000 families from each county area and mailing questionaires. Incomplete list frames were used for this purpose and the response rate was about 20%. From the mail survey, the median number of stations heard during the day was estimated for each county area. These estimates, x_i, are biased due to nonresponse and incomplete coverage but are expected to have high correlation with the true median values, y_{0i}. Unbiased estimates of y_{0i} for a sample of 85 county areas were obtained through an intensive interview survey. The sample county areas were selected by first grouping the population county areas into 85 primary strata on the basis of geographical region and type of radio service available, and then selecting one county area from each stratum with probability proportional to the estimated number of families in the county. A subsample of area segments was selected from each of the sampled county areas and the families within the selected area segments were interviewed. Using the data collected from the interviews, unbiased estimates, y_i, of the median values, y_{0i}, were obtained for the sample county areas. The correlation between y_i and x_i was 0.70.

Sample regression estimates of y_{0i} for the nonsampled county areas were obtained by fitting a linear regression to the data $(y_i, x_i; i = 1, \ldots, 85)$ and then predicting from the fitted regression equation. The resulting sample regression estimates were calculated from $\tilde{y}_i = 0.52 + 0.74x_i$, using x_i for the nonsampled counties. Note that Ericksen (1974) used the fitted regression equation for both the nonsampled and sampled counties.

3.4 Dual-system Estimation of Total Population

Population censuses are aimed at complete enumeration of the total population, but they are often subject to omissions, duplications and misclassifications. Comparisons of census counts to administrative data at aggregate levels have typically shown a differential net undercount by age, sex and race and by region (Hogan (1992)). A net undercount is the difference between omissions and erroneous inclusions in the census. Post-enumeration surveys (PESs) have also shown differential net undercount. A PES is mainly used to provide data users with estimates of coverage error as measures of quality of the census. Such estimates are also useful to census planners for taking measures to reduce coverage errors in future censuses.

In this section, we consider the dual-system method of estimating the total population, using the census count in conjunction with data from a PES. The difference between the estimated total population and the census count is taken as an estimate of the net census undercount. In practice, the population is divided into several poststrata (homogenous groups) and the dual-system estimates are made for each of the poststrata and then aggregated to desired levels. The use of dual-system estimators in small area estimation will be considered in Example 8.2.1, Chapter 8.

3.4.1 Dual-system Model

Suppose that N denotes the size of the population (poststratum) U, and that N is fixed but unknown. The main problem is to estimate N using the census count in conjunction with the PES data. The census of U aims at complete enumeration but the actual census count differs from N due to omissions and false enumerations. Typically, the census count is less than N, that is, the net census undercount is positive. We denote the census as list A and the list from which the post enumeration sample is drawn as list B, following Wolter (1986). List B is also assumed to aim at complete enumeration of U. We further assume that duplicates, fictitious cases and other erroneous inclusions are eliminated from both lists.

The dual-system model conceptualizes each person in U as either in or not in list A as well as either in or not in list B. This model leads to the following 2×2 table of counts, assuming no matching errors.

		List B		
		In	Out	Total
	In	N_{11}	N_{12}	N_{1+}
List A	Out	N_{21}	N_{22}	N_{2+}
	Total	N_{+1}	N_{+2}	N

All cell counts are conceptually observable except for N_{22}. Note that the marginal count N_{1+} differs from the actual census count due to the exclusion of false enumerations from the census.

The dual-system model makes the following assumptions: (a) The population U is closed in the sense that the census reference period is well defined and that there are no changes due to birth, death, immigration or emigration during the reference period; (b) The 2×2 table of counts is the result of N independent trials with cell probabilities $\{\pi_{ijl}; j, l = 1, 2\}$ for the ith individual ($i = 1, 2, \ldots, N$); (c) The event of inclusion in list A is independent of the event of inclusion in list B for each individual i, that is, $\pi_{ijl} = \pi_{ij+}\pi_{i+l}$ ($j, l = 1, 2$), where π_{ij+} and π_{+jl} are the marginal probabilities; (d) The marginal probabilities π_{ij+} and π_{i+l} are homogenous across individuals i, that is, $\pi_{ij+} = \pi_{j+}$ and $\pi_{i+l} = \pi_{+l}$. Assumption (d) is unlikely to hold for the total population. Therefore, poststratification is commonly used to create homogeneous groups within each of which assumption (d) holds.

Under assumptions (a)–(d), the likelihood function is given by

$$L(N, \pi_{1+}, \pi_{+1})$$
$$= \frac{N!}{N_{11}!N_{12}!N_{21}!N_{22}!} \left[\pi_{1+}^{N_{1+}} \pi_{+1}^{N_{+1}} (1 - \pi_{1+})^{N - N_{1+}} (1 - \pi_{+1})^{N - N_{+1}}\right]. \quad (3.4.1)$$

Note that the likelihood function is the probability of realizing the counts $\{N_{jl}\}$ expressed as a function of the unknown parameters N, π_{1+} and π_{+1}. The maximum likelihood estimators of N, π_{1+} and π_{+1} are

$$\tilde{N} = \left[\frac{N_{1+}N_{+1}}{N_{11}}\right] \quad (3.4.2)$$

$$\tilde{\pi}_{1+} = N_{11}/N_{+1}, \qquad \tilde{\pi}_{+1} = N_{11}/N_{1+},$$

where $[x]$ denotes the greatest integer less than or equal to x. Note that $(N_{1+}N_{+1})/N_{11} = N_{1+}/\tilde{\pi}_{1+}$ so that $\tilde{N}$ may be interpreted as the census population, N_{1+}, divided by its estimated "capture" probability $\tilde{\pi}_{1+}$.

The dual-system estimation has a long history, dating back to Petersen (1896), who obtained $\tilde{N}$ in the context of estimating the size of fish populations. In this application, the fish captured in the initial sample (list A) of size N_{1+} are tagged or marked before taking the second sample (list B) of size N_{+1}. The number of marked fish recaptured in the second sample, N_{11}, is used to compute the estimate, $\tilde{N}$, of the total fish population. The two samples are assumed to be randomly drawn from the same population, U. Sekar and Deming (1949) used $\tilde{N}$ in the context of estimating the births and deaths in the years 1945 and 1946 in a region near Calcutta, India. They also obtained the estimated variance of $\tilde{N}$ as

$$v(\tilde{N}) = (N_{1+}N_{+1}N_{12}N_{21})/N_{11}^3,$$

using the Taylor linearization method and assuming that the two samples are large. Seber (1982) has given a comprehensive account of capture-recapture (or dual-system) methods for animal populations. Fienberg (1992) presented an annotated bibliography of the literature on capture-recapture techniques. Wolter (1986) gave various extensions and refinements of the basic dual-system model.

3.4.2 Post-enumeration Surveys

When list B is sampled in a PES, the values of N_{11} and N_{+1} in the estimator $\tilde{N}$ are unknown because list B is not enumerated completely. Even the marginal count N_{1+} is unknown since it differs from the actual census count due to false enumerations and PES data are available for only a sample. In practice, these values are estimated from a PES which, in turn, lead to an estimate of N. For the 1980 census of the United States, the PES was based on two samples: the April 1980 Current Population Survey (CPS) sample and a sample (known as the E, or enumeration, sample) of the census enumerations reenumerated to ascertain erroneous inclusions in the census. The count N_{11} was estimated by clerical matching of the CPS sample to the 1980 census enumeration and using the CPS weights. The marginal count N_{1+} was also estimated using the CPS weights. The marginal count N_{+1} was estimated by subtracting from the census count an estimate of the erroneous inclusions from the E-sample. We refer the reader to Wolter (1986) for further details of the 1980 Post Enumeration Program (PEP).

The 1990 PES of the United States was designed to produce adjusted census counts for states and local areas but the U.S. Government decided not to adjust the census. For the PES, the population was divided into 1,392 poststrata based on geography, race, origin, housing tenure, age and sex, and estimates of total population were calculated for each poststratum and then aggregated to the desired level. Poststratification was designed to ensure that the homogeneity assumption (d) of the dual-system method was approximately satisfied within each poststratum as well as to reduce the sampling variance. The 1990 PES used a stratified cluster sample design with blocks or block groups as clusters. The strata were set up to correspond closely with the poststrata, and 54 geographical areas served as the major strata. The major strata were further divided into strata according to race/Hispanic origin and tenure (renters and owners) of the block clusters. A sample of 5,300 block clusters was selected, and the same sample blocks were used for the P (population) sample and the E (enumeration) sample. The P-sample covered all people living in the sample blocks at the time of the PES interview, while the E-sample consisted of all census enumerations, correct or incorrect, in the sample blocks. The P-sample was matched to the census enumeration to estimate the number of matched persons, N_{11}, as well as the marginal count N_{+1}. As before, the E-sample was used to ascertain the number of erroneous inclusions in the census.

The estimator of N within a poststratum takes the form

$$\hat{N} = (\hat{N}_{+1}\hat{N}_{1+})/\hat{N}_{11}, \tag{3.4.3}$$

where $\hat{N}_{11}$ is the weighted estimate of P-sample matches, $\hat{N}_{+1}$ is the weighted P-sample total, and $\hat{N}_{1+}$ is obtained as

$$\hat{N}_{1+} = (N_c - I_c)(1 - \hat{E}_c/\hat{N}_c).$$

Here N_c is the actual census count, I_c is the number of "whole-person" census imputations, $\hat{E}_c$ is the weighted estimate of E-sample erroneous enumerations, $\hat{N}_c$ is the weighted E-sample total (Hogan (1992)). The difference between $\hat{N}$ and N_c estimates the net census undercount and the ratio $\hat{N}/N_c$ is an estimate of the adjustment factor N/N_c. To produce adjusted census counts for small areas, these adjustment factors are smoothed using model-based methods and then applied to the census counts by block and poststrata (see Example 8.2.1, Chapter 8).

The sampling variance of the dual-system estimator $\hat{N}$ at a large area level may be estimated using the standard Taylor linearization (or delta method). We write $\hat{N}$ as

$$\hat{N} = \tilde{N}(1 + \delta\hat{N}_{1+})(1 + \delta\hat{N}_{+1})(1 + \delta\hat{N}_{11})^{-1}$$

with $\delta\hat{N}_{1+} = (\hat{N}_{1+} - \tilde{N}_{1+})/\tilde{N}_{1+}$ and similar expressions for $\delta\hat{N}_{+1}$ and $\delta\hat{N}_{11}$, where $\tilde{N}_{1+} = E_p(\hat{N}_{1+})$ and E_p denotes expectation with respect to the PES design. For large samples, we may approximate $\hat{N} - \tilde{N}$ as

$$\hat{N} - \tilde{N} \approx \tilde{N}(\delta\hat{N}_{+1} + \delta\hat{N}_{1+} - \delta\hat{N}_{11}),$$

noting that $(1 + \delta\hat{N}_{11})^{-1} \approx 1 - \delta\hat{N}_{11}$ and neglecting terms of order two and higher in the δ's. Therefore, $E_p(\hat{N}) \approx \tilde{N}$, that is, $\hat{N}$ is approximately design-unbiased for $\tilde{N}$. Further, an estimator of variance of $\hat{N}$ is

$$\begin{aligned} v(\hat{N}) = \hat{N}^2 \Big[& v(\hat{N}_{1+})/\hat{N}_{1+}^2 + v(\hat{N}_{+1})/\hat{N}_{+1}^2 + v(\hat{N}_{11})/\hat{N}_{11}^2 \\ & + 2\operatorname{cov}(\hat{N}_{1+}, \hat{N}_{+1})/(\hat{N}_{1+}\hat{N}_{+1}) - 2\operatorname{cov}(\hat{N}_{1+}, \hat{N}_{11})/(\hat{N}_{1+}\hat{N}_{11}) \\ & - 2\operatorname{cov}(\hat{N}_{+1}, \hat{N}_{11})/(\hat{N}_{+1}\hat{N}_{11}) \Big], \end{aligned} \tag{3.4.4}$$

where $v(\cdot)$ denotes the estimated variance and $\operatorname{cov}(\cdot,\cdot)$ denotes the estimated covariance. The variance estimator $v(\hat{N})$ can be computed from the PES data, noting that the same sample blocks are used for the P-sample and the E-sample. Note that $\tilde{N}$ is an estimator of N under the dual-system model, but its model variance is negligible relative to the design variance of $\hat{N}$ since N_{+1} and N_{1+} are very large in the census context. We may therefore regard $\hat{N}$ as an estimator of the true population size, N, under the dual-system model, with estimated variance $v(\hat{N})$ given by (3.4.4).

Mulry and Spencer (1991) analyzed the total error in the dual-system estimates of the total population by decomposing the total error into sampling error, matching error and other nonsampling errors. Ding and Fienberg (1994) proposed models and methods for dual-system estimation of census undercount in the presence of matching errors.

In Canada, the Coverage Error Measurement Program for 1991 consisted of four studies: (a) Reverse Record Check (RRC), (b) Overcoverage Study (OCS), (c) Vacancy Check (VC) and (d) Temporary Resident Study (TRS). Estimates from studies (c) and (d) were included in the final census counts to account for undercoverage from these two specific sources. The RRC was designed to estimate total undercoverage including that measured by VC and TRS, while OCS was designed to estimate total overcoverage. No dual-system estimation was employed as the number of persons in neither the census nor the RRC, N_{22}, was expected to be negligible relative to N. The RRC used the following sampling frames: the 1986 census, births and immigration during the period June 3, 1986 to June 3, 1991, all cases classified as not enumerated in the 1986 RRC, nonpermanent residents and health care files of Yukon and Northwest Territories on June 4, 1991. Sampling was carried out independently within each frame and the sample design varied from frame to frame. A total of 55,912 persons were selected for the RRC. We refer the reader to Statistics Canada (1993) for further details on the RRC as well as the descriptions of the OCS, VC, and TRS.

The total population at the provincial level was estimated as

$$\hat{N} = N_c + (\hat{M} - \widehat{\text{VC}} - \widehat{\text{TR}}) - \hat{O},$$

where N_c = the published census count, $\hat{M}$ = estimate of missed persons as obtained from the RRC, $\widehat{\text{VC}}$ = estimate from VC of the number of persons who occupied dwellings classified as unoccupied, $\widehat{\text{TR}}$ = estimate of the number of temporary residents from TRC, and $\hat{O}$ = estimate of overcoverage from the OCS. The estimates $\widehat{\text{VC}}$ and $\widehat{\text{TR}}$ were subtracted from $\hat{M}$ because the published census count, N_c, already included them. The population undercoverage rate was estimated as

$$\hat{R}_u = (\hat{M} - \widehat{\text{VC}} - \widehat{\text{TR}})/\hat{N},$$

where $\hat{M} - \widehat{\text{VC}} - \widehat{\text{TR}}$ is the estimated number of persons not enumerated in the census. Statistics Canada (1993) also provided standard errors for $\hat{M} - \widehat{\text{VC}} - \widehat{\text{TR}}$ and $\hat{R}_u$ for each of the ten provinces, Yukon and Northwest Territories. Thibault, Julien and Dick (1995) described the plans for the 1996 Census

Coverage Error Measurement Program that involved several methodological and operational changes to RRC.

The direct estimates, $\hat{N}$, and the associated adjustment factors $\hat{N}/N_c$, are considered to be reliable for large areas such as provinces, larger urban centers and males in the age group 20–24 years. However, the Population Estimates Program requires estimates for each single year of age for both males and females at the provincial level. The use of small area methods for smoothing the estimated adjustment factors to get more reliable small domain estimates will be considered in Chapter 7, Example 7.1.3.

3.5 Derivation of Average MSEs

In this section, we derive the average MSEs in (3.3.5), (3.3.6), (3.3.7) and (3.3.9). We consider the estimator (3.3.8) for an arbitrary constant δ ($0 \le \delta \le 1$) and write it as

$$\hat{\mathbf{Y}}(\delta) = \tilde{\mathbf{Y}} + \delta(\hat{\mathbf{Y}} - \tilde{\mathbf{Y}}),$$

that is,

$$\hat{\mathbf{Y}}(\delta) - \mathbf{Y} = (1-\delta)(\tilde{\mathbf{Y}} - \mathbf{Y}) + \delta(\hat{\mathbf{Y}} - \mathbf{Y}).$$

Further, $\hat{\mathbf{Y}} - \mathbf{Y} = \mathbf{e}$ and $\tilde{\mathbf{Y}} - \mathbf{Y} = \mathbf{He} - (\mathbf{I} - \mathbf{H})\mathbf{v}$, where $\mathbf{H} = \mathbf{X}(\mathbf{X}^T\mathbf{X})^{-1}\mathbf{X}^T$ is the projection matrix. Noting that $\mathbf{v}$ and $\mathbf{e}$ are uncorrelated and that $\mathbf{H}$ is idempotent (that is, $\mathbf{H}^2 = \mathbf{H}$), we get the average MSE of $\hat{\mathbf{Y}}(\delta)$ as

$$\begin{aligned}
&\frac{1}{k}E_m\left[(\hat{\mathbf{Y}}(\delta) - \mathbf{Y})^T(\hat{\mathbf{Y}}(\delta) - \mathbf{Y})\right] \\
&= \frac{1}{k}\left[(1-\delta)^2\{(k-p-1)\sigma_v^2 + (p+1)\sigma_e^2\}\right. \\
&\qquad \left. + \delta^2(k\sigma_e^2) + 2\delta(1-\delta)(p+1)\sigma_e^2\right] \\
&= \left(1 - \frac{p+1}{k}\right)\left[(1-\delta)^2\sigma_v^2 + \delta^2\sigma_e^2\right] + \frac{p+1}{k}\sigma_e^2 \qquad (3.5.1)
\end{aligned}$$

noting that $E[\mathbf{v}^T(\mathbf{I}-\mathbf{H})\mathbf{v}] = \sigma_v^2\text{tr}(\mathbf{I}-\mathbf{H})$, $E(\mathbf{e}^T\mathbf{He}) = \sigma_e^2\text{tr}(\mathbf{H})$, $\text{tr}(\mathbf{H}) = p+1$, and $\text{tr}(\mathbf{I} - \mathbf{H}) = k - p - 1$, where tr denotes the trace operator. Now setting $\delta = 1, 0$ and δ_0 in (3.5.1) we get (3.3.5), (3.3.6) and (3.3.9), respectively. Fuller and Battese (1981) derived (3.5.1).

To derive (3.3.7), the average MSE of $\mathbf{Y}^*$, we write model (3.3.1) for the k sample areas as $\mathbf{U} = \mathbf{Z}\gamma + \mathbf{u}$, where $\mathbf{U} = (U_1, \ldots, U_k)^T$ and $\mathbf{u} = (u_1, \ldots, u_k)^T$. Then we have $\mathbf{Y}^* = \mathbf{X}\hat{\gamma}$ with $\hat{\gamma} = (\mathbf{Z}^T\mathbf{Z})^{-1}\mathbf{Z}^T\mathbf{U}$. We may therefore write $\mathbf{Y}^* - \mathbf{Y}$ as

$$\mathbf{Y}^* - \mathbf{Y} = \mathbf{X}(\gamma - \boldsymbol{\beta}) + \mathbf{X}(\mathbf{Z}^T\mathbf{Z})^{-1}\mathbf{Z}^T\mathbf{u} - \mathbf{v}.$$

It now follows that

$$\frac{1}{k}E_m(\mathbf{Y}^* - \mathbf{Y})^T(\mathbf{Y}^* - \mathbf{Y}) = \frac{1}{k}[\{\text{tr}\mathbf{Z}(\mathbf{Z}^T\mathbf{Z})^{-1}\mathbf{X}^T\mathbf{X}(\mathbf{Z}^T\mathbf{Z})^{-1}\mathbf{Z}^T\}\sigma_u^2 + (\boldsymbol{\gamma} - \boldsymbol{\beta})^T\mathbf{X}^T\mathbf{X}(\boldsymbol{\gamma} - \boldsymbol{\beta}) + k\sigma_v^2] \tag{3.5.2}$$

noting that $\mathbf{u}$ and $\mathbf{v}$ are assumed to be uncorrelated. Now by using $\text{tr}(\mathbf{AB}) = \text{tr}(\mathbf{BA})$, we see that (3.5.2) reduces to (3.3.7).

Chapter 4

Indirect Domain Estimation

4.1 Introduction

In Chapter 2 we studied direct estimators for domains with sufficiently large sample sizes. We also introduced a "survey regression" estimator of a domain total that borrows strength for estimating the regression coefficient, but it is essentially a direct estimator. In the context of small area estimation, direct estimators lead to unacceptably large standard errors due to unduly small samples from the small areas of interest; in fact, no sample units may be selected from some small domains. This makes it necessary to find indirect estimators that increase the effective sample size and thus decrease the standard error. In this chapter we study indirect domain estimators based on implicit models that provide a link to related small areas. Such estimators include synthetic estimators (Section 4.2), composite estimators (Section 4.3) and James-Stein (or shrinkage) estimators (Section 4.4); James-Stein estimators have attracted much attention in mainstream statistics. We study their statistical properties in the design-based framework outlined in Chapter 2. Explicit small area models that account for local variation will be presented in Chapter 5, and model-based indirect estimators and associated inferences will be studied in subsequent chapters. Indirect estimators studied in Chapter 4 and later chapters are largely based on sample survey data in conjunction with auxiliary population data, whereas the demographic indirect estimators studied in Chapter 3 typically use only administrative and census data and sampling is not involved.

4.2 Synthetic Estimation

An estimator is called a synthetic estimator if a reliable direct estimator for a large area, covering several small areas, is used to derive an indirect estimator for a small area under the assumption that the small areas have the same characteristics as the large area (Gonzalez (1973)). The National Center for Health Statistics (1968) in the United States pioneered the use of synthetic estimation for developing state estimates of disability and other health characteristics from the National Health Interview Survey (NHIS). Sample sizes in most states were too small to provide reliable direct state estimates.

4.2.1 No Auxiliary Information

Suppose that auxiliary population information is not available and that we are interested in estimating the small area mean $\bar{Y}_i$; for example, the proportion P_i of persons in poverty in small area i ($\bar{Y}_i = P_i$). In this case a synthetic estimator of $\bar{Y}_i$ is given by

$$\hat{\bar{Y}}_{is} = \hat{\bar{Y}} = \frac{\hat{Y}}{\hat{N}}, \tag{4.2.1}$$

where $\hat{\bar{Y}}$ is the direct estimator of the overall population mean $\bar{Y}$, $\hat{Y} = \sum_s w_j y_j$ and $\hat{N} = \sum_s w_j$. The p-bias of $\hat{\bar{Y}}_{is}$ is approximately equal to $\bar{Y} - \bar{Y}_i$ which is small relative to $\bar{Y}_i$ if $\bar{Y}_i \approx \bar{Y}$. If the latter implicit model that the small area mean is approximately equal to the overall mean is satisfied, then the synthetic estimator will be very efficient because its mean squared error (MSE) will be small. On the other hand, it can be heavily biased for areas exhibiting strong individual effects which, in turn, can lead to large MSE. The condition $\bar{Y}_i \approx \bar{Y}$ may be relaxed to $\bar{Y}_i \approx \bar{Y}(r)$ where $\bar{Y}(r)$ is the mean of a larger area (region) covering the small area. In this case, we use $\hat{\bar{Y}}_{is} = \hat{\bar{Y}}(r)$ where $\hat{\bar{Y}}(r)$ is the regional direct estimator. The p-bias of $\hat{\bar{Y}}(r)$ is approximately equal to $\bar{Y}(r) - \bar{Y}_i$ which is negligible relative to $\bar{Y}_i$ under the weaker condition, and the MSE of $\hat{\bar{Y}}(r)$ will be small.

4.2.2 Auxiliary Information Available

If domain-specific auxiliary information is available in the form of known totals $\mathbf{X}_i$, then the regression-synthetic estimator $\mathbf{X}_i^T\hat{\mathbf{B}}$, can be used as an estimator of domain total Y_i:

$$\hat{Y}_{i\text{GRS}} = \mathbf{X}_i^T\hat{\mathbf{B}}, \tag{4.2.2}$$

where $\hat{\mathbf{B}}$ is given by (2.3.7). The p-bias of $\hat{Y}_{i\text{GRS}}$ is approximately equal to $\mathbf{X}_i^T\mathbf{B} - Y_i$, where $\mathbf{B} = (\sum_{\text{U}} \mathbf{x}_j\mathbf{x}_j^T/c_j)^{-1} (\sum_{\text{U}} \mathbf{x}_j y_j/c_j)$ is the population regression coefficient. This p-bias will be small relative to Y_i if the domain-specific regression coefficient $\mathbf{B}_i = (\sum_{\text{U}_i} \mathbf{x}_j\mathbf{x}_j^T/c_j)^{-1}(\sum_{\text{U}_i} \mathbf{x}_j y_j/c_j)$ is close to $\mathbf{B}$ and

$Y_i = \mathbf{X}_i^T \mathbf{B}_i$. The last condition $Y_i = \mathbf{X}_i^T \mathbf{B}_i$ is satisfied if $c_j = \boldsymbol{\nu}^T \mathbf{x}_j$ for some constant column vector $\boldsymbol{\nu}$. Thus the synthetic-regression estimator will be very efficient when the small area i does not exhibit strong individual effect with respect to the regression coefficient. The synthetic estimators $\hat{Y}_{i\text{GRS}}$ add up to the direct GREG estimator $\hat{Y}_{\text{GR}} = \mathbf{X}^T \hat{\mathbf{B}}$ when $c_j = \boldsymbol{\nu}^T \mathbf{x}_j$. Note that the GREG estimator at a large area level is considered to be reliable.

A special case of (4.2.2) is the ratio-synthetic estimator in the case of a single auxiliary variable x. It is obtained by letting $c_j = x_j$ in (4.2.2) and it is given by

$$\hat{Y}_{i\text{RS}} = X_i \frac{\hat{Y}}{\hat{X}}.$$

The p-bias of $\hat{Y}_{i\text{RS}}$ relative to Y_i is approximately equal to $X_i(R - R_i)/Y_i$ which will be small if the area-specific ratio $R_i = Y_i/X_i$ is close to the overall ratio $R = Y/X$. Note that $B_i = R_i$, $B = R$ and $Y_i = X_i B_i$. For the ratio-synthetic estimator, $\hat{Y}_{i\text{RS}}$ add up to the direct ratio estimator $\hat{Y}_R = (\hat{Y}/\hat{X})X$.

If domain-specific poststrata counts N_{ig} are known for poststrata $g = 1, \ldots, G$, then a count-synthetic estimator is obtained as a special case of $\hat{Y}_{i\text{GRS}}$ by setting $\mathbf{x}_j = (x_{1j}, \ldots, x_{Gj})^T$ with $x_{gj} = 1$ if $j \in U_g$ and $x_{gj} = 0$ otherwise:

$$\hat{Y}_{i\text{S/C}} = \sum_g N_{ig} \frac{\hat{Y}_{\cdot g}}{\hat{N}_{\cdot g}}, \tag{4.2.3}$$

where $\hat{Y}_{.g}$ and $\hat{N}_{.g}$ are estimators of the poststratum total $Y_{.g}$ and size $N_{.g}$ (National Center for Health Statistics (1968)). If $y_j = 1$ or 0, then a count-synthetic estimator of the proportion P_i is obtained from (4.2.3) as

$$\hat{P}_{i\text{S/C}} = \left(\sum_g N_{ig} \hat{P}_{\cdot g}\right) / \left(\sum_g N_{ig}\right), \tag{4.2.4}$$

where $\hat{P}_{\cdot g}$ is the direct estimator of the gth poststratum proportion, $P_{\cdot g}$.

More generally, a ratio-synthetic estimator is obtained if the cell totals X_{ig} of an auxiliary variable are known. By setting $x_{gj} = x_j$ if $j \in U_g$ and $x_{gj} = 0$ otherwise in (4.2.2), we get

$$\hat{Y}_{i\text{S/R}} = \sum_g X_{ig} \frac{\hat{Y}_{\cdot g}}{\hat{X}_{\cdot g}}, \tag{4.2.5}$$

where $\hat{X}_{\cdot g} = \sum_{s_{\cdot g}} w_j x_j$ (Ghangurde and Singh (1977)). The p-bias of $\hat{Y}_{i\text{S/R}}$ is approximately equal to $\sum_g X_{ig}(Y_{\cdot g}/X_{\cdot g} - Y_{ig}/X_{ig}) = \sum_g X_{ig}(R_{\cdot g} - R_{ig})$. Thus the p-bias relative to Y_i will be small if the area specific ratio R_{ig} is close to the poststratum ratio $R_{\cdot g}$ for each g. In the special case of counts, the latter implicit model is equivalent to saying that the small area mean $\bar{Y}_{ig}$ is close to the poststratum mean $\bar{Y}_{\cdot g}$ for each g. If poststrata can be formed to satisfy this model, then the count-synthetic estimator will be very efficient, provided the direct poststrata estimators $\hat{Y}_{.g}/\hat{N}_{.g}$ are reliable. Note that the estimators $\hat{Y}_{iS/C}$ and $\hat{Y}_{iS/R}$ add up to the direct poststratified estimators $\Sigma_g(N_{.g}/\hat{N}_{.g})\hat{Y}_{.g}$ and $\Sigma_g(X_{.g}/\hat{X}_{.g})\hat{Y}_{.g}$ of Y, respectively.

Alternative synthetic estimators in the poststratification context are obtained by changing $\hat{N}_{\cdot g}$ to $N_{\cdot g}$ in (4.2.3) and $\hat{X}_{\cdot g}$ to $X_{\cdot g}$ in (4.2.5)

$$\tilde{Y}_{i\mathrm{S/C}} = \sum_g N_{ig}\frac{\hat{Y}_{\cdot g}}{N_{\cdot g}} \tag{4.2.6}$$

and

$$\tilde{Y}_{i\mathrm{S/R}} = \sum_g X_{ig}\frac{\hat{Y}_{\cdot g}}{X_{\cdot g}} \tag{4.2.7}$$

(Purcell and Linacre (1976); Singh and Tessier (1976)). In large samples, the alternative synthetic estimators are less efficient than (4.2.3) and (4.2.5) when $\hat{Y}_{\cdot g}$ and $\hat{X}_{\cdot g}$ (or $\hat{N}_{\cdot g}$) are positively correlated, as in the case of the ratio estimator $\hat{Y}/\hat{X}$ compared to the expansion estimator $\hat{Y}$. The p-bias of the alternative estimator $\tilde{Y}_{i\mathrm{S/R}}$ ($\tilde{Y}_{i\mathrm{S/C}}$) in large samples remains the same as the p-bias of $\hat{Y}_{i\mathrm{S/R}}$ ($\hat{Y}_{i\mathrm{S/C}}$) but in moderate samples the p-bias will be smaller because the ratio bias is not present. The alternative synthetic estimators also satisfy the additive property by adding up to the direct estimator $\hat{Y} = \Sigma_g\hat{Y}_{\cdot g} = \Sigma_s w_j y_j$.

It is interesting to note that the synthetic method can also be used when sampling is not involved. Suppose that Y, but not Y_i, is known from some administrative source and that X_i and X are also known. Then a synthetic estimator of Y_i may be taken as $(X_i/X)Y$, whose bias relative to Y_i will be small when $R_i \approx R$. This estimator is not an estimator in the usual sense of a random quantity.

Example 4.2.1 Health Variables. The count-synthetic estimator (4.2.4) has been used to produce state estimates of proportions for certain health variables from the 1980 U.S. National Natality Survey (NNS). This survey was based on a probability sample of 9,941 live births with a four-fold oversampling of low-birth-weight infants. Information was collected from birth certificates, questionnaires sent to married mothers and hospitals. Twenty-five poststrata, g, were formed according to mother's race (white, all others), mother's age group (6 groups) and live birth order (1, 1–2, 1–3, 2, 2+, 3, 3+, 4+).

In this application (Gonzalez, Placek and Scott (1996)), a state in the United States is a small area. To illustrate the calculation of the count-synthetic estimate (4.2.4), suppose i denotes Pennsylvania, $y_j = 1$ if the jth live birth is jaundiced and $y_j = 0$ otherwise. The national estimates of percent jaundiced, $\hat{P}_g$, were obtained from the NNS for each of the 25 poststrata g. The number of hospital births in each cell, N_{ig}, were obtained from State Vital Registration data. Multiplying N_{ig} by $\hat{P}_g$ and summing over the cells g, we get the numerator of (4.2.4) as 33,806. Now dividing 33,806 by the total hospital births $\sum_g N_{ig}$ = 156,799 in Pennsylvania, the count-synthetic estimate of percent jaundiced live births in Pennsylvania is given by $(33{,}806/156{,}799) \times 100 = 21.6\%$.

External Evaluation. Gonzalez, Placek and Scott (1996) evaluated the accuracy of NNS synthetic estimates by comparing the estimates with "true" state values, P_i, of selected health variables: percent low birth weight, percent late or no prenatal care, and percent low 1-minute "Apgar" scores. Five states, covering a wide range of annual number of births (15,000 to 160,000) were selected for this purpose. True values P_i were ascertained from the State Vital Registration System. Direct state estimates were also calculated from the NNS data. Standard errors (SE) of the direct estimates were estimated using balanced repeated replication method with 20 replicate half-samples; see Rust and Rao (1996) for an overview of replication methods for estimating standard errors. The MSE of the synthetic estimator $\hat{P}_{iS/C}$ was estimated as $(\hat{P}_{iS/C} - P_i)^2$. This MSE estimator is unbiased but very unstable.

TABLE 4.1 True State Proportions, Direct and Synthetic Estimates and Associated Estimates of RRMSE

Variable/State	True	Direct Estimate		Synthetic Estimate	
	%	Est.(%)	RRMSE(%)	Est.(%)	RRMSE(%)
Low birth:					
Pennsylvania	6.5	6.6	15	6.5	0
Indiana	6.3	6.8	22	6.5	3
Tennessee	8.0	8.5	23	7.2	10
Kansas	5.8	6.8	36	6.4	10
Montana	5.6	9.2	71	6.3	13
Prenatal care:					
Pennsylvania	3.9	4.3	21	4.3	10
Indiana	3.8	2.0	21	4.7	24
Tennessee	5.4	4.7	26	5.0	7
Kansas	3.4	2.1	35	4.5	32
Montana	3.7	3.0	62	4.3	16
Apgar score:					
Pennsylvania	7.9	7.7	14	9.4	19
Indiana	10.9	9.5	16	9.4	14
Tennessee	9.6	7.3	18	9.7	1
Kansas	11.1	12.3	25	9.4	15
Montana	11.6	12.9	40	9.4	19

SOURCE: Adapted from Tables 4 and 5 in Gonzalez et al. (1996).

Table 4.1 reports the direct and synthetic estimates and the three state values as well as the estimated values of relative root mean squared error (RRMSE), where RRMSE= $\sqrt{\text{MSE}}$/(true value). It is clear from Table 4.1 that the synthetic estimates performed better than the direct estimates in terms of estimated RRMSE, especially for small states (e.g., Montana) with small numbers of sample cases. The values of estimated RRMSE ranged from 0.14 (Pennsylvania) to 0.62 (Montana) for the direct estimator, whereas those for the synthetic estimator ranged from 0.000 (Pennsylvania) to 0.32 (Kansas). The National Center for Health Statistics (NCHS) used a maximum estimated RRMSE of 25% as the standard for reliability of estimates, and most of the synthetic estimates met this criterion for reliability, unlike the direct estimates. But this conclusion should be interpreted with caution due to the instability of the MSE estimator of $\hat{P}_{iS/C}$.

Example 4.2.2 County Crop Production. Stasny, Goel and Rumsey (1991) used a regression-synthetic estimator to produce county estimates of wheat production in the state of Kansas. County estimates of farm production are often used in local decision making and by companies selling fertilizers, pesticides, crop insurance and farm equipment. They used a nonprobability sample of farms, assuming a linear regression model relating wheat production of the j th farm in the i th county, y_{ij}, to a vector of predictors $\mathbf{x}_{ij} = (1, x_{ij1}, \ldots, x_{ijp})^T$. The predictor variables $x_{ijk}(k = 1, \ldots, p)$ chosen have known county totals X_{ik} and include a measure of the size of a farm which might allow us to use the regression model to account for the fact that the sample is not a probability sample.

The regression-synthetic estimator of the i th county total Y_i is obtained as $\tilde{Y}_{iS} = \sum_j \hat{y}_{ij}$, where $\hat{y}_{ij} = \hat{\beta}_0 + \hat{\beta}_1 x_{ij1} + \ldots + \hat{\beta}_p x_{ijp}$ is the least squares predictor of y_{ij} for $j = 1, \ldots, N_i$, and N_i is the total number of farms in the i th county. The least squares estimators, $\hat{\beta}_j$, are obtained from the linear regression model $y_{ij} = \beta_0 + \beta_1 x_{ij1} + \ldots + \beta_p x_{ijp} + \varepsilon_{ij}$ with independent and identically distributed (*iid*) errors ε_{ij}, using the sample data $\{(y_{ij}, \mathbf{x}_{ij}); j = 1, \ldots, n_i; i = 1, \ldots, m\}$, where n_i is the number of sample farms from the i th county. The estimator $\tilde{Y}_{iS}$ reduces to

$$\tilde{Y}_{iS} = N_i \hat{\beta}_0 + X_{i1} \hat{\beta}_1 + \ldots + X_{ip} \hat{\beta}_p,$$

which requires only the known county totals X_{ik}. It is not necessary to know the individual values $\mathbf{x}_{ij}$ for all the farms in the i th county.

The synthetic estimates $\tilde{Y}_{iS}$ do not add up to the direct state estimate $\hat{Y}$ of wheat production obtained from a large probability sample. The state estimate $\hat{Y}$ is regarded as more accurate than the total $\sum_i \tilde{Y}_{iS}$. A simple ratio adjustment of the form

$$\tilde{Y}_{iS}(a) = \frac{\tilde{Y}_{iS}}{\sum_i \tilde{Y}_{iS}} \hat{Y}$$

was therefore used to ensure that the adjusted estimates $\tilde{Y}_{iS}(a)$ add up to the reliable direct estimate $\hat{Y}$.

The predictor variables x_{ijk} chosen for this application consist of acres planted in wheat and district indicators. A more complex model involving the interaction between acres planted and district indicators was also studied, but the two models gave similar fits.

If the sampling fractions $f_i = n_i/N_i$ are not negligible, a more efficient estimator of Y_i is given by

$$Y_{iS}^* = \sum_{j \in s_i} y_{ij} + \sum_{j \in \bar{s}_i} \hat{y}_{ij},$$

where $\bar{s}_i$ is the complement of the sample s_i (Holt, Smith and Tomberlin, (1979)). The estimator Y_{iS}^* reduces to

$$Y_{iS}^* = \sum_{j \in s_i} y_{ij} + (N_i - n_i)\hat{\beta}_0 + X_{i1}^*\hat{\beta}_1 + \ldots + X_{ip}^*\hat{\beta}_p,$$

where $X_{ik}^* = X_{ik} - \sum_{j \in s_i} x_{ijk}$ is the total of x_{ijk} for the nonsampled units $\bar{s}_i$. This estimator also requires only the county totals X_{ik}.

4.2.3 Regression-adjusted Synthetic Estimator

The regression-adjusted synthetic estimator (Levy (1971)) attempts to account for local variation by combining area-specific covariates with the synthetic estimator. Covariates $\mathbf{z}_i$ are used to model the relative bias $B_i = (\bar{Y}_i - \hat{\bar{Y}}_{is})/\hat{\bar{Y}}_{is}$ associated with the synthetic estimator $\hat{\bar{Y}}_{is}$ of the mean $\bar{Y}_i$:

$$B_i = \gamma_0 + \boldsymbol{\gamma}^T \mathbf{z}_i + \varepsilon_i,$$

where the γ's are the regression parameters and ε_i is a random error. Since B_i is not observable, the regression model is fitted by least squares to estimated bias values $\hat{B}_a = (\hat{\bar{Y}}_a - \hat{\bar{Y}}_{as})/\hat{\bar{Y}}_{as}$ for large areas $a(= 1, \ldots, A)$ using reliable direct estimators $\hat{\bar{Y}}_a$ and synthetic estimators $\hat{\bar{Y}}_{as}$. Denoting the resulting least squares estimates as $\hat{\gamma}_0$ and $\hat{\boldsymbol{\gamma}}$, we estimate B_i as $\hat{\gamma}_0 + \hat{\boldsymbol{\gamma}}^T \mathbf{z}_i$ which, in turns, leads to the regression-adjusted estimator of $\bar{Y}_i$:

$$\hat{\bar{Y}}_{is}(a) = \hat{\bar{Y}}_{is}(1 + \hat{\gamma}_0 + \hat{\boldsymbol{\gamma}}^T \mathbf{z}_i). \tag{4.2.8}$$

Note that $\hat{B}_a$ is a reliable estimator of $B_a = (\bar{Y}_a - \hat{\bar{Y}}_{as})/\hat{\bar{Y}}_{as}$. Levy (1971) obtained state level estimates by this method by fitting the bias values at the regional level a.

Gonzalez and Hoza (1978) and Nichol (1977) used a similar approach through the sample regression method (Section 3.3) by incorporating the synthetic estimator $\hat{\bar{Y}}_{is}$ as an independent variable in the regression equation along with other independent variables. This method, called the combined synthetic-regression method, showed improvement in empirical studies over both the synthetic and sample-regression methods.

4.2.4 Estimation of MSE

The p-variance of a synthetic estimator $\hat{Y}_{is}$ will be small relative to the p-variance of a direct estimator $\hat{Y}_i$ because it depends only on the precision of direct estimators at a large area level. The p-variance is readily estimated using standard design-based methods, but it is more difficult to estimate the MSE of $\hat{Y}_{iS}$. For example, the p-variance of the ratio-synthetic estimator (4.2.5) or of the count-synthetic estimator (4.2.3) can be estimated using Taylor linearization. Similarly, the p-variance of the general synthetic-regression estimator $\hat{Y}_{i\text{GRS}} = \mathbf{X}_i^T\hat{\mathbf{B}}$ can be estimated using the results of Fuller (1975) on the large sample covariance matrix of $\hat{\mathbf{B}}$ or by using a resampling method such as the jackknife. We refer the readers to Wolter (1985), Shao and Tu (1995, Chapter 6) and Rust and Rao (1996) for a detailed account of resampling methods for sample surveys.

An approximately p-unbiased estimator of MSE of $\hat{Y}_{iS}$ can be obtained using a p-unbiased direct estimator $\hat{Y}_i$. We have

$$\begin{aligned}\text{MSE}_p(\hat{Y}_{iS}) &= E_p(\hat{Y}_{iS} - Y_i)^2 \\ &= E_p(\hat{Y}_{iS} - \hat{Y}_i + \hat{Y}_i - Y_i)^2 \\ &= E_p(\hat{Y}_{iS} - \hat{Y}_i)^2 - V_p(\hat{Y}_i) + 2\,\text{Cov}_p(\hat{Y}_{iS}, \hat{Y}_i) \\ &= E_p(\hat{Y}_{iS} - \hat{Y}_i)^2 - V_p(\hat{Y}_{iS} - \hat{Y}_i) + V_p(\hat{Y}_{iS}). \end{aligned} \qquad (4.2.9)$$

It now follows from (4.2.9) that an approximately p-unbiased estimator of $\text{MSE}_p(\hat{Y}_{iS})$ is

$$\text{mse}(\hat{Y}_{iS}) = (\hat{Y}_{iS} - \hat{Y}_i)^2 - v(\hat{Y}_{iS} - \hat{Y}_i) + v(\hat{Y}_{iS}), \qquad (4.2.10)$$

where $v(\cdot)$ is a p-based estimator of $V_p(\cdot)$; for example, a jackknife estimator. The estimator (4.2.10), however, can be very unstable and it can take negative values. Consequently, it is customary to average the MSE estimator over small areas $i(=1,\ldots,m)$ belonging to a large area to get a stable estimator (Gonzalez and Waksberg (1973)). Let $\hat{\bar{Y}}_{iS} = \hat{Y}_{iS}/N_i$ so that $\text{mse}(\hat{\bar{Y}}_{iS}) = \text{mse}(\hat{Y}_{iS})/N_i^2$. We take the average of $\text{mse}(\hat{\bar{Y}}_{iS})$ over i as an estimator of $\text{MSE}(\hat{\bar{Y}}_{iS})$ so that we get $\text{mse}_a(\hat{Y}_{iS}) = N_i^2\text{mse}_a(\hat{\bar{Y}}_{iS})$ as an estimator of $\text{MSE}(\hat{Y}_{iS})$, where

$$\text{mse}_a(\hat{\bar{Y}}_{iS}) = \frac{1}{m}\sum_i \frac{1}{N_i^2}(\hat{Y}_{iS} - \hat{Y}_i)^2 - \frac{1}{m}\sum_i \frac{1}{N_i^2}v(\hat{Y}_{iS} - \hat{Y}_i) + \frac{1}{m}\sum_i \frac{1}{N_i^2}v(\hat{Y}_{iS}). \qquad (4.2.11)$$

But such a global measure of uncertainty can be misleading since it refers to the average MSE rather than to the area-specific MSE.

A good approximation to (4.2.10) is given by

$$\text{mse}(\hat{Y}_{iS}) \approx (\hat{Y}_{iS} - \hat{Y}_i)^2 - v(\hat{Y}_i), \qquad (4.2.12)$$

noting that the variance of synthetic estimator $\hat{Y}_{iS}$ is small related to the variance of the direct estimator $\hat{Y}_i$. Using the approximation (4.2.12),

$$\text{mse}_a(\hat{\bar{Y}}_{iS}) \approx \frac{1}{m}\sum_i \frac{1}{N_i^2}(\hat{Y}_{iS} - \hat{Y}_i)^2 - \frac{1}{m}\sum_i \frac{1}{N_i^2}v(\hat{Y}_i). \qquad (4.2.13)$$

Marker (1995) proposed a simple method of getting an area-specific estimator of MSE of $\hat{Y}_{is}$. It uses the assumption that the squared p-bias $B_p^2(\hat{\bar{Y}}_{is})$ is approximately equal to the average squared bias:

$$B_p^2(\hat{\bar{Y}}_{is}) \approx \frac{1}{m}\sum_i B_p^2(\hat{\bar{Y}}_{is}) = B_a^2(\hat{\bar{Y}}_{is}). \tag{4.2.14}$$

The average squared bias can be estimated as

$$b_a^2(\hat{\bar{Y}}_{is}) = \text{mse}_a(\hat{\bar{Y}}_{is}) - \frac{1}{m}\sum_i v(\hat{\bar{Y}}_{is}), \tag{4.2.15}$$

noting that average MSE = average variance + average (bias)^2. The variance estimator $v(\hat{\bar{Y}}_{is}) = v(\hat{Y}_{is})/N_i^2$ is readily obtained using traditional methods, as noted before. It now follows under the assumption (4.2.14) that $\text{MSE}_p(Y_{is})$ can be estimated as

$$\text{mse}_{\text{M}}(\hat{Y}_{is}) = v(\hat{Y}_{is}) + N_i^2 b_a^2(\hat{\bar{Y}}_{is}), \tag{4.2.16}$$

which is area-specific if $v(\hat{Y}_{is})$ depends on the area. However, the assumption (4.2.14) may not be satisfied for areas exhibiting strong individual effects. Nevertheless, (4.2.16) is an improvement over the global measure (4.2.13), provided the variance term dominates the bias term in (4.2.16). Note that both $\text{mse}_M(\hat{Y}_{is})$ and $\text{mse}_a(\hat{Y}_{is})$ require the domain sizes, N_i.

To illustrate the calculation of (4.2.16), suppose $\hat{Y}_{is}$ is the synthetic-ratio estimator $(\hat{Y}/\hat{X})X_i = \hat{R}X_i$ and $\hat{Y}_i$ is the expansion estimator $\sum_{s_i} w_j y_j$. Then $v(\hat{Y}_i) = v(y_i)$ and $v(\hat{Y}_{is}) = (X_i/\hat{X})^2 v(e)$ in the operator notation, where $e_j = y_j - \hat{R}x_j$ and y_{ij} is as defined in Chapter 2. Using these variance estimators we can compute $\text{mse}_{\text{M}}(\hat{Y}_{is})$ from (4.2.13), (4.2.15) and (4.2.16). Note that $v(y_i)$ and $v(e)$ are obtained from $v(y)$ by changing y_j to y_{ij} and e_j, respectively.

4.2.5 Structure Preserving Estimation

Structure preserving estimation (SPREE) is a generalization of synthetic estimation in the sense that it makes a fuller use of reliable direct estimates. The parameter of interest is a count such as the number of employed in a small area. SPREE uses the well-known method of iterative proportional fitting (IPF) to adjust the cell counts of a multiway table such that the adjusted counts satisfy specified margins. The cell counts are obtained from the last census while the specified margins represent reliable direct survey estimates of current margins. Thus SPREE provides intercensal estimates of small area totals of characteristics also measured in the census.

We illustrate SPREE in the context of a three-way table of census counts $\{N_{iab}\}$, where i denotes the small areas, a the categories of the variable of interest y (e.g., employed/unemployed) and b the categories of some variable closely related to y (e.g., white/nonwhite). The unknown current counts are denoted by $\{M_{iab}\}$ and the parameters of interest are the marginal counts $M_{ia\cdot} = \sum_b M_{iab}$. We assume that reliable survey estimates of some of the margins are available. In particular, we consider two cases: (1) Survey estimates $\{\hat{M}_{\cdot ab}\}$ of the margins $\{M_{\cdot ab}\}$ are available. Note that the margins correspond to a large area covering the small areas. (2) In addition to $\{\hat{M}_{\cdot ab}\}$, survey estimates $\{\hat{M}_{i\cdot\cdot}\}$ of the margins $\{M_{i\cdot\cdot}\}$ are also available. Such estimates of current small area population counts $M_{i\cdot\cdot}$ may be obtained using demographic methods considered in Chapter 3, for instance, the sample regression method (Section 3.3).

SPREE is similar to calibration estimation (Section 2.3). We seek estimates $\tilde{M}_{iab}$ of M_{iab} which minimize a distance measure between $\{N_{iab}\}$ and $\{x_{iab}\}$ subject to the constraints $\sum_i x_{iab} = \hat{M}_{\cdot ab}$ in case 1, and $\sum_i x_{iab} = \hat{M}_{\cdot ab}$ and $\sum_{a,b} x_{iab} = \hat{M}_{i\cdot\cdot}$ in case 2. The estimate of $M_{ia\cdot}$ is then obtained as $\tilde{M}_{ia\cdot} = \sum_b \tilde{M}_{iab}$.

Case 1: One-Step SPREE. Under chi-squared distance, we minimize

$$\phi = \sum_{iab}(N_{iab} - x_{iab})^2/c_{iab} - \sum_{ab}\lambda_{ab}\left(\sum_i x_{iab} - \hat{M}_{\cdot ab}\right)$$

with respect to x_{iab}, where the c_{iab} are some prespecified weights and λ_{ab} are the Lagrange multipliers. If we choose $c_{iab} = N_{iab}$, then we get the "optimal" value of x_{iab} as

$$\tilde{M}_{iab} = \frac{N_{iab}}{N_{\cdot ab}}\hat{M}_{\cdot ab}.$$

The resulting estimate of $M_{ia\cdot}$ is

$$\tilde{M}_{ia\cdot} = \sum_b \frac{N_{iab}}{N_{\cdot ab}}\hat{M}_{\cdot ab}, \tag{4.2.17}$$

which has the same form as the synthetic estimator (4.2.6). Note that the structure preserving estimates $\tilde{M}_{ia\cdot}$ satisfy the additive property when summed over i: $\tilde{M}_{\cdot a\cdot} = \hat{M}_{\cdot a\cdot}$.

The same estimates $\tilde{M}_{iab}$ are obtained if we use the discrimination information measure

$$D(N_{iab}, x_{iab}) = \sum_{iab} N_{iab}\log\frac{N_{iab}}{x_{iab}} \tag{4.2.18}$$

instead of the chi-squared distance.

The estimates $\tilde{M}_{iab}$ preserve the *association structure* in the three-way table of counts $\{N_{iab}\}$ without interfering with the current information in the *allocation structure* $\{\hat{M}_{\cdot ab}\}$; that is, the interactions as defined by cross-product ratios of the cell counts are preserved. It is easy to verify that $\{\tilde{M}_{iab}\}$ preserves area effects, that is,

$$\frac{\tilde{M}_{iab}}{\tilde{M}_{i'ab}} = \frac{N_{iab}}{N_{i'ab}} \quad \text{for all } i,\ i'; \tag{4.2.19}$$

the two-way interaction of area and variable of interest, that is, the cross-product ratios remain the same:

$$\frac{\tilde{M}_{iab}\tilde{M}_{i'a'b}}{\tilde{M}_{i'ab}\tilde{M}_{ia'b}} = \frac{N_{iab}N_{i'a'b}}{N_{i'ab}N_{ia'b}} \qquad \text{for all } i,\ i',\ a,\ a';$$

the two-way interaction of area and associated variable, and the three-way interaction of area, variable of interest and associated variable, that is, the ratio of cross-product ratios, remain the same. Because of this structure preserving property, the method is called SPREE. This property is desirable because one would expect the association structure to remain fairly stable over the intercensal period. If in fact the association structures of $\{N_{iab}\}$ and $\{M_{iab}\}$ are identical, then SPREE gives an exactly p-unbiased estimator of $M_{ia\cdot}$, provided $\hat{M}_{\cdot ab}$ is p-unbiased for $M_{\cdot ab}$. This follows by noting that

$$E_p(\tilde{M}_{ia\cdot}) = \sum_b \frac{N_{iab}}{N_{\cdot ab}} M_{\cdot ab}$$

and that the condition (4.2.19) applied to M_{iab} gives

$$M_{iab}N_{i'ab} = N_{iab}M_{i'ab},$$

which, when summed over i', leads to $N_{iab}/N_{\cdot ab} = M_{iab}/M_{\cdot ab}$.

Case 2: Two-Step SPREE. In case 2, the chi-squared distance does not preserve the association, unlike the discrimination information measure (4.2.18). We therefore consider only the latter measure and minimize $D(N_{iab}, x_{iab})$ subject to the constraints $\sum_i x_{iab} = \hat{M}_{\cdot ab}$ for all a, b and $\sum_{a,b} x_{iab} = \hat{M}_{i\cdot\cdot}$ for all i. The "optimal" solution $\tilde{M}_{iab}$ cannot be obtained in a closed form but the well-known method of IPF (Deming and Stephan (1940)) can be used to obtain $\tilde{M}_{iab}$ iteratively. The IPF procedure involves a sequence of iteration cycles each consisting of two steps. At the kth iteration cycle, the counts $\{\tilde{M}_{iab}^{(k-1)}\}$ at the end of $(k-1)$th iteration cycle are adjusted to the first set of constraints, $\sum_i x_{iab} = \hat{M}_{\cdot ab}$, and then to the second set of constraints $\sum_{a,b} x_{iab} = \hat{M}_{i\cdot\cdot}$ as follows:

$${}_1\tilde{M}_{iab}^{(k)} = \frac{\tilde{M}_{iab}^{(k-1)}}{\tilde{M}_{\cdot ab}^{(k-1)}} \hat{M}_{\cdot ab}$$

and

$$\tilde{M}_{iab}^{(k)} = \frac{{}_1\tilde{M}_{iab}^{(k)}}{{}_1\tilde{M}_{i\cdot\cdot}^{(k)}} \hat{M}_{i\cdot\cdot}.$$

The starting values for the iteration, $\tilde{M}_{iab}^{(0)}$, are set equal to the census counts N_{iab} specifying the initial association structure. If all the counts N_{iab} are strictly positive, then the sequence $\{\tilde{M}_{iab}^{(k)}\}$ converges to the solution $\{\tilde{M}_{iab}\}$

as $k \to \infty$ (see Ireland and Kullback (1968)). The resulting estimator of $M_{ia\cdot}$ is again given by the marginal value $\tilde{M}_{ia\cdot}$. The estimator $\tilde{M}_{ia\cdot}$ will be approximately p-unbiased for $M_{ia\cdot}$ if the association structure remains stable over the intercensal period.

Chambers and Feeney (1977) proposed SPREE for small area estimation. They have also given an estimator of the asymptotic covariance matrix of the estimators $\tilde{M}_{ia\cdot}$ in a general form without details needed for computation. Purcell and Kish (1980) studied SPREE under different types of marginal constraints, including cases 1 and 2 and the case when the full association structure $\{N_{iab}\}$ is not available. In the latter case, a model is used to define a dummy association structure. For example, if only the marginal counts $\{N_{i\cdot b}\}$ are known as well as the current estimates $\{\hat{M}_{\cdot ab}\}$, we may assume proportionality across the a-categories and use

$$\tilde{M}_{ia\cdot} = \sum_b \frac{N_{i\cdot b}}{N_{\cdot\cdot b}} \hat{M}_{\cdot ab} \tag{4.2.20}$$

as an estimator of $M_{ia\cdot}$.

The two-step (or one iteration cycle) estimator $\tilde{M}_{ia\cdot}^{(1)}$ is simpler than the SPREE $\tilde{M}_{ia\cdot}$. It also makes effective use of the current marginal information $\{\hat{M}_{\cdot ab}\}$ and $\{\hat{M}_{i\cdot\cdot}\}$, although both the marginal constraints may not be exactly satisfied. Rao (1986) derived a Taylor linearization variance estimator of the two-step estimator.

Example 4.2.2. Vital Statistics. Purcell and Kish (1980) made an evaluation of the one-step and the two-step SPREE estimators by comparing the estimates to true counts obtained from the Vital Statistics registration system. In this study, SPREE estimates of mortality due to each of four different causes (a) and for each state (i) in the United States were calculated for five individual years ranging over the postcensal period 1960–70. Here the categories b denote 36 age-sex-race groups, $\{N_{iab}\}$ the 1960 census counts and $\{\hat{M}_{\cdot ab} = M_{\cdot ab}\}$, $\{\hat{M}_{i\cdot\cdot} = M_{i\cdot\cdot}\}$ the known current counts.

Table 4.2 reports the medians of the state percent absolute relative errors ARE = |estimate − true value|/true value of the one-step and the two-step SPREE estimators. It is clear from Table 4.2 that the two-step estimator performs significantly better than the one-step estimator (4.2.20) in terms of median ARE. Thus it is important to incorporate, through the allocation structure, the maximum available current data into SPREE.

Purcell and Kish (1980) also computed the SPREE estimates (4.2.17) assuming that only $\{N_{i\cdot b}\}$ and $\{\hat{M}_{\cdot ab} = M_{\cdot ab}\}$ are known. As expected, these estimates were inferior to the corresponding estimates based on the full association structure $\{N_{iab}\}$.

TABLE 4.2 Medians of Percent ARE of SPREE Estimates

Cause of Death	Year	One-Step	Two-Step
Malignant	1961	1.97	1.85
Neoplasms	1964	3.50	2.21
	1967	5.58	3.22
	1970	8.18	2.75
Major CVR	1961	1.47	0.73
Diseases	1964	1.98	1.03
	1967	3.47	1.20
	1970	4.72	2.22
Suicides	1961	5.56	6.49
	1964	8.98	8.64
	1967	7.76	6.32
	1970	13.41	8.52
Total Others	1961	1.92	1.39
	1964	3.28	2.20
	1967	4.89	3.36
	1970	6.65	3.85

SOURCE: Adapted from Table 3 in Purcell and Kish (1980).

4.3 Composite Estimation

A natural way to balance the potential bias of a synthetic estimator, say $\hat{Y}_{i2}$, against the instability of a direct estimator, say $\hat{Y}_{i1}$, is to take a weighted average of $\hat{Y}_{i1}$ and $\hat{Y}_{i2}$. Such composite estimators of the small area total Y_i may be written as

$$\hat{Y}_{iC} = \phi_i \hat{Y}_{i1} + (1 - \phi_i)\hat{Y}_{i2} \tag{4.3.1}$$

for a suitably chosen weight $\phi_i (0 \leq \phi_i \leq 1)$. Many of the estimators proposed in the literature, both design-based and model-based, have the composite form (4.3.1). Model-based composite estimators under realistic small area models that account for local variation will be studied in later chapters.

4.3.1 Optimal Estimator

The design MSE of the composite estimator is given by

$$\begin{aligned}\text{MSE}_p(\hat{Y}_{iC}) = {} & \phi_i^2 \text{MSE}_p(\hat{Y}_{i1}) + (1 - \phi_i)^2 \text{MSE}_p(\hat{Y}_{i2}) \\ & +2\, \phi_i(1 - \phi_i)E_p(\hat{Y}_{i1} - Y_i)(\hat{Y}_{i2} - Y_i).\end{aligned} \tag{4.3.2}$$

By minimizing (4.3.2) with respect to ϕ_i, we get the optimal weight ϕ_i as

$$\phi_i^* = \frac{\text{MSE}_p(\hat{Y}_{i2}) - E_p(\hat{Y}_{i1} - Y_i)(\hat{Y}_{i2} - Y_i)}{\text{MSE}_p(\hat{Y}_{i1}) + \text{MSE}_p(\hat{Y}_{i2}) - 2E_p(\hat{Y}_{i1} - Y_i)(\hat{Y}_{i2} - Y_i)}$$
$$\approx \text{MSE}_p(\hat{Y}_{i2})/[\text{MSE}_p(\hat{Y}_{i1}) + \text{MSE}_p(\hat{Y}_{i2})], \quad (4.3.3)$$

assuming that the covariance term $E_p(\hat{Y}_{i1} - Y_i)(\hat{Y}_{i2} - Y_i)$ is small relative to $\text{MSE}_p(\hat{Y}_{i2})$. The approximate optimal ϕ_i^*, given by (4.3.3), lies in the interval $[0, 1]$.

The approximate optimal weight ϕ_i^* depends only on the ratio of the MSEs:

$$\phi_i^* = 1/(1 + F_i), \quad (4.3.4)$$

where $F_i = \text{MSE}_p(\hat{Y}_{i1})/\text{MSE}_p(\hat{Y}_{i2})$. Further, the MSE of $\hat{Y}_{iC}$ with optimal weight ϕ_i^* reduces to

$$\text{MSE}_p^*(\hat{Y}_{iC}) = \phi_i^* \text{MSE}_p(\hat{Y}_{i1}) = (1 - \phi_i^*)\text{MSE}_p(\hat{Y}_{i2}). \quad (4.3.5)$$

It now follows from (4.3.5) that the reduction in MSE achieved by the optimal estimator relative to the smaller of the MSEs of the component estimators is given by ϕ_i^* if $0 \le \phi_i^* \le 1/2$, and it equals $1 - \phi_i^*$ if $1/2 \le \phi_i^* \le 1$. Thus the maximum reduction of 50 percent is achieved when $\phi_i^* = 1/2$ (or $F_i = 1$).

The ratio of $\text{MSE}_p(\hat{Y}_{iC})$ with a fixed weight ϕ_i and $\text{MSE}_p(\hat{Y}_{i2})$ may be expressed as

$$\frac{\text{MSE}_p(\hat{Y}_{iC})}{\text{MSE}_p(\hat{Y}_{i2})} = (F_i + 1)\phi_i^2 - 2\phi_i + 1. \quad (4.3.6)$$

Schaible (1978) studied the behavior of the MSE ratio (4.3.6) as a function of ϕ_i for selected values of F_i $(= 1, 2, 6)$. His results suggest that sizable deviations from the optimal weight ϕ_i^* do not produce a significant increase in the MSE of the composite estimator, that is, the curve (4.3.6) is fairly flat in the neighborhood of the optimal weight. Moreover, the reduction in MSE and the range of ϕ_i for which the composite estimator has a smaller MSE than either component estimators both depend on the size of F_i. When F_i is close to one we get the most advantage in terms of both situations.

It is easy to show that $\hat{Y}_{iC}$ is better than either component estimator in terms of MSE when $\max(0, 2\phi_i^* - 1) \le \phi_i \le \min(2\phi_i^*, 1)$. The latter interval reduces to the whole range $0 \le \phi_i \le 1$ when $F_i = 1$, and it becomes narrower as F_i deviates from one. The optimal weight ϕ_i^* will be close to zero or one when one of the component estimators has a much larger MSE than the other, that is, when F_i is either large or small. In this case, the estimator with larger MSE adds little information and therefore it is better to use the component estimator with smaller MSE in preference to the composite estimator.

In practice, we use either a prior guess of the optimal value ϕ_i^* or estimate ϕ_i^* from the sample data. Assuming that the direct estimator $\hat{Y}_{i1}$ is either

p-unbiased or approximately p-unbiased as the overall sample size increases, we can estimate the optimal weight (4.3.3) using (4.2.12). We substitute the estimator $\text{mse}(\hat{Y}_{i2})$ given by (4.2.12) for the numerator $\text{MSE}_p(\hat{Y}_{i2})$ and $(\hat{Y}_{i2} - \hat{Y}_{i1})^2$ for the denominator $\text{MSE}_p(\hat{Y}_{i1}) + \text{MSE}_p(\hat{Y}_{i2})$:

$$\hat{\phi}_i^* = \frac{\text{mse}(\hat{Y}_{i2})}{(\hat{Y}_{i2} - \hat{Y}_{i1})^2} . \tag{4.3.7}$$

But this estimator of ϕ_i^* can be very unstable. One way to overcome this difficulty is to average the estimated weights $\hat{\phi}_i^*$ over several variables or "similar" areas or both. The resulting composite estimator should perform well in view of the insensitivity to deviations from the optimal weight.

Estimation of MSE of the composite estimator, even with a fixed weight, runs into difficulties similar to those for the synthetic estimator. It is possible to extend the methods in Subsection 4.2.4 to composite estimators; see Example 4.3.1.

Example 4.3.1. Labor Force Characteristics. Griffiths (1996) used a SPREE-based composite estimator to provide indirect estimates of labor force characteristics for Congressional Districts (CDs) in United States. This estimator is of the form (4.3.1) with the one-step SPREE estimator $\tilde{M}_{ia\cdot}$ (Eq. (4.2.17)) as the synthetic component, $\hat{Y}_{i2}$, and a sample-based estimator $\hat{M}_{ia\cdot}$ as the direct component, $\hat{Y}_{i1}$. To calculate the SPREE estimate, the population counts $\{N_{iab}\}$ were obtained from the 1990 Decennial Census, while the survey counts $\{\hat{M}_{\cdot ab}\}$ were obtained from the 1994 March Current Population Survey (CPS). The CPS sample is a stratified two-stage probability sample drawn independently from each of the 50 states and the District of Columbia. It was not designed to provide reliable sample-based estimates $\hat{M}_{ia\cdot}$ at the CD level; the CD sample sizes tend to be too small for direct estimation with desired reliability.

Optimal estimated weights ϕ_i^* were obtained from (4.3.3), using (4.2.10) to estimate $\text{MSE}_p(\tilde{M}_{ia\cdot})$ and $v(\hat{M}_{ia\cdot})$ to estimate $\text{MSE}_p(\hat{M}_{ia\cdot})$. Note that we replace $\hat{Y}_{iS}$ by $\tilde{M}_{ia\cdot}$ and $\hat{Y}_i$ by $\hat{M}_{ia\cdot}$ in (4.2.10). One could also use the simpler estimate, $\hat{\phi}_i^*$, given by (4.3.7), but both estimates of ϕ_i are highly unstable. Griffiths used (4.2.10) again to estimate the MSE of the SPREE-based composite estimator, by changing $\hat{Y}_{iS}$ to the composite estimator and $\hat{Y}_i$ to $\hat{M}_{ia\cdot}$. This estimate of MSE is also highly unstable.

Griffiths (1996) evaluated the efficiencies of the composite estimator and the SPREE estimator, $\tilde{M}_{ia\cdot}$, relative to the sample-based estimator $\hat{M}_{ia}$ for the five CDs in the state of Iowa. The MSE estimates based on (4.2.10) were used for this purpose. The composite estimator provided an improvement over the sample-based estimator in terms of estimated MSE. The reduction in estimated MSE, when the estimates of MSEs are averaged over the five CDs, ranged from 17% to 78%. The one-step SPREE estimator did not perform better than the direct estimator, $\hat{M}_{ia}$, for three of the variables: employed, others, household income \$10,000–\$25,000.

4.3.2 Sample Size Dependent Estimators

Sample size dependent (SSD) estimators are composite estimators with simple weights ϕ_i that depend only on the domain counts $\hat{N}_i$ and N_i or the domain totals $\hat{X}_i$ and X_i of an auxiliary variable x. These estimators were originally designed to handle domains for which the expected sample size is large enough to make the direct estimator satisfy reliability requirements whenever the realized sample size exceeds the expected sample size (Drew, Singh and Choudhry (1982)).

Drew, Singh and Choudhry (1982) proposed a SSD estimator which uses the weight

$$\phi_i(\mathrm{S1}) = \begin{cases} 1 & \text{if } \hat{N}_i/N_i \geq \delta \\ \hat{N}_i/(\delta N_i) & \text{if } \hat{N}_i/N_i < \delta, \end{cases} \tag{4.3.8}$$

where $\hat{N}_i = \sum_{s_i} w_j$ is the direct expansion estimator of N_i and δ is subjectively chosen to control the contribution of the synthetic estimator. The form of $\hat{N}_i$ suggests that it increases with the domain sample size. Another choice of ϕ_i is obtained by substituting $\hat{X}_i/X_i$ for $\hat{N}_i/N_i$ in (4.3.8). Under this choice, Drew et al. (1982) used the poststratified ratio estimator (2.4.12) as the direct estimator $\hat{Y}_{i1}$ and the synthetic ratio estimator (4.2.5) as the synthetic estimator $\hat{Y}_{i2}$. The direct estimator (2.4.12), however, suffers from the ratio bias unless the domain sample size is not small. To avoid the ratio bias, we could use the modified direct estimator $\hat{Y}_i + \sum_g (X_{ig} - \hat{X}_{ig})(\hat{Y}_g/\hat{X}_g)$ whose p-bias goes to zero as the overall sample size increases, even if the domain sample size is small. Generally, we can use the modified GREG estimator $\hat{Y}_i + (\mathbf{X}_i - \hat{\mathbf{X}}_i)^T\hat{\mathbf{B}}$ as the direct estimator and the synthetic-regression estimator $\mathbf{X}_i^T\hat{\mathbf{B}}$ as the synthetic estimator in conjunction with the weight $\phi_i(\mathrm{S1})$. A general-purpose choice of δ is $\delta = 1$. The Canadian Labour Force Survey (LFS) uses the SSD estimator with $\delta = 2/3$ to produce Census Division level estimates.

Särndal and Hidiroglou (1989) proposed the "dampened regression estimator" which is obtained from the modified GREG estimator in the form (2.5.7) by dampening the effect of the direct component $\sum_{s_i} w_j e_j$ whenever $\hat{N}_i < N_i$. It is given by

$$\hat{Y}_{i\mathrm{DR}} = \mathbf{X}_i^T\hat{\mathbf{B}} + (\hat{N}_i/N_i)^{H-1}\sum_{s_i} w_j e_j \tag{4.3.9}$$

with $H = 0$ if $\hat{N}_i \geq N_i$ and $H = h$ if $\hat{N}_i < N_i$, where h is a suitably chosen constant. This estimator can be written as a composite estimator with the improved modified GREG estimator $\mathbf{X}_i^T\hat{\mathbf{B}} + (N_i/\hat{N}_i)\sum_{s_i} w_j e_j$ as the direct estimator and $\mathbf{X}_i^T\hat{\mathbf{B}}$ as the synthetic estimator in conjunction with the weight

$$\phi_i(\mathrm{S2}) = \begin{cases} 1 & \text{if } \hat{N}_i/N_i \geq 1 \\ (\hat{N}_i/N_i)^h & \text{if } \hat{N}_i/N_i < 1. \end{cases} \tag{4.3.10}$$

A general-purpose choice of h is $h = 2$.

To study the nature of the weight $\phi_i(\text{S1})$, we consider the special case of simple random sampling. In this case, $\hat{N}_i = N(n_i/n)$. Taking $\delta = 1$ in (4.3.8), it now follows that $\phi_i(\text{S1}) = 1$ if n_i is at least as large as $E(n_i) = n(N_i/N)$. Therefore, the SSD estimator can fail to borrow strength from other domains even when $E(n_i)$ is not large enough to make the direct estimator reliable. On the other hand, when $\hat{N}_i < N_i$, the weight $\phi_i(\text{S1}) = \hat{N}_i/N_i$ decreases as n_i decreases. As a result, more weight is given to the synthetic component as n_i decreases. Thus in the case $\hat{N}_i < N_i$ the weight $\phi_i(\text{S1})$ behaves well, unlike in the case $\hat{N}_i \geq N_i$. Similar comments apply to the SSD estimator based on the weight ϕ_i (S2). Another disadvantage of SSD estimators is that the weights do not take account of the size of between area variation relative to within area variation for the characteristic of interest. That is, all the characteristics get the same weight regardless of their differences with respect to between area heterogeneity. In Chapter 7, Example 7.2.2, we demonstrate that large efficiency gains over SSD estimators can be achieved by using model-based estimators when the between area heterogeneity is relatively small.

General SSD estimators provide consistency when aggregated over characteristics because the same weight is used. However, they do not add up to a direct estimator at a large area level. A simple ratio adjustment gives

$$\hat{Y}_{i\text{C}}(a) = \frac{\hat{Y}_{i\text{C}}}{\sum_i \hat{Y}_{i\text{C}}} \hat{Y}_{\text{GR}}. \tag{4.3.11}$$

The adjusted estimators $\hat{Y}_{i\text{C}}(a)$ add up to the director estimator $\hat{Y}_{\text{GR}}$ at the large area level.

The SSD estimators with weights $\phi_i(\text{S1})$ may also be viewed as calibration estimators $\sum_s w_{ij}^* y_j$ minimizing the chi-squared distance

$$\sum_i \sum_{j \in s} c_j [w_j a_{ij} \phi_i(\text{S1}) - b_{ij}^*]^2 / w_j \tag{4.3.12}$$

subject to the constraints $\sum_{j \in s} b_{ij}^* \mathbf{x}_j = \mathbf{X}_i$, $i = 1, \ldots, m$, where a_{ij} is the domain indicator variable and c_j is a specified constant $j \in s$; that is, the "optimal" b_{ij}^* equals w_{ij}^*. Using this distance measure, we are calibrating the dampened weights $w_j a_{ij} \phi_i(\text{S1})$ rather than the original weights $w_j a_{ij}$. Singh and Mian (1995) used the calibration approach to take account of different sets of constraints simultaneously. For example, the additivity constraint $\sum_i \sum_{j \in s} b_{ij}^* y_j = \hat{Y}_{\text{GR}}$ can be introduced along with the previous constraints $\sum_{j \in s} b_{ij}^* \mathbf{x}_j = \mathbf{X}_i$, $i = 1, \ldots, m$. Note that the calibration weights w_{ij}^* for all the small areas under consideration are obtained simultaneously using this approach.

Estimation of the MSE of SSD estimators runs into difficulties similar to those for the synthetic estimator and the optimal composite estimator. One ad hoc approach to variance estimation is to use the variance estimator of the modified direct estimator $\hat{Y}_i + (\mathbf{X}_i - \hat{\mathbf{X}}_i)^T \hat{\mathbf{B}}$, namely, $v(a_i e)$, as an overestimator of the true variance of SSD estimator using either $\phi_i(\text{S1})$ or $\phi_i(\text{S2})$ as the weight attached to the direct estimator (Särndal and Hidiroglou (1989)). Another approach is to treat the weight $\phi_i(\text{S1})$ as fixed and estimate the variance as $(\phi_i(\text{S1}))^2 v(a_i e)$ noting that the variance of the synthetic component $\mathbf{X}_i^T \hat{\mathbf{B}}$ is small relative to the variance of the direct component. This variance estimator will underestimate the true variance, unlike $v(a_i e)$. Resampling methods, such as the jackknife, can also be readily used to get a variance estimator. Properties of resampling variance estimators in the context of SSD estimators have not been studied.

Example 4.3.2. Unemployed Counts. Falorsi, Falorsi and Russo (1994) compared the performances of the direct estimator, the synthetic estimator, the SSD estimator with $\delta = 1$, and the optimal composite estimator. They conducted a simulation study in which the Italian Labour Force Survey design (stratified two-stage sampling) was simulated using data from the 1981 Italian census. The "optimal" weight ϕ_i was obtained from the census data. In this study, Health Service Areas (HSAs) are the small areas (unplanned domains) that cut across design strata. The study was confined to the 14 HSAs of the Friuli region, and the sample design was based on the selection of 39 primary sampling units (PSUs) and 2,290 second stage units (SSUs); PSU is a municipality and SSU is a household. The variable of interest, y, is the number of unemployed.

The performance of the estimators was evaluated in terms of absolute relative bias (ARB) and relative root mean square error (RRMSE). The relative bias (RB) and MSE of an estimator of the total Y_i are given by

$$\text{RB} = \frac{1}{R}\sum_{r=1}^{R}\left(\frac{est_r}{Y_i} - 1\right); \text{MSE} = \frac{1}{R}\sum_{r=1}^{R}(est_r - Y_i)^2,$$

where est_r is the value of the estimator of Y_i for the rth simulation run $(r = 1, \cdots, R)$. Note that RRMSE$= \sqrt{\text{MSE}}/Y_i$ and ARB$= |\text{RB}|$. Falorsi et al. (1994) used $R = 400$ simulation runs to calculate ARB and RRMSE for each HSA and each estimator.

Table 4.3 reports the average values $\overline{\text{ARB}}$ and $\overline{\text{RRMSE}}$ of the estimators, where the average is taken over the 14 HSAs. It is clear from Table 4.3 that $\overline{\text{ARB}}$ values of the direct estimator and SSD estimator are negligible ($< 2.5\%$) whereas those of the composite and synthetic estimators are somewhat large. The synthetic estimator has the largest $\overline{\text{ARB}}\%$ (about 9%). In terms of $\overline{\text{RRMSE}}\%$, synthetic and composite estimators have the smallest values (almost one half of the value for the direct estimator) followed by the SSD estimator with approximately 30% higher value.

TABLE 4.3 Percent Average Absolute Relative Bias ($\overline{\text{ARB}}\%$) and Percent Average RRMSE ($\overline{\text{RRMSE}}\%$) of Estimators

Estimator	$\overline{\text{ARB}}\%$	$\overline{\text{RRMSE}}\%$
Direct	1.75	42.08
Synthetic	8.97	23.80
Composite	6.00	23.57
SSD	2.39	31.08

SOURCE: Adapted from Table 1 in Falorsi et al. (1994).

Falorsi et al. (1994) also examined area level values of ARB and RRMSE. Synthetic and composite estimators were found to be badly biased in small areas with low values of the ratio (population of HSA)/(population of the set of strata including the HSA), but exhibited low RRMSE compared to the other alternatives. Considering both bias and efficiency, they concluded that SSD estimator is preferable over the other estimators. It may be noted that the sampling rates were high enough leading to large enough expected domain sample sizes, a case favorable to the SSD estimator.

4.4 James-Stein Method

4.4.1 Common Weight

Another approach to composite estimation is to use a common weight, $\phi_i = \phi$, and then minimize the total MSE, $\sum_i \text{MSE}_p(\hat{Y}_{ic})$, with respect to ϕ (Purcell and Kish (1979)). This ensures good estimation for the group of small areas as a whole but not necessarily for each of the small areas in the group.

We have

$$\sum_i \text{MSE}_p(\hat{Y}_{ic}) \approx \phi^2 \sum_i \text{MSE}_p(\hat{Y}_{i1}) + (1-\phi)^2 \sum_i \text{MSE}_p(\hat{Y}_{i2}). \qquad (4.4.1)$$

Minimizing (4.4.1) with respect to ϕ gives the optimal value

$$\phi^* = \frac{\sum_i \text{MSE}_p(\hat{Y}_{i2})}{\sum_i [\text{MSE}_p(\hat{Y}_{i1}) + \text{MSE}_p(\hat{Y}_{i2})]}. \qquad (4.4.2)$$

Suppose we take $\hat{Y}_{i1}$ as the direct expansion estimator $\hat{Y}_i$ and $\hat{Y}_{i2}$ as the synthetic estimator $\hat{Y}_{iS}$. It now follows from (4.2.12) that ϕ^* may be estimated as

$$\begin{aligned}\hat{\phi}^* &= \frac{\sum_i [(\hat{Y}_{iS} - \hat{Y}_i)^2 - v(\hat{Y}_i)]}{\sum_i (\hat{Y}_{iS} - \hat{Y}_i)^2}. \\ &= 1 - \frac{\sum_i v(\hat{Y}_i)}{\sum_i (\hat{Y}_{iS} - \hat{Y}_i)^2}. \qquad (4.4.3)\end{aligned}$$

The estimator $\hat{\phi}^*$ is quite reliable, unlike $\hat{\phi}_i^*$ given by (4.3.7), because we are pooling over several small areas. However, the use of a common weight may not be reasonable if the individual variances, $V(\hat{Y}_i)$, vary considerably.

Rivest (1995) used weights of the form $\hat{\phi}^*$ in the context of adjustment for population undercount in the 1991 Canadian Census. He also obtained an approximation to the total MSE of the resulting composite estimators as well as an estimator of the total MSE. Overall performance of the composite estimators relative to the direct estimators was studied by comparing their estimated total MSE.

The composite estimator based on $\hat{\phi}^*$ is similar to the well-known James-Stein (J-S) estimator which has attracted a lot of attention in the mainstream statistical literature. We give a brief account of the James-Stein methods in this section and refer the reader to Efron and Morris (1972a, 1973, 1975) and Brandwein and Strawderman (1990) for a more extensive treatment. Efron (1975) gave an excellent expository account of the J–S methods as well as examples of their practical application, including the popular example of predicting batting averages of baseball players.

Suppose the small area means $\bar{Y}_i$ are the parameters of interest. Let $\theta_i = g(\bar{Y}_i)$ be a specified transformation of $\bar{Y}_i$ which induces normality of the corresponding estimators $\hat{\theta}_i = g(\hat{\bar{Y}}_i)$ and stabilizes the variances of $\hat{\theta}_i$. For example, if $\bar{Y}_i$ is a proportion, we can use an arc-sine transformation. Some other choices of $g(\bar{Y}_i)$ include $\theta_i = \bar{Y}_i$ and $\theta_i = \ln \bar{Y}_i$. Let $\boldsymbol{\theta} = (\theta_1, \ldots, \theta_m)^T$ and $\hat{\boldsymbol{\theta}} = (\hat{\theta}_1, \ldots, \hat{\theta}_m)^T$. We assume $\hat{\theta}_i \overset{\text{ind}}{\sim} N(\theta_i, \psi_i)$ with known variances ψ_i, $(i = 1, \ldots, m)$ where $\overset{\text{ind}}{\sim}$ denotes "independently distributed as" and $N(a, b)$ denotes a normal variable with mean a and variance b. We further assume that a prior guess of $\boldsymbol{\theta}$, say $\boldsymbol{\theta}^0$, is available or can be evaluated from the data, for example, if θ_i is linearly related to a p-vector of auxiliary variables, $\mathbf{z}_i$, then we can take the least squares predictor $\mathbf{z}_i^T(\mathbf{Z}^T\mathbf{Z})^{-1}\mathbf{Z}^T\hat{\boldsymbol{\theta}} = \mathbf{z}_i^T\hat{\boldsymbol{\beta}}_{\text{LS}}$ as θ_i^0, where $\mathbf{Z}^T = (\mathbf{z}_1, \ldots, \mathbf{z}_m)$. In the absence of such auxiliary information, we set $\mathbf{z}_i = 1$ so that $\theta_i^0 = \sum_i \hat{\theta}_i/m = \hat{\theta}_.$ for all i.

The performance of an estimator of $\boldsymbol{\theta}$, say $\tilde{\boldsymbol{\theta}}$, will be measured in terms of its total MSE (total squared error risk):

$$R(\boldsymbol{\theta}, \tilde{\boldsymbol{\theta}}) = \sum_i E_p(\tilde{\theta}_i - \theta_i)^2. \tag{4.4.4}$$

4.4.2 Equal Variances $\psi_i = \psi$

In the special case of equal sampling variances, $\psi_i = \psi$, the J–S estimator of θ_i is given by

$$\hat{\theta}_{i,\text{JS}} = \theta_i^0 + \left[1 - \frac{(m-2)\psi}{S}\right](\hat{\theta}_i - \theta_i^0), \quad m \geq 3 \tag{4.4.5}$$

assuming $\boldsymbol{\theta}^0$ is fixed, where $S = \sum_i(\hat{\theta}_i - \theta_i^0)^2$. If θ_i^0 is the least squares predictor then we replace $m-2$ by $m-p-2$ in (4.4.5), where p is the number of estimated parameters in the regression equation. Note that (4.4.5) may also be expressed as a composite estimator with weight $\hat{\phi}_{\text{JS}} = 1 - [(m-2)\psi]/S$ attached to $\hat{\theta}_i$ and $1 - \hat{\phi}_{\text{JS}}$ to the prior guess θ_i^0. The J–S estimator is also called a shrinkage estimator because it shrinks the direct estimator $\hat{\theta}_i$ toward the guess θ_i^0.

James and Stein (1961) established the following remarkable results on the superiority of $\hat{\boldsymbol{\theta}}_{\text{JS}}$ over $\hat{\boldsymbol{\theta}}$ in terms of total MSE:

Theorem 4.4.1. Suppose the direct estimators $\hat{\theta}_i$ are independent $N(\theta_i, \psi)$ with known sampling variance ψ, and the guess θ_i^0 is fixed. Then

(a) $R(\boldsymbol{\theta}, \hat{\boldsymbol{\theta}}_{\text{JS}}) < R(\boldsymbol{\theta}, \hat{\boldsymbol{\theta}}) = m\psi$ for all $\boldsymbol{\theta}$, that is, $\hat{\boldsymbol{\theta}}_{\text{JS}}$ dominates $\hat{\boldsymbol{\theta}}$ with respect to total MSE.

(b) $R(\boldsymbol{\theta}, \hat{\boldsymbol{\theta}}_{\text{JS}}) = 2\psi$ at $\boldsymbol{\theta} = \boldsymbol{\theta}^0$.

(c) $R(\boldsymbol{\theta}, \hat{\boldsymbol{\theta}}_{\text{JS}}) \le m\psi - \dfrac{(m-2)^2\psi^2}{(m-2)\psi + \sum_i(\theta_i - \theta_i^0)^2}$

so that $R(\boldsymbol{\theta}, \hat{\boldsymbol{\theta}}_{\text{JS}})$ is minimized when $\boldsymbol{\theta} = \boldsymbol{\theta}^0$.

A proof of Theorem 4.4.1 is given in Section 4.5. It follows from Theorem 4.4.1 that $\hat{\boldsymbol{\theta}}_{\text{JS}}$ leads to large reduction in total MSE when $\boldsymbol{\theta}$ is close to the guess $\boldsymbol{\theta}^0$ and m is not very small. For example, $R(\boldsymbol{\theta}^0, \hat{\boldsymbol{\theta}}_{\text{JS}})/R(\boldsymbol{\theta}^0, \hat{\boldsymbol{\theta}}) = 2/10 = 0.2$ when $m = 10$ so that the total MSE of $\hat{\boldsymbol{\theta}}_{\text{JS}}$ is only one-fifth of the total MSE of $\hat{\boldsymbol{\theta}}$ when $\boldsymbol{\theta} = \boldsymbol{\theta}^0$. On the other hand, the reduction in total MSE will be small if the variability of the errors in guessing θ is large, that is, as $\sum_i(\theta_i - \theta_i^0)^2$ increases, $R(\boldsymbol{\theta}, \hat{\boldsymbol{\theta}}_{\text{JS}})$ tends to $R(\boldsymbol{\theta}, \hat{\boldsymbol{\theta}}) = m\psi$.

We now discuss several salient features of the James-Stein method:

(1) The J–S method is attractive to users wanting good estimation for the group of small areas as a whole because large gains in efficiency can be achieved in the traditional design-based framework without assuming a model on the small area parameters θ_i.

(2) The James-Stein estimator arises quite naturally in the empirical best linear unbiased prediction (EBLUP) approach or the empirical Bayes (EB) approach, assuming a random effects model with $\theta_i \overset{\text{ind}}{\sim} N(\mathbf{z}_i^T\boldsymbol{\beta}, \sigma_v^2)$; see Section 7.1.

(3) A "plus-rule" estimator $\hat{\theta}_{\text{JS}}^+$ is obtained from (4.4.5) by changing the factor $1 - (m-2)\psi/S$ to 0 whenever $S < (m-2)\psi$, that is,

$$\hat{\theta}_{i,\text{JS}}^+ = \begin{cases} \theta_i^0 & \text{if } S < (m-2)\psi \\ \hat{\theta}_{i,\text{JS}} & \text{if } S \ge (m-2)\psi. \end{cases} \tag{4.4.6}$$

The estimator $\hat{\boldsymbol{\theta}}_{\text{JS}}^+$ dominates $\hat{\boldsymbol{\theta}}_{\text{JS}}$ in terms of total MSE, that is,

$$R(\boldsymbol{\theta}, \hat{\boldsymbol{\theta}}_{\text{JS}}^+) < R(\boldsymbol{\theta}, \hat{\boldsymbol{\theta}}_{\text{JS}}) \quad \text{for all } \boldsymbol{\theta}$$

(Efron and Morris (1973)).

(4) The dominance property is not necessarily true for the retransform of $\hat{\theta}_{i,\text{JS}}$, that is, the estimators $g^{-1}(\hat{\theta}_{i,\text{JS}}) = \hat{\bar{Y}}_{i,\text{JS}}$ may not dominate the direct estimators $g^{-1}(\hat{\theta}_i) = \hat{\bar{Y}}_i$ in terms of total MSE. Moreover, the estimator of domain totals resulting from $\hat{\bar{Y}}_{i,\text{JS}}$, namely, $N_i\hat{\bar{Y}}_{i,\text{JS}}$, may not dominate the direct estimators $N_i\hat{\bar{Y}}_i$ even when $g(\bar{Y}_i) = \bar{Y}_i$. Users may find the above lack of invariance undesirable.

(5) The assumption $E_p(\hat{\theta}_i) = \theta_i$ for nonlinear $g(\cdot)$ may not hold if the small area sample size is very small. Moreover, the common sampling variance ψ is not known in practice. If an independent estimator $\hat{\psi}$ is available such that $\nu\hat{\psi}/\psi$ is a χ^2 variable with ν degrees of freedom (df), then we modify the estimator (4.4.5) by changing ψ to $\nu\hat{\psi}/(\nu+2)$. The modified estimators retain the dominance property. For example, suppose that $\theta_i = \bar{Y}_i$ and we draw simple random samples $\{y_{ij}\}$ independently from each small area i $(j = 1,\ldots,\bar{n})$. Under the assumption $y_{ij} \overset{\text{ind}}{\sim} N(\bar{Y}_i, \sigma^2)$, we have $\nu = m(\bar{n}-1)$ and $\bar{n}\nu\hat{\psi} = \sum_i\sum_j(y_{ij} - \bar{y}_i)^2$, where $\hat{\theta}_i = \bar{y}_i$ is the sample mean in the ith area. If no guess of $\bar{Y}_i$ is available, we take θ_i^0 as the overall mean $\bar{\bar{y}}$ and change $m-2$ to $m-3$ $(m > 3)$:

$$\hat{\bar{Y}}_{i,\text{JS}} = \bar{\bar{y}} + \left[1 - \frac{(m-3)\nu\hat{\psi}}{(\nu+2)S}\right](\bar{y}_i - \bar{\bar{y}}), \tag{4.4.7}$$

where $S = \sum_i(\bar{y}_i - \bar{\bar{y}})^2$. The estimators (4.4.7) dominate the direct estimator $\bar{y}_i$ (C.R. Rao (1976)). The normality assumption on $\{y_{ij}\}$ is used in establishing the dominance result. The assumptions of normality, equal sample sizes and common variance are quite restrictive in the sample survey context.

(6) We have assumed that the direct estimators $\hat{\theta}_i$ are independent, but this may not hold when the small areas cut across the design strata and clusters. In this case, we assume that $\hat{\boldsymbol{\theta}}$ is m-variate normal, $N_m(\boldsymbol{\theta}, \boldsymbol{\Psi})$ and that an independent estimator of $\boldsymbol{\Psi}$ is based on $\mathbf{B}$ which is distributed as $W_m(\boldsymbol{\Psi}, \nu)$, a Wishart distribution with ν df. In this case, the J–S estimator of θ_i is given by

$$\hat{\theta}_{i,\text{JS}} = \theta_i^0 + \left(1 - \frac{(m-2)}{(\nu - m + 3)Q}\right)(\hat{\theta}_i - \theta_i^0), \tag{4.4.8}$$

where $Q = (\hat{\boldsymbol{\theta}} - \boldsymbol{\theta}^0)^T\mathbf{B}^{-1}(\hat{\boldsymbol{\theta}} - \boldsymbol{\theta}^0)$; see James and Stein (1961) and Bilodeau and Srivastava (1988).

(7) The J–S method assumes no specific relationship between the θ_i's, such as $\theta_i \overset{\text{iid}}{\sim} N(\theta_i^0, \sigma_v^2)$, and yet uses data other than $\hat{\theta}_i$ to estimate θ_i. Thus the total MSE is reduced even when the different θ_i refer to obviously disjoint problems (for example, some θ_i refer to batting averages of baseball players and the rest to per capita income of small areas in the United States). But the total MSE has no practical relevance in such situations.

(8) The J–S method may perform poorly in estimating those components θ_i with unusually large or small values of $\theta_i - \theta_i^0$. In fact, for large m the maximum MSE for an individual θ_i over all choices of θ can be as large as $(m/4)\psi$, using $\hat{\theta}_{i,\text{JS}}$; for example, for $m = 16$ this maximum MSE is exactly equal to 4.41 times the MSE of $\hat{\theta}_i$ (Efron and Morris (1972a)). To reduce this undesirable effect, Efron and Morris (1972a) proposed "limited translation estimators" which offer a compromise between the J–S estimator and the direct estimator. These estimators have both good ensemble and good individual properties, unlike $\hat{\theta}_{i,\text{JS}}$. A straightforward compromise estimator is obtained by restricting the amount by which $\hat{\theta}_{i,\text{JS}}$ differs from $\hat{\theta}_i$ to a multiple of the standard error of $\hat{\theta}_i$:

$$\hat{\theta}^*_{i,\text{JS}} = \begin{cases} \hat{\theta}_{i,\text{JS}} & \text{if } \hat{\theta}_i - c\psi_i^{1/2} \le \hat{\theta}_{i,\text{JS}} \le \hat{\theta}_i + c\psi_i^{1/2} \\ \hat{\theta}_i - c\psi_i^{1/2} & \text{if } \hat{\theta}_{i,\text{JS}} < \hat{\theta}_i - c\psi_i^{1/2} \\ \hat{\theta}_i + c\psi_i^{1/2} & \text{if } \hat{\theta}_{i,\text{JS}} > \hat{\theta}_i + c\psi_i^{1/2}, \end{cases} \tag{4.4.9}$$

where $c > 0$ is a suitably chosen constant. The choice $c = 1$, for example, ensures that $\text{MSE}_p(\hat{\theta}^*_{i,\text{JS}}) < 2\psi = 2\text{MSE}_p(\hat{\theta}_i)$, while retaining more than 80% of the gain of $\hat{\boldsymbol{\theta}}_{\text{JS}}$ over $\hat{\boldsymbol{\theta}}$ in terms of total MSE. We refer the reader to Efron and Morris (1972a) for further details on limited translation estimators.

Example 4.4.1. Batting Averages. Efron (1975) gave an amusing example of batting averages of major league baseball players in the United States to illustrate the superiority of J-S estimators over direct estimators. Table 4.4 gives the batting averages, $\hat{P}_i$, of $m = 18$ players after their first 45 times at bat during the 1970 season. These estimates are taken as the direct estimates $\hat{\theta}_i = \hat{P}_i$. The J-S estimates were calculated from (4.4.5) using $\theta_i^0 = \sum \hat{P}_i/18 = 0.265 = \hat{P}_.$ and $\psi = \hat{P}_.(1 - \hat{P}_.)/45 = 0.0043$, the binomial variance. Note that $\hat{P}_i$ is treated as $N(P_i, \psi)$. The compromise J-S estimates were obtained from (4.4.9) using $c = 1$. To compare the accuracies of the estimates, the batting average for each player i during the remainder of the season (about 370 more times at bat on average) was taken as the true $\theta_i = P_i$. Table 4.4 also reports the values of the J-S estimator $\hat{P}_{i,\text{JS}}$ and the compromise J-S estimator $\hat{P}^*_{i,\text{JS}}$.

Because the true values P_i are assumed to be known, we can compare the relative overall accuracies using the ratios

$$R_1 = \sum_i (\hat{P}_i - P_i)^2 / \sum_i (\hat{P}_{i,\text{JS}} - P_i)^2$$

and

$$R_2 = \sum_i (\hat{P}_i - P_i)^2 / \sum_i (\hat{P}^*_{i,\text{JS}} - P_i)^2.$$

We have $R_1 = 3.50$ so that the J-S estimates outperform the direct estimates by a factor of 3.50. Also, $R_2 = 4.09$, so that the compromise J-S estimates perform even better than the J-S estimates in this example. It may be noted that the compromise J-S estimator protects player 1's (Roberto Clemente's) proportion $\hat{P}_1 = 0.400$ from overshrinking toward the common proportion $\hat{P}_{\cdot} = 0.265$.

TABLE 4.4 Batting Averages for 18 Baseball Players

Player	Direct Estimate	True Value	J-S Estimate	Compromise J-S Estimate
1	0.400	0.346	0.293	0.334
2	0.378	0.298	0.289	0.312
3	0.356	0.276	0.284	0.290
4	0.333	0.221	0.279	0.279
5	0.311	0.273	0.275	0.275
6	0.311	0.270	0.275	0.275
7	0.289	0.263	0.270	0.270
8	0.267	0.210	0.265	0.265
9	0.244	0.269	0.261	0.261
10	0.244	0.230	0.261	0.261
11	0.222	0.264	0.256	0.256
12	0.222	0.256	0.256	0.256
13	0.222	0.304	0.256	0.256
14	0.222	0.264	0.256	0.256
15	0.222	0.226	0.256	0.256
16	0.200	0.285	0.251	0.251
17	0.178	0.319	0.247	0.243
18	0.156	0.200	0.242	0.221

SOURCE: Adapted from Table 1 in Efron (1975).

We treated $\hat{P}_i \overset{\text{ind}}{\sim} N(P_i, \psi)$ in calculating the J-S estimates in Table 4.4. But it may be more reasonable to assume that $n\hat{P}_i \overset{\text{ind}}{\sim} Bin(n, P_i)$ with $n = 45$. Under this assumption, the variance of $n\hat{P}_i$ depends on P_i. Efron and Morris (1975) used the well-known arc-sine transformation to stabilize the variance of a binomial distribution. This transformation leads to $\hat{\theta}_i = \sqrt{n}\arcsin(2\hat{P}_i - 1) = g(\hat{P}_i)$ and $\theta_i = \sqrt{n}\arcsin(2P_i - 1) = g(P_i)$, and we have approximately $\hat{\theta}_i \overset{\text{ind}}{\sim} N(\theta_i, 1)$. The J-S estimate of θ_i was calculated from (4.4.5) using $\theta_i^0 = \sum \hat{\theta}_i/18 = \hat{\theta}_{\cdot\cdot}$. The resulting estimates $\hat{\theta}_{i,JS}$ were retransformed to provide estimates $g^{-1}(\hat{\theta}_{i,JS})$ of P_i's. Efron and Morris (1975) calculated the overall accuracy of $\hat{\theta}_{i,JS}$'s relative to $\hat{\theta}_i$'s as $R = 3.50$. They also noted that $\hat{\theta}_{i,JS}$ is closer to θ_i than $\hat{\theta}_i$ to θ_i for 15 batters, being worse only for batters 1, 10 and 15.

4.4.3 Estimation of Component MSE

We now turn to the estimation of MSE of $\hat{\theta}_{i,\text{JS}}$ for each i, assuming $\hat{\theta}_i \overset{\text{ind}}{\sim} N(\theta_i, \psi)$ with known sampling variance ψ. We write $\hat{\theta}_{i,\text{JS}}$ as

$$\hat{\theta}_{i,\text{JS}} = \hat{\theta}_i + h_i(\hat{\boldsymbol{\theta}}) \tag{4.4.10}$$

with

$$h_i(\hat{\boldsymbol{\theta}}) = \frac{(m-2)\psi}{S}(\theta_i^0 - \hat{\theta}_i).$$

Using the representation (4.4.10), we have

$$\begin{aligned}\text{MSE}_p(\hat{\theta}_{i,\text{JS}}) &= E_p[\hat{\theta}_i + h_i(\hat{\boldsymbol{\theta}}) - \theta_i]^2 \\ &= \psi + 2E_p[(\hat{\theta}_i - \theta_i)h_i(\hat{\boldsymbol{\theta}})] + E_p h_i^2(\hat{\boldsymbol{\theta}}).\end{aligned}$$

By Corollary 4.5.1 to Stein's Lemma given in Section 4.5, we can write

$$E_p[(\hat{\theta}_i - \theta_i)h_i(\hat{\boldsymbol{\theta}})] = \psi E_p[\partial h_i(\hat{\boldsymbol{\theta}})/\partial\hat{\theta}_i].$$

Thus

$$\text{MSE}_p(\hat{\theta}_{i,\text{JS}}) = E_p[\psi + 2\psi\partial h_i(\hat{\boldsymbol{\theta}})/\partial\hat{\theta}_i + h_i^2(\hat{\boldsymbol{\theta}})].$$

Hence, an unbiased estimator of the MSE of $\hat{\theta}_{i,\text{JS}}$ is given by

$$\text{mse}_p(\hat{\theta}_{i,\text{JS}}) = \psi + 2\psi\partial h_i(\hat{\boldsymbol{\theta}})/\partial\hat{\theta}_i + h_i^2(\hat{\boldsymbol{\theta}}). \tag{4.4.11}$$

In fact, (4.4.11) is the minimum variance unbiased estimator due to complete sufficiency of the statistic $\hat{\boldsymbol{\theta}}$ under normality. This estimator, however, can take negative values. A better estimator is given by $\text{mse}_p^+(\hat{\theta}_{i,\text{JS}}) = \max(0, \text{mse}_p(\hat{\theta}_{i,\text{JS}}))$.

If θ_i^0 is fixed, then we have

$$\frac{\partial h_i(\hat{\boldsymbol{\theta}})}{\partial\hat{\theta}_i} = -\frac{(m-2)\psi}{S}\left[1 - \frac{2(\hat{\theta}_i - \theta_i^0)^2}{S}\right]. \tag{4.4.12}$$

If θ_i^0 is taken as the mean $\hat{\theta}_.$, then we get

$$\frac{\partial h_i(\hat{\boldsymbol{\theta}})}{\partial\hat{\theta}_i} = -\left(1 - \frac{1}{m}\right)\frac{(m-3)\psi}{S}\left[1 - \frac{2(\hat{\theta}_i - \hat{\theta}_.)^2}{S}\right], \tag{4.4.13}$$

noting that

$$h_i(\hat{\boldsymbol{\theta}}) = \frac{(m-3)\psi}{S}(\hat{\theta}_. - \hat{\theta}_i).$$

The derivations for the case of a least squares predictor θ_i^0 can be obtained in a similar manner. Note that (4.4.11) is valid for general $h_i(\hat{\boldsymbol{\theta}})$, assuming $\hat{\theta}_i \overset{\text{ind}}{\sim} N(\theta_i, \psi)$.

Although the MSE estimator (4.4.11) is the minimum variance unbiased estimator, its coefficient of variation (CV) can be quite large, that is, it can be very unstable. Model-based estimators of MSE, considered in Chapter 7, are more stable although somewhat biased in the design-based framework.

Bilodeau and Srivastava (1988) considered the general case of correlated estimators $\hat{\theta}_i$ and the resulting estimator $\hat{\boldsymbol{\theta}}_{\text{JS}}$ given by (4.4.8). They obtained the minimum variance unbiased estimator of the MSE matrix $\mathbf{M}(\hat{\boldsymbol{\theta}}_{\text{JS}}) = E_p(\hat{\boldsymbol{\theta}}_{\text{JS}} - \boldsymbol{\theta})(\hat{\boldsymbol{\theta}}_{\text{JS}} - \boldsymbol{\theta})^T$ as

$$\hat{\mathbf{M}}(\hat{\boldsymbol{\theta}}_{\text{JS}}) = \left(1 - \frac{2\kappa}{Q}\right)\frac{\mathbf{B}}{\nu} + \kappa\left(\kappa + \frac{4}{\nu}\frac{(\nu+1)}{(\nu - m + 3)}\right)\frac{(\hat{\boldsymbol{\theta}}_{\text{JS}} - \boldsymbol{\theta}^0)(\hat{\boldsymbol{\theta}}_{\text{JS}} - \boldsymbol{\theta}^0)^T}{Q^2}, \tag{4.4.14}$$

where

$$\kappa = (m-2)/(\nu - m + 3).$$

The ith diagonal element of (4.4.14) is the estimator of $\text{MSE}_p(\hat{\theta}_{i,\text{JS}})$.

Rivest and Bellmonte (2000) generalized (4.4.11) to the case of $\hat{\boldsymbol{\theta}} \sim N_m(\boldsymbol{\theta}, \Psi)$ with known covariance matrix $\Psi = (\psi_{ij})$. In particular, if the $\hat{\theta}_i$s are independent with $\psi_{ij} = 0$ when $i \neq j$, then

$$\text{mse}_p(\hat{\theta}_{i,\text{JS}}) = \psi_{ii} + 2\psi_{ii}\partial h_i(\hat{\boldsymbol{\theta}})/\partial\hat{\theta}_i + h_i^2(\hat{\boldsymbol{\theta}}), \tag{4.4.15}$$

where ψ_{ii} is the ith diagonal element of Ψ. They also noted that the derivative $\partial h_i(\hat{\boldsymbol{\theta}})/\partial\hat{\theta}_i$ may be evaluated numerically when $h_i(\cdot)$ has no explicit form. For example, suppose that $\theta_i = \overline{Y}_i$ and a reliable direct estimator $\hat{Y} = \sum_i \hat{Y}_i$ of the population total is available, where $\hat{Y}_i = N_i\hat{\overline{Y}}_i$ and the domain size N_i is known. Then it is desirable to calibrate the J-S estimator $\hat{Y}_{i,\text{JS}} = N_i\hat{\overline{Y}}_{i,\text{JS}}$ to agree with the direct estimator $\hat{Y}$.

A simple ratio adjustment of the form (4.3.11) may be used to ensure that the adjusted estimators add up to $\hat{Y}$. The ratio-adjusted J-S estimator of the ith area total Y_i is given by

$$\hat{Y}_{i,\text{JS}}(a) = \frac{\hat{Y}_{i,\text{JS}}}{\sum_i \hat{Y}_{i,\text{JS}}}\hat{Y}, \tag{4.4.16}$$

where $\hat{Y}_{i,\text{JS}} = N_i(\hat{\overline{Y}}_i + h_i(\hat{\overline{Y}})) = \hat{Y}_i + N_i h_i(\hat{\overline{Y}})$. The estimator (4.4.16) may be written as $\hat{Y}_{i,\text{JS}}(a) = \hat{Y}_i + h_i^*(\hat{Y})$, where

$$h_i^*(\hat{Y}) = \frac{\sum \hat{Y}_i}{\sum \hat{Y}_{i,\text{JS}}} N_i h_i(\hat{\overline{Y}}) + \left(\frac{\sum \hat{Y}_i}{\sum \hat{Y}_{i,\text{JS}}} - 1\right)\hat{Y}_i. \tag{4.4.17}$$

Note that $\sum_i h_i^*(\hat{Y}) = 0$ which ensures that the adjusted J-S estimators add up to $\hat{Y}$. Numerical differention may be used to evaluate $\partial h_i^*(\hat{Y})/\partial\hat{Y}_i$. An estimator of MSE of $\hat{Y}_{i,\text{JS}}(a)$ may then be obtained from (4.4.15) with $h_i(\cdot)$ changed to $h_i^*(\cdot)$.

4.4.4 Unequal Variances ψ_i

We now turn to the case of unequal but known sampling variances ψ_i. A straightforward way to generalize the James-Stein method is to define $\hat{\delta}_i = \hat{\theta}_i/\sqrt{\psi_i}$ and $\delta_i = \theta_i/\sqrt{\psi_i}$ so that $\hat{\delta}_i \overset{\text{ind}}{\sim} N(\delta_i, 1)$ and $\delta_i^0 = \theta_i^0/\sqrt{\psi_i}$ is the guess of δ_i. We can now apply (4.4.5) to the transformed data and then retransform back to the original coordinates. This leads to

$$\hat{\theta}_{i,\text{JS}} = \theta_i^0 + \left(1 - \frac{m-2}{\tilde{S}}\right)(\hat{\theta}_i - \theta_i^0), \quad m \geq 3 \tag{4.4.18}$$

where $\tilde{S} = \sum_i(\hat{\theta}_i - \theta_i^0)^2/\psi_i$. The estimator (4.4.18) dominates the direct estimator $\hat{\theta}_i$ in terms of a weighted MSE with weights $1/\psi_i$, but not in terms of total MSE:

$$\sum_i \frac{1}{\psi_i}\text{MSE}_p(\hat{\theta}_{i,\text{JS}}) < \sum_i \frac{1}{\psi_i}\text{MSE}_p(\hat{\theta}_i). \tag{4.4.19}$$

Moreover, it gives the common weight $\tilde{\phi}_{\text{JS}} = 1-(m-2)/\tilde{S}$ to $\hat{\theta}_i$ and $1-\tilde{\phi}_{\text{JS}}$ to the guess θ_i^0, that is, each $\hat{\theta}_i$ is shrunk towards the guess θ_i^0 by the same factor $\tilde{\phi}_{\text{JS}}$. This is not appealing to the user as in the case of the composite estimator with weight $\hat{\phi}^*$ given by (4.4.3). One would like to have more shrinkage the larger ψ_i is. The model-based methods of Chapters 6–8 provide such unequal shrinkage.

4.4.5 Extensions

Various extensions of the J-S method have been studied in the literature. In particular, the dominance result holds under spherically symmetric or more generally, elliptical distributions for $\hat{\boldsymbol{\theta}}$ which include the normal, student-t and double exponential distributions (see Brandwein and Strawderman (1990), Srivastava and Bilodeau (1989)), and for the exponential family of distributions (Ghosh and Auer (1983)).

Efron and Morris (1972b) extended the J–S method to the case of a vector $\boldsymbol{\theta}_i$ having q components, that is, to the case of multiple characteristics. In particular, if $\hat{\boldsymbol{\theta}}_i \overset{\text{ind}}{\sim} N(\boldsymbol{\theta}_i, \boldsymbol{\Sigma})$ with known $\boldsymbol{\Sigma}$ and the composite risk of an estimator $\tilde{\boldsymbol{\theta}} = (\tilde{\boldsymbol{\theta}}_1, \ldots, \tilde{\boldsymbol{\theta}}_m)$ of $\boldsymbol{\theta} = (\boldsymbol{\theta}_1, \ldots, \boldsymbol{\theta}_m)$ is taken as

$$R(\boldsymbol{\theta}, \tilde{\boldsymbol{\theta}}) = E_p\{\text{tr}(\boldsymbol{\theta} - \tilde{\boldsymbol{\theta}})^T \boldsymbol{\Sigma}^{-1}(\boldsymbol{\theta} - \tilde{\boldsymbol{\theta}})\},$$

then the estimators

$$\hat{\boldsymbol{\theta}}_{i,\text{JS}} = \boldsymbol{\theta}_i^0 + (\mathbf{I} - (m-q-1)\boldsymbol{\Sigma}\mathbf{S}^{-1})\hat{\boldsymbol{\theta}}_i, \quad i = 1, \ldots, m \tag{4.4.20}$$

dominate the direct estimator $\hat{\boldsymbol{\theta}}_i$ in terms of the composite risk. Here tr denotes the trace operator, $\boldsymbol{\theta}_i^0$ is a guess of $\boldsymbol{\theta}_i$ and

$$\mathbf{S} = (\hat{\boldsymbol{\theta}} - \boldsymbol{\theta}^0)(\hat{\boldsymbol{\theta}} - \boldsymbol{\theta}^0)^T$$

with $\hat{\boldsymbol{\theta}} = (\hat{\boldsymbol{\theta}}_1, \ldots, \hat{\boldsymbol{\theta}}_m)$ and $\boldsymbol{\theta}^0 = (\boldsymbol{\theta}_1^0, \ldots, \boldsymbol{\theta}_m^0)$. Note that $\mathbf{S}$ is a $q \times q$ matrix. When $q = 1$, (4.4.20) reduces to the J–S estimator (4.4.5).

4.5 Proofs

To prove Theorem 4.4.1, we need the following lemma attributed to Stein (1981); see also Brandwein and Strawderman (1990).

Lemma 4.5.1. Let $Z \sim N(\mu, 1)$, then $Eh(Z)(Z-\mu) = Eh'(Z)$, provided the expectations exist and

$$\lim_{z\to\pm\infty} h(z)\exp[-\tfrac{1}{2}(z-\mu)^2] = 0, \tag{4.5.1}$$

where $h'(Z) = \partial h(Z)/\partial Z$.

Proof:

$$\begin{aligned} E[h(Z)(Z-\mu)] &= \frac{1}{\sqrt{2\pi}}\int_{-\infty}^{\infty} h(z)(z-\mu)\exp\{-\tfrac{1}{2}(z-\mu)^2\}dz \\ &= -\frac{1}{\sqrt{2\pi}}\int_{-\infty}^{\infty} h(z)\frac{d}{dz}[\exp\{-\tfrac{1}{2}(z-\mu)^2\}]dz. \end{aligned}$$

Integration by parts now yields

$$\begin{aligned} E[h(Z)(Z-\mu)] &= -\frac{1}{\sqrt{2\pi}}h(z)\exp[-\tfrac{1}{2}(z-\mu)^2]\Big|_{-\infty}^{\infty} \\ &\quad + \frac{1}{\sqrt{2\pi}}\int_{-\infty}^{\infty} h'(z)\exp[-\tfrac{1}{2}(z-\mu)]^2 dz \\ &= Eh'(Z), \end{aligned}$$

using (4.5.1).

Corollary 4.5.1. Let $Z_i \overset{\text{ind}}{\sim} N(\mu_i, 1)$. If condition (4.5.1) holds, then $E[h(\mathbf{Z})(Z_i - \mu_i)] = E[\partial h(\mathbf{Z})/\partial Z_i]$, where $\mathbf{Z} = (Z_1, \ldots, Z_m)^T$.

It is sufficient to consider the canonical case $Z_i \overset{\text{ind}}{\sim} N(\mu_i, 1)$ to establish Theorem 4.4.1.

Theorem 4.5.1. Let $Z_i \overset{\text{ind}}{\sim} N(\mu_i, 1)$, $i = 1, \ldots, m(\geq 3)$ and

$$\hat{\boldsymbol{\mu}}_{\text{JS}}(a) = \left(1 - \frac{a}{\|\mathbf{Z}\|^2}\right)\mathbf{Z}, \quad 0 < a < 2(m-2),$$

where $\|\mathbf{Z}\|^2 = \mathbf{Z}^T\mathbf{Z}$. Then
(a) $\hat{\boldsymbol{\mu}}_{\text{JS}}(a)$ dominates $\mathbf{Z}$ for $0 < a < 2(m-2)$ in terms of total MSE, and the optimal choice of a is $m-2$ so that

$$R(\boldsymbol{\mu}, \hat{\boldsymbol{\mu}}_{\text{JS}}) < R(\boldsymbol{\mu}, \mathbf{Z}) = m \quad \text{for all } \boldsymbol{\mu},$$

where $\hat{\boldsymbol{\mu}}_{\text{JS}}$ is $\hat{\boldsymbol{\mu}}_{\text{JS}}(a)$ evaluated at $a = m-2$.
(b) $R(\boldsymbol{\mu}, \hat{\boldsymbol{\mu}}_{\text{JS}}) = 2$ at $\boldsymbol{\mu} = \mathbf{0}$.

(c) $R(\boldsymbol{\mu}, \hat{\boldsymbol{\mu}}_{\text{JS}}) \leq m - \dfrac{(m-2)^2}{(m-2) + \|\boldsymbol{\mu}\|^2}$

so that $R(\boldsymbol{\mu}, \hat{\boldsymbol{\mu}}_{\text{JS}})$ is minimized when $\boldsymbol{\mu} = 0$.

Proof: (a) We follow Brandwein and Strawderman (1990). We have

$$\begin{aligned} R(\boldsymbol{\mu}, \hat{\boldsymbol{\mu}}_{\text{JS}}(a)) &= E\|\hat{\boldsymbol{\mu}}_{\text{JS}}(a) - \boldsymbol{\mu}\|^2 \\ &= E\|\mathbf{Z} - \boldsymbol{\mu}\|^2 + a^2 E\frac{1}{\mathbf{Z}^T\mathbf{Z}} - 2aE\frac{\mathbf{Z}^T(\mathbf{Z} - \boldsymbol{\mu})}{\mathbf{Z}^T\mathbf{Z}} \\ &= m + a^2 E\frac{1}{\mathbf{Z}^T\mathbf{Z}} - 2a\sum_i E\frac{Z_i(Z_i - \mu_i)}{\mathbf{Z}^T\mathbf{Z}} \\ &= m + a^2 E\frac{1}{\mathbf{Z}^T\mathbf{Z}} - 2a\sum_i E\left[\frac{\partial}{\partial Z_i}\left(\frac{Z_i}{\mathbf{Z}^T\mathbf{Z}}\right)\right] \end{aligned}$$

by Corollary 4.5.1 with $h(\mathbf{Z}) = Z_i/(\mathbf{Z}^T\mathbf{Z})$. Further simplification gives

$$\begin{aligned} R(\boldsymbol{\mu}, \hat{\boldsymbol{\mu}}_{\text{JS}}(a)) &= m + a^2 E\frac{1}{\mathbf{Z}^T\mathbf{Z}} - 2a\sum_i E\frac{\mathbf{Z}^T\mathbf{Z} - 2Z_i^2}{(\mathbf{Z}^T\mathbf{Z})^2} \\ &= m + a^2 E\frac{1}{\mathbf{Z}^T\mathbf{Z}} - 2a(m-2)E\frac{1}{\mathbf{Z}^T\mathbf{Z}} \\ &= m + [a^2 - 2a(m-2)]E\frac{1}{\mathbf{Z}^T\mathbf{Z}}. \end{aligned} \tag{4.5.2}$$

The quadratic function $a^2 - 2a(m-2)$ is negative in the range $0 < a < 2(m-2)$ so that $\hat{\boldsymbol{\mu}}_{\text{JS}}(a)$ dominates $\mathbf{Z}$ in that range, noting that $R(\boldsymbol{\mu}, \mathbf{Z}) = m$. Further, it attains its minimum at $a = m - 2$ so that the optimum choice of a is $m - 2$.

(b) At $\boldsymbol{\mu} = \mathbf{0}$, $\mathbf{Z}^T\mathbf{Z}$ is a chi-squared variable with m df so that $E(1/\mathbf{Z}^T\mathbf{Z}) = 1/(m-2)$ and (4.5.2) reduces to

$$R(\boldsymbol{\mu}, \hat{\boldsymbol{\mu}}_{\text{JS}}) = m - (m-2) = 2.$$

(c) We follow Casella and Hwang (1982). We note that $\mathbf{Z}^T\mathbf{Z}$ is a noncentral chi-squared variable with m df and noncentrality parameter $\tau = \|\boldsymbol{\mu}\|^2/2$, denoted by $\chi_m^2(\tau)$. Further, we use the well-known fact that a noncentral chi-squared distribution is the infinite sum of central chi-squared distributions with Poisson weights. It now follows that

$$E\frac{1}{\mathbf{Z}^T\mathbf{Z}} = E\frac{1}{\chi_m^2(\tau)} = E\left[E\left(\frac{1}{\chi_{m+2K}^2}\middle|K\right)\right],$$

where χ_{m+2K}^2 is a (central) chi-squared variable with $m + 2K$ df and K is a Poisson random variable with mean τ. Therefore,

$$E\frac{1}{\mathbf{Z}^T\mathbf{Z}} = E\frac{1}{m + 2K - 2} \geq \frac{1}{m - 2 + 2\tau} = \frac{1}{m - 2 + \|\boldsymbol{\mu}\|^2},$$

by Jensen's inequality. It now follows from (4.5.2) that

$$R(\boldsymbol{\mu}, \hat{\boldsymbol{\mu}}_{\text{JS}}) = m - (m-2)^2 E \frac{1}{\mathbf{Z}^T\mathbf{Z}} \leq m - \frac{(m-2)^2}{(m-2) + \|\boldsymbol{\mu}\|^2} \, .$$

Proof of Theorem 4.4.1. Let $Z_i = (\hat{\theta}_i - \theta_i^0)/\sqrt{\psi}$ in Theorem 4.5.1 so that $\mu_i = (\theta_i - \theta_i^0)/\sqrt{\psi}$. Also,

$$R(\boldsymbol{\mu}, \mathbf{Z}) = \frac{1}{\psi} R(\boldsymbol{\theta}, \hat{\boldsymbol{\theta}})$$

and

$$R(\boldsymbol{\mu}, \hat{\boldsymbol{\mu}}_{\text{JS}}) = \frac{1}{\psi} R(\boldsymbol{\theta}, \hat{\boldsymbol{\theta}}_{\text{JS}}).$$

Chapter 5

Small Area Models

5.1 Introduction

Traditional methods of indirect estimation, studied in Chapters 3 and 4, are based on implicit models that provide a link to related small areas through supplementary data. We now turn to explicit small area models that make specific allowance for between area variation. In particular, we introduce mixed models involving random area-specific effects that account for between area variation beyond that explained by auxiliary variables included in the model. The use of explicit models offers several advantages: (1) Model diagnostics can be used to find suitable model(s) that fit the data well. Such model diagnostics include residual analysis to detect departures from the assumed model, selection of auxiliary variables for the model, and case-deletion diagnostics to detect influential observations. (2) Area-specific measures of precision can be associated with each small area estimate, unlike the global measures (averaged over small areas) often used with synthetic estimates (Chapter 4, subsection 4.2.4). (3) Linear mixed models as well as nonlinear models, such as logistic regression models and generalized linear models with random area effects, can be entertained. Complex data structures, such as spatial dependence and time series structures, can also be handled. (4) Recent methodological developments for random effects models can be utilized to achieve accurate small area inferences (Chapters 6–9).

Although we present a variety of models for small area estimation in this chapter, it is important to note that the subject matter specialists or end users should have influence on the choice of models, particularly on the choice of auxiliary variables. Also, the success of any model-based method depends on the availability of good auxiliary data. More attention should therefore be given to the compilation of auxiliary variables that are good predictors of the study variables.

We present models that permit empirical best linear unbiased prediction (EBLUP) and empirical Bayes (EB) inferences (Chapters 6–9). Additional assumptions on the model parameters, in the form of prior distributions, are needed to implement the hierarchical Bayes (HB) approach (Chapter 10). The models we study may be classified into two broad types: (i) Aggregate level (or area level) models that relate the small area means to area-specific auxiliary variables. Such models are essential if unit (element) level data are not available. (ii) Unit level models that relate the unit values of the study variable to unit-specific auxiliary variables. Models involving both unit level and area level auxiliary variables will also be studied.

5.2 Basic Area Level (Type A) Model

We assume that $\theta_i = g(\bar{Y}_i)$ for some specified $g(\cdot)$ is related to area-specific auxiliary data $\mathbf{z}_i = (z_{1i}, \ldots, z_{pi})^T$ through a linear model

$$\theta_i = \mathbf{z}_i^T \boldsymbol{\beta} + b_i v_i, \quad i = 1, \ldots, m \tag{5.2.1}$$

where the b_i's are known positive constants and $\boldsymbol{\beta} = (\beta_1, \cdots, \beta_p)^T$ is the $p \times 1$ vector of regression coefficients. Further, the v_i's are area-specific random effects assumed to be independent and identically distributed (iid) with

$$E_m(v_i) = 0, \quad V_m(v_i) = \sigma_v^2 \ (\geq 0), \tag{5.2.2}$$

where E_m denotes the model expectation and V_m the model variance. We denote this assumption as $v_i \overset{\text{iid}}{\sim} (0, \sigma_v^2)$. Normality of the random effects v_i is also often used, but it is possible to make "robust" inferences by relaxing the normality assumption (Chapter 7, subsection 7.1.5). The parameter σ_v^2 is a measure of homogeneity of the areas after accounting for the covariates $\mathbf{z}_i$.

In some applications, not all areas are selected in the sample (see Example 5.2.3). Suppose that we have M areas in the population and only m areas are selected in the sample. We assume a model of the form (5.2.1) for the population, that is, $\theta_i = \mathbf{z}_i^T \boldsymbol{\beta} + b_i v_i$, $i = 1, \ldots, M$. We further assume that the sample areas obey the population model, that is, the bias in the sample selection of areas is absent so that (5.2.1) holds for the sampled areas.

For making inferences about the small area means $\bar{Y}_i$ under model (5.2.1), we assume that direct estimators $\hat{\bar{Y}}_i$ are available. As in the James-Stein method (Chapter 4, Section 4.4), we assume that

$$\hat{\theta}_i = g(\hat{\bar{Y}}_i) = \theta_i + e_i, \quad i = 1, \ldots, m \tag{5.2.3}$$

where the sampling errors e_i are independent with

$$E_p(e_i|\theta_i) = 0, \quad V_p(e_i|\theta_i) = \psi_i. \tag{5.2.4}$$

It is also customary to assume that the sampling variances, ψ_i, are known. The above assumptions may be quite restrictive in some applications. For example, the direct estimator $\hat{\theta}_i$ may be design-biased for θ_i if $g(\cdot)$ is a nonlinear function and the area sample size n_i is small. The sampling errors may not be independent if the small areas cut across the sampling design. The assumptions of known variance ψ_i can be relaxed by estimating ψ_i from the unit-level sample data and then smoothing the estimated variances $\hat{\psi}_i$ to get a more stable estimate of ψ_i. Normality of the estimator $\hat{\theta}_i$ is also often assumed, but this may not be as restrictive as the normality of the random effects, due to the central limit theorem's effect on $\hat{\theta}_i$.

Deterministic models on θ_i are obtained by setting $\sigma_v^2 = 0$, that is, $\theta_i = \mathbf{z}_i^T\boldsymbol{\beta}$. Such models lead to synthetic estimators that do not account for local variation other than the variation reflected in the auxiliary variables $\mathbf{z}_i$.

Combining (5.2.1) with (5.2.3) we obtain the model

$$\hat{\theta}_i = \mathbf{z}_i^T\boldsymbol{\beta} + b_i v_i + e_i, \quad i = 1, \ldots, m, \tag{5.2.5}$$

Note that (5.2.5) involves design-induced errors e_i as well as model errors v_i. We assume v_i and e_i are independent. Model (5.2.5) is a special case of a linear mixed model (Chapter 6).

The assumption $E_p(e_i|\theta_i) = 0$ in the sampling model (5.2.3) may not be valid if the sample size n_i in the ith area is small and θ_i is a nonlinear function of the total Y_i, even if the direct estimator $\hat{Y}_i$ is design-unbiased. A more realistic sampling model is given by

$$\hat{Y}_i = Y_i + e_i^*, \quad i = 1, \cdots, m, \tag{5.2.6}$$

with $E_p(e_i^*|Y_i) = 0$, that is, $\hat{Y}_i$ is design-unbiased for the total Y_i. In this case, the sampling and linking models are not matched. As a result, we cannot combine (5.2.6) with the linking model (5.2.1) to produce a linear mixed model of the form (5.2.5). Therefore, standard results in linear mixed model theory, presented in Chapter 6, do not apply. In Section 10.4, we use a hierarchical Bayes (HB) approach to handle unmatched sampling and linking models.

Example 5.2.1. Income for Small Places. In the context of estimating per capita income (PCI) for small places in the United States with population less than 1,000, Fay and Herriot (1979) used model (5.2.5). In their application, $\theta_i = \log \bar{Y}_i$, and $\bar{Y}_i$ is the PCI in the ith area. Model (5.2.5) is called the Fay-Herriot model in the small area literature because they were the first to use such models for small area estimation. Some details of this application are given in Example 7.1.1.

Example 5.2.2. Census Undercount. In the context of estimating the undercount in the decennial census of the United States, Ericksen and Kadane (1985) used model (5.2.5) with $b_i = 1$ and treating σ_v^2 as known. Here, $\theta_i = (T_i - C_i)/T_i$ is the census undercount for the ith state (area) where T_i is the true (unknown) count and C_i is the census count in the ith area, and $\hat{\theta}_i$ is the estimate of θ_i from the post-enumeration survey (PES). Cressie (1989) used (5.2.5) with $b_i = 1/\sqrt{C_i}$, where $\theta_i = T_i/C_i$ is the unknown adjustment factor and $\hat{\theta}_i$ is the PES estimate of θ_i. In these applications, the PES estimates could be seriously biased, as noted by Freedman and Navidi (1986).

In the context of the Canadian Census, Dick (1995) used model (5.2.5) with $\theta_i = T_i/C_i$ and $b_i = 1$, where i denotes province × age × sex combination. He used smoothed estimates of the sampling variances ψ_i, by assuming ψ_i to be proportional to some power of the census count C_i. Some details of this application are given in Example 7.1.3.

Example 5.2.3. Poverty Counts. The basic area level model (5.2.5) has been used recently to produce model-based county estimates of poor school-age children in the United States (National Research Council (2000)). Using these estimates, the U.S. Department of Education allocates annually over $7 billion of general funds to counties, and then states distribute these funds among school districts. In the past, funds were allocated on the basis of estimated counts from the previous census, but the poverty counts have changed significantly over time.

In this application, $\theta_i = \log Y_i$, where Y_i is the true poverty count of the ith county (small area). Direct estimators $\hat{Y}_i$ were calculated as a 3-year weighted average of poor school-age children (under 18) obtained from the March Supplement of the Current Population Survey (CPS). Area level predictor variables, $\mathbf{z}_i$, were obtained from administrative records. Counties with CPS sample but no poor school-age children were excluded because $\hat{Y}_i = 0$ and $\hat{\theta}_i = \log \hat{Y}_i = -\infty$. Only auxiliary variables $\mathbf{z}_i$ are available for the counties not selected in the CPS sample. Some details of this application are given in Example 7.1.3.

5.3 Basic Unit Level (Type B) Model

We assume that unit-specific auxiliary data $\mathbf{x}_{ij} = (x_{ij1}, \ldots, x_{ijp})^T$ are available for each population element j in each small area i. It is often sufficient to assume that only the population means $\overline{\mathbf{X}}_i$ are known. Further, the variable of interest, y_{ij}, is assumed to be related to $\mathbf{x}_{ij}$ through a one-fold nested error linear regression model:

$$y_{ij} = \mathbf{x}_{ij}^T \boldsymbol{\beta} + v_i + e_{ij}; \quad j = 1, \ldots, N_i, \; i = 1, \ldots m. \tag{5.3.1}$$

Here the area-specific effects v_i are assumed to be iid random variables satisfying (5.2.2), $e_{ij} = k_{ij}\tilde{e}_{ij}$ with known constants k_{ij} and $\tilde{e}_{ij}$'s are iid random variables independent of v_i's and

$$E_m(\tilde{e}_{ij}) = 0, \quad V_m(\tilde{e}_{ij}) = \sigma_e^2. \tag{5.3.2}$$

In addition, normality of the v_i's and e_{ij}'s is often assumed. The parameters of interest are the small area means $\bar{Y}_i$ or the totals Y_i. Standard regression models are obtained by setting $\sigma_v^2 = 0$ or equivalently $v_i = 0$ in (5.3.1). Such models lead to synthetic-type estimators (Chapter 4, subsection 4.2.2).

We assume that a sample, s_i of size n_i is taken from the N_i units in the ith area ($i = 1, \ldots, m$) and that the sample values also obey the assumed model (5.3.1). The latter assumption is satisfied under simple random sampling from each area or more generally for sampling designs that use the auxiliary information $\mathbf{x}_{ij}$ in the selection of the samples s_i. To see this, we write (5.3.1) in matrix form as

$$\mathbf{y}_i^P = \mathbf{X}_i^P \boldsymbol{\beta} + v_i \mathbf{1}_i^P + \mathbf{e}_i^P, \quad i = 1, \ldots, m \tag{5.3.3}$$

where $\mathbf{X}_i^P$ is $N_i \times p$, $\mathbf{y}_i^P$, $\mathbf{1}_i^P$ and $\mathbf{e}_i^P$ are $N_i \times 1$ vectors and $\mathbf{1}_i^P = (1, \ldots, 1)^T$. We next partition (5.3.3) into sampled and nonsampled parts:

$$\mathbf{y}_i^P = \begin{bmatrix} \mathbf{y}_i \\ \mathbf{y}_i^* \end{bmatrix} = \begin{bmatrix} \mathbf{X}_i \\ \mathbf{X}_i^* \end{bmatrix} \boldsymbol{\beta} + v_i \begin{bmatrix} \mathbf{1}_i \\ \mathbf{1}_i^* \end{bmatrix} + \begin{bmatrix} \mathbf{e}_i \\ \mathbf{e}_i^* \end{bmatrix}, \tag{5.3.4}$$

where the superscript $*$ denotes the nonsampled units. If the model holds for the sample, that is, if selection bias is absent, then inferences on $\boldsymbol{\psi} = (\boldsymbol{\beta}^T, \sigma_v^2, \sigma_e^2)^T$ are based on

$$f(\mathbf{y}_i | \mathbf{X}_i^P, \boldsymbol{\psi}) = \int f(\mathbf{y}_i, \mathbf{y}_i^* | \mathbf{X}_i^P, \boldsymbol{\psi}) d\mathbf{y}_i^*, \quad i = 1, \ldots, m \tag{5.3.5}$$

where $f(\mathbf{y}_i, \mathbf{y}_i^* | \mathbf{X}_i^P, \boldsymbol{\psi})$ is the assumed joint distribution of $\mathbf{y}_i$ and $\mathbf{y}_i^*$. On the other hand, letting $\mathbf{a}_i = (a_{i1}, \ldots, a_{iN_i})^T$ with $a_{ij} = 1$ if $j \in s_i$ and $a_{ij} = 0$ otherwise, the distribution of sample data $(\mathbf{y}_i, \mathbf{a}_i)$ is given by

$$\begin{aligned} f(\mathbf{y}_i, \mathbf{a}_i | \mathbf{X}_i^P, \boldsymbol{\psi}) &= \int f(\mathbf{y}_i, \mathbf{y}_i^* | \mathbf{X}_i^P, \boldsymbol{\psi}) f(\mathbf{a}_i | \mathbf{y}_i, \mathbf{y}_i^*, \mathbf{X}_i^P) d\mathbf{y}_i^* \\ &= \left[\int f(\mathbf{y}_i, \mathbf{y}_i^* | \mathbf{X}_i^P, \boldsymbol{\psi}) d\mathbf{y}_i^* \right] f(\mathbf{a}_i | \mathbf{X}_i^P), \end{aligned}$$

provided

$$f(\mathbf{a}_i | \mathbf{y}_i, \mathbf{y}_i^*, \mathbf{X}_i^P) = f(\mathbf{a}_i | \mathbf{X}_i^P),$$

that is, the sample selection probabilities do not depend on $\mathbf{y}_i^P$ but may depend on $\mathbf{X}_i^P$. In this case, selection bias is absent and we may assume that the sample values also obey the assumed model, that is, use $f(\mathbf{y}_i | \mathbf{X}_i^P, \boldsymbol{\psi})$ for inferences on $\boldsymbol{\psi}$ (Smith (1983)).

If the sample selection probabilities depend on an auxiliary variable, say $\mathbf{z}_i^P$, which is not included in $\mathbf{X}_i^P$, then the distribution of sample data $(\mathbf{y}_i, \mathbf{a}_i)$ is

$$f(\mathbf{y}_i, \mathbf{a}_i | \mathbf{X}_i^P, \mathbf{z}_i^P, \boldsymbol{\psi}) = \left[\int f(\mathbf{y}_i, \mathbf{y}_i^* | \mathbf{X}_i^P, \mathbf{z}_i^P, \boldsymbol{\psi}) d\mathbf{y}_i^* \right] f(\mathbf{a}_i | \mathbf{z}_i^P, \mathbf{X}_i^P).$$

In this case, inference on $\boldsymbol{\psi}$ is based on $f(\mathbf{y}_i | \mathbf{X}_i^P, \mathbf{z}_i^P, \boldsymbol{\psi})$ which is different from (5.3.5) unless $\mathbf{z}_i^P$ is unrelated to $\mathbf{y}_i^P$ given $\mathbf{X}_i^P$. In this case we have sample selection bias and therefore we cannot assume that the model (5.3.3) holds for the sample values. We could extend model (5.3.3) by including $\mathbf{z}_i^P$ and then test for the significance of the associated regression coefficient using the sample data. If the null hypothesis is not rejected, then we could assume that the original model (5.3.3) also holds for the sample values (Skinner (1994)).

The model (5.3.3) is also not appropriate under two-stage cluster sampling within small areas because random cluster effects are not incorporated. But we can extend the model to account for such features (subsection 5.5.3).

We write the small area mean $\bar{Y}_i$ as

$$\bar{Y}_i = f_i \bar{y}_i + (1 - f_i)\bar{Y}_i^* \tag{5.3.6}$$

with $f_i = n_i/N_i$ and $\bar{y}_i$ and $\bar{Y}_i^*$ denoting the means of the sampled and nonsampled elements, respectively. It follows from (5.3.6) that the estimation of small area mean $\bar{Y}_i$ is equivalent to estimating the realization of the random variable $\bar{Y}_i^*$ given the sample data $\{\mathbf{y}_i\}$ and auxiliary data $\{\mathbf{X}_i^P\}$.

If the population size N_i is large, then we can take the small area means as

$$\bar{Y}_i = \bar{\mathbf{X}}_i^T \boldsymbol{\beta} + v_i \tag{5.3.7}$$

noting that $\bar{Y}_i = \bar{\mathbf{X}}_i^T \boldsymbol{\beta} + v_i + \bar{E}_i$ and $\bar{E}_i \approx 0$, where $\bar{E}_i$ is the mean of the N_i errors e_{ij} and $\bar{\mathbf{X}}_i$ is the known mean of $\mathbf{X}_i^P$. It follows from (5.3.7) that the estimation of $\bar{Y}_i$ is equivalent to the estimation of a linear combination of $\boldsymbol{\beta}$ and the realization of the random variable v_i.

We now give some examples of model (5.3.1). Some details of these applications are given in Chapter 7, subsection 7.2.6.

Example 5.3.1. County Crop Areas. Battese, Harter and Fuller (1988) used the nested error regression model (5.3.1) to estimate county crop areas using sample survey data in conjunction with satellite information. In particular, they were interested in estimating the area under corn and soybeans for each of $m = 12$ counties in North-Central Iowa using farm-interview data as $\{\mathbf{y}_i\}$ and LANDSAT satellite data as $\{\mathbf{X}_i\}$. Each county was divided into area segments and the areas under corn and soybeans were ascertained for a sample of segments by interviewing farm operators. The number of sampled segments in a county, n_i, ranged from 1 to 6. Auxiliary data in the form of numbers of pixels (a term used for "picture elements" of about 0.45 hectares) classified as corn and soybeans were also obtained for all the area segments, including the sampled segments, in each county using the LANDSAT satellite readings. Battese et al. (1988) proposed the model

$$y_{ij} = \beta_0 + \beta_1 x_{ij1} + \beta_2 x_{ij2} + v_i + \tilde{e}_{ij}, \tag{5.3.8}$$

which is a special case of model (5.3.1) with $k_{ij} = 1$, $\mathbf{x}_{ij} = (1, x_{ij1}, x_{ij2})^{\mathrm{T}}$ and $\boldsymbol{\beta} = (\beta_0, \beta_1, \beta_2)^{\mathrm{T}}$. Here $y_{ij} =$ number of hectares of corn (or soybeans), $x_{ij1} =$ number of pixels classified as corn and $x_{ij2} =$ number of pixels classified as soybeans, in the jth area segment of the ith county. Some details of this application are given in Examples 7.2.1 and 7.2.3.

Example 5.3.2. Wages and Salaries. Rao and Choudhry (1995) studied the population of unincorporated tax filers from the province of Nova Scotia,

Canada (Example 2.4.1). They proposed the model

$$y_{ij} = \beta_0 + \beta_1 x_{ij} + v_i + x_{ij}^{1/2}\tilde{e}_{ij}, \tag{5.3.9}$$

which is a special case of (5.3.1) with $k_{ij} = x_{ij}^{1/2}$. Here y_{ij} and x_{ij} denote the total wages and salaries and gross business income for the j th firm in the i th area. Simple random sampling from the overall population was used to estimate the small area totals Y_i or the means $\bar{Y}_i$. Some details of a simulation study based on this data are given in Example 7.2.2.

5.4 Extensions: Type A Models

We now consider various extensions of the basic area level (type A) model (5.2.5).

5.4.1 Multivariate Fay-Herriot Model

Suppose we have an $r \times 1$ vector of survey estimators $\hat{\boldsymbol{\theta}}_i = (\hat{\theta}_{i1}, \ldots, \hat{\theta}_{ir})^T$ and

$$\hat{\boldsymbol{\theta}}_i = \boldsymbol{\theta}_i + \mathbf{e}_i, \quad i = 1, \ldots, m \tag{5.4.1}$$

where $\boldsymbol{\theta}_i = (\theta_{i1}, \ldots, \theta_{ir})^T$ with $\theta_{ij} = g_j(\bar{Y}_{ij})$, $j = 1, \ldots, r$ and the sampling errors $\mathbf{e}_i = (e_{i1}, \cdots, e_{ir})^T$ are independent r-variate normal, $N_r(\mathbf{0}, \boldsymbol{\Psi}_i)$, with mean $\mathbf{0}$ and known covariance matrices $\boldsymbol{\Psi}_i$ conditional on $\boldsymbol{\theta}_i$. Here $\mathbf{0}$ is the $r \times 1$ null vector and $\bar{Y}_{ij}$ is the i th small area mean for the j th characteristic. We further assume that $\boldsymbol{\theta}_i$ is related to area-specific auxiliary data $\{\mathbf{z}_{ij}\}$ through a linear model

$$\boldsymbol{\theta}_i = \mathbf{Z}_i \boldsymbol{\beta} + \mathbf{v}_i, \quad i = 1, \ldots, m \tag{5.4.2}$$

where the area-specific random effects $\mathbf{v}_i$ are independent $N_r(\mathbf{0}, \boldsymbol{\Sigma}_v)$, $\mathbf{Z}_i$ is an $r \times rp$ matrix with j th row given by $(\mathbf{0}^T, \ldots, \mathbf{0}^T, \mathbf{z}_{ij}^T, \mathbf{0}^T, \ldots, \mathbf{0}^T)$ and $\boldsymbol{\beta}$ is the rp-vector of regression coefficients. Here $\mathbf{0}$ is the $p \times 1$ null vector and $\mathbf{z}_{ij}^T$ occurs in the j th position of the row vector (j th row).

Combining (5.4.1) with (5.4.2), we obtain a multivariate mixed linear model

$$\hat{\boldsymbol{\theta}}_i = \mathbf{Z}_i \boldsymbol{\beta} + \mathbf{v}_i + \mathbf{e}_i. \tag{5.4.3}$$

The model (5.4.3) is a natural extension of the Fay-Herriot model (5.2.5) with $b_i = 1$. Fay (1987) and Datta, Fay and Ghosh (1991) proposed the multivariate extension (5.4.3) and demonstrated that it can lead to more efficient estimators of the small area means $\bar{Y}_{ij}$ because it takes advantage of the correlations between the components of $\hat{\boldsymbol{\theta}}_i$ unlike the univariate model (5.2.5).

Example 5.4.1. Median Income. Datta, Fay and Ghosh (1991) applied the multivariate model (5.4.3) to estimate the current median income for four-person families in each of the American states (small areas). These estimates are used in a formula to determine the eligibility for a program of energy assistance to low-income families administered by the U.S. Department of Health and Human Services. In this application, $\boldsymbol{\theta}_i = (\theta_{i1}, \theta_{i2})^T$ with θ_{i1} = population median income of four-person families in state i and $\theta_{i2} = \frac{3}{4}$(population median income of five-person families in state i)+$\frac{1}{4}$(population median income of three-person families in state i). Direct estimates $\hat{\boldsymbol{\theta}}_i$ and the associated covariance matrix $\hat{\boldsymbol{\Psi}}_i$ were obtained from the Current Population Survey, and $\boldsymbol{\Psi}_i$ was treated as known by letting $\boldsymbol{\Psi}_i = \hat{\boldsymbol{\Psi}}_i$ and ignoring the variability of the estimate $\hat{\boldsymbol{\Psi}}_i$. Here θ_{i2} is not a parameter of interest but the associated direct estimator $\hat{\theta}_{i2}$ is strongly related to $\hat{\theta}_{i1}$. By taking advantage of this association, an improved estimator of the parameter of interest θ_{i1} can be obtained through the multivariate model (5.4.3).

The auxiliary data $\{\mathbf{z}_{ij}\}$ was based on census data: $\mathbf{z}_{ij} = (1 \ z_{ij1} \ z_{ij2})^T$, $j = 1, 2$, where z_{i11} and z_{i12} denote, respectively, the adjusted census median income for the current year and the base-year census median income for four-person families in the ith state, and z_{i21} and z_{i22} denote, respectively, the weighted average (with $\frac{3}{4}$ and $\frac{1}{4}$ as weights) of adjusted census median incomes for three- and five-person families and the corresponding weighted average (with the same weights) of the base-year census medians in the ith state. The adjusted census median incomes for the current year were obtained by multiplying the census median incomes by adjustment factors produced by the Bureau of Economic Analysis of the U.S. Department of Commerce. Some details of this application are given in Examples 8.1.1 and 10.9.1.

5.4.2 Model with Correlated Sampling Errors

A natural extension of the Fay-Herriot model (5.2.5) is to the case of correlated sampling errors e_i. Let $\hat{\boldsymbol{\theta}} = (\hat{\theta}_1, \ldots, \hat{\theta}_m)^T$, $\boldsymbol{\theta} = (\theta_1, \ldots, \theta_m)^T$ and $\mathbf{e} = (e_1, \ldots, e_m)^T$, and assume that

$$\hat{\boldsymbol{\theta}} = \boldsymbol{\theta} + \mathbf{e} \tag{5.4.4}$$

with $\mathbf{e}|\boldsymbol{\theta} \sim N_m(\mathbf{0}, \boldsymbol{\Psi})$, where the sampling error covariance matrix $\boldsymbol{\Psi} = (\psi_{ij})$ is known. Combining (5.4.4) with the model (5.2.1) for the θ_i's, we obtain a generalization of the Fay-Herriot model (5.2.5). If $b_i = 1$ for all i in (5.2.1) then the combined model may be written as

$$\hat{\boldsymbol{\theta}} = \mathbf{Z}\boldsymbol{\beta} + \mathbf{v} + \mathbf{e}, \tag{5.4.5}$$

where $\mathbf{v} = (v_1, \ldots, v_m)^T$ and $\mathbf{Z}$ is an $m \times p$ matrix with ith row equal to $\mathbf{z}_i^T$. In practice, $\boldsymbol{\Psi}$ is replaced by a survey estimator $\hat{\boldsymbol{\Psi}}$ or a smoothed estimator, but the variability associated with the estimator is often ignored.

Example 5.4.2. Census Undercount. In the context of estimating the undercount in the 1990 Census of the United States, the population was divided into 357 poststrata composed of 51 poststratum groups, each of which was subdivided into 7 age-sex categories. The 51 poststratum groups were defined on the basis of race/ethnicity, tenure (owner, renter), type of area and region. Dual-system estimates $\hat{\theta}_i$ for each poststratum $i = 1, \cdots, 357$ were obtained using the data from the 1990 Post Enumeration Survey (PES). Here $\hat{\theta}_i$ is the estimated adjustment factor for the ith poststratum. A model of the form (5.4.5) was employed to obtain smoothed estimates of the adjustment factors θ_i (Isaki, Tsay and Fuller (2000)). In a previous study (Isaki, Hwang and Tsay (1991), 1392 poststrata were employed and smoothed estimates of adjustment factors were obtained. Some details of this application are given in Example 8.2.1.

5.4.3 Time Series and Cross-sectional Models

Many sample surveys are repeated in time with partial replacement of the sample elements. For example, in the monthly U.S. Current Population Survey (CPS) an individual household remains in the sample for four consecutive months, then drops out of the sample for the eight succeeding months and comes back for another four consecutive months. In the monthly Canadian Labour Force Survey (LFS) an individual household remains in the sample for six consecutive months and then drops out of sample . For such repeated surveys considerable gain in efficiency can be achieved by borrowing strength across both small areas and time.

Rao and Yu (1992, 1994) proposed an extension of the basic Fay-Herriot model (5.2.5) to handle time series and cross-sectional data. Their model consists of a sampling error model

$$\hat{\theta}_{it} = \theta_{it} + e_{it}, \quad t = 1, \ldots, T;\ i = 1, \ldots, m \tag{5.4.6}$$

and a linking model

$$\theta_{it} = \mathbf{z}_{it}^T\boldsymbol{\beta} + v_i + u_{it}. \tag{5.4.7}$$

Here $\hat{\theta}_{it}$ is the direct survey estimator for small area i at time t, $\theta_{it} = g(\bar{Y}_{it})$ is a function of the small area mean $\bar{Y}_{it}$, the e_{it}'s are sampling errors normally distributed, given the θ_{it}'s, with zeros means and a known block diagonal covariance matrix $\boldsymbol{\Psi}$ with blocks $\boldsymbol{\Psi}_i$, and $\mathbf{z}_{it}$ is a vector of area-specific covariates some of which may change with t, for example, administrative data. Further, $v_i \overset{\text{iid}}{\sim} N(0, \sigma_v^2)$ and the u_{it}'s are assumed to follow a common first order autoregressive process for each i, that is,

$$u_{it} = \rho u_{i,t-1} + \varepsilon_{it}, \quad |\rho| < 1 \tag{5.4.8}$$

with $\varepsilon_{it} \overset{\text{iid}}{\sim} N(0, \sigma^2)$. The errors $\{e_{it}\}$, $\{v_i\}$ and $\{\varepsilon_{it}\}$ are also assumed to be independent of each other. Models of the form (5.4.7) and (5.4.8) have been extensively used in the econometrics literature (Anderson and Hsiao (1981)), ignoring sampling errors e_{it}.

The model (5.4.7) on the θ_{it}'s depends on both area-specific effects v_i and the area-by-time specific effects u_{it} which are correlated across time for each

i. We can also express (5.4.7) as a distributed-lag model

$$\theta_{it} = \rho\theta_{i,t-1} + (\mathbf{z}_{it} - \rho\mathbf{z}_{i,t-1})^T\boldsymbol{\beta} + (1-\rho)v_i + \varepsilon_{it}. \tag{5.4.9}$$

The alternative form (5.4.9) relates θ_{it} to the previous period mean $\theta_{i,t-1}$, the values of the auxiliary variables for the time points t and $t-1$, the random small area effects v_i and area-by-time effects ε_{it}. More complex models on the u_{it}'s than (5.4.8) can be formulated by assuming an autoregressive moving average (ARMA) process, but the resulting efficiency gains relative to (5.4.8) are unlikely to be significant.

Ghosh and Nangia (1993) and Ghosh, Nangia and Kim (1996) also proposed a time series cross-sectional model for small area estimation. Their model is of the form

$$\hat{\theta}_{it}|\theta_{it} \overset{\text{ind}}{\sim} N(\theta_{it}, \psi_{it}), \tag{5.4.10}$$

$$\theta_{it}|\boldsymbol{\alpha}_t \overset{\text{ind}}{\sim} N(\mathbf{z}_{it}^T\boldsymbol{\beta} + \mathbf{w}_{it}^T\boldsymbol{\alpha}_t, \sigma_t^2), \tag{5.4.11}$$

and

$$\boldsymbol{\alpha}_t|\boldsymbol{\alpha}_{t-1} \overset{\text{ind}}{\sim} N_r(\mathbf{H}_t\boldsymbol{\alpha}_{t-1}, \boldsymbol{\Delta}). \tag{5.4.12}$$

Here $\mathbf{z}_{it}$ and $\mathbf{w}_{it}$ are vectors of area specific covariates, the ψ_{it}'s are sampling variances assumed to be known, $\boldsymbol{\alpha}_t$ is a $r \times 1$ vector of time-specific random effects, and $\mathbf{H}_t$ is a known $r \times r$ matrix. The dynamic (or state space) model (5.4.12) in the univariate case ($r = 1$) with $H_t = 1$ reduces to the well-known random walk model. The above model suffers from two major limitations: (i) The direct estimators $\hat{\theta}_{it}$ are assumed to be independent over time for each i. This assumption is not realistic in the context of repeated surveys with overlapping samples, such as the CPS and the LFS. (ii) Area-specific random effects are not included in the model which leads to excessive shrinkage of small area estimators similar to synthetic estimators.

Datta, Lahiri and Maiti (2002) and You (1999) used the Rao-Yu sampling and linking models (5.4.6) and (5.4.7) but replaced the AR(1) model (5.4.8) on the u_{it} by a random walk model given by (5.4.8) with $\rho = 1 : u_{it} = u_{it-1} + \varepsilon_{it}$. Datta, Lahiri, Maiti and Lu (1999) considered a similar model but added extra terms to the linking model to reflect seasonal variation in their application.

Pfeffermann and Burck (1990) proposed a general model involving area-by-time specific random effects. Their model is of the form

$$\hat{\theta}_{it} = \theta_{it} + e_{it}, \tag{5.4.13}$$

$$\theta_{it} = \mathbf{z}_{it}^T\boldsymbol{\beta}_{it}, \tag{5.4.14}$$

where the coefficients $\boldsymbol{\beta}_{it} = (\beta_{it0}, \ldots, \beta_{itp})^T$ are allowed to vary cross-sectionally and over time, and the sampling errors e_{it} for each area i are assumed to be serially uncorrelated with mean 0 and variance ψ_{it}. The variation of $\boldsymbol{\beta}_{it}$ over time is specified by the following state-space model:

$$\begin{bmatrix} \beta_{itj} \\ \beta_{ij} \end{bmatrix} = \mathbf{T}_j \begin{bmatrix} \beta_{i,t-1,j} \\ \beta_{ij} \end{bmatrix} + \begin{bmatrix} 1 \\ 0 \end{bmatrix} v_{itj}, \quad j = 0, 1, \ldots, p. \tag{5.4.15}$$

Here the β_{ij}'s are fixed coefficients, $\mathbf{T}_j$ is a known 2×2 matrix with $(0, 1)$ as the second row, and the model errors $\{v_{itj}\}$ for each i are uncorrelated over time with mean 0 and covariances $E_m(v_{itj}v_{itl}) = \sigma_{vjl}$; $j, l = 0, 1, \ldots, p$.

The formulation (5.4.15) covers several useful models. First, the choice $\mathbf{T}_j = \begin{bmatrix} 0 & 1 \\ 0 & 1 \end{bmatrix}$ gives the well-known random coefficient regression model $\beta_{itj} = \beta_{ij} + v_{itj}$ (Swamy, (1971)). The familiar random walk model $\beta_{itj} = \beta_{i,t-1,j} + v_{itj}$ is obtained by choosing $\mathbf{T}_j = \begin{bmatrix} 1 & 0 \\ 0 & 1 \end{bmatrix}$. In this case the coefficient β_{ij} is redundant and should be omitted so that $\mathbf{T}_j = 1$. The choice $\mathbf{T}_j = \begin{bmatrix} \rho & 1-\rho \\ 0 & 1 \end{bmatrix}$ gives an AR(1) model: $\beta_{itj} - \beta_{ij} = \rho(\beta_{i,t-1,j} - \beta_{ij}) + v_{itj}$. The state space model (5.4.15) is quite general, but the assumption of serially uncorrelated sampling errors e_{it} in (5.4.13) is restrictive in the context of repeated surveys with overlapping samples.

Example 5.4.3. Median Income. Ghosh, Nangia and Kim (1996) applied the model (5.4.10)–(5.4.12) to estimate the median income of four-person families for the fifty American states and the District of Columbia. The U.S. Department of Health and Human Services uses these estimates to formulate its energy assistance program for low-income families. They used CPS data for $T = 9$ years (1981–1989) to estimate θ_{iT}, the median incomes for 1989 $(i = 1, \ldots, 51)$. Here $z_{it} = (1, z_{it1})^T$ with z_{it1} denoting the "adjusted" census median income for year t and area i which is obtained by adjusting the base year (1979) census median income by the proportional growth in per capita income (PCI). Datta, Lahiri and Maiti (2002) applied their model to estimate median income of four-person families, using the same data. Some details of this application are given in Example 8.3.3.

Example 5.4.4. U.S. Unemployment Rates. Datta, Lahiri, Maiti and Lu (1999) applied their model to estimate monthly unemployment rates for forty-nine American states (excluding the state of New York) and the District of Columbia ($m = 50$) which are used by various federal agencies for the allocation of funds and policy formulation. They considered the period January 1985–December 1988 and used the CPS estimates as $\hat{\theta}_{it}$ and the unemployment insurance UI claims rate: (percentage of unemployed workers claiming UI benefits among the total nonagricultural employment) as auxiliary data, $\mathbf{z}_{it}$. Seasonal variation in monthly unemployment rates was accounted for by introducing random month and year effects into the model. Some details of this application are given in Example 10.8.2.

Example 5.4.5. Canadian Unemployment Rates. You, Rao and Gambino (2001) applied the Rao-Yu model to estimate monthly unemployment rates for Census Metropolitan Areas (CMAs: cities with population more than 100,000) and Census Agglomerations (CAs: other urban centers) in Canada. Reliable estimates at the CMA and CA levels are used by the Employment Insurance (EI) program to determine the rules used to administer the program. Direct estimates of unemployment rates from the Canadian Labour Force Survey (LFS) are reliable for the nation and the provinces, but many CMAs and CAs do not have a large enough sample to produce reliable direct estimates from the LFS.

EI beneficiary rates were used as auxiliary data, z_{it}, in the linking model (5.4.7). Both AR(1) and random walk models on the area-by-time specific random effects, u_{it}, were considered. Some details of this application are given in Example 10.8.1.

5.4.4 Spatial Models

The basic Fay-Herriot model (5.2.5) assumes iid small area effects v_i, but in some applications it may be more realistic to entertain models that allow correlations among the v_i's. Spatial models on the v_i's are used when "neighboring" areas of each area can be defined. Such models induce correlations among the v_i's, for example, correlations that depend on geographical proximity in the context of estimating local disease and mortality rates. Cressie (1991) used a spatial model for small area estimation in the context of U.S. census undercount.

If A_i denotes a set of "neighboring" areas of the i th area, then a conditional autoregression (CAR) spatial model assumes that the conditional distribution of $b_i v_i$ given $\{v_l : \ l \neq i\}$ is given by

$$b_i v_i | \{v_l : l \neq i\} \sim N\left(\rho \sum_{l \in A_i} q_{il} b_l v_l, \ b_i^2 \sigma_v^2\right). \tag{5.4.16}$$

Here $\{q_{il}\}$ are known constants satisfying $q_{il} b_l^2 = q_{li} b_i^2$ $(i < l)$, and $\boldsymbol{\delta} = (\rho, \sigma_v^2)^T$ is the unknown parameter vector. The model (5.4.16) implies that

$$\mathbf{B}^{1/2}\mathbf{v} \sim N_m(\mathbf{0}, \ \boldsymbol{\Gamma}(\boldsymbol{\delta}) = \sigma_v^2(\mathbf{I} - \rho\mathbf{Q})^{-1}\mathbf{B}), \tag{5.4.17}$$

where $\mathbf{B} = \text{diag}(b_1^2, \ldots, b_m^2)$ and $\mathbf{Q} = (q_{il})$ is an $m \times m$ matrix with $q_{il} = 0$ whenever $l \notin A_i$ (including $q_{ii} = 0$) and $\mathbf{v} = (v_1, \ldots, v_m)^T$ (see Besag (1974)). Using (5.4.17) in (5.2.5) we obtain a spatial small area model. Note that $\boldsymbol{\delta}$ appears nonlinearly in $\boldsymbol{\Gamma}(\boldsymbol{\delta})$.

In the geostatistics literature, covariance structures of the form (i) $\boldsymbol{\Gamma}(\delta) = \sigma_v^2(\delta_1 \mathbf{I} + \delta_2 \mathbf{D})$ and (ii) $\boldsymbol{\Gamma}(\delta) = \sigma_v^2[\delta_1 \mathbf{I} + \delta_2 \mathbf{D}(\delta_3)]$ have been used, where $\mathbf{D} = (e^{-d_{il}})$ and $\mathbf{D}(\delta_3) = (\delta_3^{d_{il}})$ are $m \times m$ matrices with d_{il} denoting a "distance" (not necessarily Euclidean) between small areas i and l. Note that in case (i) the parameters δ_1 and δ_2 appear linearly in $\boldsymbol{\Gamma}(\boldsymbol{\delta})$, whereas in case (ii) δ_2 and δ_3 appear nonlinearly in $\boldsymbol{\Gamma}(\boldsymbol{\delta})$.

A drawback of the spatial model (5.4.17) is that it depends on how the neighborhoods A_i are defined. Therefore, it introduces some subjectivity (Marshall, (1991)).

Example 5.4.6. U.S. Census Undercount. Cressie (1991) extended his model with $b_i = 1/\sqrt{C_i}$ for estimating the U.S. census undercount (Example

5.2.2), by allowing spatial dependence through the CAR model (5.4.17). By exploratory spatial data analysis, he defined

$$q_{il} = \begin{cases} \sqrt{C_l/C_i} & \text{if } d_{il} \leq 700 \text{ miles}, \; i \neq l \\ 0 & \text{otherwise,} \end{cases}$$

where C_i is the census count in the ith state and d_{il} is the distance between the centers of gravity of the ith and lth states (small areas). Other choices of q_{il}, not necessarily distance-based, may be chosen. Cressie (1991) noted that the sociologist's/ethnographer's map of the small areas may be quite different from the geographer's map. For example, New York City and the rest of New York State may not be "neighbors" for the purpose of undercount estimation whereas it is more reasonable to regard other big cities like Detroit, Chicago and Los Angeles as "neighbors" of New York City. Some details of this application are given in Example 8.4.1.

5.5 Extensions: Type B Models

We now consider various extensions of the basic type B (unit level) model (5.3.1)

5.5.1 Multivariate Nested Error Regression Model

As in Section 5.3, we assume that unit-specific auxiliary data $\mathbf{x}_{ij}$ are available for all the population elements j in each small area i. We further assume that an $r \times 1$ vector of variables of interest, $\mathbf{y}_{ij}$, are related to $\mathbf{x}_{ij}$ through a multivariate nested error regression model (Fuller and Harter, (1987)):

$$\mathbf{y}_{ij} = \mathbf{B}\mathbf{x}_{ij} + \mathbf{v}_i + \mathbf{e}_{ij}; \quad j = 1, \ldots, N_i; \; i = 1, \ldots, m. \tag{5.5.1}$$

Here $\mathbf{B}$ is $r \times p$ matrix of regression coefficients, $\mathbf{v}_i$ are area-specific effects assumed to be iid random vectors with mean $\mathbf{0}$ and covariance matrix $\mathbf{\Sigma}_v$, and the random error vectors $\mathbf{e}_{ij}$ are iid with mean $\mathbf{0}$ and covariance matrix $\mathbf{\Sigma}_e$ and independent of $\mathbf{v}_i$. In addition, normality of the $\mathbf{v}_i$'s and $\mathbf{e}_{ij}$'s is often assumed. The model (5.5.1) is a natural extension of the (univariate) nested error regression model (5.3.1) with $k_{ij} = 1$.

The parameters of interest are the small area mean vectors $\bar{\mathbf{Y}}_i$ which can be approximated by $\boldsymbol{\mu}_i = \mathbf{B}\overline{\mathbf{X}}_i + \mathbf{v}_i$ if the population size N_i is large. In the latter case, it follows that the estimation of $\boldsymbol{\mu}_i$ is equivalent to the estimation of a linear combination of $\mathbf{B}$ and the realization of the random vector $\mathbf{v}_i$. As in the multivariate Fay-Herriot model, the unit level model (5.5.1) can lead to more efficient estimators by taking advantage of the correlations between the components of $\mathbf{y}_{ij}$, unlike the univariate model (5.3.1).

Example 5.5.1. County Crop Areas. In Example 5.3.1, we could take $\mathbf{y}_{ij} = (y_{ij1}, y_{ij2})$ with $y_{ij1} =$ number of hectares of corn and $y_{ij2} =$ number of hectares of soybeans, and retain the same $\mathbf{x}_{ij} = (1, x_{ij1}, x_{ij2})^T$ with $x_{ij1} =$

number of pixels classified as corn and x_{ij2} = number of pixels classified as soybeans. By taking advantage of the correlation between y_{ij1} and y_{ij2}, an improved estimator of $\boldsymbol{\mu}_i$ can be obtained through the multivariate model (5.5.1). Some details of a simulation study based on this data are given in Example 8.5.1.

5.5.2 Random Error Variance Linear Model

Consider the error-components model (5.3.1) when auxiliary data $\{\mathbf{x}_{ij}\}$ are not available: $y_{ij} = \beta + v_i + e_{ij}$ with $v_i \stackrel{\text{iid}}{\sim} N(0, \sigma_v^2)$ and $e_{ij} \stackrel{\text{iid}}{\sim} N(0, \sigma_e^2)$. The assumption of equal error variances can be relaxed by letting $e_{ij}|\sigma_{ei}^2 \stackrel{\text{ind}}{\sim} N(0, \sigma_{ei}^2)$ and assuming the error variances σ_{ei}^2 to be nonnegative iid random variables with mean σ_e^2 and variance δ_e (say) and independent of v_i. Aragon (1984) used such a random error variance model with inverse Gaussian (IG) error variances σ_{ei}^2. Kleffe and Rao (1992) used the foregoing simple model for small area estimation, and Arora and Lahiri (1997) extended it to the regression case (5.3.1) with $\mathbf{x}_{ij} = \mathbf{x}_i$, $k_{ij} = 1$, $e_{ij} = \tilde{e}_{ij}$ and $e_{ij}|\sigma_{ei}^2 \stackrel{\text{iid}}{\sim} N(0, \sigma_{ei}^2)$.

Example 5.5.1. Consumer Expenditure. Arora and Lahiri (1997) applied the random error variance model to estimate the true average weekly consumer expenditures on various items, goods and services, for $m = 43$ publication areas (small areas) throughout the United States. They actually reduced the model to an area level model by considering the direct survey estimates $\bar{y}_{iw} = \sum_j w_{ij} y_{ij} / \sum_j w_{ij}$ in place of $\{y_{ij}\}$, where w_{ij} are the survey weights. The model on $\{y_{ij}\}$ with $\mathbf{x}_{ij} = \mathbf{x}_i$ implies that $\bar{y}_{iw} = \mathbf{x}_i^T \beta + v_i + \bar{e}_{iw}$, with $\bar{e}_{iw} = \sum_j w_{ij} e_{ij} / \sum_j w_{ij} | \sigma_{ei}^2 \stackrel{\text{ind}}{\sim} N(0, K_i \sigma_{ei}^2)$, where $K_i = (\sum_j w_{ij}^2)/(\sum_j w_{ij})^2$. In their application, $\mathbf{x}_i^T \boldsymbol{\beta} = \beta_l$ if small area i belongs to the lth major area ($l = 1, \ldots, 8$). The random error variances σ_{ei}^2 are assumed to obey a specified distribution; in particular, $\sigma_{ei}^{-2} \stackrel{\text{iid}}{\sim} G(a, b)$, a gamma distribution with parameters a and b ($a > 0, b > 0$).

5.5.3 Two-fold Nested Error Regression Model

Suppose that the ith small area contains M_i primary units (or clusters) and that the jth primary unit (cluster) in the ith area contains N_{ij} subunits (elements). Let $(y_{ijl}, \mathbf{x}_{ijl})$ be the y- and $\mathbf{x}$-values for the lth elements in the jth primary unit from the ith area ($l = 1, \ldots, N_{ij}$; $j = 1, \ldots, M_i$; $i = 1, \ldots, m$). Under this population structure, it is a common practice to employ two-stage cluster sampling in each small area: a sample, s_i, of m_i primary units is selected from the ith area and if the jth cluster is sampled then a subsample, s_{ij}, of n_{ij} elements is selected from the jth cluster and the associated y-and $\mathbf{x}$-values are observed.

The foregoing population structure is reflected by a two-fold nested error regression model (Stukel and Rao (1999)):

$$y_{ijl} = \mathbf{x}_{ijl}^T\boldsymbol{\beta} + v_i + u_{ij} + e_{ijl}; \; l = 1, \ldots, N_{ij}, \; j = 1, \ldots, M_i, \quad i = 1, \ldots, m. \tag{5.5.2}$$

Here the area effects $\{v_i\}$, the cluster effects $\{u_{ij}\}$ and the residual errors $\{e_{ijl}\}$ with $e_{ijl} = k_{ijl}\tilde{e}_{ijl}$ and known constants k_{ijl} are assumed to be mutually independent. Further, $v_i \overset{\text{iid}}{\sim} (0, \sigma_v^2)$, $u_{ij} \overset{\text{iid}}{\sim} (0, \sigma_u^2)$ and $\tilde{e}_{ijl} \overset{\text{iid}}{\sim} (0, \sigma_e^2)$; normality of the random components v_i, u_{ij} and $\tilde{e}_{ijl}$ is also often assumed. We assume that the sample values also obey the assumed model (5.5.2) which is satisfied under simple random sampling of clusters and subunits within sampled clusters or more generally for sampling designs that use the auxiliary information $\mathbf{x}_{ijl}$ in the selection of the sample. Datta and Ghosh (1991) used the model (5.5.2) for the special case of cluster-specific covariates, that is, $\mathbf{x}_{ijl} = \mathbf{x}_{ij}$. Ghosh and Lahiri (1988) studied the case of no auxiliary information, that is, $\mathbf{x}_{ijl}^T\boldsymbol{\beta} = \beta$.

The parameters of interest are the small area means

$$\overline{Y}_i = \frac{1}{N_i}\left[\sum_{j\in s_i}\sum_{l\in s_{ij}} y_{ijl} + \sum_{j\in s_i}\sum_{l\in s_{ij}^c} y_{ijl}^* + \sum_{j\in s_i^c}\sum_{l=1}^{N_{ij}} y_{ijl}^*\right], \tag{5.5.3}$$

where y_{ijl}^* are the nonsampled y-values, s_i^c and s_{ij}^c denote the nonsampled clusters and subunits. If the number of primary units, N_i, is large, then $\bar{Y}_i$ may be approximated as

$$\bar{Y}_i \approx \bar{\mathbf{X}}_i^T\boldsymbol{\beta} + v_i, \tag{5.5.4}$$

noting that $\bar{Y}_i = \bar{\mathbf{X}}_i^{\mathrm{T}}\boldsymbol{\beta} + v_i + \bar{U}_i + \bar{E}_i$ and $\bar{U}_i \approx 0$, $\bar{E}_i \approx 0$, where $\bar{U}_i$ and $\bar{E}_i$ are the area means of u_{ij} and e_{ijl} and $\bar{\mathbf{X}}_i$ is the known mean of $\mathbf{x}_{ijl}$'s. It follows from (5.5.4) that the estimation of $\bar{Y}_i$ is equivalent to the estimation of a linear combination of $\boldsymbol{\beta}$ and the realization of the random variable v_i.

5.5.4 Two-level Model

The basic unit level model (5.3.1) with an intercept term β_1 may be expressed as a model with random intercept term $\beta_{1i} = \beta_1 + v_i$ and common slopes $\beta_2, \ldots, \beta_p$: $y_{ij} = \beta_{1i} + \beta_2 x_{ij2} + \cdots + \beta_p x_{ijp} + e_{ij}$. This suggests a more general model that allows differences between slopes as well as the intercepts across small area. We introduce random coefficients $\boldsymbol{\beta}_i = (\beta_{i1}, \ldots, \beta_{ip})^T$ and then model $\boldsymbol{\beta}_i$ in terms of area level covariates $\tilde{\mathbf{Z}}_i$ to arrive at a two-level small area model (Moura and Holt (1999)):

$$y_{ij} = \mathbf{x}_{ij}^T\boldsymbol{\beta}_i + e_{ij}, \quad j = 1, \ldots, N_i; \; i = 1, \ldots, m \tag{5.5.5}$$

and

$$\boldsymbol{\beta}_i = \tilde{\mathbf{Z}}_i\boldsymbol{\alpha} + \mathbf{v}_i, \tag{5.5.6}$$

where $\tilde{\mathbf{Z}}_i$ is a $p \times q$ matrix, $\boldsymbol{\alpha}$ is a $q \times 1$ vector of regression parameters, $\mathbf{v}_i \overset{\text{iid}}{\sim} (\mathbf{0}, \boldsymbol{\Sigma}_v)$ and $e_{ij} = k_{ij}\tilde{e}_{ij}$ with $\tilde{e}_{ij} \overset{\text{iid}}{\sim} (\mathbf{0}, \sigma_e^2)$. We may write (5.5.5) in a matrix form

$$\mathbf{y}_i^P = \mathbf{X}_i^P \boldsymbol{\beta}_i + \mathbf{e}_i^P. \tag{5.5.7}$$

The two-level model (5.5.6)–(5.5.7) effectively integrates the use of unit level and area level covariates into a single model:

$$\mathbf{y}_i^P = \mathbf{X}_i^P \tilde{\mathbf{Z}}_i \boldsymbol{\alpha} + \mathbf{X}_i^P \mathbf{v}_i + \mathbf{e}_i^P. \tag{5.5.8}$$

Further, the use of random slopes, $\boldsymbol{\beta}_i$, permits greater flexibility in modelling. The sample values $\{(y_{ij}, \mathbf{x}_{ij}); j = 1, \ldots, n_i; i = 1, \ldots, m\}$ are assumed to obey the model (5.5.8), that is, there is no sample selection bias. If N_i is large, we can express the mean $\bar{Y}_i$ under (5.5.8) as

$$\bar{Y}_i \approx \bar{\mathbf{X}}_i^T \tilde{\mathbf{Z}}_i \boldsymbol{\alpha} + \bar{\mathbf{X}}_i^T \mathbf{v}_i. \tag{5.5.9}$$

It follows from (5.5.9) that the estimation of $\bar{Y}_i$ is equivalent to the estimation of a linear combination of $\boldsymbol{\beta}$ and the realization of the random vector $\mathbf{v}_i$ with unknown covariance matrix $\boldsymbol{\Sigma}_v$.

The model (5.5.8) is a special case of a general linear mixed model used extensively for longitudinal data (Laird and Ware, (1982)). This model allows arbitrary matrices $\mathbf{X}_{1i}^P$ and $\mathbf{X}_{2i}^P$ to be associated with $\boldsymbol{\alpha}$ and $\mathbf{v}_i$:

$$\mathbf{y}_i^P = \mathbf{X}_{1i}^P \boldsymbol{\alpha} + \mathbf{X}_{2i}^P \mathbf{v}_i + \mathbf{e}_i^P. \tag{5.5.10}$$

The choice $\mathbf{X}_{1i}^P = \mathbf{X}_i^P \mathbf{Z}_i$ and $\mathbf{X}_{2i}^P = \mathbf{X}_i^P$ gives the two-level model (5.5.8). This model covers many of the small area models considered in the literature.

5.5.5 General Linear Mixed Model

Datta and Ghosh (1991) considered a general linear mixed model which covers the univariate unit-level models as special cases:

$$\mathbf{y}^P = \mathbf{X}^P \boldsymbol{\beta} + \mathbf{Z}^P \mathbf{v} + \mathbf{e}^P. \tag{5.5.11}$$

Here $\mathbf{e}^P$ and $\mathbf{v}$ are independent with $\mathbf{e}^P \sim N(\mathbf{0}, \sigma^2 \boldsymbol{\Psi}^P)$ and $\mathbf{v} \sim N(\mathbf{0}, \sigma^2 \mathbf{D}(\boldsymbol{\lambda}))$, where $\boldsymbol{\Psi}^P$ is a known positive definite (p.d.) matrix and $\mathbf{D}(\boldsymbol{\lambda})$ is a p.d. matrix which is structurally known except for some parameters $\boldsymbol{\lambda}$ typically involving ratios of variance components of the form σ_i^2/σ^2. Further, $\mathbf{X}^P$ and $\mathbf{Z}^P$ are known design matrices and $\mathbf{y}^P$ is the $N \times 1$ vector of population y-values.

We can partition (5.5.11), similar to (5.3.4), as

$$\mathbf{y}^P = \begin{bmatrix} \mathbf{y} \\ \mathbf{y}^* \end{bmatrix} = \begin{bmatrix} \mathbf{X} \\ \mathbf{X}^* \end{bmatrix} \boldsymbol{\beta} + \begin{bmatrix} \mathbf{Z} \\ \mathbf{Z}^* \end{bmatrix} \mathbf{v} + \begin{bmatrix} \mathbf{e} \\ \mathbf{e}^* \end{bmatrix}, \tag{5.5.12}$$

where the asterisk denotes nonsampled units. The vector of small area totals Y_i is of the form $\mathbf{Ay} + \mathbf{Cy}^*$ with $\mathbf{A} = \oplus_{i=1}^m \mathbf{1}_{n_i}^T$ and $\mathbf{C} = \oplus_{i=1}^m \mathbf{1}_{N_i - n_i}^T$, where $\oplus$ denotes the direct sum, that is, $\oplus_{i=1}^m \mathbf{A}_u = \text{blockdiag}(\mathbf{A}_1, \ldots, \mathbf{A}_m)$.

Datta and Ghosh (1991) mentioned a cross-classification model which is covered by the general model (5.5.11) but not by the "longitudinal" model (5.5.10). Suppose the units in a small area are classified into C subgroups (e.g., age, socio-economic class) labeled $j = 1, \ldots, C$ and the area-by-subgroup cell sizes N_{ij} are known. A cross-classification model is then given by

$$y_{ijk} = \mathbf{x}_{ijk}^T \boldsymbol{\beta} + v_i + a_j + u_{ij} + e_{ijk}, \qquad k = 1, \ldots, N_{ij};\ j = 1, \ldots, C;\ i = 1, \ldots, m \tag{5.5.13}$$

where $\{v_i\}$, $\{a_j\}$ and $\{u_{ij}\}$ are mutually independent with $e_{ijk} \overset{\text{iid}}{\sim} N(0, \sigma^2)$, $v_i \overset{\text{iid}}{\sim} N(0, \lambda_1\sigma^2)$, $a_j \overset{\text{iid}}{\sim} N(0, \lambda_2\sigma^2)$ and $u_{ij} \overset{\text{iid}}{\sim} N(0, \lambda_3\sigma^2)$. Lui and Cumberland (1989) considered a model of the form (5.5.13) with $\lambda_1 = \lambda_3 = 0$, that is, v_i and u_{ij} are degenerate at zero.

5.6 Generalized Linear Mixed Models

We now consider generalized linear mixed models which are especially suited for binary and count y-values.

5.6.1 Logistic Regression Models

Suppose y_{ij} is binary, that is, $y_{ij} = 0$ or 1, and the parameters of interest are the small area proportions $\bar{Y}_i = P_i = \sum_j y_{ij}/N_i$. MacGibbon and Tomberlin (1989) used a logistic regression model with random area-specific effects to estimate P_i. Given the p_{ij}'s, the y_{ij}'s are assumed to be independent Bernoulli (p_{ij}) variables, and the p_{ij}'s obey the following logistic regression model with random area effects v_i:

$$\text{logit}(p_{ij}) = \log \frac{p_{ij}}{1 - p_{ij}} = \mathbf{x}_{ij}^T \boldsymbol{\beta} + v_i, \tag{5.6.1}$$

where $v_i \overset{\text{iid}}{\sim} N(0, \sigma_v^2)$ and the $\mathbf{x}_{ij}$ are unit-specific covariates. The model-based estimator of P_i is of the form $(\sum_{j \in s_i} y_{ij} + \sum_{j \in s_i^c} \hat{p}_{ij})/N_i$, where $\hat{p}_{ij}$ is obtained from (5.6.1) by estimating $\boldsymbol{\beta}$ and the realization of v_i, using EB or HB methods.

Malec, Sedransk, Moriarity and LeClere (1997) considered a different logistic regression model with random regression coefficients. Suppose the units are grouped into classes j in each small area i, and given p_{ij}, the values y_{ijl} $(l = 1, \ldots, N_{ij})$ in the (i, j)th cell are independent Bernoulli variables with the common probability p_{ij}. Further, assume that

$$\theta_{ij} = \text{logit}(p_{ij}) = \mathbf{x}_j^T \boldsymbol{\beta}_i \tag{5.6.2}$$

and

$$\boldsymbol{\beta}_i = \mathbf{Z}_i \boldsymbol{\alpha} + \mathbf{v}_i \tag{5.6.3}$$

with $\mathbf{v}_i \stackrel{\text{iid}}{\sim} N(\mathbf{0}, \mathbf{\Sigma}_v)$, where $\mathbf{Z}_i$ is a $p \times q$ matrix of area level covariates as in subsection 5.5.4 and $\mathbf{x}_j$ is a class-specific covariate vector.

Example 5.6.1. Visits to Physicians. Malec et al. (1997) applied the model given by (5.6.2) and (5.6.3) to estimate the proportion of persons in a state or substate who have visited a physician in the past year, using the data from the U.S. National Health Interview Survey (NHIS). Here "i" denotes a county and "j" a demographic class; $\mathbf{x}_j$ is a vector of covariates that characterizes the demographic class j. Some details of this application are given in Example 10.11.3.

5.6.2 Models for Mortality and Disease Rates

Mortality and disease rates of small areas in a region or a county are often used to construct disease maps such as cancer atlases. Such maps are used to display geographical variability of a disease and identify high-rate areas warranting intervention. A simple small area model is obtained by assuming that the observed small area counts, y_i, are independent Poisson variables with conditional mean $E(y_i|\lambda_i) = n_i\lambda_i$ and that $\lambda_i \stackrel{\text{iid}}{\sim} \text{gamma}(\alpha, \nu)$. Here λ_i and n_i are the true rate and number exposed in the ith area, and (α, ν) are the scale and shape parameters of the gamma distribution. Under this model, smoothed estimates of λ_i are obtained using EB or HB methods (Clayton and Kaldor (1987), Datta, Ghosh and Waller (2000)). CAR spatial models of the form (5.4.16) on log rates $\theta_i = \log(\lambda_i)$ have also been proposed (Clayton and Kaldor (1987)). The model on λ_i can be extended to incorporate area level covariates $\mathbf{z}_i$, for example, $\theta_i = \mathbf{z}_i^T\boldsymbol{\beta} + v_i$ with $v_i \stackrel{\text{iid}}{\sim} N(0, \sigma_v^2)$. Nandram, Sedransk and Pickle (1999) studied regression models on age-specific log rates $\theta_{ij} = \log \lambda_{ij}$ involving random slopes, where j denotes age.

Joint mortality rates (y_{1i}, y_{2i}) can also be modeled by assuming that (y_{1i}, y_{2i}) are independently distributed conditional on $(\lambda_{1i}, \lambda_{2i})$ and $\theta_i = (\log \lambda_{1i}, \log \lambda_{2i})^T \stackrel{\text{iid}}{\sim} N_2(\boldsymbol{\mu}, \mathbf{\Sigma})$. Further, y_{1i} and y_{2i} are assumed to be conditionally independent Poisson variables with $E(y_{1i}|\lambda_{1i}) = n_{1i}\lambda_{1i}$ and $E(y_{2i}|\lambda_{2i}) = n_{2i}\lambda_{2i}$. As an example of this bivariate model, y_{1i} and y_{2i} denote the number of deaths due to cancer at sites 1 and 2 and (n_{1i}, n_{2i}) the populations at risk at sites 1 and 2. DeSouza (1992) showed that the bivariate model leads to improved estimates of the rates $(\lambda_{1i}, \lambda_{2i})$ compared to estimates based on separate univariate models.

Example 5.6.2. Lip Cancer. Maiti (1998) modeled $\beta_i = \log \lambda_i$ as iid $N(\mu, \sigma^2)$. He also considered a CAR spatial model on the β_i's which relates each β_i to a set of neighborhood areas of area i. He developed model-based estimates of lip cancer incidence in Scotland for each of 56 counties. Some details of this application are given in Examples 9.5.1 and 10.10.1.

5.6.3 Exponential Family Models

Ghosh, Natarajan, Stroud and Carlin (1998) proposed generalized linear models with random area effects. Conditional on the θ_{ij}'s, the sample statistics y_{ij} $(j = 1, \ldots, n_i;\ i = 1, \ldots, m)$ are assumed to be independently distributed with probability density function belonging to exponential family with canonical parameters θ_{ij}, that is,

$$f(y_{ij}|\theta_{ij}) = \exp\left[\frac{1}{\phi_{ij}}(\theta_{ij}y_{ij} - a(\theta_{ij})) + b(y_{ij}, \phi_{ij})\right] \qquad (5.6.4)$$

for known ϕ_{ij} (> 0) and functions $a(\cdot)$ and $b(\cdot)$. The exponential family (5.6.4) covers well-known distributions including the normal, binomial and Poisson distributions. For example, $\theta_{ij} = \text{logit}(p_{ij})$ and $\phi_{ij} = 1$ if y_{ij} is Binomial (n_{ij}, p_{ij}), and $\theta_{ij} = \log(\lambda_{ij})$ and $\phi_{ij} = 1$ if y_{ij} is Poisson (λ_{ij}). The θ_{ij}'s are modeled as

$$\theta_{ij} = \mathbf{x}_{ij}^T\boldsymbol{\beta} + v_i + u_{ij}, \qquad (5.6.5)$$

where v_i and u_{ij} are mutually independent with $v_i \overset{\text{iid}}{\sim} N(0, \sigma_v^2)$ and $u_{ij} \overset{\text{iid}}{\sim} N(0, \sigma_u^2)$.

The objective here is to make inferences on the small area parameters θ_{ij}. For example, $\theta_{ij} = \text{logit}(p_{ij})$ and p_{ij} denotes the proportion associated with a binary variable in the jth age-sex category in the ith region.

Ghosh, Natarajan, Waller and Kim (1999) extended the linking model (5.6.5) to handle spatial data, and applied the model to disease mapping.

5.6.4 Semi-parametric Models

Semi-parametric models based only on the specification of the first two moments of the responses y_{ij}, conditional on the small area means μ_i, and of the μ_i's have also been proposed. In the absence of covariates, Ghosh and Lahiri (1987) assumed the following model: (i) For each i, conditional on the θ_i's, the y_{ij}s are iid with mean θ_i and variance $\mu_2(\theta_i)$, denoted $y_{ij}|\theta_i \overset{\text{iid}}{\sim} (\theta_i, \mu_2(\theta_i))$, $j = 1, \ldots, N_i;\ i = 1, \ldots, m$; (ii) $\theta_i \overset{\text{iid}}{\sim} (\mu, \sigma_v^2)$; (iii) $0 < \sigma_e^2 = E\mu_2(\theta_i) < \infty$.

Raghunathan (1993) incorporated area level covariate information $\mathbf{z}_i$ as follows: (i) conditional on θ_i's, $y_{ij} \overset{\text{iid}}{\sim} (\theta_i, b_1(\phi, \theta_i, a_{ij}))$ where $b_1(\cdot)$ is a known positive function of a "dispersion" parameter ϕ, small area means θ_i and known constants a_{ij}; (ii) $\theta_i \overset{\text{ind}}{\sim} (\tau_i = h(\mathbf{z}_i^T\boldsymbol{\beta}), b_2(\psi, \tau_i, a_i))$ where $h(\cdot)$ is a known function and $b_2(\cdot)$ is a known positive function of a "dispersion" parameter ψ, the mean τ_i and a known constant a_i.

The "longitudinal" model (5.5.10) with unit-level covariates can be generalized by letting

$$E(y_{ij}|\mathbf{v}_i) = \mu_{ij}, \quad V(y_{ij}|\mathbf{v}_i) = \phi b(\mu_{ij}) \qquad (5.6.6)$$

and

$$h(\mu_{ij}) = \mathbf{x}_{ij1}^T \boldsymbol{\beta} + \mathbf{x}_{ij2}^T \mathbf{v}_i, \quad \mathbf{v}_i \overset{\text{iid}}{\sim} (\mathbf{0}, \boldsymbol{\Sigma}_v); \tag{5.6.7}$$

that is, $\mathbf{v}_i$ are independent and identically distributed with mean $\mathbf{0}$ and covariance matrix $\boldsymbol{\Sigma}_v$ (Breslow and Clayton (1993)).

Example 5.6.3. Hospital Admissions. Raghunathan (1993) obtained model-based estimates of county-specific mean number of hospital admissions for cancer chemotherapy per 1,000 individuals in the state of Washington, using 1987 hospital discharge data. His model for these data is of the form

$$E(y_{ij}|\theta_i) = \theta_i, \quad V(y_{ij}|\theta_i) = \theta_i + \phi\theta_i^2$$

and

$$E(\theta_i) = \beta, \quad V(\theta_i) = \psi,$$

where y_{ij} is the number of cancer admissions for individual j in county i, restricting to individuals ages 18 years or older.

Chapter 6

Empirical Best Linear Unbiased Prediction (EBLUP): Theory

6.1 Introduction

In Chapter 5 we presented several small area models that may be regarded as special cases of a general linear mixed model involving fixed and random effects. Moreover, small area means or totals can be expressed as linear combinations of fixed and random effects. Best linear unbiased prediction (BLUP) estimators of such parameters can be obtained in the classical frequentist framework, by appealing to general results on BLUP estimation. BLUP estimators minimize the MSE among the class of linear unbiased estimators and do not depend on normality of the random effects. But they depend on the variances (and covariances) of random effects which can be estimated by the method of fitting constants or moments. Alternatively, maximum likelihood (ML) or restricted maximum likelihood (REML) methods can be used to estimate the variance and covariance components, assuming normality. Using these estimated components in the BLUP estimator we obtain a two-stage estimator which is referred to as the empirical BLUP or EBLUP estimator (Harville (1991)), in analogy with the empirical Bayes (EB) estimator (Chapter 9).

In this chapter we present general results on EBLUP estimation. We also consider the more difficult problem of estimating the MSE of EBLUP estimators, taking account of the variability in the estimated variance and covariance components. Results for the special case of a linear mixed model with block diagonal covariance structure are spelled out. This model covers many commonly used small area models.

6.2 General Linear Mixed Model

Suppose that the sample data obey the general linear mixed model

$$\mathbf{y} = \mathbf{X}\boldsymbol{\beta} + \mathbf{Z}\mathbf{v} + \mathbf{e}. \tag{6.2.1}$$

Here $\mathbf{y}$ is the $n \times 1$ vector of sample observations, $\mathbf{X}$ and $\mathbf{Z}$ are known $n \times p$ and $n \times h$ matrices of full rank, and $\mathbf{v}$ and $\mathbf{e}$ are independently distributed with means $\mathbf{0}$ and covariance matrices $\mathbf{G}$ and $\mathbf{R}$ depending on some variance parameters $\boldsymbol{\delta} = (\delta_1, \ldots, \delta_q)^T$. We assume that $\boldsymbol{\delta}$ belongs to a specified subset of Euclidean q-space such that $\text{Var}(\mathbf{y}) = \mathbf{V} = \mathbf{V}(\boldsymbol{\delta}) = \mathbf{R} + \mathbf{Z}\mathbf{G}\mathbf{Z}^T$ is nonsingular for all $\boldsymbol{\delta}$ belonging to the subset, where $\text{Var}(\mathbf{y})$ denotes the variance-covariance matrix of $\mathbf{y}$.

We are interested in estimating a linear combination, $\mu = \mathbf{l}^T\boldsymbol{\beta} + \mathbf{m}^T\mathbf{v}$, of the regression parameters $\boldsymbol{\beta}$ and the realization of $\mathbf{v}$, for specified vectors, $\mathbf{l}$ and $\mathbf{m}$, of constants. A linear estimator of μ is of the form $\hat{\mu} = \mathbf{a}^T\mathbf{y} + b$ for known $\mathbf{a}$ and b. It is model-unbiased for μ if

$$E(\hat{\mu}) = E(\mu), \tag{6.2.2}$$

where E denotes the expectation with respect to the model (6.2.1). The MSE of $\hat{\mu}$ is given by

$$\text{MSE}(\hat{\mu}) = E(\hat{\mu} - \mu)^2, \tag{6.2.3}$$

which reduces to the variance of the error $\hat{\mu} - \mu$:

$$\text{MSE}(\hat{\mu}) = \text{Var}(\hat{\mu} - \mu)$$

if $\hat{\mu}$ is unbiased for μ. Valliant, Dorfman and Royall (2000, p.27) denote $E(\hat{\mu} - \mu)^2$ as the error variance (or prediction variance) under a model. We are interested in finding the BLUP estimator which minimizes the MSE in the class of linear unbiased estimators $\hat{\mu}$.

6.2.1 BLUP Estimator

For known $\boldsymbol{\delta}$, the BLUP estimator of μ is given by

$$\tilde{\mu}^H = t(\boldsymbol{\delta}, \mathbf{y}) = \mathbf{l}^T\tilde{\boldsymbol{\beta}} + \mathbf{m}^T\tilde{\mathbf{v}} = \mathbf{l}^T\tilde{\boldsymbol{\beta}} + \mathbf{m}^T\mathbf{G}\mathbf{Z}^T\mathbf{V}^{-1}(\mathbf{y} - \mathbf{X}\tilde{\boldsymbol{\beta}}), \tag{6.2.4}$$

where

$$\tilde{\boldsymbol{\beta}} = \tilde{\boldsymbol{\beta}}(\boldsymbol{\delta}) = (\mathbf{X}^T\mathbf{V}^{-1}\mathbf{X})^{-1}\mathbf{X}^T\mathbf{V}^{-1}\mathbf{y} \tag{6.2.5}$$

is the best linear unbiased estimator (BLUE) of $\boldsymbol{\beta}$,

$$\tilde{\mathbf{v}} = \tilde{\mathbf{v}}(\boldsymbol{\delta}) = \mathbf{G}\mathbf{Z}^T\mathbf{V}^{-1}(\mathbf{y} - \mathbf{X}\tilde{\boldsymbol{\beta}}), \tag{6.2.6}$$

and the superscript H on $\tilde{\mu}$ stands for Henderson, who proposed (6.2.4); see Henderson (1950). A direct proof that (6.2.4) is the BLUP estimator is given in subsection 6.4.1, following Henderson (1963). Robinson (1991) gave an alternative proof by writing $\hat{\mu}$ as $\hat{\mu} = t(\boldsymbol{\delta}, \mathbf{y}) + \mathbf{c}^T\mathbf{y}$ with $E(\mathbf{c}^T\mathbf{y}) = 0$, that is, $\mathbf{X}^T\mathbf{c} = \mathbf{0}$, and then showing that $E[\mathbf{l}^T(\tilde{\boldsymbol{\beta}} - \boldsymbol{\beta})\mathbf{y}^T\mathbf{c}] = 0$ and $E[\mathbf{m}^T(\tilde{\mathbf{v}} - \mathbf{v})\mathbf{y}^T\mathbf{c}] = 0$. This leads to

$$\text{MSE}(\hat{\mu}) = \text{MSE}[t(\boldsymbol{\delta}, \mathbf{y})] + E(\mathbf{c}^T\mathbf{y})^2 \geq \text{MSE}[t(\boldsymbol{\delta}, \mathbf{y})].$$

Robinson's proof assumes the knowledge of $t(\boldsymbol{\delta}, \mathbf{y})$.

Henderson, Kempthorne, Searle and von Krosigk (1959) assumed normality of $\mathbf{v}$ and $\mathbf{e}$ and maximized the joint density of $\mathbf{y}$ and $\mathbf{v}$ with respect to $\boldsymbol{\beta}$ and $\mathbf{v}$. This is equivalent to maximizing

$$\phi = -\frac{1}{2}(\mathbf{y} - \mathbf{X}\boldsymbol{\beta} - \mathbf{Z}\mathbf{v})^T\mathbf{R}^{-1}(\mathbf{y} - \mathbf{X}\boldsymbol{\beta} - \mathbf{Z}\mathbf{v}) - \frac{1}{2}\mathbf{v}^T\mathbf{G}^{-1}\mathbf{v}, \tag{6.2.7}$$

which leads to the following "mixed model" equations:

$$\begin{bmatrix} \mathbf{X}^T\mathbf{R}^{-1}\mathbf{X} & \mathbf{X}^T\mathbf{R}^{-1}\mathbf{Z} \\ \mathbf{Z}^T\mathbf{R}^{-1}\mathbf{X} & \mathbf{Z}^T\mathbf{R}^{-1}\mathbf{Z} + \mathbf{G}^{-1} \end{bmatrix} \begin{bmatrix} \boldsymbol{\beta}^* \\ \mathbf{v}^* \end{bmatrix} = \begin{bmatrix} \mathbf{X}^T\mathbf{R}^{-1}\mathbf{y} \\ \mathbf{Z}^T\mathbf{R}^{-1}\mathbf{y} \end{bmatrix}. \tag{6.2.8}$$

The solution of (6.2.8) is identical to the BLUP estimators of $\boldsymbol{\beta}$ and $\mathbf{v}$, that is, $\boldsymbol{\beta}^* = \tilde{\boldsymbol{\beta}}$ and $\mathbf{v}^* = \tilde{\mathbf{v}}$. This follows by noting that

$$\mathbf{R}^{-1} - \mathbf{R}^{-1}\mathbf{Z}(\mathbf{Z}^T\mathbf{R}^{-1}\mathbf{Z} + \mathbf{G}^{-1})^{-1}\mathbf{Z}^T\mathbf{R}^{-1} = \mathbf{V}^{-1}$$

and

$$(\mathbf{Z}^T\mathbf{R}^{-1}\mathbf{Z} + \mathbf{G}^{-1})^{-1}\mathbf{Z}^T\mathbf{R}^{-1} = \mathbf{G}\mathbf{Z}^T\mathbf{V}^{-1}.$$

The mixed model equations (6.2.8) are often computationally simpler than (6.2.5) and (6.2.6) if $\mathbf{G}$ and $\mathbf{R}$ are easily invertible (e.g., diagonal) and $\mathbf{V}$ has no simple inverse.

In view of the equivalence of $(\boldsymbol{\beta}^*, \mathbf{v}^*)$ and $(\tilde{\boldsymbol{\beta}}, \tilde{\mathbf{v}})$, the BLUP estimators are often called "joint maximum likelihood estimates" but the function being maximized, ϕ, is not a log likelihood in the usual sense because $\mathbf{v}$ is nonobservable (Robinson (1991)). It is called a "penalized likelihood" because a "penalty" $-\frac{1}{2}\mathbf{v}^T\mathbf{G}^{-1}\mathbf{v}$ is added to the log likelihood when $\mathbf{v}$ is regarded as fixed.

The best prediction (BP) estimator of μ is given by its conditional expectation $E(\mu|\mathbf{y})$ in the sense that

$$E[d(\mathbf{y}) - \mu]^2 \geq E[E(\mu|\mathbf{y}) - \mu]^2$$

for any estimator $d(\mathbf{y})$ of μ, not necessarily linear or unbiased. This result follows by noting that

$$\begin{aligned} E[(d(\mathbf{y}) - \mu)^2|\mathbf{y}] &= E[(d(\mathbf{y}) - E(\mu|\mathbf{y}) + E(\mu|\mathbf{y}) - \mu)^2|\mathbf{y}] \\ &= [d(\mathbf{y}) - E(\mu|\mathbf{y})]^2 + E[(E(\mu|\mathbf{y}) - \mu)^2|\mathbf{y}] \\ &\geq E[(E(\mu|\mathbf{y}) - \mu)^2|\mathbf{y}] \end{aligned}$$

with equality if and only if $d(\mathbf{y}) = E(\mu|\mathbf{y})$. Under normality, the BP estimator $E(\mu|\mathbf{y})$ reduces to the BLUP estimator (6.2.5) with $\tilde{\boldsymbol{\beta}}$ replaced by $\boldsymbol{\beta}$, that is, it depends on the unknown $\boldsymbol{\beta}$. In particular,

$$E(\mathbf{m}^T\mathbf{v}|\mathbf{y}) = \mathbf{m}^T\mathbf{G}\mathbf{Z}^T\mathbf{V}^{-1}(\mathbf{y} - \mathbf{X}\boldsymbol{\beta}).$$

This estimator is also the best linear prediction (BLP) estimator of $\mathbf{m}^T\mathbf{v}$ without assuming normality.

Since we do not want the estimator to depend on $\boldsymbol{\beta}$, we transform $\mathbf{y}$ to all error contrasts $\mathbf{A}^T\mathbf{y}$ with mean $\mathbf{0}$, that is, $\mathbf{A}^T\mathbf{X} = \mathbf{0}$ where $\mathbf{A}$ is any $n \times (n-p)$ matrix of full rank orthogonal to the $n \times p$ model matrix $\mathbf{X}$. For the transformed data, the best predictor $E(\mathbf{m}^T\mathbf{v}|\mathbf{A}^T\mathbf{y})$ in fact reduces to the BLUP estimator $\mathbf{m}^T\tilde{\mathbf{v}}$ (see subsection 6.4.2). This result provides an alternative justification of BLUP without linearity and unbiasedness, but assuming normality. It is interesting to note that the transformed data $\mathbf{A}^T\mathbf{y}$ are used to obtained the restricted (or residual) maximum likelihood (REML) estimators of the variance parameters $\boldsymbol{\delta}$ (see subsection 6.2.3).

We have considered the BLUP estimation of a single linear combination $\mu = \mathbf{l}^T\boldsymbol{\beta} + \mathbf{m}^T\mathbf{v}$, but the method readily extends to simultaneous estimation of $r(\geq 2)$ linear combinations, $\boldsymbol{\mu} = \mathbf{L}\boldsymbol{\beta} + \mathbf{M}\mathbf{v}$, where $\boldsymbol{\mu} = (\mu_1, \ldots, \mu_r)^T$. The BLUP estimator of $\boldsymbol{\mu}$ is

$$\mathbf{t}(\boldsymbol{\delta}, \mathbf{y}) = \mathbf{L}\tilde{\boldsymbol{\beta}} + \mathbf{M}\tilde{\mathbf{v}} = \mathbf{L}\tilde{\boldsymbol{\beta}} + \mathbf{M}\mathbf{G}\mathbf{Z}^T\mathbf{V}^{-1}(\mathbf{y} - \mathbf{X}\tilde{\boldsymbol{\beta}}). \quad (6.2.9)$$

The estimator $\mathbf{t}(\boldsymbol{\delta}, \mathbf{y})$ is optimal in the sense that for any other linear unbiased estimator $\mathbf{t}^*(\mathbf{y})$ of $\boldsymbol{\mu}$, the matrix $E(\mathbf{t}^* - \boldsymbol{\mu})(\mathbf{t}^* - \boldsymbol{\mu})^T - E(\mathbf{t} - \boldsymbol{\mu})(\mathbf{t} - \boldsymbol{\mu})^T$ is positive semi-definite (psd). Note that $E(\mathbf{t} - \boldsymbol{\mu})(\mathbf{t} - \boldsymbol{\mu})^T$ is the dispersion matrix of $\mathbf{t} - \boldsymbol{\mu}$.

6.2.2 MSE of BLUP

The BLUP estimator $t(\boldsymbol{\delta}, \mathbf{y})$ may be expressed as

$$t(\boldsymbol{\delta}, \mathbf{y}) = t^*(\boldsymbol{\delta}, \boldsymbol{\beta}, \mathbf{y}) + \mathbf{d}^T(\tilde{\boldsymbol{\beta}} - \boldsymbol{\beta}),$$

where $t^*(\boldsymbol{\delta}, \boldsymbol{\beta}, \mathbf{y})$ is the BLUP estimator when $\boldsymbol{\beta}$ is known:

$$t^*(\boldsymbol{\delta}, \boldsymbol{\beta}, \mathbf{y}) = \mathbf{l}^T\boldsymbol{\beta} + \mathbf{b}^T(\mathbf{y} - \mathbf{X}\boldsymbol{\beta}), \quad (6.2.10)$$

with

$$\mathbf{b}^T = \mathbf{m}^T\mathbf{G}\mathbf{Z}^T\mathbf{V}^{-1},$$

and

$$\mathbf{d}^T = \mathbf{l}^T - \mathbf{b}^T\mathbf{X}.$$

It now follows that $t^*(\boldsymbol{\delta}, \boldsymbol{\beta}, \mathbf{y}) - \mu$ and $\mathbf{d}^T(\tilde{\boldsymbol{\beta}} - \boldsymbol{\beta})$ are uncorrelated, noting that

$$E[(\mathbf{b}^T(\mathbf{Z}\mathbf{v} + \mathbf{e}) - \mathbf{m}^T\mathbf{v})(\mathbf{v}^T\mathbf{Z}^T + \mathbf{e}^T)\mathbf{V}^{-1}] = \mathbf{0}.$$

Therefore,

$$\text{MSE}[t(\boldsymbol{\delta},\mathbf{y})] = \text{MSE}[t^*(\boldsymbol{\delta},\boldsymbol{\beta},\mathbf{y})] + \text{Var}[\mathbf{d}^T(\tilde{\boldsymbol{\beta}}-\boldsymbol{\beta})] = g_1(\boldsymbol{\delta}) + g_2(\boldsymbol{\delta}), \qquad (6.2.11)$$

where

$$g_1(\boldsymbol{\delta}) = \text{Var}[t^*(\boldsymbol{\delta},\boldsymbol{\beta},\mathbf{y}) - \mu] = \mathbf{m}^T(\mathbf{G} - \mathbf{G}\mathbf{Z}^T\mathbf{V}^{-1}\mathbf{Z}\mathbf{G})\mathbf{m} \qquad (6.2.12)$$

and

$$g_2(\boldsymbol{\delta}) = \mathbf{d}^T(\mathbf{X}^T\mathbf{V}^{-1}\mathbf{X})^{-1}\mathbf{d}. \qquad (6.2.13)$$

The second term, $g_2(\boldsymbol{\delta})$, in (6.2.11) accounts for the variability in the estimator $\tilde{\boldsymbol{\beta}}$.

Henderson (1975) used the mixed model equations (6.2.8) to obtain an alternative formula for $\text{MSE}[t(\boldsymbol{\delta},\mathbf{y})]$:

$$\text{MSE}[t(\boldsymbol{\delta},\mathbf{y})] = (\mathbf{l}^T, \mathbf{m}^T)\begin{bmatrix} \mathbf{C}_{11} & \mathbf{C}_{12} \\ \mathbf{C}_{21} & \mathbf{C}_{22} \end{bmatrix}\begin{pmatrix} \mathbf{l} \\ \mathbf{m} \end{pmatrix}, \qquad (6.2.14)$$

where the matrix $\mathbf{C}$ with blocks $\mathbf{C}_{ij}$ $(i,j = 1,2)$ is the inverse of the coefficient matrix of the mixed model equations. This form is computationally simpler than (6.2.11) when $\mathbf{V}$ is not easily invertible.

6.2.3 EBLUP Estimator

The BLUP estimator $t(\boldsymbol{\delta},\mathbf{y})$ given by (6.2.4) depends on the variance parameters $\boldsymbol{\delta}$ which are unknown in practical applications. Replacing $\boldsymbol{\delta}$ by an estimator $\hat{\boldsymbol{\delta}} = \hat{\boldsymbol{\delta}}(\mathbf{y})$, we obtain a two-stage estimator $\hat{\mu}^H = t(\hat{\boldsymbol{\delta}},\mathbf{y})$, which is referred to as the EBLUP estimator. For convenience, we also write $t(\hat{\boldsymbol{\delta}},\mathbf{y})$ and $t(\boldsymbol{\delta},\mathbf{y})$ as $t(\hat{\boldsymbol{\delta}})$ and $t(\boldsymbol{\delta})$.

The two-stage estimator $t(\hat{\boldsymbol{\delta}})$ remains unbiased for μ, that is, $E[t(\hat{\boldsymbol{\delta}}) - \mu] = 0$, provided (i) $E[t(\hat{\boldsymbol{\delta}})]$ is finite; (ii) $\hat{\boldsymbol{\delta}}$ is any even translation-invariant estimator of $\boldsymbol{\delta}$, that is, $\hat{\boldsymbol{\delta}}(-\mathbf{y}) = \hat{\boldsymbol{\delta}}(\mathbf{y})$ and $\hat{\boldsymbol{\delta}}(\mathbf{y} - \mathbf{X}\mathbf{b}) = \hat{\boldsymbol{\delta}}(\mathbf{y})$ for all $\mathbf{y}$ and $\mathbf{b}$; (iii) The distributions of $\mathbf{v}$ and $\mathbf{e}$ are both symmetric around $\mathbf{0}$ (not necessarily normal). A proof of unbiasedness of $t(\hat{\boldsymbol{\delta}})$, due to Kacker and Harville (1981), uses the following results: (a) $\hat{\boldsymbol{\delta}}(\mathbf{y}) = \hat{\boldsymbol{\delta}}(\mathbf{Z}\mathbf{v} + \mathbf{e}) = \hat{\boldsymbol{\delta}}(-\mathbf{Z}\mathbf{v} - \mathbf{e})$; (b) $t(\hat{\boldsymbol{\delta}}) - \mu = \phi(\mathbf{v},\mathbf{e}) - \mathbf{m}^T\mathbf{v}$, where $\phi(\mathbf{v},\mathbf{e})$ is an odd function of $\mathbf{v}$ and $\mathbf{e}$, that is, $\phi(-\mathbf{v},-\mathbf{e}) = -\phi(\mathbf{v},\mathbf{e})$. Result (b) implies that $E\phi(\mathbf{v},\mathbf{e}) = E\phi(-\mathbf{v},-\mathbf{e}) = -E\phi(\mathbf{v},\mathbf{e})$ or $E\phi(\mathbf{v},\mathbf{e}) = 0$ so that $E[t(\hat{\boldsymbol{\delta}}) - \mu] = 0$.

Kackar and Harville (1981) have also shown that standard procedures for estimating $\boldsymbol{\delta}$ yield even translation invariant estimators; in particular, ML, REML and the method of fitting constants (also called Henderson's method 3). We refer the reader to Searle, Casella and McCulloch (1992) and P.S.R.S. Rao (1997) for details of these methods for the analysis of variance (ANOVA) model, which is a special case of the general linear mixed model (6.2.1). The ANOVA model is given by

$$\mathbf{y} = \mathbf{X}\boldsymbol{\beta} + \mathbf{Z}_1\mathbf{v}_1 + \cdots + \mathbf{Z}_r\mathbf{v}_r + \mathbf{e}, \qquad (6.2.15)$$

where $\mathbf{v}_1, \ldots, \mathbf{v}_r$ and $\mathbf{e}$ are independently distributed with means $\mathbf{0}$ and covariance matrices $\sigma_1^2\mathbf{I}_{h_1}, \ldots, \sigma_r^2\mathbf{I}_{h_r}$ and $\sigma_e^2\mathbf{I}_n$. The parameters $\boldsymbol{\delta} = (\sigma_0^2, \ldots, \sigma_r^2)^T$ with $\sigma_i^2 \geq 0$ $(i = 1, \ldots, r)$ and $\sigma_0^2 = \sigma_e^2 > 0$ are the variance components. Note that $\mathbf{G}$ is now block diagonal with blocks $\sigma_i^2\mathbf{I}_{h_i}$, $\mathbf{R} = \sigma_e^2\mathbf{I}_n$ and $\mathbf{V} = \sigma_e^2\mathbf{I}_n + \sum \sigma_i^2\mathbf{Z}_i\mathbf{Z}_i^T$, which is a special case of covariance matrix with linear structure: $\mathbf{V} = \sum \boldsymbol{\delta}_i\mathbf{H}_i$ for known symmetric matrices $\mathbf{H}_i$.

6.2.4 ML and REML Estimators

We now provide formulas for the ML and REML estimators of $\boldsymbol{\beta}$ and $\boldsymbol{\delta}$ under the general linear mixed model (6.2.1) and the associated asymptotic covariance matrices, assuming normality (Cressie (1992)). Under normality, the partial derivative of the log likelihood function, $l(\boldsymbol{\beta}, \boldsymbol{\delta})$, with respect to $\boldsymbol{\delta}$ is given by $\mathbf{s}(\boldsymbol{\beta}, \boldsymbol{\delta})$ with jth element

$$s_j(\boldsymbol{\beta}, \boldsymbol{\delta}) = \partial l(\boldsymbol{\beta}, \boldsymbol{\delta})/\partial \delta_j = -\frac{1}{2}\text{tr}(\mathbf{V}^{-1}\mathbf{V}_{(j)}) - \frac{1}{2}(\mathbf{y} - \mathbf{X}\boldsymbol{\beta})^T\mathbf{V}^{(j)}(\mathbf{y} - \mathbf{X}\boldsymbol{\beta}),$$

where $\mathbf{V}_{(j)} = \partial\mathbf{V}/\partial\delta_j$ and $\mathbf{V}^{(j)} = \partial\mathbf{V}^{-1}/\partial\delta_j = -\mathbf{V}^{-1}\mathbf{V}_{(j)}\mathbf{V}^{-1}$. Note that $\mathbf{V} = \mathbf{V}(\boldsymbol{\delta})$. Also, the matrix of expected second derivatives of $-l(\boldsymbol{\beta}, \boldsymbol{\delta})$ with respect to $\boldsymbol{\delta}$ is given by $\mathcal{I}(\boldsymbol{\delta})$ with (j, k)th element

$$\mathcal{I}_{jk}(\boldsymbol{\delta}) = \frac{1}{2}\text{tr}(\mathbf{V}^{-1}\mathbf{V}_{(j)}\mathbf{V}^{-1}\mathbf{V}_{(k)}). \tag{6.2.16}$$

The ML estimator of $\boldsymbol{\delta}$ is obtained iteratively using the "scoring" algorithm:

$$\boldsymbol{\delta}^{(a+1)} = \boldsymbol{\delta}^{(a)} + [\mathcal{I}(\boldsymbol{\delta}^{(a)})]^{-1}\mathbf{s}[\tilde{\boldsymbol{\beta}}(\boldsymbol{\delta}^{(a)}), \boldsymbol{\delta}^{(a)}], \tag{6.2.17}$$

where the superscript (a) denotes that the specified terms are evaluated at $\boldsymbol{\delta} = \boldsymbol{\delta}^{(a)}$ and $\tilde{\boldsymbol{\beta}} = \tilde{\boldsymbol{\beta}}(\boldsymbol{\delta}^{(a)})$, the values of $\boldsymbol{\delta}$ and $\tilde{\boldsymbol{\beta}} = \tilde{\boldsymbol{\beta}}(\boldsymbol{\delta})$ at the ath iteration $(a = 0, 1, 2, \ldots)$. At convergence of the iterations (6.2.17), we get ML estimators $\hat{\boldsymbol{\delta}}_{\text{ML}}$ of $\boldsymbol{\delta}$ and $\hat{\boldsymbol{\beta}}_{\text{ML}} = \tilde{\boldsymbol{\beta}}(\hat{\boldsymbol{\delta}}_{\text{ML}})$ of $\boldsymbol{\beta}$. The asymptotic covariance matrix of $\hat{\boldsymbol{\beta}}_{\text{ML}}$ and $\hat{\boldsymbol{\delta}}_{\text{ML}}$ has a block diagonal structure, $\text{diag}[\overline{\mathbf{V}}(\hat{\boldsymbol{\beta}}_{\text{ML}}), \overline{\mathbf{V}}(\hat{\boldsymbol{\delta}}_{\text{ML}})]$, with

$$\overline{\mathbf{V}}(\hat{\boldsymbol{\beta}}_{\text{ML}}) = (\mathbf{X}^T\mathbf{V}^{-1}\mathbf{X})^{-1}; \overline{\mathbf{V}}(\hat{\boldsymbol{\delta}}_{\text{ML}}) = \mathcal{I}^{-1}(\boldsymbol{\delta}). \tag{6.2.18}$$

A drawback of the ML estimator of $\boldsymbol{\delta}$ is that it does not take account of the loss in degrees of freedom (df) due to estimating $\boldsymbol{\beta}$. For example, when $y_1, \ldots, y_n$ are iid $N(\mu, \sigma^2)$, the ML estimator $\hat{\sigma}^2 = [(n-1)/n]s^2$ is not equal to the customary unbiased estimator $s^2 = \sum_i(y_i - \overline{y})^2/(n-1)$ of σ^2. The REML method takes account of the loss in df by using the transformed data $\mathbf{y}^* = \mathbf{A}^T\mathbf{y}$, where $\mathbf{A}$ is any $n \times (n-p)$ matrix of full rank orthogonal to the $n \times p$ matrix $\mathbf{X}$. Noting that $\mathbf{y}^* = \mathbf{A}^T\mathbf{y}$ is $(n-p)$-variate normal with mean $\mathbf{0}$ and covariance matrix $\mathbf{A}^T\mathbf{V}\mathbf{A}$, the partial derivative of the restricted log likelihood function $l_R(\boldsymbol{\delta})$ with respect to $\boldsymbol{\delta}$ is given by $\mathbf{s}_R(\boldsymbol{\delta})$ with jth element

$$s_{Rj}(\boldsymbol{\delta}) = \partial l_R(\boldsymbol{\delta})/\partial\delta_j = -\frac{1}{2}\text{tr}[\mathbf{P}\mathbf{V}_{(j)}] + \frac{1}{2}\mathbf{y}^T\mathbf{P}\mathbf{V}_{(j)}\mathbf{P}\mathbf{y},$$

where

$$\mathbf{P} = \mathbf{V}^{-1} - \mathbf{V}^{-1}\mathbf{X}(\mathbf{X}^T\mathbf{V}^{-1}\mathbf{X})^{-1}\mathbf{X}^T\mathbf{V}^{-1}$$

and $l_R(\boldsymbol{\delta})$ is based on y^*. Note that $\mathbf{Py} = \mathbf{V}^{-1}(\mathbf{y} - \mathbf{X}\tilde{\boldsymbol{\beta}})$, Also, the matrix of expected second derivatives of $-l_R(\boldsymbol{\delta})$ with respect to $\boldsymbol{\delta}$ is given by $\mathcal{I}_R(\boldsymbol{\delta})$ with (j,k)th element

$$\mathcal{I}_{R,jk}(\boldsymbol{\delta}) = \frac{1}{2}\text{tr}[\mathbf{PV}_{(j)}\mathbf{PV}_{(k)}]. \tag{6.2.19}$$

Note that both $\mathbf{s}_R(\boldsymbol{\delta})$ and $\mathcal{I}_R(\boldsymbol{\delta})$ are invariant to the choice of $\mathbf{A}$.

The REML estimator of $\boldsymbol{\delta}$ is obtained iteratively from (6.2.17) by replacing $\mathcal{I}(\boldsymbol{\delta}^{(a)})$ and $\mathbf{s}[\tilde{\boldsymbol{\beta}}(\boldsymbol{\delta})^{(a)}, \boldsymbol{\delta}^{(a)}]$ by $\mathcal{I}_R(\boldsymbol{\delta}^{(a)})$ and $\mathbf{s}_R(\boldsymbol{\delta}^{(a)})$, respectively. At convergence of the iterations, we get REML estimators $\hat{\boldsymbol{\delta}}_{\text{RE}}$ and $\hat{\boldsymbol{\beta}}_{\text{RE}} = \tilde{\boldsymbol{\beta}}(\hat{\boldsymbol{\delta}}_{\text{RE}})$. Asymptotically, $\overline{\mathbf{V}}(\hat{\boldsymbol{\delta}}_{\text{RE}}) \approx \overline{\mathbf{V}}(\hat{\boldsymbol{\delta}}_{\text{ML}}) = \mathcal{I}^{-1}(\boldsymbol{\delta})$ and $\overline{\mathbf{V}}(\hat{\boldsymbol{\beta}}_{\text{RE}}) \approx \overline{\mathbf{V}}(\hat{\boldsymbol{\beta}}_{\text{ML}}) = (\mathbf{X}^T\mathbf{V}^{-1}\mathbf{X})^{-1}$, provided p is fixed.

For the ANOVA model (6.2.15), the ML estimator of $\boldsymbol{\delta}$ can be obtained iteratively as follows, using the BLUP estimators $\tilde{\boldsymbol{\beta}}$ and $\tilde{\mathbf{v}}$ (Hartley and Rao (1967); Henderson (1973)):

$$\sigma_i^{2(a+1)} = \frac{1}{h_i}[\tilde{\mathbf{v}}_i^{(a)T}\tilde{\mathbf{v}}_i^{(a)} + \sigma_i^{2(a)}\text{tr } \mathbf{T}^{*(a)}_{ii}] \tag{6.2.20}$$

and

$$\sigma_e^{2(a+1)} = \mathbf{y}^T(\mathbf{y} - \mathbf{X}\tilde{\boldsymbol{\beta}}^{(a)} - \mathbf{Z}\tilde{\mathbf{v}}^{(a)})/n, \tag{6.2.21}$$

where

$$\mathbf{T}^*_{ii} = (\mathbf{I} + \mathbf{Z}^T\mathbf{R}^{-1}\mathbf{Z}\mathbf{G})^{-1}\mathbf{F}_{\text{ii}}$$

with $tr\mathbf{T}^*{}_{ii} > 0$, where $\mathbf{F}_{ii}$ is given by $\mathbf{G}$ with unity in place of σ_i^2 and zero in place of σ_j^2 $(j \neq i)$. The values of $\tilde{\boldsymbol{\beta}}$ and $\tilde{\mathbf{v}}$ for a specified $\boldsymbol{\delta}$ are readily obtained from the mixed model equations (6.2.8), without evaluating $\mathbf{V}^{-1}$. The algorithm given by (6.2.20) and (6.2.21) is similar to the EM algorithm (Dempster, Laird and Rubin (1977)). The asymptotic covariance matrices of $\hat{\boldsymbol{\beta}}_{\text{ML}}$ and $\hat{\boldsymbol{\delta}}_{\text{ML}}$ are given by (6.2.18) using $\mathbf{V}_{(j)} = \mathbf{Z}_j\mathbf{Z}_j^T, (j = 0, 1, \ldots, r)$ and $\mathbf{Z}_0 = \mathbf{I}_n$ in (6.2.16).

Anderson (1973) suggested the following iterative algorithm to obtain $\hat{\boldsymbol{\delta}}_{\text{ML}}$:

$$\boldsymbol{\delta}^{(a+1)} = [\mathcal{I}(\boldsymbol{\delta}^{(a)})]^{-1}\mathbf{b}(\boldsymbol{\delta}^{(a)}), \tag{6.2.22}$$

where the ith element $\mathbf{b}(\boldsymbol{\delta})$ is given by

$$b_i(\boldsymbol{\delta}) = \frac{1}{2}\mathbf{y}^T\mathbf{PZ}_i\mathbf{Z}_i^T\mathbf{Py}. \tag{6.2.23}$$

This algorithm is equivalent to the scoring algorithm for solving ML equations (J.N.K. Rao (1974)). The algorithm is also applicable to any covariance matrix with linear structure, $\mathbf{V} = \sum \boldsymbol{\delta}_i \mathbf{H}_i$. We simply replace $\mathbf{Z}_i\mathbf{Z}_i^T$ by $\mathbf{H}_i$ to define $\mathcal{I}(\boldsymbol{\delta})$ and $\mathbf{b}(\boldsymbol{\delta})$ using (6.2.16) and (6.2.23).

For the ANOVA model (6.2.15), the REML estimator of $\boldsymbol{\delta}$ can be obtained iteratively from (6.2.20) and (6.2.21) by changing n to $n-p$ in (6.2.21) and $\mathbf{T}_{ii}^*$ to $\mathbf{T}_{ii}$ in (6.2.19), where

$$\mathbf{T}_{ii} = (\mathbf{I} + \mathbf{Z}^T\mathbf{Q}\mathbf{Z}\mathbf{G})^{-1}\mathbf{F}_{ii}$$

with

$$\mathbf{Q} = \mathbf{R}^{-1} - \mathbf{R}^{-1}\mathbf{X}(\mathbf{X}^T\mathbf{R}^{-1}\mathbf{X})^{-1}\mathbf{X}^T\mathbf{R}^{-1}$$

(Harville (1977)). The elements of the information matrix, $\mathcal{I}_R(\boldsymbol{\delta})$, are given by (6.2.19) using $\mathbf{V}_{(j)} = \mathbf{Z}_j\mathbf{Z}_j^T$.

For the REML, an iterative algorithm similar to (6.2.22) is given by

$$\boldsymbol{\delta}^{(a+1)} = [\mathcal{I}_R(\boldsymbol{\delta}^{(a)})]^{-1}\mathbf{b}(\boldsymbol{\delta}^{(a)}). \tag{6.2.24}$$

The algorithm (6.2.24) is also equivalent to the scoring algorithm for solving REML equations (Hocking and Kutner (1975)).

C.R. Rao (1971) proposed the method of minimum norm quadratic unbiased (MINQU) estimation that does not require the normality assumption, unlike ML and the REML. The MINQU estimators depend on a preassigned value $\boldsymbol{\delta}_0$ for $\boldsymbol{\delta}$, and they are identical to the first iterative solution, $\boldsymbol{\delta}^{(1)}$, of REML iteration (6.2.24) using $\boldsymbol{\delta}_0$ as the starting value, that is, $\boldsymbol{\delta}^{(0)} = \boldsymbol{\delta}_0$. This result suggests that REML (or ML) estimators of $\boldsymbol{\delta}$, derived under normality, may perform well even under nonnormal distributions. In fact, Jiang (1996) established asymptotic consistency of the REML estimator of $\boldsymbol{\delta}$ for the ANOVA model (6.2.15) when normality may not hold.

The method of fitting constants and the method of moments are also used to estimate $\boldsymbol{\delta}$. We study these methods for particular cases of the general linear mixed model or the ANOVA model (Chapters 7 and 8).

Henderson's (1973) iterative algorithms for computing ML and REML estimates of variance components σ_e^2 and σ_i^2 $(i = 1, \ldots, r)$ in the ANOVA model (6.2.15) are not affected by the constraints on the parameter space: $\sigma_e^2 > 0$; $\sigma_i^2 \geq 0$, $i = 1, \ldots, r$ (Harville (1977)). If the starting values $\sigma_e^{2(0)}$ and $\sigma_i^{2(0)}$ are strictly positive, then at every iteration a the values $\sigma_e^{2(a+1)}$ and $\sigma_i^{2(a+1)}$ remain positive, although it is possible for some of them to be arbitrarily close to 0. The scoring method for general linear mixed models (6.2.1) and the equivalent Anderson's method for models with linear covariance structures do not enjoy this nice property of Henderson's algorithm. Modifications are needed to accommodate constraints on $\boldsymbol{\delta} = (\delta_1, \ldots, \delta_q)^T$. We refer the reader to Harville (1977) for details. MINQUE and the method of fitting constants or the method of moments also require modifications to account for the constraints on $\boldsymbol{\delta}$.

6.2.5 MSE of EBLUP

The error in the EBLUP estimator $t(\hat{\boldsymbol{\delta}})$ may be decomposed as

$$t(\hat{\boldsymbol{\delta}}) - \mu = [t(\boldsymbol{\delta}) - \mu] + [t(\hat{\boldsymbol{\delta}}) - t(\boldsymbol{\delta})].$$

Therefore,

$$\text{MSE}[t(\hat{\boldsymbol{\delta}})] = \text{MSE}[t(\boldsymbol{\delta})] + E[t(\hat{\boldsymbol{\delta}}) - t(\boldsymbol{\delta})]^2 + 2E[t(\boldsymbol{\delta}) - \mu][t(\hat{\boldsymbol{\delta}}) - t(\boldsymbol{\delta})]. \quad (6.2.25)$$

Under normality of the random effects $\mathbf{v}$ and $\mathbf{e}$, the cross-product term in (6.2.25) is zero provided $\hat{\boldsymbol{\delta}}$ is translation invariant (see Section 6.4) so that

$$\text{MSE}[t(\hat{\boldsymbol{\delta}})] = \text{MSE}[t(\boldsymbol{\delta})] + E[t(\hat{\boldsymbol{\delta}}) - t(\boldsymbol{\delta})]^2. \quad (6.2.26)$$

It is clear from (6.2.26) that the MSE of the EBLUP estimator is always larger than that of the BLUP estimator $t(\boldsymbol{\delta})$, under normality. The common practice of approximating MSE$[t(\hat{\boldsymbol{\delta}})]$ by MSE$[t(\boldsymbol{\delta})]$ could therefore lead to significant underestimation, especially in cases where $t(\boldsymbol{\delta})$ varies with $\boldsymbol{\delta}$ to a significant extent and where the variability of $\hat{\boldsymbol{\delta}}$ is not small.

The last term of (6.2.26) is generally intractable except in special cases, such as the balanced one-way model $y_{ij} = \mu + v_i + e_{ij}$, $i = 1, \ldots, m$, $j = 1, \ldots, \bar{n}$ (Peixoto and Harville (1986)). It is therefore necessary to obtain an approximation to this term. We present a heuristic approximation along the lines of Kackar and Harville (1984), which can be justified rigorously for particular small area models (see Section 6.3). By a Taylor approximation, we have

$$t(\hat{\boldsymbol{\delta}}) - t(\boldsymbol{\delta}) \approx \mathbf{d}(\boldsymbol{\delta})^T(\hat{\boldsymbol{\delta}} - \boldsymbol{\delta}) \quad (6.2.27)$$

with $\mathbf{d}(\boldsymbol{\delta}) = \partial t(\boldsymbol{\delta})/\partial \boldsymbol{\delta}$, assuming that the terms involving higher powers of $\hat{\boldsymbol{\delta}} - \boldsymbol{\delta}$ are of lower order relative to $\mathbf{d}(\boldsymbol{\delta})^T(\hat{\boldsymbol{\delta}} - \boldsymbol{\delta})$. Further, under normality,

$$\mathbf{d}(\boldsymbol{\delta}) \approx \partial t^*(\boldsymbol{\delta}, \boldsymbol{\beta})/\partial \boldsymbol{\delta} = (\partial \mathbf{b}^T/\partial \boldsymbol{\delta})(\mathbf{y} - \mathbf{X}\boldsymbol{\beta}) = \mathbf{d}^*(\boldsymbol{\delta}),$$

noting that the terms involving the derivatives of $\tilde{\boldsymbol{\beta}} - \boldsymbol{\beta}$ with respect to $\boldsymbol{\delta}$ are of lower order, where $t^*(\boldsymbol{\delta}, \boldsymbol{\beta})$ is given by (6.2.10). Thus

$$E[\mathbf{d}(\boldsymbol{\delta})^T(\hat{\boldsymbol{\delta}} - \boldsymbol{\delta})]^2 \approx E[\mathbf{d}^*(\boldsymbol{\delta})^T(\hat{\boldsymbol{\delta}} - \boldsymbol{\delta})]^2. \quad (6.2.28)$$

Further,

$$\begin{aligned} E[\mathbf{d}^*(\boldsymbol{\delta})^T(\hat{\boldsymbol{\delta}} - \boldsymbol{\delta})]^2 &\approx \text{tr}[E(\mathbf{d}^*(\boldsymbol{\delta})\mathbf{d}^*(\boldsymbol{\delta})^T)\overline{\mathbf{V}}(\hat{\boldsymbol{\delta}})] \\ &= \text{tr}[(\partial \mathbf{b}^T/\partial \boldsymbol{\delta})\mathbf{V}(\partial \mathbf{b}^T/\partial \boldsymbol{\delta})^T\overline{\mathbf{V}}(\hat{\boldsymbol{\delta}})] =: g_3(\boldsymbol{\delta}), \quad (6.2.29) \end{aligned}$$

where the neglected terms are of lower order, $\overline{\mathbf{V}}(\hat{\boldsymbol{\delta}})$ is the asymptotic covariance matrix of $\hat{\boldsymbol{\delta}}$ and $A =: B$ means that B is defined to be equal to A. It now follows from (6.2.27), (6.2.28) and (6.2.29) that

$$E[t(\hat{\boldsymbol{\delta}}) - t(\boldsymbol{\delta})]^2 \approx g_3(\boldsymbol{\delta}). \quad (6.2.30)$$

Combining (6.2.11) with (6.2.30), we get a second-order approximation to the MSE of $t(\hat{\boldsymbol{\delta}})$ as

$$\mathrm{MSE}[t(\hat{\boldsymbol{\delta}})] \approx g_1(\boldsymbol{\delta}) + g_2(\boldsymbol{\delta}) + g_3(\boldsymbol{\delta}). \tag{6.2.31}$$

The terms $g_2(\boldsymbol{\delta})$ and $g_3(\boldsymbol{\delta})$, due to estimating $\boldsymbol{\beta}$ and $\boldsymbol{\delta}$, are of lower order than the leading term $g_1(\boldsymbol{\delta})$.

6.2.6 Estimation of MSE of EBLUP

For practical applications, we need an estimator of $\mathrm{MSE}[t(\hat{\boldsymbol{\delta}})]$ as a measure of variability associated with $t(\hat{\boldsymbol{\delta}})$. A naive approach approximates $\mathrm{MSE}[t(\hat{\boldsymbol{\delta}})]$ by $\mathrm{MSE}[t(\boldsymbol{\delta})]$ and then substitutes $\hat{\boldsymbol{\delta}}$ for $\boldsymbol{\delta}$. The resulting estimator of MSE is given by

$$\mathrm{mse}_N[t(\hat{\boldsymbol{\delta}})] = g_1(\hat{\boldsymbol{\delta}}) + g_2(\hat{\boldsymbol{\delta}}). \tag{6.2.32}$$

Another estimator of MSE is obtained by substituting $\hat{\boldsymbol{\delta}}$ for $\boldsymbol{\delta}$ in the MSE approximation (6.2.31):

$$\mathrm{mse}_1[t(\hat{\boldsymbol{\delta}})] = g_1(\hat{\boldsymbol{\delta}}) + g_2(\hat{\boldsymbol{\delta}}) + g_3(\hat{\boldsymbol{\delta}}). \tag{6.2.33}$$

We have $Eg_2(\hat{\boldsymbol{\delta}}) \approx g_2(\boldsymbol{\delta})$ and $Eg_3(\hat{\boldsymbol{\delta}}) \approx g_3(\boldsymbol{\delta})$ to the desired order of approximation, but $g_1(\hat{\boldsymbol{\delta}})$ is not the correct estimator of $g_1(\boldsymbol{\delta})$ because its bias is generally of the same order as $g_2(\boldsymbol{\delta})$ and $g_3(\boldsymbol{\delta})$.

To evaluate the bias of $g_1(\hat{\boldsymbol{\delta}})$, we make a Taylor expansion of $g_1(\hat{\boldsymbol{\delta}})$ around $\boldsymbol{\delta}$:

$$\begin{aligned} g_1(\hat{\boldsymbol{\delta}}) &= g_1(\boldsymbol{\delta}) + (\hat{\boldsymbol{\delta}} - \boldsymbol{\delta})^T \nabla g_1(\boldsymbol{\delta}) + \frac{1}{2}(\hat{\boldsymbol{\delta}} - \boldsymbol{\delta})^T \nabla^2 g_1(\boldsymbol{\delta})(\hat{\boldsymbol{\delta}} - \boldsymbol{\delta}) \\ &= g_1(\boldsymbol{\delta}) + \Delta_1 + \Delta_2, \end{aligned}$$

say, where $\nabla g_1(\boldsymbol{\delta})$ is the vector of first-order derivatives of $g_1(\boldsymbol{\delta})$ with respect to $\boldsymbol{\delta}$ and $\nabla^2 g_1(\boldsymbol{\delta})$ is the matrix of second-order derivatives of $g_1(\boldsymbol{\delta})$ with respect to $\boldsymbol{\delta}$. If $\hat{\boldsymbol{\delta}}$ is unbiased for $\boldsymbol{\delta}$, then $E(\Delta_1) = 0$. In general, if $E(\Delta_1) \approx \mathbf{b}_{\hat{\boldsymbol{\delta}}}^T(\boldsymbol{\delta}) \nabla g_1(\boldsymbol{\delta})$ is of lower order than $E(\Delta_2)$, then

$$Eg_1(\hat{\boldsymbol{\delta}}) \approx g_1(\boldsymbol{\delta}) + \frac{1}{2}\mathrm{tr}[\nabla^2 g_1(\boldsymbol{\delta}) \bar{\mathbf{V}}(\hat{\boldsymbol{\delta}})], \tag{6.2.34}$$

where $\mathbf{b}_{\hat{\boldsymbol{\delta}}}(\boldsymbol{\delta})$ is an approximation to the bias $E(\hat{\boldsymbol{\delta}}) - \boldsymbol{\delta}$. Further, if the covariance matrix $\mathbf{V}$ has a linear structure, (6.2.34) reduces to

$$Eg_1(\hat{\boldsymbol{\delta}}) \approx g_1(\boldsymbol{\delta}) - g_3(\boldsymbol{\delta}). \tag{6.2.35}$$

It now follows from (6.2.32), (6.2.33) and (6.2.35) that the biases of $\mathrm{mse}_N[t(\hat{\boldsymbol{\delta}})]$ and $\mathrm{mse}_1[t(\hat{\boldsymbol{\delta}})]$ are

$$B_N \approx -2g_3(\boldsymbol{\delta}), \quad B_1 \approx -g_3(\boldsymbol{\delta}).$$

A correct estimator of $\mathrm{MSE}[t(\hat{\boldsymbol{\delta}})]$ to the desired order of approximation is given by

$$\mathrm{mse}[t(\hat{\boldsymbol{\delta}})] \approx g_1(\hat{\boldsymbol{\delta}}) + g_2(\hat{\boldsymbol{\delta}}) + 2g_3(\hat{\boldsymbol{\delta}}), \tag{6.2.36}$$

noting that $E[g_1(\hat{\boldsymbol{\delta}}) + g_3(\hat{\boldsymbol{\delta}})] \approx g_1(\boldsymbol{\delta})$ from (6.2.35). Consequently,

$$E\,\mathrm{mse}[t(\hat{\boldsymbol{\delta}})] \approx \mathrm{MSE}[t(\hat{\boldsymbol{\delta}})].$$

Formula (6.2.36) holds for the REML estimator, $\hat{\boldsymbol{\delta}}_{\mathrm{RE}}$, and some moment estimators.

If $E(\Delta_1)$ is of the same order as $E(\Delta_2)$, as in the case of the ML estimator $\hat{\boldsymbol{\delta}}_{\mathrm{ML}}$, then an extra term $\mathbf{b}^T_{\hat{\boldsymbol{\delta}}}(\hat{\boldsymbol{\delta}})\nabla g_1(\hat{\boldsymbol{\delta}})$ is subtracted from (6.2.36):

$$\mathrm{mse}_*[t(\hat{\boldsymbol{\delta}})] \approx g_1(\hat{\boldsymbol{\delta}}) - \mathbf{b}^T_{\hat{\boldsymbol{\delta}}}(\hat{\boldsymbol{\delta}})\nabla g_1(\hat{\boldsymbol{\delta}}) + g_2(\hat{\boldsymbol{\delta}}) + 2g_3(\hat{\boldsymbol{\delta}}). \tag{6.2.37}$$

The term $\mathbf{b}^T_{\hat{\boldsymbol{\delta}}}(\hat{\boldsymbol{\delta}})\nabla g_1(\hat{\boldsymbol{\delta}})$ is spelled out in subsection 6.3.2 for the special case of block diagonal covariance matrix $\mathbf{V} = \mathbf{V}(\boldsymbol{\delta})$ and $\hat{\boldsymbol{\delta}} = \hat{\boldsymbol{\delta}}_{\mathrm{ML}}$.

Prasad and Rao (1990) derived the MSE estimator (6.2.36) for special cases covered by the general linear mixed model with a block diagonal covariance structure (Section 6.3 and Chapter 7). Following Prasad and Rao (1990), Harville and Jeske (1992) proposed (6.2.36) for the general linear mixed model (6.2.1), assuming $E(\hat{\boldsymbol{\delta}}) = \boldsymbol{\delta}$, and referred to (6.2.36) as the Prasad-Rao estimator.

Das, Jiang and Rao (2001) provide rigorous proofs of the approximations (6.2.36) and (6.2.37) for the REML and ML methods, respectively.

6.2.7 Software

PROC MIXED (SAS/STAT Users Guide (1999), Chapter 41) implements ML and REML estimation of model parameters $\boldsymbol{\beta}$ and $\boldsymbol{\delta}$ for the general linear mixed model, using the Newton-Raphson method. The inverse of the observed information matrix is used to estimate the covariance matrix of parameter estimates $\hat{\boldsymbol{\beta}}$ and $\hat{\boldsymbol{\delta}}$. The SCORING option in PROC MIXED uses Fisher's scoring method to estimate $\boldsymbol{\beta}$ and $\boldsymbol{\delta}$, and the inverse of the expected information matrix to estimate the covariance matrix of parameter estimates. PROC MIXED also gives the EBLUP estimate $\hat{\mu} = t(\hat{\boldsymbol{\delta}}, \mathbf{y})$ of a specified $\mu = l^T\boldsymbol{\beta} + \mathbf{m}^T\mathbf{v}$ along with the naive MSE estimate, $\mathrm{mse}_N(\hat{\mu})$, given by (6.2.32). An option called DDFM=KENWARDROGER also computes the nearly unbiased MSE estimate (6.2.36), using the estimated covariance matrix of $\hat{\boldsymbol{\beta}}$ and $\hat{\boldsymbol{\delta}}$ based on the observed information matrix.

PROC MIXED also implements model selection using Akaike's information criterion (AIC) and Schwarz's Bayes criterion (BIC). The AIC based on REML is given by $\mathrm{AIC} = l_R(\hat{\boldsymbol{\delta}}_{\mathrm{RE}}) - q^*$, where $l_R(\hat{\boldsymbol{\delta}}_{\mathrm{RE}})$ is the restricted log likelihood function evaluated at $\hat{\boldsymbol{\delta}} = \hat{\boldsymbol{\delta}}_{\mathrm{RE}}$ and q^* = effective number of variance parameters $\boldsymbol{\delta}$ (that is, those not estimated to be on a boundary constraint). If ML is used, then $l_R(\hat{\boldsymbol{\delta}}_{\text{RE}})$ is replaced by $l(\hat{\boldsymbol{\beta}}_{\text{ML}}, \hat{\boldsymbol{\delta}}_{\text{ML}})$ in the AIC criterion. The AIC compares models with the same fixed effects but different random effects. Using AIC, models with larger AIC–values are preferred. The BIC based on REML is given by BIC$= l_R(\hat{\boldsymbol{\delta}}_{\text{RE}}) - \frac{1}{2}q^* \log N^*$, where $n^* = n - p$. If ML is used, then $l_R(\hat{\boldsymbol{\delta}}_{\text{RE}})$ and N^* are replaced by $l(\hat{\boldsymbol{\beta}}_{\text{ML}}, \hat{\boldsymbol{\delta}}_{\text{ML}})$ and n, respectively in the BIC. The BIC permits comparison of models with different random effects as well as fixed effects, unlike the AIC. Using the BIC, models with larger BIC-values are preferred. Note that the BIC penalizes models with a greater number of variance parameters $\boldsymbol{\delta}$ more than the AIC does. As a result, the two criteria may lead to different models. PROC MIXED warns against the mechanical use of model selection criteria by noting that "subject matter considerations and objectives are of great importance when selecting a model."

The S-PLUS function lme (S-PLUS 6 for Windows Guide to Statistics (2001)) fits general linear mixed models. It, like SAS PROC MIXED, uses ML or REML to estimate the regression parameters, $\boldsymbol{\beta}$, and the variance parameters, $\boldsymbol{\delta}$, and it provides values for the AIC and BIC for the model fit. In lme, the within-group (or area) errors can have unequal variances or be correlated. Naive MSE estimates for the EBLUP can be calculated from information returned by the function. A number of diagnostic plots are available to check model assumptions, including plots of within-area residuals (differences between the observed values and the EBLUPs) and normal probability plots of the estimated random effects. More information on lme is given in Pinheiro and Bates (2000).

Jiang and J.S. Rao (2002) studied the selection of fixed and random effects for the general linear mixed model (6.2.1). They studied two cases: (i) selection of fixed covariates, $\mathbf{X}$, from a set of candidate covariates when the random effects are not subject to selection; (ii) selection of both covariates (or fixed effects) and random effects. Normality of the random effects, $\mathbf{v}$, and errors, $\mathbf{e}$, is not assumed. For case (i), suppose $\mathbf{x}_1, \ldots, \mathbf{x}_q$ denote the $n \times 1$ candidate vectors of covariates from which the columns of the $n \times p$ matrix $\mathbf{X}$ are to be selected ($p \leq q$). Let $\mathbf{X}(a)$ denote the matrix for a subset, a, of the q column vectors, and let $\hat{\boldsymbol{\beta}}(a) = [\mathbf{X}(a)^T\mathbf{X}(a)]^{-1}\mathbf{X}(a)^T\mathbf{y}$ be the ordinary least squares estimator of $\boldsymbol{\beta}(a)$ for the model $\mathbf{y} = \mathbf{X}(a)\boldsymbol{\beta}(a) + \zeta$ with $\zeta = \mathbf{Z}\mathbf{v} + \mathbf{e}$. A generalized information criterion (GIC) for the selection of covariates is then given by

$$C_n(a) = (\mathbf{y} - \mathbf{X}(a)\hat{\boldsymbol{\beta}}(a))^T(\mathbf{y} - \mathbf{X}(a)\hat{\boldsymbol{\beta}}(a)) + \lambda_n|a|,$$

where $|a|$ is the cardinality of the subset a and λ_n is a positive number satisfying certain conditions. GIC selects the subset $\hat{a}$ that minimizes $C_n(a)$ over all (or specified) subsets of the q column vectors. Jiang and J.S. Rao (2002) proved that GIC is consistent in the sense of $P(\hat{a} \neq a_0) \to 0$ as $n \to \infty$, where a_0 is the true subset (or model). Note that $C_n(a)$ does not require estimates of variance components and uses only the least squares estimates $\hat{\boldsymbol{\beta}}(a)$. This feature is computationally attractive because the number of candidate models can be very large.

The AIC criterion uses $\lambda_n = 2$, while the BIC uses $\lambda_n = \log n$. If the true underlying model is based on a single random effect, as in the case of the basic area level model (5.2.5) and the basic unit level model (5.3.4), the BIC choice, $\lambda_n = \log n$, satisfies the conditions needed for consistency, but not the AIC choice $\lambda_n = 2$.

GIC does not work for the case (ii) involving the selection of both fixed and random effects. Jiang and J.S. Rao (2002) proposed an alternative method for the general ANOVA model (6.2.15). This method divides the random effect factors into several groups and then applies different model selection procedures for different groups simultaneously.

6.3 Block Diagonal Covariance Structure

6.3.1 EBLUP Estimator

A special case of the general linear mixed model (6.2.1) covers many small area models considered in the literature. For this model

$$\mathbf{y} = \text{col}_{1\le i\le m}(\mathbf{y}_i) = (\mathbf{y}_1^T, \ldots, \mathbf{y}_m^T)^T, \quad \mathbf{X} = \text{col}_{1\le i\le m}(\mathbf{X}_i),$$
$$\mathbf{Z} = \text{diag}_{1\le i\le m}(\mathbf{Z}_i), \quad \mathbf{v} = \text{col}_{1\le i\le m}(\mathbf{v}_i), \quad \mathbf{e} = \text{col}_{1\le i\le m}(\mathbf{e}_i),$$

where m is the number of small areas, $\mathbf{X}_i$ is $n_i \times p$, $\mathbf{Z}_i$ is $n_i \times h_i$ and $\mathbf{y}_i$ is an $n_i \times 1$ vector $(\sum n_i = n, \sum h_i = h)$. Further,

$$\mathbf{R} = \text{diag}_{1\le i\le m}(\mathbf{R}_i), \quad \mathbf{G} = \text{diag}_{1\le i\le m}(\mathbf{G}_i)$$

so that $\mathbf{V}$ has a block diagonal structure:

$$\mathbf{V} = \text{diag}_{1\le i\le m}(\mathbf{V}_i)$$

with

$$\mathbf{V}_i = \mathbf{R}_i + \mathbf{Z}_i\mathbf{G}_i\mathbf{Z}_i^T.$$

The model, therefore, may be decomposed into m submodels

$$\mathbf{y}_i = \mathbf{X}_i\boldsymbol{\beta} + \mathbf{Z}_i\mathbf{v}_i + \mathbf{e}_i, \quad i = 1, \ldots, m. \tag{6.3.1}$$

We are interested in estimating linear combinations $\mu_i = \mathbf{l}_i^T\boldsymbol{\beta} + \mathbf{m}_i^T\mathbf{v}_i$, $i = 1, \ldots, m$.

It follows from (6.2.4) that the BLUP estimator of μ_i reduces to

$$\tilde{\mu}_i^H = t_i(\boldsymbol{\delta}, \mathbf{y}_i) = \mathbf{l}_i^T\tilde{\boldsymbol{\beta}} + \mathbf{m}_i^T\tilde{\mathbf{v}}_i, \tag{6.3.2}$$

where

$$\tilde{\mathbf{v}}_i = \mathbf{G}_i\mathbf{Z}_i^T\mathbf{V}_i^{-1}(\mathbf{y}_i - \mathbf{X}_i\tilde{\boldsymbol{\beta}}), \tag{6.3.3}$$

and

$$\tilde{\beta} = \left(\sum_i \mathbf{X}_i^T \mathbf{V}_i^{-1} \mathbf{X}_i\right)^{-1} \left(\sum_i \mathbf{X}_i^T \mathbf{V}_i^{-1} \mathbf{y}_i\right). \tag{6.3.4}$$

The MSE of the BLUP estimator, given by (6.2.11), reduces to

$$\text{MSE}(\tilde{\mu}_i^H) = g_{1i}(\boldsymbol{\delta}) + g_{2i}(\boldsymbol{\delta}) \tag{6.3.5}$$

with

$$g_{1i}(\boldsymbol{\delta}) = \mathbf{m}_i^T(\mathbf{G}_i - \mathbf{G}_i \mathbf{Z}_i^T \mathbf{V}_i^{-1} \mathbf{Z}_i \mathbf{G}_i)\mathbf{m}_i \tag{6.3.6}$$

and

$$g_{2i}(\boldsymbol{\delta}) = \mathbf{d}_i^T \left(\sum_i \mathbf{X}_i^T \mathbf{V}_i^{-1} \mathbf{X}_i\right)^{-1} \mathbf{d}_i, \tag{6.3.7}$$

where

$$\mathbf{d}_i^T = \mathbf{l}_i^T - \mathbf{b}_i^T \mathbf{X}_i$$

with

$$\mathbf{b}_i^T = \mathbf{m}_i^T \mathbf{G}_i \mathbf{Z}_i^T \mathbf{V}_i^{-1}.$$

Replacing $\boldsymbol{\delta}$ by an estimator of $\hat{\boldsymbol{\delta}}$ in (6.3.2), we get the EBLUP estimator

$$\hat{\mu}_i^H = t_i(\hat{\boldsymbol{\delta}}, \mathbf{y}_i) = \mathbf{l}_i^T \hat{\boldsymbol{\beta}} + \mathbf{m}_i^T \hat{\mathbf{v}}_i. \tag{6.3.8}$$

6.3.2 Estimation of MSE

The second-order MSE approximation (6.2.31) reduces to

$$\text{MSE}(\hat{\mu}_i^H) \approx g_{1i}(\boldsymbol{\delta}) + g_{2i}(\boldsymbol{\delta}) + g_{3i}(\boldsymbol{\delta}) \tag{6.3.9}$$

with

$$g_{3i}(\boldsymbol{\delta}) = \text{tr}[(\partial \mathbf{b}_i^T/\partial \boldsymbol{\delta}) \mathbf{V}_i (\partial \mathbf{b}_i^T/\partial \boldsymbol{\delta})^T \overline{\mathbf{V}}(\hat{\boldsymbol{\delta}})]. \tag{6.3.10}$$

Neglected terms in the approximation (6.3.9) are of order $o(m^{-1})$. The estimator of MSE, given by (6.2.36), reduces to

$$\text{mse}(\hat{\mu}_i^H) \approx g_{1i}(\hat{\boldsymbol{\delta}}) + g_{2i}(\hat{\boldsymbol{\delta}}) + 2\, g_{3i}(\hat{\boldsymbol{\delta}}) \tag{6.3.11}$$

for large m. The estimator of MSE given by (6.2.37) reduces to

$$\text{mse}_*(\hat{\mu}_i^H) = g_{1i}(\hat{\boldsymbol{\delta}}) - \mathbf{b}_{\hat{\boldsymbol{\delta}}}^T(\hat{\boldsymbol{\delta}}) \nabla g_{1i}(\hat{\boldsymbol{\delta}}) + g_{2i}(\hat{\boldsymbol{\delta}}) + 2 g_{3i}(\hat{\boldsymbol{\delta}}) \tag{6.3.12}$$

for large m. If $\hat{\boldsymbol{\delta}} = \hat{\boldsymbol{\delta}}_{\text{ML}}$, then

$$\mathbf{b}_{\hat{\boldsymbol{\delta}}_{\text{ML}}}(\boldsymbol{\delta}) = \frac{1}{2m}\left\{\mathcal{I}^{-1}(\boldsymbol{\delta})\ \text{col}_{1\le j\le m}\ \text{tr}\left[\sum_i (\mathbf{X}_i^T\mathbf{V}_i^{-1}\mathbf{X}_i)^{-1}\left(\sum_i \mathbf{X}_i^T\mathbf{V}_i^{(j)}\mathbf{X}_i\right)\right]\right\} \tag{6.3.13}$$

with

$$\mathbf{V}_i^{(j)} = \partial\mathbf{V}_i^{-1}/\partial\delta_j = -\mathbf{V}_i^{-1}(\partial\mathbf{V}_i/\partial\delta_j)\mathbf{V}_i^{-1}$$

and

$$\mathcal{I}_{jk}(\boldsymbol{\delta}) = \frac{1}{2}\sum_i \text{tr}(\mathbf{V}_i^{-1}\partial\mathbf{V}_i/\partial\delta_j)(\mathbf{V}_i^{-1}\partial\mathbf{V}_i/\partial\delta_k).$$

For ML and REML, Datta and Lahiri (2000) justified the second-order approximation (6.3.9) to MSE($\hat{\mu}_i^H$) and the estimators of MSE, (6.3.11) and (6.3.12), under the following regularity conditions:

(i) The elements of $\mathbf{X}_i$ and $\mathbf{Z}_i$ are uniformly bounded such that $\sum_i \mathbf{X}_i^T\mathbf{V}_i^{-1}\mathbf{X}_i = O(m)$.

(ii) $\sup_{i\ge 1} n_i < \infty$ and $\sup_{i\ge 1} h_i < \infty$, where h_i is the number of columns in $\mathbf{Z}_i$.

(iii) Covariance matrices $\mathbf{G}_i$ and $\mathbf{R}_i$ have linear structures of the form $\mathbf{G}_i = \sum_{j=0}^q \delta_j\mathbf{A}_{ij}\mathbf{A}_{ij}^T$ and $\mathbf{R}_i = \sum_{j=0}^q \delta_j\mathbf{B}_{ij}\mathbf{B}_{ij}^T$, where $\delta_0 = 1$, $\mathbf{A}_{ij}$ and $\mathbf{B}_{ij}(i = 1,\ldots,m; j = 0,\ldots,q)$ are known matrices of order $n_i \times h_i$ and $h_i \times h_i$ respectively, and the elements of $\mathbf{A}_{ij}$ and $\mathbf{B}_{ij}$ are uniformly bounded such that $\mathbf{G}_i$ and $\mathbf{R}_i$ are positive definite matrices for $i = 1,\ldots,m$.

The neglected terms in the second-order approximation (6.3.9) to MSE($\hat{\mu}_i^H$) are of order $o(m^{-1})$ for large m. The MSE estimators (6.3.11) for REML and (6.3.12) for ML are approximately unbiased in the sense $E[\text{mse}(\hat{\mu}_i^H)] - \text{MSE}(\hat{\mu}_i^H) = o(m^{-1})$.

The MSE estimators (6.3.11) and (6.3.12) are not area-specific in the sense that they do not depend directly on the area-specific data $\mathbf{y}_i$. But it is easy to find other choices, using the form (6.3.10) for $g_{3i}(\boldsymbol{\delta})$, that yield area-specific MSE estimators. For example, noting that $\tilde{\mathbf{V}}_i(\boldsymbol{\delta},\mathbf{y}_i) = (\mathbf{y}_i - \mathbf{X}_i^T\tilde{\boldsymbol{\beta}})(\mathbf{y}_i - \mathbf{X}_i^T\tilde{\boldsymbol{\beta}})^T$ is approximately unbiased for $\mathbf{V}_i$, we get an alternative area-specific estimator of $g_{3i}(\boldsymbol{\delta})$ as $g_{3i}^*(\hat{\boldsymbol{\delta}},\mathbf{y}_i)$, where

$$g_{3i}^*(\boldsymbol{\delta},\mathbf{y}_i) = \text{tr}[(\partial\mathbf{b}_i^T/\partial\boldsymbol{\delta})\tilde{\mathbf{V}}_i(\boldsymbol{\delta},\mathbf{y}_i)(\partial\mathbf{b}_i^T/\partial\boldsymbol{\delta})^T\overline{\mathbf{V}}(\hat{\boldsymbol{\delta}})]. \tag{6.3.14}$$

This choice leads to two alternative area-specific versions of (6.3.11):

$$\text{mse}_1(\hat{\mu}_i^H) \approx g_{1i}(\hat{\boldsymbol{\delta}}) + g_{2i}(\hat{\boldsymbol{\delta}}) + 2g_{3i}^*(\hat{\boldsymbol{\delta}},\mathbf{y}_i) \tag{6.3.15}$$

and

$$\text{mse}_2(\hat{\mu}_i^H) \approx g_{1i}(\hat{\boldsymbol{\delta}}) + g_{2i}(\hat{\boldsymbol{\delta}}) + g_{3i}(\hat{\boldsymbol{\delta}}) + g_{3i}^*(\hat{\boldsymbol{\delta}},\mathbf{y}_i). \tag{6.3.16}$$

Similarly, area-specific versions of (6.3.12) are given by

$$\text{mse}_{*1}(\hat{\mu}_i^H) \approx g_{1i}(\hat{\boldsymbol{\delta}}) - \mathbf{b}_{\hat{\boldsymbol{\delta}}}^T(\boldsymbol{\delta})\nabla g_{1i}(\hat{\boldsymbol{\delta}}) + g_{2i}(\hat{\boldsymbol{\delta}}) + 2g_{3i}^*(\hat{\boldsymbol{\delta}}, \mathbf{y}_i) \tag{6.3.17}$$

and

$$\text{mse}_{*2}(\hat{\mu}_i^H) \approx g_{1i}(\hat{\boldsymbol{\delta}}) - \mathbf{b}_{\hat{\boldsymbol{\delta}}}^T(\boldsymbol{\delta})\nabla g_{1i}(\hat{\boldsymbol{\delta}}) + g_{2i}(\hat{\boldsymbol{\delta}}) + g_{3i}(\hat{\boldsymbol{\delta}}) + g_{3i}^*(\hat{\boldsymbol{\delta}}, \mathbf{y}_i). \tag{6.3.18}$$

The use of the area-specific matrix $\tilde{\mathbf{V}}_i(\boldsymbol{\delta}, \mathbf{y}_i)$, based on the residuals $\mathbf{y}_i - \mathbf{X}_i\tilde{\boldsymbol{\beta}}$, induces instability in the MSE estimators, but its effect should be small because only the $O(m^{-1})$ term, $2g_{3i}(\hat{\boldsymbol{\delta}})$, is changed.

6.3.3 Extension

The BLUP estimator (6.3.2) readily extends to the vector case $\boldsymbol{\mu}_i = \mathbf{L}_i\boldsymbol{\beta} + \mathbf{M}_i\mathbf{v}_i$ for specified matrices $\mathbf{L}_i$ and $\mathbf{M}_i$. It is given by

$$\begin{aligned}\tilde{\boldsymbol{\mu}}_i^H &= \mathbf{t}_i(\boldsymbol{\delta}, \mathbf{y}) = \mathbf{L}_i\tilde{\boldsymbol{\beta}} + \mathbf{M}_i\tilde{\mathbf{v}}_i \\ &= \mathbf{L}_i\tilde{\boldsymbol{\beta}} + \mathbf{M}_i\mathbf{G}_i\mathbf{Z}_i^T\mathbf{V}_i^{-1}(\mathbf{y}_i - \mathbf{X}_i\tilde{\boldsymbol{\beta}}).\end{aligned} \tag{6.3.19}$$

The covariance matrix of $\tilde{\boldsymbol{\mu}}_i^H - \boldsymbol{\mu}_i$ follows from (6.3.5) by changing $\mathbf{l}_i^T$ and $\mathbf{m}_i^T$ to $\mathbf{L}_i$ and $\mathbf{M}_i$, respectively:

$$\begin{aligned}\text{MSE}(\tilde{\boldsymbol{\mu}}_i^H) &= E(\tilde{\boldsymbol{\mu}}_i^H - \boldsymbol{\mu}_i)(\tilde{\boldsymbol{\mu}}_i^H - \boldsymbol{\mu}_i)^T \\ &= \mathbf{M}_i(\mathbf{G}_i - \mathbf{G}_i\mathbf{Z}_i^T\mathbf{V}_i^{-1}\mathbf{Z}_i\mathbf{G}_i)\mathbf{M}_i^T \\ &\quad + \mathbf{D}_i\Big(\sum_i \mathbf{X}_i^T\mathbf{V}_i^{-1}\mathbf{X}_i\Big)^{-1}\mathbf{D}_i^T,\end{aligned} \tag{6.3.20}$$

where

$$\mathbf{D}_i = \mathbf{L}_i - \mathbf{M}_i\mathbf{G}_i\mathbf{Z}_i^T\mathbf{V}_i^{-1}\mathbf{X}_i.$$

EBLUP estimator is given by $\hat{\boldsymbol{\mu}}_i^H = \mathbf{t}_i(\hat{\boldsymbol{\delta}}, \mathbf{y})$. An estimator of $\text{MSE}(\hat{\boldsymbol{\mu}}_i^H)$ that accounts for the estimation of $\boldsymbol{\delta}$ may be obtained along the lines of subsection 6.3.2, but details are omitted here for simplicity.

6.3.4 Model Diagnostics

In this subsection, we give a brief account of some model diagnostics for the linear mixed model (6.3.1) with block diagonal covariance structure.

(i) *Transformation Method*

Let $\mathbf{V}_i = \sigma_e^2\mathbf{A}_i$ where σ_e^2 denotes the error variance. Using $\mathbf{A}_i$, we transform the model (6.3.1) to $\tilde{\mathbf{y}}_i = \tilde{\mathbf{X}}_i\boldsymbol{\beta} + \tilde{\mathbf{e}}_i$, $i = 1, \ldots, m$, where $\tilde{\mathbf{y}}_i = \mathbf{A}_i^{-1/2}\mathbf{y}_i$, $\tilde{\mathbf{X}}_i = \mathbf{A}_i^{-1/2}\mathbf{X}_i$ and $\tilde{\mathbf{e}}_i = \mathbf{A}_i^{-1/2}(\mathbf{Z}_i\mathbf{v}_i + \mathbf{e}_i)$. We may express the transformed model as a standard linear regression model, $\tilde{\mathbf{y}} = \tilde{\mathbf{X}}\boldsymbol{\beta} + \tilde{\mathbf{e}}$ with iid errors $\tilde{\mathbf{e}}$, that is, $\tilde{\mathbf{e}}$ has mean $\mathbf{0}$ and covariance matrix $\sigma_e^2\mathbf{I}_n$. Therefore, we can apply standard linear regression methods for model selection and validation such as the selection of fixed effects, residual analysis and influence analysis. In practice, we replace the unknown variance components in $\mathbf{A}_i$ by suitable estimates. The transformation method looks appealing, but it focuses only on the fixed effects; random effects are also of interest in practice. Moreover, the method is sensitive to the estimates of variance components used in $\mathbf{A}_i$.

(ii) *Influence diagnostics*

Cook's (1977) distance is widely used in standard linear regression to study the effect of deleting a "case" on the estimate of $\boldsymbol{\beta}$. Banerjee and Frees (1997) extended Cook's distance to the model (6.3.1) with block diagonal covariance structure to study the effect of deleting a block (small area) on the estimate of $\boldsymbol{\beta}$.

Let $\hat{\boldsymbol{\beta}}_{(i)}$ be the estimator of $\boldsymbol{\beta}$ after dropping block i. The influence of block i on the estimate $\hat{\boldsymbol{\beta}} = (\Sigma_i \mathbf{X}_i^T \hat{\mathbf{V}}_i^{-1}\mathbf{X}_i)^{-1}(\Sigma_i \mathbf{X}_i^T \hat{\mathbf{V}}_i^{-1}\mathbf{y}_i)$ may be measured by a Mahalanobis distance given by

$$B_i(\hat{\boldsymbol{\beta}}) = (\hat{\boldsymbol{\beta}} - \hat{\boldsymbol{\beta}}_{(i)})^T(\Sigma_i \mathbf{X}_i^T \hat{\mathbf{V}}_i^{-1}\mathbf{X}_i)(\hat{\boldsymbol{\beta}} - \hat{\boldsymbol{\beta}}_{(i)})/p. \tag{6.3.21}$$

Note that $B_i(\hat{\boldsymbol{\beta}})$ is the squared distance from $\hat{\boldsymbol{\beta}}$ to $\hat{\boldsymbol{\beta}}_{(i)}$ relative to the estimated covariance matrix, $(\Sigma_i \mathbf{X}_i^T \hat{\mathbf{V}}_i^{-1}\mathbf{X}_i)^{-1}$, of $\hat{\boldsymbol{\beta}}$. The measure $B_i(\hat{\boldsymbol{\beta}})$ may be simplified to

$$B_i(\hat{\boldsymbol{\beta}}) = (\mathbf{y}_i - \mathbf{X}_i\hat{\boldsymbol{\beta}})^T(\hat{\mathbf{V}}_i - \mathbf{H}_i)^{-1}\mathbf{H}_i(\hat{\mathbf{V}}_i - \mathbf{H}_i)^{-1}(\mathbf{y}_i - \mathbf{X}_i\hat{\boldsymbol{\beta}})/p, \tag{6.3.22}$$

where

$$\mathbf{H}_i = \mathbf{X}_i(\Sigma_i \mathbf{X}_i^T \hat{\mathbf{V}}_i^{-1}\mathbf{X}_i)^{-1}\mathbf{X}_i^T. \tag{6.3.23}$$

A dot plot of the measure $B_i(\hat{\boldsymbol{\beta}})$ for $i = 1, \ldots, m$ is useful in identifying influential blocks (or small areas). Note that we are assuming that $\mathbf{V}_i$ is correctly specified.

Christiansen, Pearson and Johnson (1992) studied case deletion diagnostics for mixed ANOVA models. They extended Cook's distance to measure influence on the fixed effects as well as the variance components. Computational procedures are also provided. These measures are useful for identifying influential observations (or cases).

Cook (1986) developed a "local" influence approach to diagnostics for assessing how slight perturbations to the model can influence ML inferences. Beckman, Nachtsheim and Cook (1987) applied this approach to linear mixed ANOVA models. Let $\boldsymbol{\omega}$ be a q-vector of perturbations and $l_{\boldsymbol{\omega}}(\boldsymbol{\beta}, \boldsymbol{\delta})$ be the corresponding log likelihood function. For example, to check the assumption of homogeneity of error variances in the ANOVA model, that is, $\mathbf{R} = \sigma_e^2\mathbf{I}_n$, we introduce perturbation of the form $\mathbf{R}_{\boldsymbol{\omega}} = \sigma_e^2\mathbf{D}(\boldsymbol{\omega})$, where $\mathbf{D}(\boldsymbol{\omega}) = \text{diag}(\omega_1, \ldots, \omega_n)$. We denote the "null" perturbations that yield the original model by $\boldsymbol{\omega}_0$ so that $l_{\boldsymbol{\omega}_0}(\boldsymbol{\beta}, \boldsymbol{\delta}) = l(\boldsymbol{\beta}, \boldsymbol{\delta})$, the original log likelihood function. Note that $\boldsymbol{\omega}_0 = \mathbf{1}$ in the above example.

The influence of the perturbation $\boldsymbol{\omega}$ can be assessed by using the "likelihood displacement"

$$\text{LD}(\boldsymbol{\omega}) = 2\left[l(\hat{\boldsymbol{\beta}}, \hat{\boldsymbol{\delta}}) - l(\hat{\boldsymbol{\beta}}_{\boldsymbol{\omega}}, \hat{\boldsymbol{\delta}}_{\boldsymbol{\omega}})\right], \tag{6.3.24}$$

where $(\hat{\boldsymbol{\beta}}_{\boldsymbol{\omega}}, \hat{\boldsymbol{\delta}}_{\boldsymbol{\omega}})$ are the ML estimators of $\boldsymbol{\beta}$ and $\boldsymbol{\delta}$ under the perturbed model. Note that $\text{LD}(\boldsymbol{\omega})$ is nonnegative and achieves its minimum value of 0 when $\boldsymbol{\omega} = \boldsymbol{\omega}_0$. Large values of $\text{LD}(\boldsymbol{\omega})$ indicate that $(\hat{\boldsymbol{\beta}}_{\boldsymbol{\omega}}, \hat{\boldsymbol{\delta}}_{\boldsymbol{\omega}})$ differ considerably from $(\hat{\boldsymbol{\beta}}, \hat{\boldsymbol{\delta}})$ relative to the contours of the unperturbed log likelihood $l(\boldsymbol{\beta}, \boldsymbol{\delta})$. Rather than calculating LD($\boldsymbol{\omega}$), we examine the behavior of LD($\boldsymbol{\omega}$) as a function of $\boldsymbol{\omega}$ for values that are "local" to $\boldsymbol{\omega}_0$; in particular, we examine curvatures of LD($\boldsymbol{\omega}$). Let $\boldsymbol{\omega}_0 + a\boldsymbol{\nu}$ be a vector through $\boldsymbol{\omega}_0$ in the direction $\boldsymbol{\nu}$, then the "normal curvature" of the surface $(\boldsymbol{\omega}^T, \text{LD}(\boldsymbol{\omega}))$ in the direction of $\boldsymbol{\nu}$ is given by

$$C_{\boldsymbol{\nu}} = \partial^2 \text{LD}(\boldsymbol{\omega}_0 + a\boldsymbol{\nu})/\partial a^2|_{a=0}. \tag{6.3.25}$$

Large values of $C_{\boldsymbol{\nu}}$ indicate sensitivity to the induced perturbations in the direction $\boldsymbol{\nu}$. We find the largest curvature $C_{\max}$ and its corresponding direction $\boldsymbol{\nu}_{\max}$. A plot of the elements of the normalized vector $\boldsymbol{\nu}_{\max}/||\boldsymbol{\nu}_{\max}||$ against the indices $1, \ldots, q$ is useful for identifying influential perturbations; large elements indicate influential perturbations. We refer to Beckman et al. (1987) for computational details. Hartless, Booth and Littell (2000) applied the above method to the county crop areas data of Example 5.3.1. By perturbing the error variances, that is, using $\mathbf{R}_{\boldsymbol{\omega}} = \sigma_e^2 \mathbf{D}(\boldsymbol{\omega})$, they identified an erroneous observation in the original data (see Example 7.2.1 of Chapter 7).

6.4 Proofs

6.4.1 Derivation of BLUP

A linear estimator $\hat{\mu} = \mathbf{a}^T\mathbf{y} + b$ is unbiased for $\mu = \mathbf{l}^T\boldsymbol{\beta} + \mathbf{m}^T\mathbf{v}$ under the linear mixed model, that is, $E(\hat{\mu}) = E(\mu)$, if and only if $\mathbf{a}^T\mathbf{X} = \mathbf{l}^T$ and $b = 0$. The MSE of a linear unbiased estimator $\hat{\mu}$ is given by

$$\text{MSE}(\hat{\mu}) = V(\hat{\mu} - \mu) = \mathbf{a}^T\mathbf{V}\mathbf{a} - 2\mathbf{a}^T\mathbf{Z}\mathbf{G}\mathbf{m} + \mathbf{m}^T\mathbf{G}\mathbf{m}.$$

We now minimize $V(\hat{\mu} - \mu)$ subject to the unbiasedness condition $\mathbf{a}^T\mathbf{X} = \mathbf{l}^T$ using Lagrange multipliers $2\boldsymbol{\lambda}$. We get

$$\mathbf{V}\mathbf{a} + \mathbf{X}\boldsymbol{\lambda} = \mathbf{Z}\mathbf{G}\mathbf{m}$$

or

$$\mathbf{a} = -\mathbf{V}^{-1}\mathbf{X}\boldsymbol{\lambda} + \mathbf{V}^{-1}\mathbf{Z}\mathbf{G}\mathbf{m}. \tag{6.4.1}$$

Substituting (6.4.1) for $\mathbf{a}$ in the constraint $\mathbf{a}^T\mathbf{X} = \mathbf{l}^T$, we solve for $\boldsymbol{\lambda}$:

$$\boldsymbol{\lambda} = -(\mathbf{X}^T\mathbf{V}^{-1}\mathbf{X})^{-1}\mathbf{l} + (\mathbf{X}^T\mathbf{V}^{-1}\mathbf{X})^{-1}\mathbf{X}^T\mathbf{V}^{-1}\mathbf{Z}\mathbf{G}\mathbf{m}. \qquad (6.4.2)$$

Again, substituting (6.4.2) for $\boldsymbol{\lambda}$ in (6.4.1), we obtain

$$\mathbf{a}^T\mathbf{y} = \mathbf{l}^T\tilde{\boldsymbol{\beta}} + \mathbf{m}^T\mathbf{G}\mathbf{Z}^T\mathbf{V}^{-1}(\mathbf{y} - \mathbf{X}\tilde{\boldsymbol{\beta}}),$$

which is identical to the BLUP given by (6.2.4), where $\tilde{\boldsymbol{\beta}} = (\mathbf{X}^T\mathbf{V}^{-1}\mathbf{X})^{-1}$ $\times\mathbf{X}^T\mathbf{V}^{-1}\mathbf{y}$.

6.4.2 Equivalence of BLUP and Best Predictor $E(\mathbf{m}^T\mathbf{v}|\mathbf{A}^T\mathbf{y})$

For the transformed data $\mathbf{A}^T\mathbf{y}$ with $\mathbf{A}^T\mathbf{X} = \mathbf{0}$, we show here that the best predictor, $E(\mathbf{m}^T\mathbf{v}|\mathbf{A}^T\mathbf{y})$, of $\mathbf{m}^T\mathbf{v}$ reduces to the BLUP estimator $\mathbf{m}^T\tilde{\mathbf{v}}$, where $\tilde{\mathbf{v}}$ is given by (6.2.6). Under normality, it is easy to verify that the conditional distribution of $\mathbf{m}^T\mathbf{v}$ given $\mathbf{A}^T\mathbf{y}$ is normal with mean $E(\mathbf{m}^T\mathbf{v}|\mathbf{A}^T\mathbf{y}) = \mathbf{C}\mathbf{A}(\mathbf{A}^T\mathbf{V}\mathbf{A})^{-1}\mathbf{A}^T\mathbf{y}$, where $\mathbf{C} = \mathbf{m}^T\mathbf{G}\mathbf{Z}^T$ and $\mathbf{A}^T\mathbf{X} = 0$. Since $\mathbf{V}^{1/2}\mathbf{A}$ and $\mathbf{V}^{-1/2}\mathbf{X}$ are orthogonal to each other, and $\text{rank}(\mathbf{V}^{1/2}\mathbf{A}) + \text{rank}(\mathbf{V}^{-1/2}\mathbf{X}) = n$, the following decomposition of projections holds:

$$\mathbf{I} = \mathbf{P}_{l(\mathbf{V}^{1/2}\mathbf{A})} + \mathbf{P}_{l(\mathbf{V}^{-1/2}\mathbf{X})},$$

where $l(\mathbf{B})$ denotes the linear space spanned by the columns of $\mathbf{B}$, and $\mathbf{P}$ denotes the projection. Hence

$$\mathbf{I} = \mathbf{V}^{1/2}\mathbf{A}(\mathbf{A}^T\mathbf{V}\mathbf{A})^{-1}\mathbf{A}^T\mathbf{V}^{1/2} + \mathbf{V}^{-1/2}\mathbf{X}(\mathbf{X}^T\mathbf{V}^{-1}\mathbf{X})^{-1}\mathbf{X}^T\mathbf{V}^{-1/2}. \qquad (6.4.3)$$

It now follows from (6.4.3) that

$$\begin{aligned} &E(\mathbf{m}^T\mathbf{v}|\mathbf{A}^T\mathbf{y}) \\ &= \mathbf{C}\left[\mathbf{V}^{-1} - \mathbf{V}^{-1}\mathbf{X}(\mathbf{X}^T\mathbf{V}^{-1}\mathbf{X})^{-1}\mathbf{X}^T\mathbf{V}^{-1}\right]\mathbf{y} = \mathbf{C}\mathbf{V}^{-1}(\mathbf{y} - \mathbf{X}\tilde{\boldsymbol{\beta}}), \end{aligned}$$

which is equal to the BLUP estimator $\mathbf{m}^T\tilde{\mathbf{v}}$. This proof is due to Jiang (1997).

6.4.3 Derivation of the Decomposition (6.2.26)

The result of subsection 6.4.2 may be used to provide a simple proof of the MSE decomposition (6.2.26): $\text{MSE}[t(\hat{\boldsymbol{\delta}})] = \text{MSE}[t(\boldsymbol{\delta})] + E[t(\hat{\boldsymbol{\delta}}) - t(\boldsymbol{\delta})]^2$, where $\hat{\boldsymbol{\delta}}$ is a function of $\mathbf{A}^T\mathbf{y}$. Write $\mathbf{V}(\hat{\boldsymbol{\delta}}) = \hat{\mathbf{V}}$, $\hat{\boldsymbol{\beta}} = \tilde{\boldsymbol{\beta}}(\hat{\boldsymbol{\delta}})$ and $\tilde{\mathbf{v}}(\hat{\boldsymbol{\delta}}) = \hat{\mathbf{v}}$. Also, write $t(\hat{\boldsymbol{\delta}}) - \mu = t(\hat{\boldsymbol{\delta}}) - t(\boldsymbol{\delta}) + t(\boldsymbol{\delta}) - \mu$. We first note that $t(\hat{\boldsymbol{\delta}}) - t(\boldsymbol{\delta}) = \mathbf{l}'(\hat{\boldsymbol{\beta}} - \tilde{\boldsymbol{\beta}}) + \mathbf{m}^T(\hat{\mathbf{v}} - \tilde{\mathbf{v}})$ is a function of $\mathbf{A}^T\mathbf{y}$. This follows by writing

$$\hat{\boldsymbol{\beta}} - \tilde{\boldsymbol{\beta}} = \left[(\mathbf{X}^T\hat{\mathbf{V}}^{-1}\mathbf{X})^{-1}\mathbf{X}\hat{\mathbf{V}}^{-1} - (\mathbf{X}^T\mathbf{V}^{-1}\mathbf{X})^{-1}\mathbf{X}^T\mathbf{V}^{-1}\right](\mathbf{y} - \mathbf{X}\tilde{\boldsymbol{\beta}})$$

and noting that both $\hat{\mathbf{V}}$ and $\mathbf{y} - \mathbf{X}\tilde{\boldsymbol{\beta}}$ are functions of $\mathbf{A}^T\mathbf{y}$. Similarly, $\hat{\mathbf{v}} - \tilde{\mathbf{v}}$ also depends only on $\mathbf{A}^T\mathbf{y}$. Further, the first term on the right-hand side of

$t(\boldsymbol{\delta}) - \mu = \mathbf{l}^T(\tilde{\boldsymbol{\beta}} - \boldsymbol{\beta}) + \mathbf{m}^T(\tilde{\mathbf{v}} - \mathbf{v})$ is independent of $\mathbf{A}^T\mathbf{y}$ because of the condition $\mathbf{A}^T\mathbf{X} = 0$, and the last term equals $\mathbf{m}^T[E(\mathbf{v}|\mathbf{A}^T\mathbf{y}) - \mathbf{v}]$, using the result of subsection 6.4.2. Hence, we have

$$\begin{aligned} \text{MSE}[t(\hat{\boldsymbol{\delta}})] &= \text{MSE}[t(\boldsymbol{\delta})] + E[t(\hat{\boldsymbol{\delta}}) - t(\boldsymbol{\delta})]^2 + 2E\left[(t(\hat{\boldsymbol{\delta}}) - t(\boldsymbol{\delta}))\mathbf{l}^T E(\tilde{\boldsymbol{\beta}} - \boldsymbol{\beta}|\mathbf{A}^T\mathbf{y})\right] \\ &\quad + 2E\left[(t(\hat{\boldsymbol{\delta}}) - t(\boldsymbol{\delta}))\mathbf{m}^T E(E(\mathbf{v}|\mathbf{A}^T\mathbf{y}) - \mathbf{v}|\mathbf{A}^T\mathbf{y})\right] \\ &= \text{MSE}[t(\boldsymbol{\delta})] + E[t(\hat{\boldsymbol{\delta}}) - t(\boldsymbol{\delta})]^2, \end{aligned} \tag{6.4.4}$$

noting that $E(\tilde{\boldsymbol{\beta}} - \boldsymbol{\beta}|\mathbf{A}^T\mathbf{y}) = E(\tilde{\boldsymbol{\beta}} - \boldsymbol{\beta}) = 0$.

The above proof of (6.4.4) is due to Jiang (2001). Kackar and Harville (1984) gave a somewhat different proof of (6.4.4).

Chapter 7

Empirical Best Linear Unbiased Prediction (EBLUP): Basic Models

We presented the EBLUP theory in Chapter 6 under a general linear mixed model. We also studied the special case of a linear mixed model with a block diagonal covariance structure. This model covers many small area models used in practice. In this chapter, we apply the EBLUP results in Section 6.3 to the basic area level (type A) and the basic unit level (type B) models, described in Sections 5.2 and 5.3 of Chapter 5. EBLUP estimation and the associated MSE estimation are spelled out in Section 7.1 for the basic area level model. Details of the applications in Examples 5.2.1–5.2.3, Chapter 5, are also given. Section 7.2 deals with the basic unit level model, and EBLUP estimation and the associated MSE estimation are presented. Details of the applications in Examples 5.3.1 and 5.3.2, Chapter 5, are also given. Pseudo-EBLUP estimators that depend on the design weights and satisfy the design-consistency property are studied in subsection 7.2.7. Those estimators also satisfy the benchmarking property without any post-adjustment in the sense that they add up automatically to a reliable direct estimator when aggregated over the small areas.

7.1 Basic Area Level Model

In this section we consider the basic area level (type A) model (5.2.5) and spell out EBLUP estimation using the results in Section 6.3, Chapter 6 for the general linear mixed model with block diagonal covariance structure.

7.1.1 BLUP Estimator

The basic area level model is given by

$$\hat{\theta}_i = \mathbf{z}_i^T \boldsymbol{\beta} + b_i v_i + e_i, \ i = 1, \ldots, m, \tag{7.1.1}$$

where $\mathbf{z}_i$ is a $p \times 1$ vector of area level covariates, $v_i \overset{\text{iid}}{\sim} (0, \sigma_v^2)$ and independent of the sampling error $e_i \overset{\text{ind}}{\sim} (0, \psi_i)$ with known variance ψ_i, $\hat{\theta}_i$ is a direct estimator of ith area parameter $\theta_i = g(\bar{Y}_i)$, and b_i is a known positive constant. Model (7.1.1) is a special case of the general linear mixed model with block diagonal covariance structure, given by (6.3.1). We have

$$\mathbf{y}_i = \hat{\theta}_i, \quad \mathbf{X}_i = \mathbf{z}_i^T, \quad \mathbf{Z}_i = b_i$$

and

$$\mathbf{v}_i = v_i, \quad \mathbf{e}_i = e_i, \quad \boldsymbol{\beta} = (\beta_1, \ldots, \beta_p)^T.$$

Further,

$$\mathbf{G}_i = \sigma_v^2, \quad \mathbf{R}_i = \psi_i$$

so that

$$\mathbf{V}_i = \psi_i + \sigma_v^2 b_i^2.$$

Also, $\mu_i = \theta_i = \mathbf{z}_i^T \boldsymbol{\beta} + b_i v_i$ so that $\mathbf{l}_i = \mathbf{z}_i$ and $\mathbf{m}_i = b_i$.

Making the above substitutions in the general formula (6.3.2) for the BLUP estimator of μ_i, we get the BLUP estimator of θ_i as

$$\tilde{\theta}_i^H = \mathbf{z}_i^T \tilde{\boldsymbol{\beta}} + \gamma_i (\hat{\theta}_i - \mathbf{z}_i^T \tilde{\boldsymbol{\beta}}) \tag{7.1.2}$$

$$= \gamma_i \hat{\theta}_i + (1 - \gamma_i) \mathbf{z}_i^T \tilde{\boldsymbol{\beta}}, \tag{7.1.3}$$

where

$$\gamma_i = \sigma_v^2 b_i^2 / (\psi_i + \sigma_v^2 b_i^2) \tag{7.1.4}$$

and

$$\tilde{\boldsymbol{\beta}} = \tilde{\boldsymbol{\beta}}(\sigma_v^2) = \left[\sum_{i=1}^m \mathbf{z}_i \mathbf{z}_i^T / (\psi_i + \sigma_v^2 b_i^2) \right]^{-1} \left[\sum_{i=1}^m \mathbf{z}_i \hat{\theta}_i / (\psi_i + \sigma_v^2 b_i^2) \right].$$

It is clear from (7.1.3) that the BLUP estimator, $\tilde{\theta}_i^H$, can be expressed as a weighted average of the direct estimator $\hat{\theta}_i$ and the regression-synthetic estimator $\mathbf{z}_i^T \tilde{\boldsymbol{\beta}}$, where the weight γ_i $(0 \le \gamma_i \le 1)$, given by (7.1.4), measures the uncertainty in modeling the θ_i's, namely, the model variance $\sigma_v^2 b_i^2$ relative to the total variance $\psi_i + \sigma_v^2 b_i^2$. Thus $\tilde{\theta}_i^H$ takes proper account of the between area variation relative to the precision of the direct estimator. If the model variance $\sigma_v^2 b_i^2$ is relatively small, then γ_i will be small and more weight is attached to the synthetic estimator. Similarly, more weight is attached to the direct estimator if the design variance ψ_i is relatively small or γ_i is large. The form (7.1.2) for $\tilde{\theta}_i^H$ suggests that it adjusts the synthetic estimator $\mathbf{z}_i^T \tilde{\boldsymbol{\beta}}$ to account for model uncertainty.

It is important to note that $\tilde{\theta}_i^H$ is valid for general sampling designs because we are modeling only the $\hat{\theta}_i$'s and not the individual elements in the population, unlike type B models, and the direct estimator $\hat{\theta}_i$ uses the design weights. Further, $\tilde{\theta}_i^H$ is design-consistent because $\gamma_i \to 1$ as the sampling variance $\psi_i \to 0$. The design-bias of $\tilde{\theta}_i^H$ is given by

$$B_p(\tilde{\theta}_i^H) \approx (1-\gamma_i)(\mathbf{z}_i^T \boldsymbol{\beta}^* - \theta_i), \tag{7.1.5}$$

where $\boldsymbol{\beta}^* = E_2(\tilde{\boldsymbol{\beta}})$ is the conditional expectation of $\tilde{\boldsymbol{\beta}}$ given $\boldsymbol{\theta} = (\theta_1, \ldots, \theta_m)^T$. It follows from (7.1.5) that the design-bias relative to θ_i tends to zero as $\psi_i \to 0$ or $\gamma_i \to 1$. Note that $E_m(\mathbf{z}_i^T \boldsymbol{\beta}^*) = E_m(\theta_i)$ so that the average bias is zero when the linking model $\theta_i = \mathbf{z}_i^T \boldsymbol{\beta} + b_i v_i$ holds.

The MSE of the BLUP estimator $\tilde{\theta}_i^H$ is easily obtained either from the general result (6.3.5) or by direct calculation. It is given by

$$\text{MSE}(\tilde{\theta}_i^H) = E(\tilde{\theta}_i^H - \theta_i)^2 = g_{1i}(\sigma_v^2) + g_{2i}(\sigma_v^2), \tag{7.1.6}$$

where

$$g_{1i}(\sigma_v^2) = \sigma_v^2 b_i^2 \psi_i / (\psi_i + \sigma_v^2 b_i^2) = \gamma_i \psi_i \tag{7.1.7}$$

and

$$g_{2i}(\sigma_v^2) = (1-\gamma_i)^2 \mathbf{z}_i^T \left[\sum_i \mathbf{z}_i \mathbf{z}_i^T / (\psi_i + \sigma_v^2 b_i^2) \right]^{-1} \mathbf{z}_i. \tag{7.1.8}$$

The first term, $g_{1i}(\sigma_v^2)$ in (7.1.6), is of order $O(1)$, whereas the second term, $g_{2i}(\sigma_v^2)$, due to estimating $\boldsymbol{\beta}$, is of order $O(m^{-1})$ for large m, assuming the following regularity conditions:

(i) ψ_i and b_i are uniformly bounded. (7.1.9)

(ii) $\sup_i \tilde{h}_{ii} = O(m^{-1})$, where $\tilde{h}_{ii} = \tilde{\mathbf{z}}_i^T \left(\sum_i \tilde{\mathbf{z}}_i \tilde{\mathbf{z}}_i^T \right)^{-1} \tilde{\mathbf{z}}_i$ (7.1.10)

with $\tilde{\mathbf{z}}_i = \mathbf{z}_i / b_i$. Condition (ii) is a standard condition in linear regression analysis (Wu (1986)).

Comparing the leading term $\gamma_i \psi_i$ with ψ_i, the MSE of the direct estimator $\hat{\theta}_i$, it is clear that $\tilde{\theta}_i^H$ leads to large gains in efficiency when γ_i is small, that is, when the variability of the model error $b_i v_i$ is small relative to the total variability. Note that ψ_i is also the design variance of $\hat{\theta}_i$.

The BLUP estimator (7.1.3) depends on the variance component σ_v^2 which is unknown in practical applications. Replacing σ_v^2 by an estimator $\hat{\sigma}_v^2$, we obtain an EBLUP estimator $\hat{\theta}_i^H$:

$$\hat{\theta}_i^H = \hat{\gamma}_i \hat{\theta}_i + (1 - \hat{\gamma}_i)\mathbf{z}_i^T \hat{\boldsymbol{\beta}}, \tag{7.1.11}$$

where $\hat{\gamma}_i$ and $\hat{\boldsymbol{\beta}}$ are the values of γ_i and $\tilde{\boldsymbol{\beta}}$ when σ_v^2 is replaced by $\hat{\sigma}_v^2$.

Fay and Herriot (1979) recommended the use of a compromise EBLUP estimator, $\hat{\theta}_{i*}^H$, similar to the compromise James-Stein (J-S) estimator (4.4.9) in Chapter 4. That is, (i) use $\hat{\theta}_i^H$ if $\hat{\theta}_i^H$ lies in the interval $[\hat{\theta}_i - c\sqrt{\psi_i}, \hat{\theta}_i + c\sqrt{\psi_i}]$ for a specified constant c (typically $c = 1$); (ii) use $\hat{\theta}_i^H - c\sqrt{\psi_i}$ if $\hat{\theta}_i^H$ is less than $\hat{\theta}_i - c\sqrt{\psi_i}$; (iii) use $\hat{\theta}_i^H + c\sqrt{\psi_i}$ if $\hat{\theta}_i^H$ is greater than $\hat{\theta}_i + c\sqrt{\psi_i}$. The compromise estimator $\hat{\theta}_{i*}^H$ (or $\hat{\theta}_i^H$) is transformed back to the original scale to obtain an estimator of the i th area mean $\bar{Y}_i$ as $g^{-1}(\hat{\theta}_{i*}^H)$ (or $g^{-1}(\hat{\theta}_i^H)$).

7.1.2 Estimation of σ_v^2

A method of moments estimator $\hat{\sigma}_{vm}^2$ can be obtained by noting that

$$E\left[\sum_i (\hat{\theta}_i - \mathbf{z}_i^T \tilde{\boldsymbol{\beta}})^2 / (\psi_i + \sigma_v^2 b_i^2)\right] = E[h(\sigma_v^2)] = m - p,$$

where $\tilde{\boldsymbol{\beta}} = \tilde{\boldsymbol{\beta}}(\sigma_v^2)$. It follows that $\hat{\sigma}_{vm}^2$ is obtained by solving

$$h(\sigma_v^2) = m - p$$

iteratively and letting $\hat{\sigma}_{vm}^2 = 0$ when no positive solution exists. Fay and Herriot (1979) suggested the following iterative solution: starting with $\sigma_v^{2(0)} = 0$, define

$$\sigma_v^{2(a+1)} = \sigma_v^{2(a)} + \frac{1}{h_*'(\sigma_v^{2(a)})}[m - p - h(\sigma_v^{2(a)})] \tag{7.1.12}$$

constraining $\sigma_v^{2(a+1)} \geq 0$, where

$$h_*'(\sigma_v^2) = -\sum_i b_i^2 (\hat{\theta}_i - \mathbf{z}_i^T \tilde{\boldsymbol{\beta}})^2 / (\psi_i + \sigma_v^2 b_i^2)^2$$

is an approximation to the derivative of $h(\sigma_v^2)$. Convergence of the iteration (7.1.12) is rapid, generally requiring less than 10 iterations.

Alternatively, a simple moment estimator is given by $\hat{\sigma}_{vs}^2 = \max(\tilde{\sigma}_{vs}^2, 0)$, where

$$\tilde{\sigma}_{vs}^2 = \frac{1}{m-p}\left[\sum_i (b_i^{-1}\hat{\theta}_i - \tilde{\mathbf{z}}_i^T \hat{\boldsymbol{\beta}}_{\mathrm{WLS}})^2 - \sum_i \frac{\psi_i}{b_i^2}(1 - \tilde{h}_{ii})\right] \tag{7.1.13}$$

and

$$\hat{\beta}_{\text{WLS}} = \left(\sum_i \tilde{\mathbf{z}}_i \tilde{\mathbf{z}}_i^T\right)^{-1} \left(\sum_i \tilde{\mathbf{z}}_i \hat{\theta}_i / b_i\right)$$

is a weighted least squares estimator of $\boldsymbol{\beta}$. If $b_i = 1$, then (7.1.13) reduces to the formula of Prasad and Rao (1990). Neither moment estimator of σ_v^2 requires normality, and both lead to consistent estimators as $m \to \infty$.

The scoring algorithm (6.2.17) for ML estimation of σ_v^2 reduces to

$$\sigma_v^{2(a+1)} = \sigma_v^{2(a)} + [\mathcal{I}(\sigma_v^{2(a)})]^{-1} s(\tilde{\boldsymbol{\beta}}^{(a)}, \sigma_v^{2(a)})$$

where

$$\mathcal{I}(\sigma_v^2) = \frac{1}{2} \sum_{i=1}^{m} \frac{b_i^4}{(\sigma_v^2 b_i^2 + \psi_i)^2} \tag{7.1.14}$$

and

$$s(\tilde{\boldsymbol{\beta}}, \sigma_v^2) = -\frac{1}{2} \sum_{i=1}^{m} \frac{b_i^2}{\sigma_v^2 b_i^2 + \psi_i} + \frac{1}{2} \sum_{i=1}^{m} b_i^2 \frac{(\hat{\theta}_i - \mathbf{z}_i^T \tilde{\boldsymbol{\beta}})^2}{(\sigma_v^2 b_i^2 + \psi_i)^2}.$$

Similarly, the scoring algorithm for REML estimation of σ_v^2 reduces to

$$\sigma_v^{2(a+1)} = \sigma_v^{2(a)} + [\mathcal{I}_R(\sigma_v^{2(a)})]^{-1} s_R(\sigma_v^{2(a)}),$$

where

$$\mathcal{I}_R(\sigma_v^2) = \frac{1}{2} \text{tr}[\mathbf{PBPB}] \tag{7.1.15}$$

and

$$s_R(\sigma_v^2) = -\frac{1}{2} \text{tr}[\mathbf{PB}] + \frac{1}{2} \mathbf{y}^T \mathbf{PBPy},$$

where $\mathbf{B} = \text{diag}(b_1^2, \ldots, b_m^2)$ and $\mathbf{P}$ is defined in subsection 6.2.4; see Cressie (1992). Asymptotically, $\mathcal{I}(\sigma_v^2)/\mathcal{I}_R(\sigma_v^2) \to 1$ as $m \to \infty$.

The EBLUP estimator $\hat{\theta}_i^H$, based on a moment, ML or REML estimator of σ_v^2, remains model-unbiased if the errors v_i and e_i are symmetrically distributed around 0. In particular, $\hat{\theta}_i^H$ is model-unbiased for θ_i if v_i and e_i are normally distributed.

For the special case of $b_i = 1$ and equal sampling variances $\psi_i = \psi$, the BLUP estimator (7.1.3) reduces to

$$\tilde{\theta}_i^H = \gamma \hat{\theta}_i + (1 - \gamma) \mathbf{z}_i^T \hat{\boldsymbol{\beta}}_{\text{LS}}$$

with $1 - \gamma = \psi/(\psi + \sigma_v^2)$, where $\hat{\boldsymbol{\beta}}_{\text{LS}}$ is the least squares estimator of $\boldsymbol{\beta}$. Under normality, we can obtain an unbiased estimator $1 - \gamma^*$ of $1 - \gamma$ by noting that

$S/(\psi + \sigma_v^2)$ is a χ^2 variable with $m - p$ df, where $S = \sum_i(\hat{\theta}_i - \mathbf{z}_i^T\hat{\boldsymbol{\beta}}_{\text{LS}})^2$ is the residual sum of squares. We have

$$1 - \gamma^* = \psi(m - p - 2)/S,$$

and an EBLUP estimator of θ_i is therefore given by

$$\hat{\theta}_i^H = \gamma^*\hat{\theta}_i + (1 - \gamma^*)\mathbf{z}_i^T\hat{\boldsymbol{\beta}}_{\text{LS}}.$$

This estimator is identical to the James-Stein estimator, studied in subsection 4.4.2, with guess $\theta_i^0 = \mathbf{z}_i^T\hat{\boldsymbol{\beta}}_{\text{LS}}$. Note that the simple moment estimators, $\hat{\sigma}_{vs}^2$ and $\hat{\sigma}_{vm}^2$, both lead to

$$1 - \hat{\gamma} = \psi(m - p)/S,$$

which is approximately equal to $1 - \gamma^*$ for large m and fixed p.

7.1.3 Relative Efficiency of Estimators of σ_v^2

Asymptotic variances (as $m \to \infty$) of ML and REML estimators are equal:

$$\bar{V}(\hat{\sigma}_{v\text{ML}}^2) = \bar{V}(\hat{\sigma}_{v\text{RE}}^2) = [\mathcal{I}(\sigma_v^2)]^{-1} = 2\left[\sum_{i=1}^m b_i^4/(\sigma_v^2 b_i^2 + \psi_i)^2\right]^{-1}, \quad (7.1.16)$$

where the information number, $\mathcal{I}(\sigma_v^2)$, is given by (7.1.14). The asymptotic variance of the simple moment estimator $\hat{\sigma}_{vs}^2$ is given by

$$\bar{V}(\hat{\sigma}_{vs}^2) = 2m^{-2}\sum_{i=1}^m(\sigma_v^2 b_i^2 + \psi_i)^2/b_i^4 \quad (7.1.17)$$

(Prasad and Rao (1990)). Datta, Rao and Smith (2002) derived the asymptotic variance of the Fay-Herriot moment estimator, $\hat{\sigma}_{vm}^2$, as

$$\bar{V}(\hat{\sigma}_{vm}^2) = 2m\left[\sum_{i=1}^m b_i^2/(\sigma_v^2 b_i^2 + \psi_i)\right]^{-2}. \quad (7.1.18)$$

Using the Cauchy-Schwarz inequality and the fact that the arithmetic mean is greater than or equal to the harmonic mean, we get

$$\bar{V}(\hat{\sigma}_{v\text{RE}}^2) = \bar{V}(\hat{\sigma}_{v\text{ML}}^2) \le \bar{V}(\hat{\sigma}_{vm}^2) \le \bar{V}(\hat{\sigma}_{vs}^2). \quad (7.1.19)$$

Equality in (7.1.19) holds if $\psi_i = \psi$ and $b_i = 1$. The Fay-Herriot moment estimator, σ_{vm}^2, becomes significantly more efficient relative to $\hat{\sigma}_{vs}^2$ as the variability of the terms $\sigma_v^2 + \psi_i/b_i^2$ increases. On the other hand, the loss in efficiency of σ_{vm}^2 relative to the REML (ML) estimator is relatively small as it depends on the variability of the terms $(\sigma_v^2 + \psi_i/b_i^2)^{-1}$.

7.1.4 Examples

We now provide some details of the applications in Examples 5.2.1–5.2.3, Chapter 5.

Example 7.1.1. Income for Small Places. The U.S. Bureau of the Census is required to provide the Treasury Department with the estimates of per capita income (PCI) and other statistics for state and local governments receiving funds under the General Revenue Sharing Program. Those statistics are then used by the Treasury Department to determine allocations to the local government units (places) within the different states by dividing the corresponding state allocations. Initially, the Census Bureau determined the current PCI estimate for a place by multiplying the 1970 census estimate of PCI in 1969 (based on a 20% sample) by the ratio of an administrative estimate of PCI in the current year and a similarly derived estimate for 1969. But the sampling error of the PCI estimates turned out to be quite large for places having fewer than 500 persons in 1970, with coefficient of variation (CV) of about 13% for a place of 500 persons and 30% for a place with 100 persons. As a result, the Census Bureau initially decided to set aside the census estimates for these small places and to substitute the corresponding county estimates in their place. But this solution turned out to be unsatisfactory because the census estimates for many small places differed significantly from the corresponding county estimates after accounting for sampling errors.

Using the EBLUP estimator (7.1.11) with $b_i = 1$ (equivalent to the empirical Bayes (EB) estimator under normality of the errors v_i and e_i; see Chapter 9), Fay and Herriot (1979) presented empirical evidence that the EBLUP estimates for small places have average error smaller than either the census estimates or the county estimates. The EBLUP estimator used by them is a weighted average of the census estimator $\hat{\theta}_i$ and a regression synthetic estimator $\mathbf{z}_i^T\hat{\boldsymbol{\beta}}$, obtained by fitting a linear regression equation to $(\hat{\theta}_i, \mathbf{z}_i^T), i = 1, \ldots, m$, where $\mathbf{z}_i = (z_{1i} = 1, z_{2i}, \ldots, z_{pi})^T$ and the independent variables $z_{2i}, \ldots, z_{pi}$ are based on the associated county PCI, tax return data for 1969 and data on housing from the 1970 census. The Fay-Herriot method was adopted by the Census Bureau in 1974 to form updated PCI estimates for small places. This was the largest application (prior to 1990) of EBLUP (EB) methods in a U.S. Federal Statistical Program.

We now present some details of the Fay-Herriot application and the results of an external evaluation. First, based on past studies, the CV of the sample estimate, $\hat{\hat{Y}}_i$, of PCI was taken as $3.0/\hat{N}_i^{1/2}$ for the ith small place, where $\hat{N}_i$ is the weighted sample count; $\hat{\hat{Y}}_i$ and $\hat{N}_i$ were available for almost all places. This suggested the use of logarithmic transformation, $\hat{\theta}_i = \log(\hat{\hat{Y}}_i)$, with $\mathrm{Var}(\hat{\theta}_i) \approx [\mathrm{CV}(\hat{\hat{Y}}_i)]^2 = 9/\hat{N}_i = \psi_i$. Secondly, four separate regression models were evaluated to determine a suitable combined model, treating the sampling variances, ψ_i, as known. The independent variables, $\mathbf{z}$, for the four models are respectively given by (1) $z_1 = 1, z_2 = \log(\text{county PCI})$: $p = 2$; (2) $z_1, z_2, z_3 = \log(\text{value of owner-occupied housing for the place})$, $z_4 = \log(\text{value}$

of owner-occupied housing for the county): $p = 4$; (3) $z_1, z_2, z_5 = \log$(adjusted gross income per exemption from the 1969 tax returns for the place), $z_6 = \log$(adjusted gross income per exemption from the 1969 tax returns for the county): $p = 4$; (4) $z_1, z_2, \ldots, z_6 : p = 6$.

Fay and Herriot (1979) calculated the values of $\hat{\sigma}_{vm}^2$ for each of the four models, using the iterative solution (7.1.12). The values of $\hat{\sigma}_{vm}^2$ provide a measure of the average fit of the regression models to the sample data, after allowing for the sample errors in the $\hat{\theta}_i$'s. A value for $\hat{\sigma}_{vm}^2$ of 0.045 corresponds to $\psi_i = 9/200 = 0.045$ for a place of size 200 and equal weighting ($\hat{\gamma}_{im} = 1/2$) of the direct estimate, $\hat{\theta}_i$, and the regression synthetic estimate, $\mathbf{z}_i^T\hat{\beta}$, where $\hat{\gamma}_{im}$ is the value of γ_i when σ_v^2 is replaced by $\hat{\sigma}_{vm}^2$. The resulting MSE, based on the leading term $g_{1i}(\hat{\sigma}_{vm}^2) = \hat{\sigma}_{vm}^2\hat{\gamma}_{im}$, is one-half of the sampling variance ψ_i; that is, the EBLUP estimate for a place of 200 persons roughly has the same precision as the direct estimate for a place of 400 persons. Fay and Herriot used the values of $\hat{\sigma}_v^2$ relative to ψ_i as the criterion for model selection.

Table 7.1 reports the values of $\hat{\sigma}_{vm}^2$ for the states with more than 500 small places (of size less than 500). It is clear from Table 7.1 that regressions involving either tax or housing data, but especially those involving both, are significantly better than the regression on the county values alone; that is, model 4 and to a lesser extent models 2 and 3 provide better fits to the data in terms of $\hat{\sigma}_{vm}^2$-values than model 1. Note that the values of $\hat{\sigma}_{vm}^2$ for model 4 are much smaller than 0.045, especially for North Dakota, Nebraska, Wisconsin and Iowa.

TABLE 7.1 Values of $\hat{\sigma}_{vm}^2$ for States with More Than 500 Small Places

	Model			
State	(1)	(2)	(3)	(4)
Illinois	0.036	0.032	0.019	0.017
Iowa	0.029	0.011	0.017	0.000
Kansas	0.064	0.048	0.016	0.020
Minnesota	0.063	0.055	0.014	0.019
Missouri	0.061	0.033	0.034	0.017
Nebraska	0.065	0.041	0.019	0.000
North Dakota	0.072	0.081	0.020	0.004
South Dakota	0.138	0.138	0.014	*
Wisconsin	0.042	0.025	0.025	0.004

* Not fitted because of too few points.
SOURCE: Adapted from Table 1 in Fay and Herriot (1979).

Fay and Herriot obtained compromise EBLUP estimates, $\hat{\theta}_{i*}^H$, from $\hat{\theta}_i^H$ and then transformed $\hat{\theta}_{i*}^H$ back to the original scale. The estimate of $\bar{Y}_i$ is therefore given by $\exp(\hat{\theta}_{i*}^H)$. The latter estimates were then subjected to a two-step raking (benchmarking) to ensure consistency with following aggregate sample estimates: (i) For each of the classes ($< 500, 500 - 1000$ and > 1000), the total estimated income for all places equals the direct estimate at the state level. (ii) The total estimated income for all places in a county equals the direct county estimate of total income.

External Evaluation. Fay and Herriot (1979) also conducted an external evaluation by comparing the 1972 estimates to "true" values obtained from a special complete census of a random sample of places in 1973. The 1972 estimates for each place were obtained by multiplying the 1970 estimates by an updating factor derived from administrative sources. Table 7.2 reports the values of percentage absolute relative error ARE = (|estimate−true value|/true value)×100, for the special census area using direct, county and EBLUP estimates. Values of average ARE are also reported. Table 7.2 shows that the EBLUP estimates exhibit smaller average errors and a lower incidence of extreme errors than either the direct estimates or the county estimates: the average ARE for places with population less than 500 is 22% compared to 28.6% (for the direct estimates) and 31.6% (for the county estimates). The EBLUP estimates were consistently higher than the special census values. But missing income was not imputed in the special census (unlike in the 1970 census). As a result, the special census values, which are based on only completed cases, may be subject to a downward bias.

TABLE 7.2 Values of Percentage Absolute Relative Error of Estimates from True Values: Places with Population Less Than 500

Special Census Area	Direct Estimate	EBLUP Estimate	County Estimate
1	10.2	14.0	12.9
2	4.4	10.3	30.9
3	34.1	26.2	9.1
4	1.3	8.3	24.6
5	34.7	21.8	6.6
6	22.1	19.8	14.6
7	14.1	4.1	18.7
8	18.1	4.7	25.9
9	60.7	78.7	99.7
10	47.7	54.7	95.3
11	89.1	65.8	86.5
12	1.7	9.1	12.7
13	11.4	1.4	6.6
14	8.6	5.7	23.5
15	23.6	25.3	34.3
16	53.6	10.5	11.7
17	51.4	14.4	23.7
Average	28.6	22.0	31.6

SOURCE: Adapted from Table 3 in Fay and Herriott (1979)).

Example 7.1.2. Poverty Counts. Current county estimates of school-age children in poverty in the United States are used by the U.S. Department of Education to allocate over \$7 billion of funds, called Title I funds, annually to counties, and then states distribute the funds among school districts. The allocated funds support compensatory education programs to meet the needs of educationally disadvantaged children.

The EBLUP estimator (7.1.11) under the basic area level model with $b_i = 1$ has been used recently to produce model-based county estimates (National Research Council (2000)). In this application, $\hat{\theta}_i = \log(\hat{Y}_i)$, where $\hat{Y}_i$ is the 3-year weighted average count of poor school-age children (under age 18) in county i obtained from the March Income Supplement of the Current Population Survey (CPS). Over one-half of the more than 3000 counties do not have CPS samples, and sample sizes in the remaining counties are too small to provide reliable direct estimates $\hat{Y}_i$ of true poverty counts Y_i.

To fit the county model, $\hat{\theta}_i = \mathbf{z}_i^T\boldsymbol{\beta}+v_i+e_i$, the following area level predictor variables, obtained from administrative records, were used: For county i, $z_{1i} = 1, z_{2i} =$ log(number of child exemptions reported by families in poverty on tax returns), $z_{3i} =$ log(number of people receiving food stamps), $z_{4i} =$ log (estimated population under age 18), $z_{5i} =$ log(number of child exemptions on tax returns), $z_{6i} =$ log(number of poor school-age children estimated in the previous census). The predictors $\mathbf{z}_i$ were available for all the counties. Counties with CPS samples but no poor school-age children (i.e., $\hat{Y}_i = 0$) were excluded in fitting the county model because $\hat{\theta}_i$ is not defined if $\hat{Y}_i = 0$.

The difficulty with the unknown sampling variances ψ_i was handled by (i) using a county model of the same form as above for the census year 1990 for which reliable estimates $\hat{\psi}_{ic}$ of sampling variances, ψ_{ic}, of census county estimates $\hat{\theta}_{ic}$ were obtained through generalized variance functions (see Wolter (1985), Chapter 5), and (ii) assuming the census model errors v_{ic} follow the same distribution as the current model errors v_i (i.e., normal with mean 0 and variance $\sigma_{vc}^2 = \sigma_v^2$). Under assumption (ii), a ML estimate of σ_v^2 was obtained from the census sample estimates $\hat{\theta}_{ic}$ assuming $\hat{\psi}_{ic} = \psi_{ic}$, and then used in the current model, assuming $\psi_i = \sigma_e^2/n_i$, to get an estimate $\tilde{\sigma}_e^2$ of σ_e^2, where n_i is the sample size in the ith county. The resulting estimate, $\tilde{\psi}_i = \tilde{\sigma}_e^2/n_i$, was treated as the true ψ_i in calculating the EBLUP estimate, $\hat{\theta}_i^H$, of $\theta_i = \log Y_i$. The assumption $\psi_i = \sigma_e^2/n_i$ is somewhat restrictive because it ignores the design effect.

The total Y_i for the counties with CPS sample and positive estimates $\hat{Y}_i$ may be estimated as $\hat{Y}_i^H = \exp(\hat{\theta}_i^H)$, but a more refined method based on the mean of a lognormal distribution was used. For the remaining counties, synthetic estimates $\hat{\theta}_{iS} = \mathbf{z}_i^T\hat{\boldsymbol{\beta}}$ were computed using the known auxiliary data $\mathbf{z}_i$ and then transformed back to obtain synthetic estimates of totals Y_i. All county estimates were then ratio-adjusted to agree with model-based state estimates obtained from a separate state model; the state estimates were ratio-adjusted to agree with the direct national CPS estimate.

The state model is of the form $p_t^* = \alpha_1 + \alpha_2 z_{2t}^* + \alpha_3 z_{3t}^* + \alpha_4 z_{4t}^* + \alpha_5 z_{5t}^* + v_t^* + e_t^*$, where p_t^* is the proportion of poor school-age children estimated from the current year CPS sample, v_t^* is the model error and e_t^* is the sampling error associated with state t and the z^*-variables are state level predictor variables. The following predictor variables were used: $z_{2t}^* =$ proportion of child exemptions reported by families in poverty on their tax returns, $z_{3t}^* =$ proportion of people receiving food stamps, $z_{4t}^* =$ proportion of people under

65 who did not file income tax returns, z^*_{5t} = residual from a linear regression of the proportion of poor school-age children from the most recent census on the other three predictor variables.

For the state model, reliable estimates of the sampling variances, ψ^*_t, were obtained through generalized variance functions and, treating these estimates as true variances ψ^*_t, the REML estimate of the model variance σ^{*2}_v for the state model was obtained. The resulting EBLUP estimates of the totals $Y^*_t = N^*_t p^*_t$ were ratio-adjusted to obtain the final model-based state estimates, where N^*_t is the current estimate of school-age children in state t derived from the Census Bureau's program of population estimates.

Finally, school district estimates of poor school-age children were computed, using simple synthetic estimation, due to lack of suitable data for fitting school district level models. Let $\hat{p}^{(c)}_{ia}$ be the census estimated proportion of poor school-age children in school district a wholly within county i. Then the synthetic estimate of poor school-age children in school district a is given by $\hat{p}^{(c)}_{ia}\hat{Y}^H_i$, where $\hat{Y}^H_i = \exp(\hat{\theta}^H_i)$ and $\hat{\theta}^H_i$ is the EBLUP estimate of θ_i for the current year. If a school district crossed county lines, then the synthetic estimate is given by the sum of the synthetic estimates of school district parts.

Evaluations. Both internal and external evaluations were conducted for model choice and for checking the validity of the chosen model. In an internal evaluation, the validity of the underlying assumptions and features of the model are examined. On the other hand, an external evaluation compares the estimates derived from a model to "true" values that were not used in the development of the model.

For internal evaluation, the following features of the model were examined: (a) linearity of the regression; (b) constancy of regression coefficients over time; (c) choice of predictor variables; (d) normality of the standardized residuals through quantile-quantile (q–q) plots; (e) homogeneity of the variances of standardized residuals; (f) residual analysis to detect outliers. No significant departures from the assumed county model, $\hat{\theta}_i = \beta_1 + \beta_2 z_{2i} + \ldots + \beta_6 z_{6i} + v_i + e_i$, were observed. We denote this model as log number (under 18). Random effects in the model were not taken into account in examining the above model features.

For external evaluation, several candidate area level models were fitted to the 1989 CPS estimates and predictor variables, and EBLUP estimates of county poverty counts were obtained. The 1989 county estimates obtained from the 1990 census were used to estimate the model variance σ^2_v. Models studied included log number (under 18), log number (under 21), log rate (under 18) and log rate (under 21). Log rate models used estimated log poverty ratio as the dependent variable, and the predictors were also log ratios. In the models with "under 21" in the brackets, the predictor variable z_{4i} was changed to z^*_{4i} = log(estimated population under 21). Two simple synthetic estimates, called stable shares and stable rates, were also included in the external evaluation. Stable shares estimates were obtained by assuming that county shares within a state were the same as those in the 1980 census and then calibrating the county estimates from the 1980 census to current state estimates for 1990. On the other hand, stable rates estimates were obtained by assuming that the county ratios (poor/population) within a state were the same as those in the 1980 census, multiplying the 1980 census ratio by the current population estimate and then calibrating to current state estimates. The state estimates were derived from the state model described earlier. Note that both stable shares and stable rates estimates relied heavily on the 1980 census values.

By treating the 1990 census estimates as true values, the values of average ARE, $M^{-1}\sum_i |\text{est}_i - \text{true}_i|/\text{true}_i$, were computed for the model-based estimates and stable shares and stable rates estimates, where M is the total number of counties and the summation is over the counties, i. The 1990 census estimates that were used in calculating the average proportional absolute difference (%) were ratio adjusted by a common factor to make the census national estimate of poor school-age children equal to the 1989 CPS national estimate. This was done to account for the differences between CPS and census in the measurement of income and poverty. The values of average ARE (%) for the model-based estimates were as follows: log number (under 21): 15.4, log number (under 18): 16.4, log rate (under 21): 17.5, log rate (under 18): 18.8, compared to 27.1 for stable shares estimate and 26.2 for stable rates estimate. Thus the model-based estimates performed much better than the stable shares or stable rates estimates. Also, log number models were better than the log rate models. Apart from the overall comparisons, "algebraic differences" for subgroups were also examined for various types of subgroups. The formula for algebraic difference in subgroups is defined as $\sum_i(\text{est}_{ij} - \text{true}_{ij})/\sum_i \text{true}_{ij}$, where the summation is over the counties i in subgroup j and $\text{est}_{ij}(\text{true}_{ij})$ denote the estimate (true value) for the ith county in the jth subgroup. This measure was intended to identify cases of potential bias in model-based estimates. For example, the estimates derived from a model might be consistently larger (smaller) than the true values in a particular subgroup relative to other subgroups. Examination of algebraic differences revealed that the use of z_{4i} was better than z_{4i}^* as a predictor variable even though the latter model (i.e., log number (under 21) model) led to a slightly smaller average proportional absolute difference (%): 16.4 versus 15.4. For example, the log number (under 21) model did not perform well relative to the log number (under 18) model for counties with large proportions of people under 21 living in group quarters: algebraic difference (%) of 14.2 versus −3.2.

Based on extensive internal and external evaluations, the log number (under 18) model was selected to produce county estimates of poor school-age children. The U.S. Census Bureau has an active research program to improve the county model as well as the school district estimates. Some of the topics under investigation include: (a) Modeling of CPS county sample variances; (b) discrete variable models that make use of counties with no sampled poor school-age achildren; (c) improvements to the county model through addition of prediction variables or modification of the model; (d) evaluation of food stamp and other input data; (e) reducing the variance of census school district estimates of poor school-age children. We refer the readers to National Research Council (2000, Chapter 9) for further details.

Example 7.1.3. Canadian Census Undercoverage. We have already noted in Example 5.2.2, Chapter 5, that the basic area level model (7.1.1) has been used to estimate the undercount in the decennial census of the United States and in the Canadian census. We now provide a brief account of the application to the 1991 Canadian census (Dick (1995)).

The objective here is to estimate $m = 96$ adjustment factors $\theta_i = T_i/C_i$ corresponding to sex (2) $\times$ age (4) $\times$ province (12) domains, where T_i is the true (unknown) count and C_i is the census count in the ith domain. The net undercoverage rate in the ith domain is given by $U_i = 1 - \theta_i^{-1}$. Direct estimates $\hat{\theta}_i$ were obtained from a post-enumeration survey (see subsection 2.4.2, Chapter 2). The associated sampling variances, ψ_i, were derived through smoothing of the estimated variances. In particular, the variance of the estimated number of missing persons, $\hat{M}_i$, was assumed to be proportional to a power of the census count C_i, and a linear regression was fitted to $\{\log(v(\hat{M}_i)), \log(C_i); i = 1, \ldots, m\}$, where $v(\hat{M}_i)$ is the estimated variance of $\hat{M}_i$. Using this relationship, the sampling variances were calculated from $\log(\psi_i) = -6.13 - 0.28 \log(C_i)$, and the resulting ψ_i were treated as the true ψ_i.

Auxiliary (predictor) variables $\mathbf{z}$ for building the model were selected from a set of 42 variables by backward stepwise regression (Draper and Smith (1981), Chapter 6). Note that the random area effects v_i are not taken into account in stepwise regression. Internal evaluation of the resulting model (7.1.1) with $b_i = 1$ was then performed, by treating the standard residuals $r_i = (\hat{\theta}_i^H - \mathbf{z}_i^T\hat{\beta})/(\hat{\sigma}_v^2 + \psi_i)^{1/2}$ as iid $N(0,1)$ variables, where $\hat{\beta}$ and $\hat{\sigma}_v^2$ are the REML estimates of β and σ_v^2. No significant departures from the assumed model were observed. In particular, to check for the normality of the standardized residuals r_i and to detect outlier r_i's, a normal q–q plot of the r_i versus $\Phi^{-1}[F_m(r_i)]$, $i = 1, \ldots, m$ was examined, where $\Phi(x)$ is the cumulative distribution function (CDF) of $N(0,1)$ variable and $F_m(x)$ is the empirical CDF of the r_i's, that is, $F_m(x) = m^{-1}\Sigma_{i=1}^m I(x - r_i)$, where $I(x - r_i) = 1$ if $x \geq r_i$ and 0 otherwise. Dempster and Ryan (1989) proposed a weighted normal q–q plot that is more sensitive to departures from normality than the unweighted normal q–q plot. This plot uses a weighted empirical CDF

$$F_m^*(x) = \left[\sum_{i=1}^m (\hat{\sigma}_v^2 + \psi_i)^{-1} I(x - r_i)\right] \Big/ \left[\sum_{i=1}^m (\hat{\sigma}_v^2 + \psi_i)^{-1}\right]$$

$$= \sum_{i=1}^m \hat{\gamma}_i I(x - r_i) \Big/ \sum_{i=1}^m \hat{\gamma}_i$$

for the special case $b_i = 1$, instead of $F_m(x)$. Note that $F_m^*(x)$ assigns greater weight to those areas for which $\hat{\sigma}_v^2$ accounts for a larger part of the total variance $\hat{\sigma}_v^2 + \psi_i$.

The EBLUP adjustment factors, $\hat{\theta}_i^H$, were converted to estimates of missing persons, $\hat{M}_i^H$. These estimates were then subjected to two-step raking (see subsection 4.2.5, Chapter 4) to ensure consistency with the reliable direct estimates of marginal totals, $\hat{M}_{p+}$ and $\hat{M}_{+a}$, where "p" denotes a province, "a" denotes an age-sex group and $M_i = M_{pa}$. The raked EBLUP estimates, $\hat{M}_{pa}^{RH}$, were used as the final estimates. These estimates were further divided into single year of age estimates by using simple synthetic estimation:

$$\hat{M}_{pa}^S(q) = \hat{M}_{pa}^{RH}[C_{pa}(q)/C_{pa}],$$

where q denotes a sub-age group and $C_{pa}(q)$ is the associated census count.

7.1.5 MSE Estimation

The second-order MSE approximation (6.3.9) is valid for the Fay-Herriot model, $\hat{\theta}_i = \mathbf{z}_i^T\beta + b_i v_i + e_i$, under regularity conditions (7.1.9) and (7.1.10) and normality of the errors v_i and e_i. It reduces to

$$\text{MSE}(\hat{\theta}_i^H) \approx g_{1i}(\sigma_v^2) + g_{2i}(\sigma_v^2) + g_{3i}(\sigma_v^2), \tag{7.1.20}$$

where $g_{1i}(\sigma_v^2)$ and $g_{2i}(\sigma_v^2)$ are given by (7.1.7) and (7.1.8), and

$$g_{3i}(\sigma_v^2) = \psi_i^2 b_i^4(\psi_i + \sigma_v^2 b_i^2)^{-3}\bar{V}(\hat{\sigma}_v^2), \tag{7.1.21}$$

where $\bar{V}(\hat{\sigma}_v^2)$ is the asymptotic variance of an estimator, $\hat{\sigma}_v^2$, of σ_v^2. We have $\bar{V}(\hat{\sigma}_v^2) = \bar{V}(\hat{\sigma}_{v\text{ML}}^2) = \bar{V}(\hat{\sigma}_{v\text{RE}}^2)$, given by (7.1.16), if we use $\hat{\sigma}_{v\text{ML}}^2$ or $\hat{\sigma}_{v\text{RE}}^2$ to estimate σ_v^2. If the simple moment estimator $\hat{\sigma}_{vs}^2$ is used, then $\bar{V}(\hat{\sigma}_v^2) = \bar{V}(\hat{\sigma}_{vs}^2)$ given by (7.1.17). For the Fay-Herriot moment estimator, $\hat{\sigma}_{vm}^2$, we have $\bar{V}(\hat{\sigma}_v^2) = \bar{V}(\hat{\sigma}_{vm}^2)$ given by (7.1.18). It follows from (7.1.21) that $g_{3i}(\sigma_v^2)$ is the smallest for ML and REML estimators of σ_v^2 followed by $\hat{\sigma}_{vm}^2$. The g_{3i} term for $\hat{\sigma}_{vm}^2$ is significantly smaller than the g_{3i}-term for $\hat{\sigma}_{vs}^2$ if the variability of the terms $\sigma_v^2 + \psi_i/b_i^2$ is substantial.

We now turn to the estimation of $\text{MSE}(\hat{\theta}_i^H)$. The estimator of MSE given by (6.3.11) is valid for REML and the simple moment estimator of σ_v^2 under regularity conditions (7.1.9) and (7.1.10) and normality of the errors v_i and e_i. It reduces to

$$\text{mse}(\hat{\theta}_i^H) = g_{1i}(\hat{\sigma}_v^2) + g_{2i}(\hat{\sigma}_v^2) + 2g_{3i}(\hat{\sigma}_v^2) \tag{7.1.22}$$

if $\hat{\sigma}_v^2$ is chosen to be $\hat{\sigma}_{v\text{RE}}^2$ or $\hat{\sigma}_{vs}^2$. The corresponding area-specific versions, $\text{mse}_1(\hat{\theta}_i^H)$ and $\text{mse}_2(\hat{\theta}_i^H)$, are obtained from (6.3.15) and (6.3.16) by changing $g_{3i}(\delta, \mathbf{y}_i)$ to $g_{3i}^*(\sigma_v^2, \hat{\theta}_i)$, where $g_{3i}^*(\sigma_v^2, \hat{\theta}_i)$ is obtained from (6.3.14) as

$$g_{3i}^*(\sigma_v^2, \hat{\theta}_i) = [b_i^4\psi_i^2/(\psi_i + \sigma_v^2 b_i^2)^4](\hat{\theta}_i - \mathbf{z}_i^T\tilde{\beta})^2\overline{V}(\hat{\sigma}_v^2). \tag{7.1.23}$$

We have

$$\text{mse}_1(\hat{\theta}_i^H) = g_{1i}(\hat{\sigma}_v^2) + g_{2i}(\hat{\sigma}_v^2) + 2g_{3i}^*(\hat{\sigma}_v^2, \hat{\theta}_i) \tag{7.1.24}$$

and

$$\text{mse}_2(\hat{\theta}_i^H) = g_{1i}(\hat{\sigma}_v^2) + g_{2i}(\hat{\sigma}_v^2) + g_{3i}(\hat{\sigma}_v^2) + g_{3i}^*(\hat{\sigma}_v^2, \hat{\theta}_i). \tag{7.1.25}$$

Rao (2001) obtained the area-specific MSE estimators (7.1.24) and (7.1.25) for the special case of $b_i = 1$.

For the ML estimator $\hat{\sigma}_{v\text{ML}}^2$ and the Fay-Herriot estimator $\hat{\sigma}_{vm}^2$, the MSE estimator (6.3.12) is valid. It reduces to

$$\text{mse}_*(\hat{\theta}_i^H) = g_{1i}(\hat{\sigma}_v^2) - b_{\hat{\sigma}_v^2}(\hat{\sigma}_v^2)\nabla g_{1i}(\hat{\sigma}_v^2) + g_{2i}(\hat{\sigma}_v^2) + 2g_{3i}(\hat{\sigma}_v^2), \tag{7.1.26}$$

where

$$\nabla g_{1i}(\hat{\sigma}_v^2) = b_i^2(1-\gamma_i)^2. \tag{7.1.27}$$

The bias term $b_{\hat{\sigma}_v^2}(\sigma_v^2)$ for the ML estimator is obtained from (6.3.13), which reduces to

$$b_{\hat{\sigma}_{v\text{ML}}^2}(\sigma_v^2) = -[2\mathcal{I}(\sigma_v^2)]^{-1}\text{tr}\left[\left\{\sum_i (\psi_i + \sigma_v^2 b_i^2)^{-1}\mathbf{z}_i\mathbf{z}_i^T\right\}^{-1} \times \left\{\sum_i b_i^2(\psi_i + \sigma_v^2 b_i^2)^{-2}\mathbf{z}_i\mathbf{z}_i^T\right\}\right], \tag{7.1.28}$$

where $\mathcal{I}(\sigma_v^2)$ is given by (7.1.14). It follows from (7.1.27) and (7.1.28) that the term $-b_{\hat{\sigma}_{v\text{ML}}^2}(\hat{\sigma}_{v\text{ML}}^2)\ \nabla g_{1i}(\hat{\sigma}_{v\text{ML}}^2)$ in (7.1.26) is positive. Therefore, ignoring this term and using (7.1.22) with $\hat{\sigma}_v^2 = \hat{\sigma}_{v\text{ML}}^2$ would lead to underestimation of MSE approximation given by (7.1.20).

The bias term $b_{\hat{\sigma}_v^2}(\sigma_v^2)$ for the Fay-Herriot moment estimator $\hat{\sigma}_{vm}^2$ is given by

$$b_{\hat{\sigma}_{vm}^2}(\sigma_v^2) = \frac{2\left[m\sum_i(\psi_i + \sigma_v^2 b_i^2)^{-2} - \{\sum_i(\psi_i + \sigma_v^2 b_i^2)^{-1}\}^2\right]}{[\sum_i(\psi_i + \sigma_v^2 b_i^2)^{-1}]^3} \tag{7.1.29}$$

(Datta, Rao and Smith (2002)). It follows from (7.1.29) that the bias of $\hat{\sigma}_{vm}^2$ is positive (unlike the bias of $\hat{\sigma}_{v\text{ML}}^2$) and it reduces to 0 if $b_i = 1$ and $\psi_i = \psi$ for all i. As a result, the term $-b_{\hat{\sigma}_{vm}^2}(\hat{\sigma}_{vm}^2)\nabla g_{1i}(\hat{\sigma}_{vm}^2)$ in (7.1.26) is negative. Therefore, ignoring this term and using (7.1.22) with $\hat{\sigma}_v^2 = \hat{\sigma}_{vm}^2$ would lead to overestimation of the MSE approximation given by (7.1.20).

Area-specific versions of (7.1.26) for the ML estimator, $\hat{\sigma}_{v\text{ML}}^2$, and the moment estimator, $\hat{\sigma}_{vm}^2$, are obtained from (6.3.17) and (6.3.18):

$$\text{mse}_{*1}(\hat{\theta}_i^H) = g_{1i}(\hat{\sigma}_v^2) - b_{\hat{\sigma}_v^2}(\hat{\sigma}_v^2)\nabla g_{1i}(\hat{\sigma}_v^2) + g_{2i}(\hat{\sigma}_v^2) + 2g_{3i}^*(\hat{\sigma}_v^2, \hat{\theta}_i) \tag{7.1.30}$$

and

$$\text{mse}_{*2}(\hat{\theta}_i^H) = g_{1i}(\hat{\sigma}_v^2) - b_{\hat{\sigma}_v^2}(\hat{\sigma}_v^2)\nabla g_{1i}(\hat{\sigma}_v^2) + g_{2i}(\hat{\sigma}_v^2) + g_{3i}(\hat{\sigma}_v^2) + g_{3i}^*(\hat{\sigma}_v^2, \hat{\theta}_i), \tag{7.1.31}$$

where $\hat{\sigma}_v^2$ is chosen as $\hat{\sigma}_{v\text{ML}}^2$ or $\hat{\sigma}_{vm}^2$.

The above MSE estimators are approximately unbiased in the sense of bias lower order than m^{-1}, for large m. We assumed normality of the random effects, v_i, in deriving the MSE estimators of the EBLUP estimator $\hat{\theta}_i^H$. Lahiri and Rao (1995), however, showed that the MSE estimator (7.1.22) is also valid under nonnormal v_i with $E|v_i|^{8+\delta} < \infty$ for $0 < \delta < 1$. This robustness result was established using the simple moment estimator $\hat{\sigma}_{vs}^2$, assuming normality of the sampling errors e_i. The latter assumption, however, is not restrictive, unlike the normality of v_i, due to the central limit theorem effect on the direct estimator $\hat{\theta}_i$. The moment condition $E|v_i|^{8+\delta} < \infty$ is satisfied by many continuous distributions including the double exponential, "shifted" exponential and lognormal distributions with $E(v_i) = 0$. The proof of the validity of (7.1.22) under nonnormal v_i is highly technical. We refer the reader to the Appendix in Lahiri and Rao (1995) for details of the proof. It may be noted that MSE($\hat{\theta}_i^H$) is affected by the nonnormality of the v_i's. In fact, it depends on the fourth moment of v_i, and the cross-product term $E(\hat{\theta}_i^H - \theta_i)(\hat{\theta}_i^H - \tilde{\theta}_i^H)$ in the decomposition of the MSE (see (6.2.25)) is nonzero, unlike in the case of normal v_i. However, the formula (7.1.22) for the MSE estimator remains valid in the sense of $E(\text{mse}(\hat{\theta}_i^H)) = \text{MSE}(\hat{\theta}_i^H) + o(m^{-1})$. It is not known if this robustness property of the MSE estimator holds in the case of ML, REML and the Fay-Herriot estimators of σ_v^2. However, simulation results (see Example 7.1.4) indicate robustness.

Example 7.1.4. Simulation Study Datta, Rao and Smith (2002) studied the relative bias of MSE estimators through simulation. They used the basic area level model (7.1.1) with $b_i = 1$ and no covariates, that is, $\hat{\theta}_i = \mu + v_i + e_i$. Since the MSE of an EBLUP is translation invariant, they took $\mu = 0$ without loss of generality. However, to account for the estimation of unknown regression parameters in practical applications, this zero mean was estimated from each simulation run. The simulation runs consisted of R = 100,000 data sets of $\hat{\theta}_i = v_i + e_i$, $i = 1, \ldots, 15$ generated from $v_i \overset{\text{iid}}{\sim} N(0, \sigma_v^2 = 1)$ and $e_i \overset{\text{ind}}{\sim} N(0, \psi_i)$ for specified sampling variances ψ_i. In particular, three different ψ_i–patterns, with five groups $G_1, \ldots, G_5$ for each pattern and equal number of areas and equal ψ_i's within each group G_t, were chosen: pattern (a): 0.7, 0.6, 0.5, 0.4, 0.3; pattern (b) 2.0, 0.6, 0.5, 0.4, 0.2; pattern (c): 4.0, 0.6, 0.5, 0.4, 0.1. Note that pattern (a) is nearly balanced while pattern (c) has the largest variability ($\max \psi_i / \min \psi_i = 40$) and pattern (b) intermediate variability ($\max \psi_i / \min \psi_i = 10$). Patterns similar to (c) can occur when the sample sizes for some areas are significantly larger than those for the remaining areas.

The MSEs of the EBLUP estimators $\hat{\theta}_i^H$ were calculated from the simulated data sets, using MSE$= R^{-1}\Sigma_{r=1}^{R}[\hat{\theta}_i^H(r) - \theta_i(r)]^2$, where $\hat{\theta}_i^H(r)$ and $\theta_i(r)$ denote the values of $\hat{\theta}_i^H$ and $\theta_i = v_i$ for the rth simulation run. The MSE values were calculated using the estimators of σ_v^2 discussed in subsection 7.1.2: the Prasad-Rao (PR) moment estimator $\hat{\sigma}_{vs}^2$, the Fay-Herriot (FH) moment estimator

$\hat{\sigma}^2_{vm}$ and the ML and REML estimators. The relative bias (RB) of a MSE estimator was calculated as RB $= [R^{-1}\Sigma_{r=1}^{R}(\text{mse}_r - \text{MSE})]/\text{MSE}$, where mse_r is the value of a MSE estimator for the rth run. The MSE estimator (7.1.22) associated with PR and REML and the MSE estimator (7.1.26) associated with ML and FH were compared with respect to relative bias.

For the nearly balanced pattern (a), all four methods performed well with average relative bias (ARB) less than 2%. However, for the extreme pattern (c), PR led to large ARB when ψ_i/σ_v^2 was small: about 80% for G_4 and 700% for G_5. This overestimation of MSE decreased for $m = 30$: 140% for G_5 (Datta and Lahiri 2000)). The remaining three methods performed well for patterns (b) and (c) with ARB less than 13% for REML and less than 10% for FH and ML. Datta, Rao and Smith (2002) also calculated ARB values for two nonnormal v_i's with mean 0 and variance 1: double exponential (DE) and location exponential (LE). FH also performed well for DE and LE with ARB less than 10% (ARB for ML less than 16%). On the other hand, REML led to larger ARB for group G_5: 26% for DE and 44% for LE under pattern (c). Overall, FH performed better than the other methods in terms of ARB.

7.1.6 Conditional MSE

In subsection 7.1.5 we studied the estimation of unconditional MSE of the EBLUP estimator $\hat{\theta}_i^H$. It is more appealing to survey practitioners to consider the estimation of the conditional MSE, $\text{MSE}_p(\hat{\theta}_i^H) = E_p(\hat{\theta}_i^H - \theta_i)^2$, where the expectation, E_p, is with respect to the sampling design only, treating the small area means, θ_i, as fixed unknown parameters. Following subsection 4.4.3, Chapter 4, $\hat{\theta}_i^H$ may be expressed as $\hat{\theta}_i + h_i(\hat{\theta})$, where

$$h_i(\hat{\theta}) = -(1-\gamma_i)(\hat{\theta}_i - \mathbf{z}_i^T\hat{\boldsymbol{\beta}}).$$

Assuming $\hat{\theta}_i|\theta_i \overset{\text{ind}}{\sim} N(\theta_i, \psi_i)$ and appealing to the general formula (4.4.15), a design-unbiased estimator, $\text{mse}_p(\hat{\theta}_i^H)$, of the conditional MSE is obtained. Rivest and Bellmonte (2000) gave an explicit formula for $\text{mse}_p(\hat{\theta}_i^H)$ applicable to the simple moment estimator $\hat{\sigma}^2_{vs}$ given by (7.1.13). When $\hat{\sigma}^2_{vs} = 0$, the formula for $\text{mse}_p(\hat{\theta}_i^H)$ reduces to

$$\text{mse}_p(\hat{\theta}_i^H) = (\hat{\theta}_i - \mathbf{z}_i^T\hat{\boldsymbol{\beta}})^2 - \psi_i + 2\mathbf{z}_i^T\left(\sum_i \mathbf{z}_i\mathbf{z}_i^T/\psi_i\right)^{-1}\mathbf{z}_i. \qquad (7.1.32)$$

As in the case of the James-Stein MSE estimator, (4.4.15), $\text{mse}_p(\hat{\theta}_i^H)$ can take negative values. We therefore use the truncated estimator $\text{mse}_p^*(\hat{\theta}_i^H) = \max(0, \text{mse}_p(\hat{\theta}_i^H))$. The estimator $\text{mse}_p(\hat{\theta}_i^H)$ is also unbiased for the unconditional MSE, but it can be highly variable especially when the weight (or shrinking factor), γ_i, attached to the direct estimator is small.

Rivest and Bellmonte (2000) studied the properties of $\text{mse}_p(\hat{\theta}_i^H)$ in the simple case where all the parameters, $(\boldsymbol{\beta}, \sigma_v^2)$, of the linking model are assumed to be known. In this case, assuming $b_i = 1$,

$$\text{mse}_p(\hat{\theta}_i^H) = \frac{\psi_i \sigma_v^2}{\psi_i + \sigma_v^2} + \left(\frac{\psi_i}{\psi_i + \sigma_v^2}\right)^2 [(\hat{\theta}_i - \mathbf{z}_i^T \boldsymbol{\beta})^2 - \psi_i - \sigma_v^2] \tag{7.1.33}$$

and

$$\text{mse}(\hat{\theta}_i^H) = g_{1i}(\sigma_v^2) = \frac{\psi_i \sigma_v^2}{\psi_i + \sigma_v^2} = \psi_i \gamma_i. \tag{7.1.34}$$

When γ_i is close to zero, the probability of getting a negative estimate $\text{mse}_p(\hat{\theta}_i^H)$ is close to 0.5. The relative performance of $\text{mse}_p(\hat{\theta}_i^H)$ and $\text{mse}(\hat{\theta}_i^H)$ in estimating the conditional MSE, $\text{MSE}_p(\hat{\theta}_i^H)$, may be studied by letting $\psi_i = \psi$ for all i, and calculating the ratio, R, of $E_m[\overline{\text{MSE}}_p(\text{mse}_p(\hat{\theta}_i^H))]$ to $E_m[\overline{\text{MSE}}_p(\text{mse}(\hat{\theta}_i^H))]$, where $\overline{\text{MSE}}_p$ denotes the average of MSE_p over the m areas and E_m denotes the expectation with respect to the linking model. The ratio R measures the average efficiency of $\text{mse}(\hat{\theta}_i^H)$ relative to $\text{mse}_p(\hat{\theta}_i^H)$: $R = (\psi^2 + 2\psi\sigma_v^2)/\sigma_v^4$. We have $R > 1$ if $\sigma_v^2/\psi < 2.4$. When shrinking is appreciable, that is, when γ_i is small, $\text{mse}(\hat{\theta}_i^H)$ is a much better estimator of $\text{MSE}_p(\hat{\theta}_i^H)$ than $\text{mse}_p(\hat{\theta}_i^H)$. For example, if $\gamma = 1/2$ or $\sigma_v^2/\psi = 1$, we have $R = 3$. Note that the EBLUP estimator, $\hat{\theta}_i^H$, leads to significant gain in efficiency relative to the direct estimator, $\hat{\theta}_i$, only when the shrinking factor, γ, is small.

Hwang and Rao (1987; unpublished work) conducted a small simulation study by first generating a large number of simulation runs, $\hat{\boldsymbol{\theta}}$, from the sampling model, $\hat{\theta}_i = \theta_i + e_i$, for specified ψ_i. The EBLUP estimator, $\hat{\theta}_i^H$, was based on the simple linking model $\theta_i = \beta + v_i$. The simulation results suggested that the model-based MSE estimator, $\text{mse}(\hat{\theta}_i^H)$, tracks the conditional MSE, $\text{MSE}_p(\hat{\theta}_i^H)$, quite well even under moderate violations of the assumed linking model. Moreover, $\text{mse}(\hat{\theta}_i^H)$ was much less variable than $\text{mse}_p(\hat{\theta}_i^H)$. Only in extreme cases, such as with a large outlier value θ_i, did $\text{mse}(\hat{\theta}_i^H)$ perform poorly compared to $\text{mse}_p(\hat{\theta}_i^H)$.

Fuller (1989) proposed a compromise measure of conditional MSE. This measure is given by $\text{MSE}_c(\hat{\theta}_i^H) = E[(\hat{\theta}_i^H - \theta_i)^2 | \hat{\theta}_i]$, where the expectation is conditional on the observed $\hat{\theta}_i$ for the ith area. For the special case of $\psi_i = \psi$, it leads to an estimator of MSE_c that is closely related to the area-specific estimator of the unconditional MSE given by (7.1.24).

7.1.7 Mean Product Error of Two Estimators

The small area estimators $\hat{\theta}_i^H$ are often aggregated to obtain an estimator for a larger area. In this case, we also need the mean product error (MPE) of two estimators $\hat{\theta}_i^H$ and $\hat{\theta}_t^H$ $(i \neq t)$ to get the MSE of the larger area estimator. We have

$$\text{MPE}(\hat{\theta}_i^H, \hat{\theta}_t^H) = E(\hat{\theta}_i^H - \theta_i)(\hat{\theta}_t^H - \theta_t) = \text{MPE}(\tilde{\theta}_i^H, \tilde{\theta}_t^H) + \text{ lower order terms.} \tag{7.1.35}$$

The leading term in (7.1.35) is given by

$$\text{MPE}(\tilde{\theta}_i^H, \tilde{\theta}_t^H) = (1-\gamma_i)(1-\gamma_t)\mathbf{z}_i^T \left[\sum_i \mathbf{z}_i \mathbf{z}_i^T/(\psi_i + \sigma_v^2 b_i^2)\right]^{-1} \mathbf{z}_t$$

$$:= g_{2it}(\sigma_v^2), \qquad (7.1.36)$$

which is of order $o(m^{-1})$, unlike $\text{MSE}(\tilde{\theta}_i^H)$, which is $O(1)$. It follows from (7.1.35) that

$$\text{MPE}(\hat{\theta}_i^H, \hat{\theta}_t^H) \approx \text{MPE}(\tilde{\theta}_i^H, \tilde{\theta}_t^H) = g_{2it}(\sigma_v^2) \qquad (7.1.37)$$

is correct to terms of order $o(m^{-1})$. Further, an estimator of $\text{MPE}(\hat{\theta}_i^H, \hat{\theta}_t^H)$ is given by

$$\text{mpe}(\hat{\theta}_i^H, \hat{\theta}_t^H) = g_{2it}(\hat{\sigma}_v^2), \qquad (7.1.38)$$

which is correct to terms of order $o(m^{-1})$, that is,

$$E[\text{mpe}(\hat{\theta}_i^H, \hat{\theta}_t^H)] = \text{MPE}(\hat{\theta}_i^H, \hat{\theta}_t^H) + o(m^{-1}). \qquad (7.1.39)$$

7.1.8 Estimation of Small Area Means

We have considered the estimation of $\theta_i = g(\bar{Y}_i)$ in the previous subsections. If $g(\bar{Y}_i) = \bar{Y}_i$, then the estimator of the small area mean $\bar{Y}_i$ is $\hat{\bar{Y}}_i^H = \hat{\theta}_i^H$ and $\text{mse}(\hat{\bar{Y}}_i^H) = \text{mse}(\hat{\theta}_i^H)$. On the other hand, for nonlinear $g(\cdot)$ as in Example 7.1.1, we use

$$\hat{\bar{Y}}_i^H = g^{-1}(\hat{\theta}_i^H) = h(\hat{\theta}_i^H). \qquad (7.1.40)$$

Note that $\hat{\bar{Y}}_i^H$ does not retain the optimality of $\hat{\theta}_i^H$.

Using a Taylor expansion, we can approximate the MSE of $\hat{\bar{Y}}_i^H$ as

$$\text{MSE}(\hat{\bar{Y}}_i^H) \approx [h'(\theta_i)]^2 \text{MSE}(\hat{\theta}_i^H). \qquad (7.1.41)$$

An estimator of $\text{MSE}(\hat{\bar{Y}}_i^H)$ is then obtained as

$$\text{mse}(\hat{\bar{Y}}_i^H) = [h'(\hat{\theta}_i^H)]^2 \text{mse}(\hat{\theta}_i^H), \qquad (7.1.42)$$

where $h'(\hat{\theta}_i^H)$ is the value of $h'(\theta_i)$ evaluated at $\theta_i = \hat{\theta}_i^H$. This estimator is not correct to terms of order $o(m^{-1})$, unlike $\text{mse}(\hat{\theta}_i^H)$. In fact, the Taylor approximation may not be justifiable in the small area context because $\hat{\theta}_i^H - \theta_i$ is of order $O(1)$ for large m, and hence its higher powers cannot be neglected unless the small area sample size is also large. The EB and HB approaches (Chapters 9 and 10) are better suited for handling nonlinear cases, $h(\theta_i)$.

7.1.9 Weighted Estimator

The EBLUP estimator $\hat{\theta}_i^H$ runs into difficulties when $\hat{\sigma}_v^2 = 0$. In this case, it reduces to the synthetic estimator $\mathbf{z}_i^T\hat{\boldsymbol{\beta}}$ regardless of the sampling variance, ψ_i, of the direct estimator $\hat{\theta}_i$. For example, in the state model of Example 7.1.2 (poverty counts of school-age children), $\hat{\sigma}_{v\mathrm{ML}}^2 = \hat{\sigma}_{v\mathrm{RE}}^2 = 0$ in 1992. As a result, the EBLUP attached zero weight to the direct estimates $\hat{\theta}_i$ regardless of the CPS sample sizes n_i (number of households); see Example 10.3.1, Chapter 10 for HB estimation. Also the leading term, $g_{1i}(\hat{\sigma}_v^2) = \hat{\gamma}_i\psi_i$, of the MSE estimate becomes zero when $\hat{\sigma}_v^2 = 0$. One way to get around this problem is to use a weighted combination of $\hat{\theta}_i$ and $\mathbf{z}_i^T\hat{\boldsymbol{\beta}}$ with fixed weights a_i and $1 - a_i$:

$$\hat{\theta}_i(a_i) = a_i\hat{\theta}_i + (1 - a_i)\mathbf{z}_i^T\hat{\boldsymbol{\beta}}. \tag{7.1.43}$$

A prior guess of σ_v^2, say σ_{v0}^2, available from past studies, may be used to construct the weight a_i as $a_i = \sigma_{v0}^2 b_i^2/(\psi_i + \sigma_{v0}^2 b_i^2)$. Also, the synthetic estimator $\mathbf{z}_i^T\hat{\boldsymbol{\beta}}$ is a special case of the weighted estimator $\hat{\theta}_i(a_i)$ with $a_i = 0$. Note that $\hat{\theta}_i(a_i)$ remains model-unbiased for θ_i for any fixed weight a_i, provided the linking model holds, that is, $E(\theta_i) = \mathbf{z}_i^T\boldsymbol{\beta}$.

Datta, Kubakowa and Rao (2002) derived an approximately unbiased estimator of the MSE of $\hat{\theta}_i(a_i)$, assuming $b_i = 1$ in the basic area level model (7.1.1). It is given by

$$\mathrm{mse}[\hat{\theta}_i(a_i)] = g_{1i}(\hat{\sigma}_v^2) + g_{2i}(\hat{\sigma}_v^2) + g_{ai}^*(\hat{\sigma}_v^2) - b_{\hat{\sigma}_v^2}(\hat{\sigma}_v^2), \tag{7.1.44}$$

where

$$g_{ai}^*(\sigma_v^2) = (a_i - \gamma_i)^2\left[\sigma_v^2 + \psi_i - g_{2i}(\sigma_v^2)/(1-\gamma_i)^2\right] \tag{7.1.45}$$

and $b_{\hat{\sigma}_v^2}(\sigma_v^2)$ is the bias of $\hat{\sigma}_v^2$. The bias term $b_{\hat{\sigma}_v^2}(\sigma_v^2)$ is zero to terms of order m^{-1} for the simple moment estimator $\hat{\sigma}_{vs}^2$ and the REML estimator $\hat{\sigma}_{v\mathrm{RE}}^2$. The bias terms for the ML estimator $\hat{\sigma}_{v\mathrm{ML}}^2$ and the Fay-Herriot estimator $\hat{\sigma}_{vm}^2$ are given by (7.1.28) and (7.1.29), respectively. Note that $E[\mathrm{mse}(\hat{\theta}_i(a_i)] - \mathrm{MSE}[\hat{\theta}_i(a_i)] = o(m^{-1})$ for any fixed weight a_i $(0 \le a_i \le 1)$.

The MSE estimator of the synthetic estimator $\mathbf{z}_i^T\hat{\boldsymbol{\beta}}$ is obtained by letting $a_i = 0$ in (7.1.44). Also, if $\hat{\sigma}_v^2 = 0$ the leading term of order $O(1)$ in (7.1.44) reduces to $a_i^2\psi_i$ noting that $\hat{\gamma}_i = 0$, whereas the leading term, $g_{1i}(\hat{\sigma}_v^2) = \hat{\gamma}_i\psi_i$, of $\mathrm{mse}(\hat{\theta}_i^H)$ is zero. The leading term, $a_i^2\psi_i$ for $a_i > 0$, decreases as the sampling variance decreases, and this is a desirable property. We use the synthetic estimates $\mathbf{z}_i^T\hat{\boldsymbol{\beta}}$ for the areas with no samples. A nearly unbiased MSE estimator of $\mathbf{z}_i^T\hat{\boldsymbol{\beta}}$ is obtained from (7.1.44) by letting $a_i = 0$.

7.2 Basic Unit Level Model

In this section we consider the basic unit level (type B) model (5.3.4) and spell out EBLUP estimation, using the general results in Section 6.3, Chapter 6 for the general linear mixed model with block diagonal covariance structure.

As noted in Section 5.3, Chapter 5, we can take the ith small area mean as $\mu_i = \bar{\mathbf{X}}_i^T\boldsymbol{\beta} + v_i$ if the population sizes, N_i, of the small areas are sufficiently large. In this case, we can use the sample part of the model (5.3.4), namely, $y_{ij} = \mathbf{x}_{ij}^T\beta + v_i + e_{ij}$, $j = 1,\ldots,n_i$; $i = 1,\ldots,m$ which may be written in matrix notation as

$$\mathbf{y}_i = \mathbf{X}_i\boldsymbol{\beta} + v_i\mathbf{1}_{n_i} + \mathbf{e}_i, \quad i = 1,\ldots,m \tag{7.2.1}$$

to make inference on $\bar{Y}_i$, by appealing to the general results in Section 6.3. The case of nonnegligible sampling fractions, n_i/N_i, will be handled by appealing to the population model (5.3.4); see subsection 7.2.5. Note that n_i/N_i is negligible if N_i is large because n_i is assumed to be small.

7.2.1 BLUP Estimator

The model (7.2.1) is a special case of the general model (6.3.1) with block diagonal covariance structure. We have

$$\mathbf{y}_i = \mathbf{y}_i, \quad \mathbf{X}_i = \mathbf{X}_i, \quad \mathbf{Z}_i = \mathbf{1}_{n_i},$$
$$\mathbf{v}_i = v_i, \quad \mathbf{e}_i = \mathbf{e}_i, \quad \boldsymbol{\beta} = (\beta_1,\ldots,\beta_p)^T,$$

where $\mathbf{y}_i$ is the $n_i \times 1$ vector of sample observations y_{ij} from the ith area. Further,

$$\mathbf{G}_i = \sigma_v^2, \quad \mathbf{R}_i = \sigma_e^2\,\mathrm{diag}_{1\le j\le n_i}(k_{ij}^2)$$

so that

$$\mathbf{V}_i = \mathbf{R}_i + \sigma_v^2\mathbf{1}_{n_i}\mathbf{1}_{n_i}^T.$$

Also, $\mu_i = \theta_i = \bar{\mathbf{X}}_i^T\boldsymbol{\beta} + v_i$ so that $\mathbf{l}_i = \bar{\mathbf{X}}_i$ and $\mathbf{m}_i = 1$. The matrix $\mathbf{V}_i$ can be inverted explicitly as

$$\mathbf{V}_i^{-1} = \frac{1}{\sigma_e^2}\left[\mathrm{diag}_j(a_{ij}) - \frac{\gamma_i}{a_{i\cdot}}\mathbf{a}_i\mathbf{a}_i^T\right] \tag{7.2.2}$$

using the following standard result on matrix inversion:

$$(\mathbf{A} + \mathbf{u}\mathbf{v}^T)^{-1} = \mathbf{A}^{-1} - \mathbf{A}^{-1}\mathbf{u}\mathbf{v}^T\mathbf{A}^{-1}/(1 + \mathbf{v}^T\mathbf{A}^{-1}\mathbf{u}). \tag{7.2.3}$$

Here we have

$$a_{ij} = k_{ij}^{-2}, \quad a_{i\cdot} = \sum\nolimits_j a_{ij}, \quad \mathbf{a}_i = (a_{i1},\ldots,a_{in_i})^T$$

and

$$\gamma_i = \sigma_v^2/(\sigma_v^2 + \sigma_e^2/a_{i\cdot}). \tag{7.2.4}$$

Making the above substitutions in the general formula (6.3.2) and noting that $(\sigma_v^2/\sigma_e^2)(1-\gamma_i) = \gamma_i/a_{i\cdot}$, we get the BLUP estimator of μ_i as

$$\tilde{\mu}_i^H = \bar{\mathbf{X}}_i^T\tilde{\boldsymbol{\beta}} + \gamma_i(\bar{y}_{ia} - \bar{\mathbf{x}}_{ia}^T\tilde{\boldsymbol{\beta}}), \tag{7.2.5}$$

where $\bar{y}_{ia}$ and $\bar{\mathbf{x}}_{ia}$ are weighted means given by

$$\bar{y}_{ia} = \sum_j a_{ij}y_{ij}/a_{i\cdot}, \quad \bar{\mathbf{x}}_{ia} = \sum_j a_{ij}\mathbf{x}_{ij}/a_{i\cdot},$$

and $\tilde{\boldsymbol{\beta}}$ is the BLUE of $\boldsymbol{\beta}$:

$$\tilde{\boldsymbol{\beta}} = \left(\sum_i \mathbf{X}_i^T\mathbf{V}_i^{-1}\mathbf{X}_i\right)^{-1}\left(\sum_i \mathbf{X}_i^T\mathbf{V}_i^{-1}\mathbf{y}_i\right), \tag{7.2.6}$$

where

$$\mathbf{X}_i^T\mathbf{V}_i^{-1}\mathbf{X}_i = \mathbf{A}_i = \sigma_e^{-2}\left(\sum_j a_{ij}\mathbf{x}_{ij}\mathbf{x}_{ij}^T - \gamma_i a_{i\cdot}\bar{\mathbf{x}}_{ia}\bar{\mathbf{x}}_{ia}^T\right) \tag{7.2.7}$$

and

$$\mathbf{X}_i^T\mathbf{V}_i^{-1}\mathbf{y}_i = \sigma_e^{-2}\left(\sum_j a_{ij}\mathbf{x}_{ij}y_{ij} - \gamma_i a_{i\cdot}\bar{\mathbf{x}}_{ia}\bar{y}_{ia}\right). \tag{7.2.8}$$

The BLUP estimator (7.2.5) can also be expressed as a weighted average of the "survey regression" estimator $\bar{y}_{ia} + (\bar{\mathbf{X}}_i - \bar{\mathbf{x}}_{ia})^T\tilde{\boldsymbol{\beta}}$ and the regression synthetic estimator $\bar{\mathbf{X}}_i^T\tilde{\boldsymbol{\beta}}$:

$$\tilde{\mu}_i^H = \gamma_i[\bar{y}_{ia} + (\bar{\mathbf{X}}_i - \bar{\mathbf{x}}_{ia})^T\tilde{\boldsymbol{\beta}}] + (1-\gamma_i)\bar{\mathbf{X}}_i^T\tilde{\boldsymbol{\beta}}. \tag{7.2.9}$$

The weight γ_i $(0 \le \gamma_i \le 1)$ measures the model variance, σ_v^2, relative to the total variance $\sigma_v^2 + \sigma_e^2/a_{i\cdot}$. If the model variance is relatively small, then γ_i will be small and more weight is attached to the synthetic component. Similarly, more weight is attached to the survey regression estimator as $a_{i\cdot}$ increases. Note that $a_{i\cdot}$ is of order $O(n_i)$ and it reduces to n_i if $k_{ij} = 1$ for all (i,j). Also, in the latter case the survey regression estimator is approximately design-unbiased for μ_i under simple random sampling, provided the total sample size $n = \sum_i n_i$ is large. In the case of general k_{ij}'s, it is model-unbiased for μ_i conditional on the realized local effect v_i, provided $\tilde{\boldsymbol{\beta}}$ is conditionally unbiased for $\boldsymbol{\beta}$. On the other hand, the BLUP estimator (7.2.9) is conditionally biased due to the presence of the synthetic component $\bar{\mathbf{X}}_i^T\tilde{\boldsymbol{\beta}}$. The sample regression estimator is therefore valid under weaker assumptions, but it fails to borrow strength as illustrated in Section 3.5, Chapter 3.

Under simple random sampling and $k_{ij} = 1$ for all (i,j), the BLUP estimator is design-consistent for $\bar{Y}_i$ as n_i increases because $\gamma_i \to 1$. For general designs with unequal survey weights w_{ij}, we consider survey weighted pseudo-BLUP estimators that are design consistent (subsection 7.2.7).

The MSE of the BLUP estimator is easily obtained either directly or from the general result (6.3.5) by letting $\boldsymbol{\delta} = (\sigma_v^2, \sigma_e^2)^T$. It is given by

$$\text{MSE}(\tilde{\mu}_i^H) = E(\tilde{\mu}_i^H - \mu_i)^2 = g_{1i}(\sigma_v^2, \sigma_e^2) + g_{2i}(\sigma_v^2, \sigma_e^2), \tag{7.2.10}$$

where

$$g_{1i}(\sigma_v^2, \sigma_e^2) = \gamma_i(\sigma_e^2/a_{i\cdot}) \tag{7.2.11}$$

and

$$g_{2i}(\sigma_v^2, \sigma_e^2) = (\bar{\mathbf{X}}_i - \gamma_i \bar{\mathbf{x}}_{ia})^T \left(\sum_i \mathbf{A}_i\right)^{-1} (\bar{\mathbf{X}}_i - \gamma_i \bar{\mathbf{x}}_{ia}) \tag{7.2.12}$$

with $\mathbf{A}_i$ given by (7.2.7). The first term, $g_{1i}(\sigma_v^2, \sigma_e^2)$, is of order $O(1)$, whereas the second term, $g_{2i}(\sigma_v^2, \sigma_e^2)$, due to estimating $\boldsymbol{\beta}$, is of order $O(m^{-1})$ for large m, assuming the following regularity conditions:

(i) k_{ij} and n_i are uniformly bounded. (7.2.13)

(ii) Elements of $\mathbf{X}_i$ are uniformly bounded such that $\mathbf{A}_i$ is of order $O(1)$. (7.2.14)

The leading term of the MSE of the BLUP estimator is given by $g_{1i}(\sigma_v^2, \sigma_e^2) = \gamma_i(\sigma_e^2/a_{i\cdot})$. Comparing this term to $\sigma_e^2/a_{i\cdot}$, the leading term of the MSE of the sample regression estimator, it is clear that the BLUP estimator provides considerable gain in efficiency over the sample regression estimator if γ_i is small. Therefore, models with smaller γ_i should be preferred, provided they provide adequate fit in terms of residual analysis and other model diagnostics. This is similar to the model choice in Example 7.1.1 for the basic area level model.

The BLUE $\tilde{\boldsymbol{\beta}}$ and its covariance matrix $(\sum_i \mathbf{X}_i^T \mathbf{V}_i^{-1} \mathbf{X}_i)^{-1}$ can be calculated using only ordinary least squares (OLS) by first transforming the model (7.2.1) with correlated errors $u_{ij} = v_i + e_{ij}$ to a model with uncorrelated errors u_{ij}^*. The transformed model is given by

$$k_{ij}^{-1}(y_{ij} - \tau_i \bar{y}_{ia}) = k_{ij}^{-1}(\mathbf{x}_{ij} - \tau_i \bar{\mathbf{x}}_{ia})^T \boldsymbol{\beta} + u_{ij}^*, \tag{7.2.15}$$

where $\tau_i = 1 - (1 - \gamma_i)^{1/2}$ and the u_{ij}^*'s have mean zero and constant variance σ_e^2 (Stukel and Rao (1997)). If $k_{ij} = 1$ for all (i, j), (7.2.15) reduces to the transformed model of Fuller and Battese (1973). Another advantage of the transformation method is that standard OLS model diagnostics may be applied to the transformed data $\{k_{ij}^{-1}(y_{ij} - \tau_i \bar{y}_{ia}), k_{ij}^{-1}(\mathbf{x}_{ij} - \tau_i \bar{\mathbf{x}}_{ia})\}$ to check the validity of the nested error regression model (7.2.1); see Example 7.2.1. In practice, τ_i is estimated from the data (subsection 7.2.2).

The BLUP estimator (7.2.9) depends on the variance ratio σ_v^2/σ_e^2, which is unknown in practice. Replacing σ_v^2 and σ_e^2 by estimators $\hat{\sigma}_v^2$ and $\hat{\sigma}_e^2$, we obtain an EBLUP estimator

$$\hat{\mu}_i^H = \hat{\gamma}_i[\bar{y}_{ia} + (\bar{\mathbf{X}}_i - \bar{\mathbf{x}}_{ia})^T \hat{\boldsymbol{\beta}}] + (1 - \hat{\gamma}_i)\bar{\mathbf{X}}_i^T \hat{\boldsymbol{\beta}}, \tag{7.2.16}$$

where $\hat{\gamma}_i$ and $\hat{\boldsymbol{\beta}}$ are the values of γ_i and $\tilde{\boldsymbol{\beta}}$ when (σ_v^2, σ_e^2) is replaced by $(\hat{\sigma}_v^2, \hat{\sigma}_e^2)$.

7.2.2 Estimation of σ_v^2 and σ_e^2

We present a simple method of estimating the variance components σ_v^2 and σ_e^2. It involves performing two ordinary least squares regressions and then using the method of moments to get unbiased estimators of σ_e^2 and σ_v^2 (Stukel and Rao (1997)). Fuller and Battese (1973) proposed this method for the special case $k_{ij} = 1$ for all (i, j).

We first calculate the residual sum of squares SSE(1) with ν_1 degrees of freedom by regressing through the origin the y-deviations $k_{ij}^{-1}(y_{ij} - \bar{y}_{ia})$ on the nonzero $\mathbf{x}$-deviations $k_{ij}^{-1}(\mathbf{x}_{ij} - \bar{\mathbf{x}}_{ia})$ for those areas with $n_i > 1$. This leads to an unbiased estimator of σ_e^2:

$$\hat{\sigma}_{em}^2 = \nu_1^{-1}\text{SSE}(1), \tag{7.2.17}$$

where $\nu_1 = n - m - p_1$ and p_1 is the number of nonzero $\mathbf{x}$-deviations. We next calculate the residual sum of squares SSE(2) by regressing y_{ij}/k_{ij} on $\mathbf{x}_{ij}/k_{ij}$. An unbiased estimator of σ_v^2 is then given by

$$\tilde{\sigma}_{vm}^2 = \eta_1^{-1}[\text{SSE}(2) - (n-p)\hat{\sigma}_e^2], \tag{7.2.18}$$

noting that

$$E[\text{SSE}(2)] = \eta_1\sigma_v^2 + (n-p)\sigma_e^2,$$

where

$$\eta_1 = \sum_i a_{i\cdot}\left[1 - a_{i\cdot}\bar{\mathbf{x}}_{ia}^T\left(\sum_i\sum_j a_{ij}\mathbf{x}_{ij}\mathbf{x}_{ij}^T\right)^{-1}\bar{\mathbf{x}}_{ia}\right]. \tag{7.2.19}$$

The estimators $\tilde{\sigma}_{vm}^2$ and $\hat{\sigma}_{em}^2$ are equivalent to those found by using the well-known "fitting-of-constants" method attributed to Henderson (1953). However, the latter method requires OLS regression on p_1+m variables, in contrast to p_1 variables for the transformation method, and thus is computationally more cumbersome as the number of small areas, m, increases.

Since $\tilde{\sigma}_{vm}^2$ can take negative values, we truncate $\tilde{\sigma}_{vm}^2$ to zero whenever it is negative. The truncated estimator $\hat{\sigma}_{vm}^2 = \max(\tilde{\sigma}_{vm}^2, 0)$ is no longer unbiased but it is consistent as m increases. For the special case of $k_{ij} = 1$ for all (i, j), Battese, Harter and Fuller (1988) proposed an alternative estimator of γ_i which is approximately unbiased for γ_i.

Assuming normality of the errors v_i and e_{ij}, ML or REML can also be employed. For example, PROC MIXED in SAS can be used to calculate ML or REML estimates of σ_v^2 and σ_e^2 and the associated EBLUP estimate $\hat{\mu}_i^H$. A naive estimate of MSE,

$$\text{mse}_N(\hat{\mu}_i^H) = g_{1i}(\hat{\sigma}_v^2, \hat{\sigma}_e^2) + g_{2i}(\hat{\sigma}_v^2, \hat{\sigma}_e^2), \tag{7.2.20}$$

can also be computed using PROC MIXED in SAS, where $(\hat{\sigma}_v^2, \hat{\sigma}_e^2)$ are the ML or REML estimators of (σ_v^2, σ_e^2).

7.2.3 MSE of EBLUP

The second-order MSE approximation (6.3.9) is valid for the nested error linear regression model (7.2.1), under regularity conditions (7.2.13) and (7.2.14) and normality of the errors v_i and e_{ij}. It reduces to

$$\text{MSE}(\hat{\mu}_i^H) \approx g_{1i}(\sigma_v^2, \sigma_e^2) + g_{2i}(\sigma_v^2, \sigma_e^2) + g_{3i}(\sigma_v^2, \sigma_e^2), \tag{7.2.21}$$

where $g_{1i}(\sigma_v^2, \sigma_e^2)$ and $g_{2i}(\sigma_v^2, \sigma_e^2)$ are given by (7.2.11) and (7.2.12), respectively. Further,

$$g_{3i}(\sigma_v^2, \sigma_e^2) = a_{i\cdot}^{-2}(\sigma_v^2 + \sigma_e^2/a_{i\cdot})^{-3} h(\sigma_v^2, \sigma_e^2) \tag{7.2.22}$$

with

$$h(\sigma_v^2, \sigma_e^2) = \sigma_e^4 \bar{V}_{vv}(\boldsymbol{\delta}) + \sigma_v^4 \bar{V}_{ee}(\boldsymbol{\delta}) - 2\sigma_e^2\sigma_v^2 \overline{V}_{ve}(\boldsymbol{\delta}), \tag{7.2.23}$$

where $\boldsymbol{\delta} = (\sigma_v^2, \sigma_e^2)^T$, $\bar{V}_{ee}(\boldsymbol{\delta})$ and $\bar{V}_{vv}(\boldsymbol{\delta})$ are the asymptotic variances of the estimators $\hat{\sigma}_e^2$ and $\hat{\sigma}_v^2$ and $\overline{V}_{ve}(\boldsymbol{\delta})$ is the asymptotic covariance of $\hat{\sigma}_v^2$ and $\hat{\sigma}_e^2$.

If the method of fitting-of-constants is used, then $\hat{\boldsymbol{\delta}} = (\hat{\sigma}_v^2, \hat{\sigma}_e^2)^T = (\hat{\sigma}_{vm}^2, \hat{\sigma}_{em}^2)^T$ with asymptotic variances

$$\bar{V}_{vvm}(\boldsymbol{\delta}) = 2\eta_1^{-2}\left[\nu_1^{-1}(n-p-\nu_1)(n-p)\sigma_e^4 + \eta_2\sigma_v^4 + 2\eta_1\sigma_e^2\sigma_v^2\right], \tag{7.2.24}$$

$$\bar{V}_{eem}(\boldsymbol{\delta}) = 2\nu_1^{-1}\sigma_e^4, \tag{7.2.25}$$

and asymptotic covariance

$$\bar{V}_{vem}(\boldsymbol{\delta}) = -2\eta_1^{-1}\nu_1^{-1}(n-p-\nu_1)\sigma_e^4, \tag{7.2.26}$$

where η_1 is given by (7.2.19), ν_1 is defined below (7.2.17) and

$$\eta_2 = \sum_i a_{i\cdot}^2\left(1 - a_{i\cdot}\bar{\mathbf{x}}_{ia}^T\mathbf{A}_1^{-1}\bar{\mathbf{x}}_{ia}\right) + \text{tr}\left(\mathbf{A}_1^{-1}\sum_i a_{i\cdot}^2\bar{\mathbf{x}}_{ia}\bar{\mathbf{x}}_{ia}^T\right)^2$$

with $\mathbf{A}_1 = \sum_i\sum_j a_{ij}\mathbf{x}_{ij}\mathbf{x}_{ij}^T$ (Stukel (1991), Chapter 2). If the errors e_{ij} have equal variance σ_e^2 (i.e., $k_{ij} = 1, a_{ij} = k_{ij}^{-2} = 1$), then (7.2.24)–(7.2.26) reduce to the formulas given by Prasad and Rao (1990).

If ML or REML is used to estimate σ_v^2 and σ_e^2, then $\hat{\boldsymbol{\delta}} = \hat{\boldsymbol{\delta}}_{\text{ML}}$ or $\hat{\boldsymbol{\delta}} = \hat{\boldsymbol{\delta}}_{\text{RE}}$ with asymptotic covariance matrix $\overline{\mathbf{V}}(\boldsymbol{\delta}) = \overline{\mathbf{V}}_{\text{ML}}(\boldsymbol{\delta}) = \overline{\mathbf{V}}_{\text{RE}}(\boldsymbol{\delta})$ given by the inverse of the 2×2 information matrix $\mathbf{I}(\delta)$ with diagonal elements

$$\mathcal{I}_{vv}(\boldsymbol{\delta}) = \frac{1}{2}\sum_i \text{tr}(\mathbf{V}_i^{-1}\partial\mathbf{V}_i/\partial\sigma_v^2)^2,$$

$$\mathcal{I}_{ee}(\boldsymbol{\delta}) = \frac{1}{2}\sum_i \text{tr}(\mathbf{V}_i^{-1}\partial\mathbf{V}_i/\partial\sigma_e^2)^2$$

and off-diagonal elements

$$\mathcal{I}_{ve}(\boldsymbol{\delta}) = \mathcal{I}_{ev}(\boldsymbol{\delta}) = \frac{1}{2}\sum_i \mathrm{tr}(\mathbf{V}_i^{-1}\partial\mathbf{V}_i/\partial\sigma_v^2)(\mathbf{V}_i^{-1}\partial\mathbf{V}_i/\partial\sigma_e^2),$$

where

$$\partial\mathbf{V}_i/\partial\sigma_v^2 = \mathbf{1}_{n_i}\mathbf{1}_{n_i}^T; \quad \partial\mathbf{V}_i/\partial\sigma_e^2 = \sigma_e^{-2}\mathbf{R}_i.$$

Using the formula (7.2.2) for $\mathbf{V}_i^{-1}$ we get, after simplification,

$$\mathcal{I}_{vv}(\boldsymbol{\delta}) = \frac{1}{2}\sum_i a_{i\cdot}^2\alpha_i^{-2}, \tag{7.2.27}$$

$$\mathcal{I}_{ee}(\boldsymbol{\delta}) = \frac{1}{2}\sum_i[(n_i-1)\sigma_e^{-4} + \alpha_i^{-2}] \tag{7.2.28}$$

and

$$\mathcal{I}_{ve}(\boldsymbol{\delta}) = \frac{1}{2}\sum_i a_{i\cdot}\alpha_i^{-2}, \tag{7.2.29}$$

where

$$\alpha_i = \sigma_e^2 + a_{i\cdot}\sigma_v^2. \tag{7.2.30}$$

In the special case of equal error variances (i.e., $k_{ij} = 1$), (7.2.27)–(7.2.29) reduce to the formulas given by Datta and Lahiri (2000).

7.2.4 MSE Estimation

The estimator of MSE given by (6.3.11) is valid for REML and the method of fitting-of-constants under regularity conditions (7.2.13) and (7.2.14) and normality of the errors v_i and e_{ij}. It reduces to

$$\mathrm{mse}(\hat{\mu}_i^H) = g_{1i}(\hat{\sigma}_v^2, \hat{\sigma}_e^2) + g_{2i}(\hat{\sigma}_v^2, \hat{\sigma}_e^2) + 2\, g_{3i}(\hat{\sigma}_v^2, \hat{\sigma}_e^2) \tag{7.2.31}$$

if $\hat{\boldsymbol{\delta}}$ is chosen as $\hat{\boldsymbol{\delta}}_{\mathrm{RE}}$ or $\hat{\boldsymbol{\delta}}_m$. The corresponding area-specific versions, $\mathrm{mse}_1(\hat{\mu}_i^H)$ and $\mathrm{mse}_2(\hat{\mu}_i^H)$, are obtained from (6.3.15) and (6.3.16), by evaluating $g_{3i}^*(\boldsymbol{\delta}, \mathbf{y}_i)$ given by (6.3.14). After some simplification, we get

$$g_{3i}^*(\boldsymbol{\delta}, \mathbf{y}_i) = a_{i\cdot}^{-2}(\sigma_v^2 + \sigma_e^2/a_{i\cdot})^{-4}h(\sigma_v^2, \sigma_e^2)(\bar{y}_{ia} - \bar{\mathbf{x}}_{ia}^T\tilde{\beta})^2. \tag{7.2.32}$$

Hence, the area-specific versions are given by

$$\mathrm{mse}_1(\hat{\mu}_i^H) = g_{1i}(\hat{\sigma}_v^2, \hat{\sigma}_e^2) + g_{2i}(\hat{\sigma}_v^2, \hat{\sigma}_e^2) + 2g_{3i}^*(\hat{\boldsymbol{\delta}}, \mathbf{y}_i) \tag{7.2.33}$$

and

$$\mathrm{mse}_2(\hat{\mu}_i^H) = g_{1i}(\hat{\sigma}_v^2, \hat{\sigma}_e^2) + g_{2i}(\hat{\sigma}_v^2, \hat{\sigma}_e^2) + g_{3i}(\hat{\sigma}_v^2, \hat{\sigma}_e^2) + g_{3i}^*(\hat{\boldsymbol{\delta}}, \mathbf{y}_i). \tag{7.2.34}$$

For the ML estimator $\hat{\boldsymbol{\delta}}_{\text{ML}}$, the MSE estimator (6.3.12) is valid. It reduces to

$$\text{mse}(\hat{\mu}_i^H) = g_{1i}(\hat{\boldsymbol{\delta}}) - \mathbf{b}_{\hat{\boldsymbol{\delta}}}^T(\hat{\boldsymbol{\delta}})\nabla g_{1i}(\hat{\boldsymbol{\delta}}) + g_{2i}(\hat{\boldsymbol{\delta}}) + 2\, g_{3i}(\hat{\boldsymbol{\delta}}), \tag{7.2.35}$$

where $\hat{\boldsymbol{\delta}} = \hat{\boldsymbol{\delta}}_{\text{ML}}$ and

$$\nabla g_{1i}(\hat{\boldsymbol{\delta}}) = \alpha_i^{-2}(\sigma_e^4, \sigma_v^4 a_{i\cdot})^T.$$

The bias term $\mathbf{b}_{\hat{\boldsymbol{\delta}}}(\boldsymbol{\delta})$ for the ML estimator $\hat{\boldsymbol{\delta}}_{\text{ML}}$ is obtained from (6.3.13) as

$$\mathbf{b}_{\hat{\boldsymbol{\delta}}_{\text{ML}}}(\boldsymbol{\delta}) = \frac{1}{2}\mathcal{I}^{-1}(\boldsymbol{\delta})\left[\text{tr}\left(\sum_i \mathbf{X}_i^T\mathbf{V}_i^{-1}\mathbf{X}_i\right)^{-1}\left(\sum_i \mathbf{X}_i^T(\partial\mathbf{V}_i^{-1}/\partial\sigma_v^2)\mathbf{X}_i\right),\right.$$

$$\left.\text{tr}\left(\sum_i \mathbf{X}_i^T\mathbf{V}_i^{-1}\mathbf{X}_i\right)^{-1}\left(\sum_i \mathbf{X}_i^T(\partial\mathbf{V}_i^{-1}/\partial\sigma_e^2)\mathbf{X}_i\right)\right]^T, \tag{7.2.36}$$

where $\mathcal{I}(\boldsymbol{\delta})$ is the 2×2 information matrix with elements given by (7.2.27)–(7.2.29), and $\mathbf{X}_i^T\mathbf{V}_i^{-1}\mathbf{X}_i$ is given by (7.2.7). Further, it follows from (7.2.7) that

$$\mathbf{X}_i^T(\partial\mathbf{V}_i^{-1}/\partial\sigma_v^2)\mathbf{X}_i = -\alpha_i^{-2}a_{i\cdot}^2\overline{\mathbf{x}}_{ia}\overline{\mathbf{x}}_{ia}^T$$

and

$$\mathbf{X}_i^T(\partial\mathbf{V}_i^{-1}/\partial\sigma_e^2)\mathbf{X}_i = -\sigma_e^{-4}\Sigma_j a_{ij}\mathbf{x}_{ij}\mathbf{x}_{ij}^T + \alpha_i^{-2}\sigma_e^{-4}\sigma_v^2(\sigma_e^2 + \alpha_i)a_{i\cdot}^2\overline{\mathbf{x}}_{ia}\overline{\mathbf{x}}_{ia}^T.$$

7.2.5 Non-negligible Sampling Rates

If the sampling rate $f_i = n_i/N_i$ is not negligible, then we cannot take the small area mean $\bar{Y}_i$ as $\overline{\mathbf{X}}\boldsymbol{\beta} + v_i$. However, we can write $\bar{Y}_i$ as

$$\bar{Y}_i = f_i\overline{y}_i + (1 - f_i)\overline{y}_i^*,$$

where $\overline{y}_i$ is the sample mean and $\overline{y}_i^*$ is the mean of nonsampled values y_{ij}^*, of the ith area. Under the population model (5.3.4), we replace the unobserved y_{ij}^* by its estimator $\mathbf{x}_{ij}^{*T}\tilde{\boldsymbol{\beta}} + \tilde{v}_i$, where $\mathbf{x}_{ij}^*$ is the $\mathbf{x}$-value associated with y_{ij}^*. The resulting BLUP estimator of $\bar{Y}_i$ is given by

$$\tilde{\bar{Y}}_i^H = f_i\bar{y}_i + (1 - f_i)(\bar{\mathbf{x}}_i^{*T}\tilde{\boldsymbol{\beta}} + \tilde{v}_i) = f_i\bar{y}_i + (1 - f_i)\tilde{\bar{y}}_i^{*H}, \tag{7.2.37}$$

where $\bar{\mathbf{x}}_i^*$ is the mean of nonsampled values $\mathbf{x}_{ij}^*$ and

$$\tilde{\bar{y}}_i^{*H} = \gamma_i[\bar{y}_{ia} + (\bar{\mathbf{x}}_i^* - \bar{\mathbf{x}}_{ia})^T)\tilde{\boldsymbol{\beta}}] + (1 - \gamma_i)\bar{\mathbf{x}}_i^{*T}\tilde{\boldsymbol{\beta}}. \tag{7.2.38}$$

Note that $\bar{\mathbf{x}}_i^* = (N_i\overline{\mathbf{X}}_i - n_i\bar{\mathbf{x}}_i)/(N_i - n_i)$ can be computed knowing only the population mean $\overline{\mathbf{X}}_i$. The BLUP property of $\tilde{\bar{Y}}_i^H$ is easily established by showing that $\text{Cov}(\mathbf{b}^T\mathbf{y}, \tilde{\bar{Y}}_i^H - \overline{Y}_i) = 0$ for every zero linear function $\mathbf{b}^T\mathbf{y}$, that is, $E(\mathbf{b}^T\mathbf{y}) = 0$ (Stukel (1991), Chapter 2).

Replacing σ_v^2 and σ_e^2 in (7.2.38) by estimators $\hat{\sigma}_v^2$ and $\hat{\sigma}_e^2$ we obtain an EBLUP estimator:

$$\hat{\bar{Y}}_i^H = f_i \bar{y}_i + (1 - f_i)\hat{\bar{y}}_i^{*H}. \tag{7.2.39}$$

The MSE of $\hat{\bar{Y}}_i^H$ is given by

$$\text{MSE}(\hat{\bar{Y}}_i^H) = E(\hat{\bar{Y}}_i^H - \bar{Y}_i)^2 = (1-f_i)^2 E(\hat{\bar{y}}_i^{*H} - \mu_i^*)^2 + (N_i^{-2}\mathbf{k}_i^{*\,T}\mathbf{k}_i^*)\sigma_e^2, \tag{7.2.40}$$

noting that

$$\hat{\bar{Y}}_i^H - \bar{Y}_i = (1 - f_i)[(\hat{\bar{y}}_i^{*H} - \mu_i^*) - \bar{e}_i^*].$$

Here $\mathbf{k}_i^*$ is the known vector of k_{ij}-values for nonsampled elements, $\bar{e}_i^*$ is the e-mean of the nonsampled elements and $\mu_i^* = \bar{\mathbf{x}}_i^{*\,T}\boldsymbol{\beta} + v_i$ for the ith area.

Now noting that $\hat{\bar{y}}_i^{*H}$ is the EBLUP estimator of μ_i^*, we can use (7.2.21) to get a second-order approximation to $E(\hat{\bar{y}}_i^{*H} - \mu_i^*)^2$ as

$$E(\hat{\bar{y}}_i^{*H} - \mu_i^*)^2 \approx g_{1i}(\sigma_v^2, \sigma_e^2) + g_{2i}^*(\sigma_v^2, \sigma_e^2) + g_{3i}(\sigma_v^2, \sigma_e^2), \tag{7.2.41}$$

where $g_{2i}^*(\sigma_v^2, \sigma_e^2)$ is obtained from $g_{2i}(\sigma_v^2, \sigma_e^2)$ by changing $\bar{\mathbf{X}}_i$ to $\bar{\mathbf{x}}_i^*$. Substituting (7.2.41) in (7.2.40) gives a second-order approximation to $\text{MSE}(\hat{\bar{Y}}_i^H)$.

Similarly, an estimator of $\text{MSE}(\hat{\bar{Y}}_i^H)$, correct to terms of order $o(m^{-1})$, is given by

$$\text{mse}(\hat{\bar{Y}}_i^H) = (1 - f_i)^2 \text{mse}(\hat{\bar{y}}_i^{*H}) + (N_i^{-2}\mathbf{k}_i^{*\,T}\mathbf{k}_i^*)\hat{\sigma}_e^2, \tag{7.2.42}$$

where

$$\text{mse}(\hat{\bar{y}}_i^{*H}) = g_{1i}(\hat{\sigma}_v^2, \hat{\sigma}_e^2) + g_{2i}^*(\hat{\sigma}_v^2, \hat{\sigma}_e^2) + 2\, g_{3i}(\hat{\sigma}_v^2, \hat{\sigma}_e^2) \tag{7.2.43}$$

and $(\hat{\sigma}_v^2, \hat{\sigma}_e^2)$ are the REML or the fitting-of-constants estimators of (σ_v^2, σ_e^2). Area-specific versions can be obtained similarly, using (7.2.33) and (7.2.34).

7.2.6 Examples

We now provide some details of the applications in Examples 5.3.1 and 5.3.2, Chapter 5.

Example 7.2.1. County Crop Areas. Battese, Harter and Fuller (1988) applied the nested error linear regression model with equal error variances (i.e., $k_{ij} = 1$) to estimate area under corn and area under soybeans for each of $m = 12$ counties in north-central Iowa, using farm interview data in conjunction with LANDSAT satellite data. Each county was divided into area segments, and the areas under corn and soybeans were ascertained for a sample of segments by interviewing farm operators. The number of sample segments,

n_i, in a county ranged from 1 to 5 ($\sum n_i = n = 37$), while the total number of segments in a county ranged from 394 to 687. Because of negligible sample fractions n_i/N_i, the small area means $\overline{Y}_i$ are taken as $\mu_i = \overline{\mathbf{X}}_i^T \beta + v_i$.

Unit level auxiliary data $\{x_{1ij}, x_{2ij}\}$ in the form of the number of pixels classified as corn, x_{1ij}, and the number of pixels classified as soybeans, x_{2ij}, were also obtained for all the area segments, including the sample segments, in each county, using the LANDSAT satellite readings. In this application, y_{ij} denotes the number of hectares of corn (soybeans) in the jth sample area segment in the ith county. Table 1 in Battese, Harter and Fuller (1988) reports the values of n_i, N_i, $\{y_{ij}, x_{1ij}, x_{2ij}\}$ as well as the area means $\overline{X}_{1i}$ and $\overline{X}_{2i}$. The sample data for the second sample segment in Hardin county was deleted from the estimation of means μ_i because the corn area for that segment looked erroneous in a preliminary analysis. It is interesting to note that model diagnostics, based on the local influence approach (see subsection 6.3.4), identified this sample data point as possibly erroneous (Hartless, Booth and Littell (2000)).

Battese et al. (1988) used model (7.2.1) with $\mathbf{x}_{ij} = (1, x_{1ij}, x_{2ij})^T$ and normally distributed errors v_i and e_{ij} with common variances σ_v^2 and σ_e^2. Estimates of σ_v^2 and σ_e^2 were obtained as $\hat{\sigma}_e^2 = \hat{\sigma}_{em}^2 = 150$, $\hat{\sigma}_v^2 = 140$ for corn and $\hat{\sigma}_e^2 = \hat{\sigma}_{em}^2 = 195$, $\hat{\sigma}_v^2 = 272$ for soybeans. The estimate $\hat{\sigma}_v^2$ is slightly different from $\hat{\sigma}_{vm}^2$, the estimate based on the fitting-of-constants method. The regression coefficients associated with x_{1ij} and x_{2ij} were both significant for the corn model, but only the coefficient associated with x_{2ij} was significant for the soybeans model. The between-county variance, σ_v^2, was significant at the 10% level for corn, and at the 1% level for soybeans.

Battese, Harter and Fuller (1988) reported some methods for validating the assumed model with $\mathbf{x}_{ij} = (1, x_{1ij}, x_{2ij})^T$. First, they introduced quadratic terms x_{1ij}^2 and x_{2ij}^2 into the model and tested the null hypothesis that the regression coefficients of the quadratic terms are zero. The null hypothesis was not rejected at the 5% level. Second, to test the normality of the error terms v_i and e_{ij}, the transformed residuals $(y_{ij} - \hat{\tau}_i \overline{y}_i) - (\mathbf{x}_{ij} - \hat{\tau}_i \overline{\mathbf{x}}_i)^T \hat{\beta}$ were computed, where $\hat{\tau}_i = 1 - (1 - \hat{\gamma}_i)^{1/2}$. Under the null hypothesis, the transformed residuals are approximately iid $N(0, \sigma_e^2)$. Hence, standard methods for checking normality can be applied to the transformed residuals. In particular, the well-known Shapiro-Wilk W statistic for the transformed residuals gave values of 0.985 and 0.957 for corn and soybeans, respectively, yielding p-values (probabilities of getting values less than those observed under normality) of 0.921 and 0.299, respectively (Shapiro and Wilk (1965)). It may be noted that small values of W correspond to departure from normality. Although $n = 37$ is not large, the p-values suggest no evidence against the hypothesis of normality. Normal quantile-quantile (q–q) plots of transformed residuals also indicated no evidence against normality for both corn and soybeans. Jiang, Lahiri and Wu (1999) proposed an alternative test of normality (or specified distributions) of the error terms v_i and e_{ij}, based on a chi-squared statistic with estimated cell frequencies.

A limitation of the transformed residuals is that the effect of individual errors, e_{ij}, may be masked. To study the effects of individual errors, we can examine the standardized EBLUP residuals $(y_{ij} - \mathbf{x}_{ij}^T\hat{\boldsymbol{\beta}} - \hat{v}_i)/\hat{\sigma}_e$. If the model is valid, the standardized residuals are approximately iid $N(0,1)$ variables. Residual plots of the standardized residuals can reveal the effects of individual errors. To check for the normality of the small area effects v_i and to detect outlier v_i's, a normal probability plot of EBLUP estimates $\hat{v}_i$, divided by their standardized errors, may be examined (Lange and Ryan (1989)).

Table 7.3 reports the EBLUP estimates, $\hat{\mu}_i^H$, of small area means, μ_i, for corn and soybeans using $\hat{\sigma}_{em}^2$ and $\hat{\sigma}_{vm}^2$. Estimated standard errors of the EBLUP estimates and the survey regression estimates, $\hat{\mu}_i^{\text{SR}} = \bar{y}_i + (\overline{\mathbf{X}}_i - \bar{\mathbf{x}}_i)^T\hat{\boldsymbol{\beta}}$, denoted by $s(\hat{\mu}_i^H)$ and $s(\hat{\mu}_i^{\text{SR}})$, are also given. It is clear from Table 7.3 that the ratio of the estimated standard error of the EBLUP estimate to that of the survey regression estimate decreases from 0.97 to 0.77 as the number of sample area segments, n_i, decreases from 5 to 1. The reduction in the estimated standard error is considerable when $n_i \leq 3$.

TABLE 7.3 EBLUP Estimates of County Means and Estimated Standard Errors of EBLUP and Survey Regression Estimates

		Corn			Soybeans		
County	n_i	$\hat{\mu}_i^H$	$s(\hat{\mu}_i^H)$	$s(\hat{\mu}_i^{\text{SR}})$	$\hat{\mu}_i^H$	$s(\hat{\mu}_i^H)$	$s(\hat{\mu}_i^{\text{SR}})$
Cerro Gordo	1	122.2	9.6	13.7	77.8	12.0	15.6
Hamilton	1	126.3	9.5	12.9	94.8	11.8	14.8
Worth	1	106.2	9.3	12.4	86.9	11.5	14.2
Humboldt	2	108.0	8.1	9.7	79.7	9.7	11.1
Franklin	3	145.0	6.5	7.1	65.2	7.6	8.1
Pocahontas	3	112.6	6.6	7.2	113.8	7.7	8.2
Winnebago	3	112.4	6.6	7.2	98.5	7.7	8.3
Wright	3	122.1	6.7	7.3	112.8	7.8	8.4
Webster	4	115.8	5.8	6.1	109.6	6.7	7.0
Hancock	5	124.3	5.3	5.7	101.0	6.2	6.5
Kossuth	5	106.3	5.2	5.5	119.9	6.1	6.3
Hardin	5	143.6	5.7	6.1	74.9	6.6	6.9

SOURCE: Adapted from Table 2 in Battese et al. (1988).

The EBLUP estimates, $\hat{\mu}_i^H$, were adjusted (calibrated) to agree with the survey regression estimate for the entire area covering the 12 counties. The latter estimate is given by $\hat{\mu}^{\text{SR}} = \sum_{l=1}^{12} W_l \hat{\mu}_l^{\text{SR}}$, where W_l is the proportion of population area segments belong to lth area, and the overall population mean $\mu = \sum W_l \mu_l$. The estimator $\hat{\mu}^{\text{SR}}$ is approximately design-unbiased for μ under simple random sampling within areas and its design standard error is relatively small. The adjusted EBLUP estimates are given by

$$\hat{\mu}_i^H(a) = \hat{\mu}_i^H + \hat{p}_i^{(1)}(\hat{\mu}^{\text{SR}} - \hat{\mu}^H), \tag{7.2.44}$$

where $\hat{\mu}^H = \sum W_l \hat{\mu}_l^H$ and

$$\hat{p}_i^{(1)} = \left[\sum W_l^2 \text{mse}(\hat{\mu}_l^H)\right]^{-1} [W_i \text{mse}(\hat{\mu}_i^H)].$$

It follows from (7.2.44) that $\sum W_i\hat{\mu}_i^H(a) = \hat{\mu}^{\text{SR}}$. A drawback of $\hat{\mu}_i^H(a)$ is that it depends on the MSE estimates $\text{mse}(\hat{\mu}_l^H)$. Pfeffermann and Barnard (1991) proposed an "optimal" adjustment of the form (7.2.44) with $\hat{p}_i^{(1)}$ changed to

$$\hat{p}_i^{(2)} = [\text{mse}(\hat{\mu}^H)]^{-1}\text{mpe}(\hat{\mu}_i^H, \hat{\mu}^H), \tag{7.2.45}$$

where $\text{mpe}(\hat{\mu}_i^H, \hat{\mu}^H)$ is an estimate of the mean product error of $\hat{\mu}_i^H$ and $\hat{\mu}^H$, $\text{MPE}(\hat{\mu}_i^H, \hat{\mu}^H) = E(\hat{\mu}_i^H - \mu_i)(\mu^H - \mu)$. The term $\hat{p}_i^{(2)}$ involves estimates of $\text{MPE}(\hat{\mu}_i^H, \hat{\mu}_l^H), l \neq i$. A simple adjustment compared to (7.2.44) and (7.2.45) is obtained from (7.2.44) by changing $\hat{p}_i^{(1)}$ to

$$\hat{p}_i^{(3)} = \left[\sum W_l^2 \text{mse}(\hat{\mu}_l^{\text{SR}})\right]^{-1} \left[W_i \text{mse}(\hat{\mu}_i^{\text{SR}})\right], \tag{7.2.46}$$

following Isaki, Tsay and Fuller (2000) and Wang (2000). An alternative is to use the simple ratio adjustment which is given by (7.2.44) with $\hat{p}_i^{(1)}$ changed to

$$\hat{p}_i^{(4)} = \hat{\mu}_i^H / \hat{\mu}^H. \tag{7.2.47}$$

Mantel, Singh and Bureau (1993) conducted an empirical study on the performance of the adjustment estimators using $\hat{p}_i^{(1)}$, $\hat{p}_i^{(2)}$ and $\hat{p}_i^{(4)}$, using a synthetic population based on Statistics Canada's Survey of Employment, Payroll and Hours. Their results indicated that the simple ratio adjustment, $\hat{p}_i^{(4)}$, often performs better than the more complex adjustments, possibly due to instability of mse- and mpe-terms involved in $\hat{p}_i^{(1)}$ and $\hat{p}_i^{(2)}$.

Example 7.2.2. Simulation Study. Rao and Choudhry (1995) studied the relative performance of some direct and indirect estimators, using real and synthetic populations. For the real population, a sample of 1,678 unincorporated tax filers (units) from the province of Nova Scotia, Canada, divided into 18 census divisions, was treated as the overall population. In each census division, units were further classified into four mutually exclusive industry groups. The objective was to estimate the total wages and salaries (Y_i) for each nonempty census division by industry group i (small areas of interest). Here we focus on the industry group "construction" with 496 units and average sample size equal to 27.5. Gross business income, available for all the units, was used as an auxiliary variable (x). The overall correlation coefficient between y and x for construction was 0.64.

To make comparisons between estimators under customary repeated sampling, $R = 500$ samples, each of size $n = 149$, from the overall population of $N = 1,678$ units were selected by simple random sampling. From each simulated sample, the following estimators were calculated: (1) Post-stratified estimator (PST): $= N_i\bar{y}_i$ if $n_i \geq 1; = 0$ if $n_i = 0$, where N_i and n_i are the population and sample sizes in the ith area (n_i is a random variable). (2) Ratio-synthetic estimator (SYN): $(\bar{y}/\bar{x})X_i$, where $\bar{y}$ and $\bar{x}$ are the overall sample means in the industry group and X_i is the x-total for the ith area.

(3) Sample size dependent estimator (SSD): $\psi_i(D)(\text{PST}) + [1 - \psi_i(D)](\text{SYN})$ with $\psi_i(D) = 1$ if $n_i/n \geq N_i/N; = (n_i/n)(N_i/N)^{-1}$ otherwise. (4) EBLUP estimator $\hat{Y}_i^H = N_i \hat{\bar{Y}}_i^H$ where $\hat{\bar{Y}}_i^H$ given by (7.2.39), using the nested error regression model (7.2.1) with $\mathbf{x}_{ij}^T\beta = \beta x_{ij}$ and $k_{ij} = x_{ij}^{1/2}$ and the fitting-of-constants estimators $\hat{\sigma}_{vm}^2$ and $\hat{\sigma}_{em}^2$. To examine the aptness of this model, the model was fitted to the 496 population pairs (y_{ij}, x_{ij}) from the construction group and the standardized EBLUP residuals $(y_{ij} - \hat{\beta}x_{ij} - \hat{v}_i)/(\hat{\sigma}_e x_{ij}^{1/2})$ were examined. A plot of these residuals against the x_{ij}'s indicated a reasonable but not good fit in the sense that the plot revealed an upward shift with several values larger than 1.0 but none below −1.0. Several variations of the model, including a model with an intercept term, did not lead to better fits.

For each estimator, the values of average absolute relative bias ($\overline{\text{ARB}}$), average relative efficiency ($\overline{\text{EFF}}$) and average absolute relative error ($\overline{\text{ARE}}$), defined as follows, were calculated:

$$\overline{\text{ARB}} = \frac{1}{m}\sum_{i=1}^{m}\left|\frac{1}{500}\sum_{r=1}^{500}(\text{est}_r/Y_i - 1)\right|,$$

$$\overline{\text{EFF}} = \left[\overline{\text{MSE}}(\text{PST})/\overline{\text{MSE}}(\text{est})\right]^{1/2},$$

$$\overline{\text{ARE}} = \frac{1}{m}\sum_{i=1}^{m}\frac{1}{500}\sum_{r=1}^{500}|\text{est}_r/Y_i - 1|,$$

where the average is taken over $m = 18$ census divisions in the industry group. Here est_r denotes the value of the estimator, est, for the rth simulated sample $(r = 1, 2, \ldots, 500)$, Y_i is the true small area total and

$$\overline{\text{MSE}}(\text{est}) = \frac{1}{m}\sum_{i=1}^{m}\frac{1}{500}\sum_{r=1}^{500}(\text{est}_r - Y_i)^2;$$

$\overline{\text{MSE}}(\text{PST})$ is obtained by changing est_r to PST_r, the value of the post-stratified estimator for the rth simulated sample. Note that $\overline{\text{ARB}}$ measures the bias of an estimator, whereas $\overline{\text{EFF}}$ and $\overline{\text{ARE}}$ both measure the accuracy of an estimator.

Table 7.4 reports the percentage values of $\overline{\text{ARB}}$, $\overline{\text{EFF}}$ and $\overline{\text{ARE}}$ for the construction group. It is clear from Table 7.4 that both SYN and EBLUP perform significantly better than PST and SSD in terms of $\overline{\text{EFF}}$ and $\overline{\text{ARE}}$, leading to larger $\overline{\text{EFF}}$ values and smaller $\overline{\text{ARE}}$ values. For example, $\overline{\text{EFF}}$ for the EBLUP estimator is 266.1% compared to 137.6% for SSD. In terms of $\overline{\text{ARB}}$, SYN has the largest value (45.7%) as expected, followed by the EBLUP estimator with $\overline{\text{ARB}}$=11.3%; PST and SSD have smaller $\overline{\text{ARB}}$: 5.4% and 2.9%, respectively. Overall, EBLUP is somewhat better than SYN: $\overline{\text{EFF}}$ value of 261.1% versus 232.8% and $\overline{\text{ARE}}$ value of 13.5% versus 16.5%. It is gratifying that the EBLUP estimator under the assumed model performed well, despite the problems found in the residual analysis.

The estimators were also compared under a synthetic population generated from the assumed model with the real population x-values. The parameter values $(\beta, \sigma_v^2, \sigma_e^2)$ used for generating the synthetic population were the estimates obtained by fitting the model to the real population pairs (y_{ij}, x_{ij}): $\beta = 0.21, \sigma_v^2 = 1.58$ and $\sigma_e^2 = 1.34$. A plot of the standardized EBLUP residuals, obtained by fitting the model to the synthetic population, showed an excellent fit as expected. Table 7.4 also reports the percentage values of $\overline{\text{ARB}}$, $\overline{\text{EFF}}$ and $\overline{\text{ARE}}$ for the synthetic population. Comparing these values to the corresponding values for the real population, it is clear that $\overline{\text{EFF}}$ increases for EBLUP and SYN, while it remains essentially unchanged for SSD. Similarly, $\overline{\text{ARE}}$ decreases for EBLUP and SYN, while it remains essentially unchanged for SSD. The value of $\overline{\text{ARB}}$ also decreases for EBLUP and SYN: 11.3% versus 8.4% for EBLUP and 15.7% versus 12.5% for SYN.

TABLE 7.4 Unconditional Comparisons of Estimators: Real and Synthetic Population

Quality Measure	Estimator			
	PST	SYN	SSD	EBLUP
		Real Population		
$\overline{\text{ARB}}$%	5.4	15.7	2.9	11.3
$\overline{\text{EFF}}$%	100.0	232.8	137.6	261.1
$\overline{\text{ARE}}$%	32.2	16.5	24.0	13.5
		Synthetic Population		
$\overline{\text{ARB}}$%	5.6	12.5	2.4	8.4
$\overline{\text{EFF}}$%	100.0	313.3	135.8	319.1
$\overline{\text{ARE}}$%	35.0	13.2	25.9	11.8

SOURCE: Adapted from Tables 27.1 and 27.3 in Rao and Choudhry (1995).

Conditional comparisons of the estimators were also made by conditioning on the realized sample sizes in the small areas. This is a more realistic approach because the domain sample sizes, n_i, are random with known distribution. To make conditional comparisons under repeated sampling, a simple random sample of size $n = 419$ was first selected to determine the sample sizes, n_i, in the small areas. Regarding the n_i's as fixed, 500 stratified random samples were then selected, treating the small areas as strata. The conditional values of $\overline{\text{ARB}}$, $\overline{\text{EFF}}$ and $\overline{\text{ARE}}$ were computed from the simulated stratified samples. The conditional performances were similar to the unconditional performances, but different when two separate values for each quality measure were computed by averaging first over areas with $n_i < 6$ only and then over areas with $n_i \geq 6$. In particular, $\overline{\text{EFF}}$ ($\overline{\text{ARE}}$) for EBLUP is much larger (smaller) than the value for SSD when $n_i < 6$.

As noted in Chapter 4, subsection 4.3.2, the SSD estimator does not take advantage of the between area homogeneity, unlike the EBLUP estimator. To demonstrate this point, a series of synthetic populations was generated, using the previous parameter values, $\beta = 0.21, \sigma_v^2 = 1.58$ and $\sigma_e^2 = 1.34$, and the model $y_i = \beta x_i + v_i \theta^{1/2} + e_{ij} x_{ij}^{1/2}$, by varying θ from 0.1 to 10.0

($\theta = 1$ corresponds to the previous synthetic population). Note that for a given ratio σ_v^2/σ_e^2, the between area homogeneity increases as θ decreases. Table 7.5 reports the unconditional values of $\overline{\text{EFF}}$ and $\overline{\text{ARE}}$ for the estimators SSD and EBLUP, as θ varies from 0.1 to 10.0. It is clear from Table 7.5 that $\overline{\text{EFF}}$ and $\overline{\text{ARE}}$ for SSD remain essentially unchanged as θ increases from 0.1 to 10.0. On the other hand, $\overline{\text{EFF}}$ for EBLUP is largest when $\theta = 0.1$ and decreases as θ increases to 10.0. Similarly, $\overline{\text{ARE}}$ for EBLUP is smallest when $\theta = 0.1$ and increases as θ increases to 10.0.

TABLE 7.5 Effect of Between Area Homogeneity on the Performance of SSD and EBLUP

	Between Area Homogeneity: θ					
Estimator	0.1	0.5	1.0	2.0	5.0	10.0
			$\overline{\text{EFF}}$%			
SSD	136.0	136.0	135.8	135.6	134.7	133.1
EBLUP	324.3	324.6	319.1	305.0	270.8	239.9
			$\overline{\text{ARE}}$%			
SSD	25.6	25.7	25.9	26.3	27.2	28.2
EBLUP	11.5	11.6	11.8	12.5	14.5	16.7

SOURCE: Adapted from Tables 27.4 and 27.5 in Rao and Choudhry (1995).

7.2.7 Pseudo-EBLUP Estimation

The EBLUP estimator $\hat{\mu}_i^H$ under the unit level model (7.2.1) does not make use of the design weights, w_{ij}, attached to the sampled elements $(i,j); j = 1,\ldots,n_i; i = 1,\ldots,m$. As a result, it is not design-consistent unless the sampling design is self-weighting within areas, that is, $w_{ij} = w_i$ for all j. On the other hand, the EBLUP estimator under the area level model is design-consistent.

As noted by Kott (1990) and Prasad and Rao (1999), it is appealing to survey practitioners to use design-consistent, model-based estimators because such estimators provide protection against model failures as the small area sample size, n_i, increases. Note that n_i could be moderately large for some of the areas under consideration in which case design-consistency becomes relevant. In this subsection, pseudo-EBLUP estimators that depend on the design weights and satisfy the design-consistency property are developed for the special case of equal error variances, that is, $k_{ij} = 1$ for all (i,j). These estimators satisfy the benchmarking property without any adjustment in the sense that they add up to the direct survey regression estimator when aggregated over the areas i (You and Rao (2002a)).

We obtain a survey-weighted area level model from the unit level model by taking a weighted average with weights $\tilde{w}_{ij} = w_{ij}/\sum_k w_{ik} = w_{ij}/w_{i\cdot}$:

$$\bar{y}_{iw} = \sum_j \tilde{w}_{ij} y_{ij} = \sum_j \tilde{w}_{ij}(\mathbf{x}_{ij}^T \boldsymbol{\beta} + v_i + e_{ij})$$
$$= \bar{\mathbf{x}}_{iw}^T \boldsymbol{\beta} + v_i + \bar{e}_{iw}, \quad (7.2.48)$$

where $\bar{e}_{iw} = \sum_j \tilde{w}_{ij} e_{ij}$ with $E(\bar{e}_{iw}) = 0$ and $V(\bar{e}_{iw}) = \sigma_e^2 \sum_j \tilde{w}_{ij}^2 = \sigma_e^2 \delta_{iw}$ and $\bar{\mathbf{x}}_{iw} = \sum_j \tilde{w}_{ij} \mathbf{x}_{ij}$. We first assume that the parameters $\boldsymbol{\beta}, \sigma_e^2$ and σ_v^2 are known in the aggregated area level model (7.2.48). Then the BLUP estimator of $\mu_i = \bar{\mathbf{X}}_i^T \boldsymbol{\beta} + v_i$ from the aggregated model is obtained as

$$\tilde{\mu}_{iw}^H = \bar{\mathbf{X}}_i^T \boldsymbol{\beta} + \gamma_{iw}(\bar{y}_{iw} - \bar{\mathbf{x}}_{iw}^T \boldsymbol{\beta}), \quad (7.2.49)$$

where $\gamma_{iw} = \sigma_v^2/(\sigma_v^2 + \sigma_e^2 \delta_{iw})$. This estimator depends on $\boldsymbol{\beta}, \sigma_e^2$ and σ_v^2. We estimate the variance components, σ_e^2 and σ_v^2, from the unit level model using the fitting-of-constants or the REML method (subsection 7.2.2).

To estimate the regression parameter $\boldsymbol{\beta}$, we first obtain the BLUP estimator of v_i from the aggregated model (7.2.48), given $(\boldsymbol{\beta}, \sigma_e^2, \sigma_v^2)$, as

$$\tilde{v}_{iw}(\boldsymbol{\beta}, \sigma_e^2, \sigma_v^2) = \gamma_{iw}(\bar{y}_{iw} - \bar{\mathbf{x}}_{iw}^T \boldsymbol{\beta}). \quad (7.2.50)$$

We then solve the following design-weighted estimating equations for $\boldsymbol{\beta}$:

$$\sum_{i=1}^{m} \sum_{j=1}^{n_i} w_{ij} \mathbf{x}_{ij} [y_{ij} - \mathbf{x}_{ij}^T \boldsymbol{\beta} - \tilde{v}_{iw}(\boldsymbol{\beta}, \sigma_e^2, \sigma_v^2)] = 0. \quad (7.2.51)$$

The solution of (7.2.51) is given by

$$\tilde{\boldsymbol{\beta}}_w(\sigma_e^2, \sigma_v^2) = \left[\sum_i \sum_j w_{ij} \mathbf{x}_{ij} (\mathbf{x}_{ij} - \gamma_{iw} \bar{\mathbf{x}}_{iw})^T \right]^{-1}$$
$$\times \left[\sum_i \sum_j w_{ij} (\mathbf{x}_{ij} - \gamma_{iw} \bar{\mathbf{x}}_{iw}) y_{ij} \right]. \quad (7.2.52)$$

Given σ_e^2 and σ_v^2, the estimator $\tilde{\boldsymbol{\beta}}_w$ is model-unbiased for $\boldsymbol{\beta}$. Now replacing σ_e^2 and σ_v^2 by the estimators $\hat{\sigma}_e^2$ and $\hat{\sigma}_v^2$ we obtain a design-weighted estimator of $\boldsymbol{\beta}$ as $\hat{\boldsymbol{\beta}}_w = \tilde{\boldsymbol{\beta}}_w(\hat{\sigma}_e^2, \hat{\sigma}_v^2)$.

A pseudo-EBLUP estimator of μ_i is obtained from (7.2.49) by replacing $(\boldsymbol{\beta}, \sigma_e^2, \sigma_v^2)$ by $(\hat{\boldsymbol{\beta}}_w, \hat{\sigma}_e^2, \hat{\sigma}_v^2)$:

$$\hat{\mu}_{iw}^H = \bar{\mathbf{X}}_i^T \hat{\boldsymbol{\beta}}_w + \hat{\gamma}_{iw}(\bar{y}_{iw} - \bar{\mathbf{x}}_{iw}^T \hat{\boldsymbol{\beta}}_w), \quad (7.2.53)$$

where $\hat{\gamma}_{iw} = \hat{\sigma}_v^2/(\hat{\sigma}_v^2 + \delta_i \hat{\sigma}_e^2)$. The estimator $\hat{\mu}_{iw}^H$ automatically satisfies the benchmarking property when aggregating over i, assuming that the weights are calibrated to agree with the known sizes N_i, that is, $\sum_j w_{ij} = w_{i\cdot} = N_i$, and that the unit level model includes the intercept term, that is, $x_{ij1} = 1$.

That is, $\sum_i N_i \hat{\mu}_{iw}^H$ equals the direct survey regression estimator $\hat{Y}_w + (\mathbf{X} - \hat{\mathbf{X}}_w)^T \hat{\boldsymbol{\beta}}_w$, where $\hat{Y}_w = \sum_i w_{i\cdot} \bar{y}_{iw} = \sum_i \sum_j w_{ij} y_{ij}$ and $\hat{\mathbf{X}}_w = \sum_i w_{i\cdot} \bar{\mathbf{x}}_{iw} = \sum_i \sum_j w_{ij} \mathbf{x}_{ij}$ are the direct estimators of the overall totals Y and $\mathbf{X}$, respectively. To prove this property, consider the first equation of (7.2.51) corresponding to the intercept term $x_{ij1} = 1$ and substitute $\hat{\boldsymbol{\beta}}_w$ and $\hat{v}_{iw} = \tilde{v}_{iw}(\hat{\boldsymbol{\beta}}_w, \hat{\sigma}_v^2, \hat{\sigma}_e^2)$ for $\boldsymbol{\beta}$ and $\tilde{v}_{iw}(\boldsymbol{\beta}, \sigma_v^2, \sigma_e^2)$ to get $\sum_i \sum_j w_{ij}(y_{ij} - \mathbf{x}_{ij}^T \hat{\boldsymbol{\beta}}_w - \hat{v}_{iw}) = 0$ or

$$\sum_i N_i \hat{v}_{iw} = \hat{Y}_w - \hat{\mathbf{X}}_w^T \hat{\boldsymbol{\beta}}_w. \tag{7.2.54}$$

Also, noting that $\hat{\mu}_{iw}^H = \bar{\mathbf{X}}_i^T \hat{\boldsymbol{\beta}}_w + \hat{v}_{iw}$ and using (7.2.53), we obtain

$$\sum_i N_i \hat{\mu}_{iw}^H = \mathbf{X}^T \hat{\boldsymbol{\beta}}_w + \sum_i N_i \hat{v}_{iw} = \hat{Y}_w + (\mathbf{X} - \hat{\mathbf{X}}_w)^T \hat{\boldsymbol{\beta}}_w.$$

Thus, the pseudo-EBLUP estimator, $\hat{\mu}_{iw}^H$, satisfies the benchmarking property without any adjustment, unlike the EBLUP estimator $\hat{\mu}_i^H$.

Under normality of the errors v_i and e_{ij}, an estimator of $\text{MSE}(\hat{\mu}_{iw}^H) = E(\hat{\mu}_{iw}^H - \mu_i)^2$ is given by

$$\text{mse}(\hat{\mu}_{iw}^H) = g_{1iw}(\hat{\sigma}_v^2, \hat{\sigma}_e^2) + g_{2iw}(\hat{\sigma}_v^2, \hat{\sigma}_e^2) + 2g_{3iw}(\hat{\sigma}_v^2, \hat{\sigma}_e^2), \tag{7.2.55}$$

where

$$g_{1iw}(\sigma_v^2, \sigma_e^2) = \gamma_{iw} \delta_{iw} \sigma_e^2, \tag{7.2.56}$$

$$g_{2iw}(\sigma_v^2, \sigma_e^2) = (\bar{\mathbf{X}}_i - \gamma_{iw} \bar{\mathbf{x}}_{iw})^T \Phi_w(\sigma_v^2, \sigma_e^2)(\bar{\mathbf{X}}_i - \gamma_{iw} \bar{\mathbf{x}}_{iw}), \tag{7.2.57}$$

and

$$g_{3iw}(\sigma_v^2, \sigma_e^2) = \gamma_{iw}(1 - \gamma_{iw})^2 \sigma_v^{-2} \sigma_e^{-2} h(\sigma_v^2, \sigma_e^2), \tag{7.2.58}$$

where $h(\sigma_v^2, \sigma_e^2)$ is given by (7.2.23). Further, $\Phi_w(\sigma_v^2, \sigma_e^2)$ is the covariance matrix of $\tilde{\boldsymbol{\beta}}_w(\sigma_v^2, \sigma_e^2)$. It is given by

$$\begin{aligned}\Phi_w(\sigma_v^2, \sigma_e^2) = &\left(\sum_i \sum_j \mathbf{x}_{ij} \mathbf{z}_{ij}^T\right)^{-1} \left(\sum_i \sum_j \mathbf{z}_{ij} \mathbf{z}_{ij}^T\right) \left[\left(\sum_i \sum_j \mathbf{x}_{ij} \mathbf{z}_{ij}^T\right)^{-1}\right]^T \sigma_e^2 \\ &+ \left(\sum_i \sum_j \mathbf{x}_{ij} \mathbf{z}_{ij}^T\right)^{-1} \left[\sum_i \left(\sum_j \mathbf{z}_{ij}\right) \left(\sum_j \mathbf{z}_{ij}\right)^T\right] \\ &\times \left[\left(\sum_i \sum_j \mathbf{x}_{ij} \mathbf{z}_{ij}^T\right)^{-1}\right]^T \sigma_v^2,\end{aligned} \tag{7.2.59}$$

where $\mathbf{z}_{ij} = w_{ij}(\mathbf{x}_{ij} - \gamma_{iw} \bar{\mathbf{x}}_{iw})$. The MSE estimator (7.2.55) is valid for the fitting-of-constants or REML estimators of σ_v^2 and σ_e^2. However, the cross-product term $E(\tilde{\mu}_{iw}^H - \mu_i)(\hat{\mu}_{iw}^H - \tilde{\mu}_{iw}^H)$ may not be zero in the survey-weighted case, and the MSE estimator (7.2.55) ignores this term. As a result, the bias of (7.2.55) may contain a term of order $O(m^{-1})$.

Example 7.2.3. County Corn Crop Areas. You and Rao (2002a) applied the pseudo-EBLUP method to the county crop areas data from Battese, Harter and Fuller (1988); see Example 7.2.1. Simple random sampling within areas was assumed so that $w_{ij} = w_i = N_i/n_i$ for all sample elements j in area i. In this case, the EBLUP estimator is also design-consistent, and the pseudo-EBLUP estimator has the same form as the EBLUP estimator except for the estimator of $\boldsymbol{\beta} = (\beta_0, \beta_1, \beta_2)^T$ under the model $y_{ij} = \beta_0 + \beta_1 x_{ij1} + \beta_2 x_{ij2} + v_i + e_{ij}$, $j = 1, \ldots, n_i$; $i = 1, \ldots, m$.

Table 7.6 reports the EBLUP and the pseudo-EBLUP estimates of mean hectares of corn for the 12 counties and the associated standard errors (square root of estimated MSEs). The fitting-of-constants method was used to estimate σ_v^2 and σ_e^2. The EBLUP estimates and associated standard errors differ slightly from those reported in Table 7.3 because of differences in the method of estimating σ_v^2 and σ_e^2, as noted in Example 7.2.1.

TABLE 7.6 EBLUP and Pseudo-EBLUP Estimates and Associated Standard Errors (s.e.): County Corn Crop Areas

		EBLUP		Pseudo-EBLUP	
County	n_i	Estimate	s.e.	Estimate	s.e.
		Corn			
Cerro Gordo	1	122.2	9.7	120.5	9.9
Hamilton	1	126.2	9.6	125.2	9.7
Worth	1	106.8	9.5	106.4	9.6
Humboldt	2	108.5	8.1	107.4	8.3
Franklin	3	144.2	6.6	143.7	6.6
Pocahontas	3	112.1	6.6	111.5	6.6
Winnebago	3	112.8	6.6	112.1	6.6
Wright	3	122.0	6.7	121.3	6.8
Webster	4	115.3	5.8	115.0	5.8
Hancock	5	124.4	5.4	124.5	5.4
Kossuth	5	106.9	5.3	106.6	5.3
Hardin	5	143.0	5.6	143.5	5.8

SOURCE: Adapted from Table 1 in You and Rao (2002a).

It is clear from Table 7.6 that the pseudo-EBLUP method compares favourably to the "optimal" EBLUP method, leading to a slight loss in efficiency. The loss in efficiency is due to the weighted method of estimating $\boldsymbol{\beta}$, but the standard errors of $\hat{\boldsymbol{\beta}}_w$ compare favourably to those of $\hat{\boldsymbol{\beta}}$. For example, the standard errors of $\hat{\beta}_1$ and $\hat{\beta}_2$ are 0.050 and 0.056 compared to 0.054 and 0.062, the standard errors of $\hat{\beta}_{1w}$ and $\hat{\beta}_{2w}$, respectively.

The pseudo-EBLUP estimates satisfy the benchmarking property; that is, $\sum_i N_i \hat{\mu}_{iw}^H = \hat{Y}_w + (\mathbf{X} - \hat{\mathbf{X}}_w)^T \hat{\boldsymbol{\beta}}_w$, the survey regression estimate of the overall total: 815,025.2 hectares.

Chapter 8

EBLUP: Extensions

In Chapter 7, we presented a detailed account of the EBLUP theory for small area estimation under the basic area level (type A) and unit level (type B) models. In this chapter, we provide a brief account of the EBLUP method for various extensions of the basic area level and unit level models. Multivariate area level models are studied in Section 8.1, and applied to EBLUP estimation of median income of four-person families for each of the American states. Models with correlated sampling errors are investigated in Section 8.2, and applied to EBLUP estimation of U.S. census undercount. Time series and cross-sectional models are studied in Section 8.2, and applied to EBLUP estimation of current median income of four-person families for each of the American states. Spatial models that include spatial correlations among the random small area effects are considered in Section 8.3, and applied to EBLUP estimation of U.S. census undercount. Unit level models with random error variances and two-fold nested error regression models are studied in Sections 8.6 and 8.7, respectively. Two-level models that effectively integrate the use of unit level and area level covariates are used in Section 8.8 for EBLUP estimation of small area means.

8.1 Multivariate Fay-Herriot Model

The multivariate Fay-Herriot model is given by (5.4.3). This model can also be expressed as a special case of the general model (6.3.1) with block diagonal covariance structure. We have

$$\mathbf{y}_i = \hat{\boldsymbol{\theta}}_i, \quad \mathbf{X}_i = \mathbf{Z}_i, \quad \mathbf{Z}_i = \mathbf{I}_r$$

and

$$\mathbf{v}_i = \mathbf{v}_i, \quad \mathbf{e}_i = \mathbf{e}_i, \quad \boldsymbol{\beta} = (\beta_1, \ldots, \beta_p)^T,$$

where $\hat{\boldsymbol{\theta}}_i = (\hat{\theta}_{i1}, \ldots, \hat{\theta}_{ir})^T$ is an r-vector of direct estimators for the ith area. Further,

$$\mathbf{G}_i = \boldsymbol{\Sigma}_v, \quad \mathbf{R}_i = \boldsymbol{\Psi}_i,$$

so that

$$\mathbf{V}_i = \boldsymbol{\Psi}_i + \boldsymbol{\Sigma}_v.$$

Also, $\boldsymbol{\mu}_i = \boldsymbol{\theta}_i = \mathbf{L}_i\boldsymbol{\beta} + \mathbf{M}_i\mathbf{v}_i$ with $\mathbf{L}_i = \mathbf{Z}_i$ and $\mathbf{M}_i = \mathbf{I}_r$.

Making the above substitutions in the general formula (6.3.19), we get the BLUP estimator of $\boldsymbol{\theta}_i$ as

$$\tilde{\boldsymbol{\theta}}_i^H = \boldsymbol{\Sigma}_v(\boldsymbol{\Psi}_i + \boldsymbol{\Sigma}_v)^{-1}\hat{\boldsymbol{\theta}}_i + \boldsymbol{\Psi}_i(\boldsymbol{\Psi}_i + \boldsymbol{\Sigma}_v)^{-1}\mathbf{Z}_i\tilde{\boldsymbol{\beta}}, \tag{8.1.1}$$

where

$$\tilde{\boldsymbol{\beta}} = \left[\sum_{i=1}^m \mathbf{Z}_i^T(\boldsymbol{\Psi}_i + \boldsymbol{\Sigma}_v)^{-1}\mathbf{Z}_i\right]^{-1}\left[\sum_{i=1}^m \mathbf{Z}_i^T(\boldsymbol{\Psi}_i + \boldsymbol{\Sigma}_v)^{-1}\hat{\boldsymbol{\theta}}_i\right]. \tag{8.1.2}$$

The estimator (8.1.1) is a natural extension of the univariate BLUP estimator (7.1.3).

The MSE of $\tilde{\boldsymbol{\theta}}_i^H$ (i.e., the covariance matrix of $\tilde{\boldsymbol{\theta}}_i^H - \boldsymbol{\theta}_i$) follows from the general formula (6.3.20) by noting that $\boldsymbol{\Sigma}_v\mathbf{V}_i^{-1}\boldsymbol{\Psi}_i = (\boldsymbol{\Psi}_i^{-1} + \boldsymbol{\Sigma}_v^{-1})^{-1}$:

$$\begin{aligned}\text{MSE}(\tilde{\boldsymbol{\theta}}_i^H) &= E(\tilde{\boldsymbol{\theta}}_i^H - \boldsymbol{\theta}_i)(\tilde{\boldsymbol{\theta}}_i^H - \boldsymbol{\theta}_i)^T \\ &= (\boldsymbol{\Psi}_i^{-1} + \boldsymbol{\Sigma}_v^{-1})^{-1} + (\boldsymbol{\Psi}_i^{-1} + \boldsymbol{\Sigma}_v^{-1})^{-1}\boldsymbol{\Sigma}_v^{-1}\mathbf{Z}_i \\ &\quad \times \left[\sum_{i=1}^m \mathbf{Z}_i^T(\boldsymbol{\Psi}_i + \boldsymbol{\Sigma}_v)^{-1}\mathbf{Z}_i\right]^{-1}\mathbf{Z}_i^T\boldsymbol{\Sigma}_v^{-1}(\boldsymbol{\Psi}_i^{-1} + \boldsymbol{\Sigma}_v^{-1})^{-1}. \end{aligned} \tag{8.1.3}$$

Formula (8.1.3) reduces to (7.1.6) in the univariate case, $r = 1$. Also, the leading term in (8.1.3) is of order $O(1)$, whereas the second-term, due to estimating $\boldsymbol{\beta}$, is of order $O(m^{-1})$ for large m. The diagonal elements of $\text{MSE}(\tilde{\boldsymbol{\theta}}_i^H)$ represent the MSEs of the components $\tilde{\theta}_{ij}^H$ $(j = 1, \ldots, r)$ of $\tilde{\boldsymbol{\theta}}_i^H$. By taking advantage of the correlations between the components $\hat{\theta}_{ij}$ of the direct estimator $\hat{\boldsymbol{\theta}}_i$, the multivariate model leads to more efficient estimators compared to those based on a univariate model for each $j = 1, \ldots, r$.

The EBLUP estimator $\hat{\boldsymbol{\theta}}_i^H$ is obtained by substituting an estimator $\hat{\boldsymbol{\Sigma}}_v$ for $\boldsymbol{\Sigma}_v$ in (8.1.1). For example, one could use the REML estimator of $\boldsymbol{\Sigma}_v$.

Example 8.1.1. Median Income. Datta, Fay and Ghosh (1991) applied the multivariate model to estimate the current median income, θ_{i1}, of four-person families in each of the American states, $i = 1, \ldots, 51$. In this application, $\theta_{i2} =$ 3/4(median income of five-person families in state i)+1/4(median income of three-person families in state i); see Example 5.4.1, Chapter 5 for details on the auxiliary variables $\mathbf{z}_{ij}$ used in this study. Direct survey estimates, $\hat{\boldsymbol{\theta}}_i$, and associated covariance matrices, $\boldsymbol{\Psi}_i$, were obtained from the 1979 Current Population Survey data. Datta et al. (1991) made an external evaluation of the estimates, $\hat{\theta}_{i1}^H$ and $\hat{\theta}_{i1}$, by comparing them to "true" values, θ_{i1}, available from the 1980 census data.

In terms of absolute relative error averaged over the states ($\overline{\text{ARE}}$), the EBLUP estimates, $\hat{\theta}_{i1}^H$, outperformed the direct estimates $\hat{\theta}_{i1}$: $\overline{\text{ARE}}$ of 2% compared to $\overline{\text{ARE}}$ of 5% for the direct estimates. Note that $\overline{\text{ARE}} = \frac{1}{51}\sum_i |\text{est}_i - \theta_{i1}|/\theta_{i1}$.

8.2 Correlated Sampling Errors

The Fay-Herriot model with correlated sampling errors is given by (5.4.5). It is a special case of the general linear mixed model (6.2.1) with $\mathbf{R} = \boldsymbol{\Psi}$, $\mathbf{Z} = \mathbf{I}$, $\mathbf{G} = \sigma_v^2\mathbf{I}_m$, $\mathbf{X} = \mathbf{Z}$ and $\mathbf{y} = \hat{\boldsymbol{\theta}} = (\hat{\theta}_1, \ldots, \hat{\theta}_m)^T$. Using these values in the general BLUP estimator (6.2.9) with $\mathbf{L} = \mathbf{Z}$ and $\mathbf{M} = \mathbf{I}$, we get the BLUP estimator of $\boldsymbol{\theta} = (\theta_1, \ldots, \theta_m)^T$ as

$$\tilde{\boldsymbol{\theta}}^H = \mathbf{Z}\tilde{\boldsymbol{\beta}} + \sigma_v^2\mathbf{V}^{-1}(\hat{\boldsymbol{\theta}} - \mathbf{Z}\tilde{\boldsymbol{\beta}}), \tag{8.2.1}$$

where

$$\tilde{\boldsymbol{\beta}} = (\mathbf{Z}^T\mathbf{V}^{-1}\mathbf{Z})^{-1}(\mathbf{Z}^T\mathbf{V}^{-1}\hat{\boldsymbol{\theta}})$$

and

$$\mathbf{V} = \boldsymbol{\Psi} + \sigma_v^2\mathbf{I}.$$

The EBLUP estimator $\hat{\boldsymbol{\theta}}^H$ is obtained by assuming known $\boldsymbol{\Psi}$ and substituting an estimator of $\hat{\sigma}_v^2$ for σ_v^2 in (8.2.1).

Datta et al. (1992) obtained a second-order approximation to the covariance matrix of $\hat{\boldsymbol{\theta}}^H - \boldsymbol{\theta}$, under normality of the errors $\mathbf{v}$ and $\mathbf{e}$, as

$$\text{MSE}(\hat{\boldsymbol{\theta}}^H) \approx \mathbf{G}_1(\sigma_v^2) + \mathbf{G}_2(\sigma_v^2) + \mathbf{G}_3(\sigma_v^2), \tag{8.2.2}$$

where

$$\mathbf{G}_1(\sigma_v^2) = \boldsymbol{\Psi} - \boldsymbol{\Psi}\mathbf{V}^{-1}\boldsymbol{\Psi},$$
$$\mathbf{G}_2(\sigma_v^2) = \boldsymbol{\Psi}\mathbf{V}^{-1}\mathbf{Z}(\mathbf{Z}^T\mathbf{V}^{-1}\mathbf{Z})^{-1}\mathbf{Z}^T\mathbf{V}^{-1}\boldsymbol{\Psi}$$

and

$$\mathbf{G}_3(\sigma_v^2) = \boldsymbol{\Psi}\mathbf{K}^3\boldsymbol{\Psi}\bar{\mathbf{V}}(\hat{\sigma}_v^2)$$

with

$$\mathbf{K} = \mathbf{V}^{-1} - \mathbf{V}^{-1}\mathbf{Z}(\mathbf{Z}^T\mathbf{V}^{-1}\mathbf{Z})^{-1}\mathbf{Z}^T\mathbf{V}^{-1}.$$

If $\hat{\sigma}_v^2$ is chosen as the REML estimator then

$$\bar{\mathbf{V}}(\hat{\sigma}_v^2) = 2/\text{tr}(\mathbf{V}^{-2}).$$

A second-order approximation to the estimator of MSE($\hat{\boldsymbol{\theta}}^H$) is obtained along the lines of (6.3.11) as

$$\text{mse}(\hat{\boldsymbol{\theta}}^H) = \mathbf{G}_1(\hat{\sigma}_v^2) + \mathbf{G}_2(\hat{\sigma}_v^2) + 2\mathbf{G}_3(\hat{\sigma}_v^2). \tag{8.2.3}$$

Isaki, Tsay and Fuller (2000) relaxed the assumption of a known sampling covariance matrix $\boldsymbol{\Psi}$. They replaced $\boldsymbol{\Psi}$ in the EBLUP estimator $\hat{\boldsymbol{\theta}}^H$ by

$$\hat{\boldsymbol{\Psi}}_\phi = \phi\hat{\boldsymbol{\Psi}}_d + (1-\phi)\hat{\boldsymbol{\Psi}}, \tag{8.2.4}$$

where $\hat{\boldsymbol{\Psi}}$ is a sample-based estimator of $\boldsymbol{\Psi}$ with diagonal elements $\hat{\psi}_{ii}, i = 1,\ldots,m$, $\hat{\boldsymbol{\Psi}}_d = \text{diag}(\psi_{ii}, i = 1,\ldots,m)$ and ϕ is a prespecified constant, $0 \le \phi \le 1$. We denote the EBLUP estimator of $\boldsymbol{\theta}$ based on $\hat{\boldsymbol{\Psi}}_\phi$ as $\hat{\boldsymbol{\theta}}^H(\phi)$. Isaki et al. (2000) adjusted $\hat{\boldsymbol{\theta}}^H(\phi)$ to agree with a direct estimator for the entire area covering the small areas. This adjustment is similar to (7.2.44) but more complex because of correlated sampling errors. For the special case of $\phi = 1$, it uses multipliers similar to $\hat{p}_i^{(3)}$ given by (7.2.46). Isaki et al. (2000) also obtained an estimator of the covariance matrix of $\hat{\boldsymbol{\theta}}_a^H(\phi) - \boldsymbol{\theta}$, where $\hat{\boldsymbol{\theta}}_a^H(\phi)$ denotes the adjusted EBLUP estimator of $\boldsymbol{\theta}$. This estimator is based on additional assumptions: (i) correlation between $\hat{\sigma}_v^2$ and $\hat{\boldsymbol{\Psi}}$ is negligible; (ii) $\hat{\boldsymbol{\Psi}}$ is distributed as a "Wishart" matrix with fairly large degrees of freedom. Note that the Wishart distribution is the multivariate version of the chi-squared distribution.

Example 8.2.1. U.S. Census Undercoverage. The adjustment of undercounts in the 1980 census attracted a lot of attention and controversy. In 1980 the U.S. Census Bureau expended vast sums of money and intellectual resources on improving coverage in the 1980 census enumeration. After the enumeration, however, several large states and cities claimed that the census had undercounted their populations. In fact, New York State filed a lawsuit against the Census Bureau in 1980, demanding that the Bureau adjust its count for New York State. E.P. Ericksen, J.B. Kadane and some other prominent statisticians appeared as the plaintiff's expert witnesses in New York State's lawsuit. They proposed EBLUP state estimates of census undercounts, $\theta_i = (T_i - C_i)/T_i$, based on a state level model, using dual-system estimates of θ_i's obtained from the 1980 post-enumeration Survey (PES) (Ericksen and Kadane (1985)). Here T_i and C_i denote the true count and the census count for the i th state, respectively. Freedman and Navidi (1986) criticized the Ericksen-Kadane proposal for not validating their model fully and for not making their assumptions explicit. They also raised other technical issues, including the effect of large biases and large sampling errors in the PES estimates, $\hat{\theta}_i$. Cressie (1992) modelled the adjustment factors $\theta_i = T_i/C_i$ and proposed improved state level models.

In 1990, dual system PES estimates of adjustment factors $\theta_i = T_i/C_i$ were produced for 1392 subdivisions (poststrata) of the total population. The PES sample contained approximately 377,000 persons in roughly 5200 sample blocks. The poststrata were defined on the basis of geographical divisions of the country, tenure (owners or renters of homes), size-of-place, race, sex and age. A poststratum level model was developed, using the dual system estimates, $\hat{\theta}_i$, and poststratum level covariates $\mathbf{z}_i$. The PES variance estimates, $\hat{\psi}_i$, were smoothed through a variance function model, while the original estimated correlations were used with the smoothed standard errors to obtain partially smoothed estimates of the sampling covariances, ψ_{il}. Treating the resulting covariance matrix as the true $\mathbf{\Psi}$, EBLUP estimates, $\hat{\boldsymbol{\theta}}^H$, were obtained, using (8.2.1) with σ_v^2 changed to $\hat{\sigma}_v^2$. The EBLUP estimates were then ratio-adjusted to agree with regional total population estimates derived from the direct estimates of regional adjustment factors. Finally, the ratio-adjusted EBLUP adjustment factors were applied to block level census counts to produce synthetic estimates of true block counts. Note that synthetic estimates are used only at the block level in contrast to state level modeling which entails synthetic estimation at all levels of geography lower than the state. Isaki, Huang and Tsay (1991) provided a detailed account of EBLUP estimation of the poststratum adjustment factors $\theta_i = T_i/C_i$.

Datta et al. (1992) conducted an evaluation of the above methodology, prior to the 1990 census, using the 1988 Census Dress Rehearsal Data from test sites in Missouri. A post-enumeration survey was also conducted as part of the 1988 study to produce direct estimates of the adjustment factors. Datta et al. (1992) used a best subset selection procedure, based on the well-known Mallow's C_p criterion, to select z-variables from a set of 22 possible explanatory variables. Both $\mathbf{y}$ and $\mathbf{Z}$ were first transformed to ensure approximately iid errors, and the best subset selection procedure was then applied to the transformed data $\{\hat{\mathbf{V}}^{-1/2}\mathbf{y}, \hat{\mathbf{V}}^{-1/2}\mathbf{Z}\}$, where $\hat{\mathbf{V}}$ is an estimate of the covariance matrix of the combined error vector, $\mathbf{v}+\mathbf{e}$. Datta et al. (1992) computed the EBLUP estimates and associated standard errors, using the estimator (8.2.1) and the MSE estimator (8.2.3). Their study revealed that for every poststratum the EBLUP estimates of adjustment factors outperformed the direct estimates.

The Census Bureau recommended the use of 1990 census adjusted counts based on the above EBLUP methodology. But Fay (1992) identified some serious deficiencies in the recommended EBLUP estimates. In particular, he noted that some of the β-coefficients in the model were very sensitive to the choice of estimated covariance matrix used to construct $\hat{\boldsymbol{\beta}}$. He attributed this difficulty to an unstable estimate of $\mathbf{\Psi}$ caused by direct estimates that were zero for many of the sample blocks. Subsequently, the Secretary of Commerce decided against using the proposed adjusted counts.

Isaki, Tsay and Fuller (2000) investigated alternative EBLUP estimates from the 1990 census and PES data, using $\hat{\mathbf{\Psi}}_\phi$ given by (8.2.4). The proposed method is less influenced by the variability of the estimated covariance matrix compared to the previous method. It was applied to a new set of 357 poststrata composed of 51 poststratum groups, each of which was subdivided into 7 age-sex categories (see Example 5.4.2, Chapter 5). Adjusted EBLUP estimates of adjustment factors and associated standard errors, as described before, were computed for each of 336 poststrata obtained by eliminating Asian and American Indian data from the EBLUP estimation. The poststratum level model contained 21 explanatory variables including the intercept. The ratio of the standard error of $\hat{\theta}_{ia}^{H}(\phi)$ to $\hat{\theta}_{ia}^{H}(0.6)$, averaged over $i = 1, \ldots, 336$, was computed for selected values of ϕ including $\phi = 0$ and 1. Those average ratios revealed that $\phi = 0.5$ or 0.6 is a good choice. The standard errors of adjusted EBLUP estimates $\hat{\theta}_{ia}^{H}(0.6)$ were much smaller than the corresponding direct estimates, $\hat{\theta}_i$, of adjustment factors $\theta_i = T_i/C_i$; the average estimated MSE efficiency of the adjusted EBLUP estimates (with $\phi = 0.6$) is about 400% relative to the direct estimates $\hat{\theta}_i$. Moreover, the set of adjusted EBLUP estimates contained fewer extreme estimates compared to the set of direct estimates.

8.3 Time Series and Cross-sectional Models

8.3.1 Rao–Yu Model

The model given by (5.4.6) and (5.4.7) with AR(1) or random walk model on the errors u_{it} provides an extension of the basic type A model to handle time series and cross-sectional data. Arranging the direct estimators $\hat{\theta}_{it}$ as $\hat{\boldsymbol{\theta}}_i = (\hat{\theta}_{i1}, \ldots, \hat{\theta}_{iT})^T$, $i = 1, \ldots, m$, we can write the model in the form (6.3.1) with block diagonal covariance structure. We have

$$\mathbf{y}_i = \hat{\boldsymbol{\theta}}_i, \quad \mathbf{X}_i = (\mathbf{z}_{i1}, \ldots, \mathbf{z}_{iT})^T, \quad \mathbf{Z}_i = (\mathbf{1}_T, \mathbf{I}_T),$$
$$\mathbf{v}_i^T = (v_i, \mathbf{u}_i^T), \quad \mathbf{e}_i = (\mathbf{e}_{i1}, \ldots, \mathbf{e}_{iT})^T$$

and $\boldsymbol{\beta} = (\beta_1, \ldots, \beta_p)^T$, where $\mathbf{1}_T$ is the $T \times 1$ vector of 1's, and $\mathbf{I}_T$ is the identity matrix of order T. Further,

$$\mathbf{G}_i = \begin{bmatrix} \sigma_v^2 & \mathbf{0}^T \\ \mathbf{0} & \sigma^2\boldsymbol{\Lambda} \end{bmatrix}, \quad \mathbf{R}_i = \boldsymbol{\Psi}_i$$

where $\boldsymbol{\Lambda}_i = \boldsymbol{\Lambda}$ is the $T \times T$ covariance matrix of $\mathbf{u}_i = (u_{i1}, \ldots, u_{iT})^T$ with (t,s)th element $\rho^{|t-s|}/(1-\rho^2)$ for the AR(1) model $u_{it} = \rho u_{i,t-1} + \epsilon_{it}, |\rho| < 1$; and (t,s)th element $\min(t,s)$ for the random walk model $u_{it} = u_{i,t-1} + \epsilon_{it}$. We may write $\mathbf{V}_i$ as

$$\mathbf{V}_i = \boldsymbol{\Psi}_i + \sigma^2\boldsymbol{\Lambda} + \sigma_v^2\mathbf{J}_T$$

with $\mathbf{J}_T = \mathbf{1}_T\mathbf{1}_T^T$ denoting a $T \times T$ matrix having every element equal to one. Further, the small area parameters θ_{iT} for the current occasion T may be expressed as $\theta_{iT} = \mu_i = \mathbf{l}_i^T\boldsymbol{\beta} + \mathbf{m}_i^T\mathbf{v}_i$ with $\mathbf{l}_i = \mathbf{z}_{iT}$ and $\mathbf{m}_i = (1, 0, \ldots, 0, 1)^T$.

Making the above substitution in the general BLUP formula (6.3.2), we get the BLUP estimator of θ_{iT} as

$$\tilde{\theta}_{iT}^{H} = \mathbf{z}_{iT}^{T}\tilde{\boldsymbol{\beta}} + (\sigma_v^2 \mathbf{1}_T + \sigma^2 \boldsymbol{\lambda}_T)^T \mathbf{V}_i^{-1}(\hat{\boldsymbol{\theta}}_i - \mathbf{X}_i\tilde{\boldsymbol{\beta}}), \tag{8.3.1}$$

where $\boldsymbol{\lambda}_T^T$ is the Tth row of $\boldsymbol{\Lambda}$ and

$$\tilde{\boldsymbol{\beta}} = \left(\sum_i \mathbf{X}_i^T \mathbf{V}_i^{-1} \mathbf{X}_i\right)^{-1} \left(\sum_i \mathbf{X}_i^T \mathbf{V}_i^{-1} \hat{\boldsymbol{\theta}}_i\right).$$

The BLUP estimator (8.3.1) may also be expressed as a weighted combination of the direct estimator of $\hat{\theta}_{iT}$, the regression synthetic estimator $\mathbf{z}_{iT}^T\tilde{\boldsymbol{\beta}}$ and the residuals $\hat{\theta}_{it} - \mathbf{z}_{it}^T\tilde{\boldsymbol{\beta}}$, $t = 1, \ldots, T-1$:

$$\tilde{\theta}_{iT}^{H} = w_{iT}^{*}\hat{\theta}_{iT} + (1 - w_{iT}^{*})\mathbf{z}_{iT}^{T}\tilde{\boldsymbol{\beta}} + \sum_{t=1}^{T-1} w_{it}^{*}(\hat{\theta}_{it} - \mathbf{z}_{it}^{T}\tilde{\boldsymbol{\beta}}), \tag{8.3.2}$$

where

$$(w_{i1}^{*}, \ldots, w_{iT}^{*}) = (\sigma_v^2 \mathbf{1}_T + \sigma^2 \boldsymbol{\Lambda}_T)^T \mathbf{V}_i^{-1}; i = 1, \ldots, m.$$

For the case of an AR(1) model with known ρ, Rao and Yu (1994) estimated σ^2 and σ_v^2 by extending the simple transformation method of Fuller and Battese (1973) considered in subsection 7.2.2, Chapter 7. This method is equivalent to the fitting-of-constants method. It uses two ordinary least squares steps to obtain unbiased estimators $\tilde{\sigma}^2(\rho)$ and $\tilde{\sigma}_v^2(\rho)$, which are then truncated to zero. The truncated estimators, $\hat{\sigma}^2(\rho)$ and $\hat{\sigma}_v^2(\rho)$, are not unbiased, but they are consistent as $m \to \infty$. The transformation method also works for the random walk model (You (1999), Chapter 8). Datta, Lahiri and Maiti (2002) used REML estimates $\hat{\sigma}^2$ and $\hat{\sigma}_v^2$ under the random walk model. REML can also be used for the AR(1) model with known ρ.

Replacing σ^2 and σ_v^2 by their estimators $\hat{\sigma}^2(\rho)$ and $\hat{\sigma}_v^2(\rho)$ in (8.3.1), we get the EBLUP estimator $\hat{\theta}_{iT}^H(\rho)$ under the AR(1) model with known ρ. Similarly, the EBLUP estimator, $\hat{\theta}_{iT}^H$, under the random walk model is obtained. The EBLUP estimator does not require normality of the errors; only symmetrically distributed errors are needed. However, normality is used to derive an estimator of MSE correct to terms of order $o(m^{-1})$ for large m.

The MSE estimators of $\hat{\theta}_{iT}^H(\rho)$ and $\hat{\theta}_{iT}^H$, correct to terms of order $o(m^{-1})$, may be obtained from the general results of subsection 6.3.2, Chapter 6. Rao and Yu (1994) used the formula (6.3.11) to obtain an MSE estimator of $\hat{\theta}_{iT}^H(\rho)$ under the transformation method of estimating σ^2 and σ_v^2. Datta, Lahiri and Maiti (2002) used the area-specific version (6.3.16) to obtain an MSE estimator of $\hat{\theta}_{iT}^H$ using REML estimators of σ^2 and σ_v^2. You (1999, Chapter 8) used (6.3.11) to obtain an MSE estimator of $\hat{\theta}_{iT}^H$ under the transformation method of estimating σ^2 and σ_v^2.

The case of unknown ρ in the AR(1) model is more difficult to handle. Rao and Yu (1994) obtained a consistent moment estimator $\hat{\rho}$, but it often takes values outside the admissible range $(-1, 1)$, especially for small T or small σ^2 relative to the sampling variation. To avoid this difficulty, they proposed a naive estimator of ρ obtained by ignoring the sampling errors. It is given by

$$\hat{\rho}_N = \left[\sum_{i=1}^{m}\sum_{t=1}^{T}\hat{\eta}_{it}(\hat{\eta}_{i,t+1} - \hat{\eta}_{i,t+2})\right]\left[\sum_{i=1}^{m}\sum_{t=1}^{T-2}\hat{\eta}_{it}(\hat{\eta}_{it} - \hat{\eta}_{i,t+1})\right]^{-1}, \qquad (8.3.3)$$

where

$$\hat{\eta}_{it} = \hat{\theta}_{it} - \mathbf{z}_{it}^T\Big(\sum_i \mathbf{Z}_i\mathbf{Z}_i^T\Big)^{-1}\Big(\sum_i \mathbf{Z}_i\hat{\boldsymbol{\theta}}_i\Big)$$

are the least squares residuals. The estimator $\hat{\rho}_N$ is inconsistent in the presence of sampling errors and typically underestimates ρ. Nevertheless, the resulting EBLUP estimator, $\hat{\theta}_{iT}^H(\hat{\rho}_N)$, remains unbiased. An estimator of the MSE of $\hat{\theta}_{iT}^H(\hat{\rho}_N)$ is obtained from the MSE estimator of $\hat{\theta}_{iT}^H(\rho)$ by substituting $\hat{\rho}_N$ for ρ. However, this MSE estimator is not correct to terms of order $o(m^{-1})$.

Example 8.3.1. Simulation Study. Rao and Yu (1994) conducted a limited simulation study on the properties of $\hat{\theta}_{iT}^H(\rho)$, for known ρ or prior guess $\rho = \rho_0$, and $\hat{\theta}_{iT}^H(\hat{\rho}_N)$; the parameters σ_v^2 and σ^2 were estimated by the transformation method. They used the following simple model:

$$y_{it} = v_i + u_{it} + e_{it}, \quad u_{it} = \rho u_{i,t-1} + \varepsilon_{it}, \quad |\rho| < 1$$

with $\rho = 0.2$ and 0.4, $e_{it} \overset{\text{iid}}{\sim} N(0, \psi = 1)$, $v_i \overset{\text{iid}}{\sim} N(0, \sigma_v^2)$ and $\varepsilon_{it} \overset{\text{iid}}{\sim} N(0, \sigma^2)$. They used $m = 40$ small areas and both small T $(= 5)$ and moderate T $(= 10)$, and generated 5000 independent samples $\{y_{it}\}$ for each selected pair (σ_v^2, σ^2). The simulated values of the gain in efficiency (GE) of an EBLUP estimator over the Fay-Herriot (FH) estimator and the relative biases (RB) of MSE estimators were then computed, using the simulated samples. The FH estimator uses only the cross-sectional data for the current period T.

Results for the known ρ case may be summarized as follows: (1) Substantial GE is achieved when the between time variation relative to the sampling variation is small ($\sigma^2 = 0.25$, 0.5), especially when the between small area variation relative to the sampling variation is substantial. For example, GE = 105% when $T = 10$, $\sigma^2 = 0.25$, $\rho = 0.4$ and $\sigma_v^2 = 1.0$. Note that the FH estimator gives more weight to the direct estimator when σ_v^2 is substantial. (2) GE increases significantly with T, especially for small σ^2, that is, the use of data for more time points improves the efficiency of the EBLUP estimator. (3) mse$[\hat{\theta}_{iT}^H(\rho)]$ performs well in terms of RB leading to slight overestimation (RB$\leq$ 3% for $T = 5$ and RB$\leq$ 5% for $T = 10$).

Since ρ is unknown in practice, Rao and Yu (1994) also computed the GE-values of $\hat{\theta}_{iT}^H(\rho_0)$ and the RB-values of mse$[\hat{\theta}_{iT}^H(\rho)]$ evaluated at $\rho = \rho_0$, using prior guesses $\rho_0 = 0.1, 0.3$ for true $\rho = 0.2$ and $\rho_0 = 0.2, 0.3, 0.5\ 0.6$ for true $\rho = 0.4$. Results may be summarized as follows: (1) The GE-values are virtually unaffected by the choice of ρ_0, that is, the EBLUP estimator $\hat{\theta}_{iT}^H(\rho_0)$ retains its efficiency even when the prior guess ρ_0 deviates substantially from ρ. (2) The MSE estimator performs well when $\rho_0 < \rho$, but it can lead to substantial overestimation (RB$\geq$ 10%) when ρ_0 is substantially larger than ρ and σ^2 is small. The MSE estimator, therefore, appears to be satisfactory provided the prior guess ρ_0 is not significantly larger than the true ρ.

Turning to $\hat{\theta}_{iT}^H(\hat{\rho}_N)$, the GE values of $\hat{\theta}_{iT}^H(\hat{\rho}_N)$ are also close to those under the true ρ, that is, the EBLUP estimator retains efficiency even when a naive estimator $\hat{\rho}_N$ is used; note that $\hat{\rho}_N$ leads to significant underestimation of ρ. Further, the MSE estimator, mse$[\hat{\theta}_{iT}^H(\rho)]$ evaluated at $\rho = \hat{\rho}_N$, performs well for $\rho = 0.2$ and 0.4, but it can lead to significant underestimation as ρ increases (RB$= -10\%$ for $T = 5$ and $\rho = 0.7$). Thus, the MSE estimator is quite satisfactory unless ρ is expected to be large. Also, it is data-based unlike mse$[\hat{\theta}_{iT}^H(\rho)]$ evaluated at $\rho = \rho_0$ which depends on a prior guess ρ_0.

Example 8.3.2. Unemployment in Census Divisions. Choudhry and Rao (1989) considered a special form of the Rao-Yu model given by (5.4.6) and (5.4.7). They treated the composite errors $a_{it} = e_{it} + u_{it}$ as an AR(1) process: $a_{it} = \tilde{\rho} a_{i,t-1} + \tilde{\varepsilon}_{it}, |\tilde{\rho}| < 1$, and $\tilde{\varepsilon}_{it} \overset{\text{iid}}{\sim} N(0, \tilde{\sigma}^2)$. The parameters of interest were taken as $\theta_{it} = \mathbf{z}_{it}^T\boldsymbol{\beta} + v_i$, that is, the time effects on the area means are reflected only through varying auxiliary variables, $\mathbf{z}_{it}$, unlike (5.4.7) which uses $\theta_{it} = \mathbf{z}_{it}^T\beta + v_i + u_{it}$. Choudhry and Rao (1989) used this simplified model to obtain EBLUP estimates of monthly unemployment for census divisions (small areas), using data from the Canadian Labour Force Survey. The parameters σ_v^2 and $\tilde{\sigma}^2$ were estimated by the transformation method of Pantula and Pollock (1989); note that the sampling error, e_{it}, is absorbed in the composite error a_{it}.

In this application, we have $m = 21$ small areas, $T = 36$ time points, $\hat{\theta}_{it}$ = log(survey estimate of proportion of population unemployed) and $\mathbf{z}_{it} = (1, z_{it1}, z_{it2})^T$ with z_{it1} = log(unemployment insurance (UI) beneficiaries/projected population 15 years and over) and z_{it2} = survey estimate of labour force participation rate, defined as the proportion of target population which is either employed or unemployed. The UI beneficiary counts were obtained from the UI system. The monthly small area estimates were needed in producing three-year average unemployment rates for census divisions which were used in conjunction with other variables to produce an index which, in turn, was used to allocate funds for industrial incentive.

Example 8.3.3. Median Income. Datta, Lahiri and Maiti (2002) applied the Rao-Yu model (5.4.6) and (5.4.7), Chapter 5, with random walk errors u_{it}, to estimate median income of four-person families for the fifty American states and the District of Columbia. They used the Current Population Survey (CPS) data for nine years (1981–1989) to produce EBLUP estimates $\hat{\theta}_{iT}^H$ of the median income, θ_{iT}, for 1989 ($i = 1, \ldots, 51$). The estimates for 1989 enabled them to conduct an external evaluation, by comparing the estimates to 1990 census estimates for 1989. The sampling error model is given by $\hat{\theta}_{it} = \theta_{it} + e_{it}$ with sampling covariance matrix $\mathbf{\Psi}_i$ for each area i estimated from the CPS data, using the jackknife method and some smoothing techniques. The linking model is given by $\theta_{it} = \beta_0 + \beta_1 z_{it1} + v_i + u_{it}$, where z_{it1} is the "adjusted" census median income for year t in area i (see Example 5.4.3, Chapter 5). The EBLUP estimates were calculated using the REML and ML estimates of model parameters σ^2 and σ_v^2.

Average absolute relative error ($\overline{\text{ARE}}$) values for the CPS direct estimates and the EBLUP estimates were calculated by treating the census values as the true values θ_{iT}. Note that $\overline{\text{ARE}} = \frac{1}{51}\sum_i |\text{est}_{iT} - \theta_{iT}|/\theta_{iT}$. They obtained the following values: $\overline{\text{ARE}}$(CPS)=7.3%, $\overline{\text{ARE}}$ (EBLUP)=2.9% using REML, and $\overline{\text{ARE}}$(EBLUP)=2.8% using ML. The EBLUP estimates outperformed the CPS estimates in terms of $\overline{\text{ARE}}$.

Datta, Lahiri and Maiti (2002) also calculated the coefficient of variation (CV) of the estimates, using the Tth diagonal element of $\mathbf{\Psi}_i$ for the direct estimate $\hat{\theta}_{iT}$, $\text{mse}_2(\hat{\theta}_{iT}^H)$ for the EBLUP estimate based on REML and $\text{mse}_{*2}(\hat{\theta}_{iT}^H)$ for the EBLUP estimate based on ML. The distribution of CVs is given in Table 8.1.

TABLE 8.1 Distribution of Coefficient of Variation (%)

	Coefficient of Variation		
Estimate	2–4%	4-6%	$\geq$ 6%
CPS	6	7	38
EBLUP (REML)	49	2	0
EBLUP (ML)	49	2	0

SOURCE: Adapted from Table 1 in Datta, Lahiri and Maiti (2002).

It is clear from Table 8.1 that EBLUP (ML) and EBLUP (REML) estimates outperformed the CPS direct estimates by bringing the CV down to the range 2–4% for 49 areas, while the CV of CPS estimates is at least 6% for 38 areas.

8.3.2 State Space Models

The area level model given by (5.4.13)–(5.4.15), Chapter 5, allows the model coefficients to vary both cross-sectionally and over time. It is a special case of the general state-space model which may be expressed in the form

$$\mathbf{y}_t = \mathbf{Z}_t\boldsymbol{\alpha}_t + \boldsymbol{\varepsilon}_t; \quad E(\boldsymbol{\varepsilon}_t) = \mathbf{0}, \quad E(\boldsymbol{\varepsilon}_t\boldsymbol{\varepsilon}_t^T) = \mathbf{\Sigma}_t \tag{8.3.4}$$

$$\boldsymbol{\alpha}_t = \mathbf{H}\boldsymbol{\alpha}_{t-1} + \mathbf{A}\boldsymbol{\eta}_t; \quad E(\boldsymbol{\eta}_t) = \mathbf{0}, \quad E(\boldsymbol{\eta}_t\boldsymbol{\eta}_t^T) = \mathbf{\Gamma} \tag{8.3.5}$$

where $\boldsymbol{\varepsilon}_t$ and $\boldsymbol{\eta}_t$ are uncorrelated contemporaneously and over time. This model is a special case of the general linear mixed model but the state-space form permits updating of the estimates over time, using the Kalman filter equations (8.3.7) and (8.3.8) below, and smoothing past estimates as new data becomes available, using an appropriate smoothing algorithm. The vector $\boldsymbol{\alpha}_t$

is known as the *state vector*, (8.3.5) as the *transition equation* and (8.3.4) as the *measurement equation.*

Let $\tilde{\boldsymbol{\alpha}}_{t-1}$ be the BLUP estimator of $\boldsymbol{\alpha}_{t-1}$ based on all data observed up to time $t-1$, so that $\tilde{\boldsymbol{\alpha}}_{t|t-1} = \mathbf{H}\tilde{\boldsymbol{\alpha}}_{t-1}$ is the BLUP of $\boldsymbol{\alpha}_t$ at time $t-1$. Further,

$$\mathbf{P}_{t|t-1} = \mathbf{H}\mathbf{P}_{t-1}\mathbf{H}^T + \mathbf{A}\boldsymbol{\Gamma}\mathbf{A}^T$$

is the covariance matrix of the prediction errors $\tilde{\boldsymbol{\alpha}}_{t|t-1} - \boldsymbol{\alpha}_t$, where $\mathbf{P}_{t-1} = E(\tilde{\boldsymbol{\alpha}}_{t-1} - \boldsymbol{\alpha}_{t-1})(\tilde{\boldsymbol{\alpha}}_{t-1} - \boldsymbol{\alpha}_{t-1})^T$ is the covariance matrix of the prediction errors at time $t-1$. This result readily follows from (8.3.5). At time t, the predictor of $\boldsymbol{\alpha}_t$ and its covariance matrix are updated using the new data $(\mathbf{y}_t, \mathbf{Z}_t)$. We have

$$\mathbf{y}_t - \mathbf{Z}_t\tilde{\boldsymbol{\alpha}}_{t|t-1} = \mathbf{Z}_t(\boldsymbol{\alpha}_t - \tilde{\boldsymbol{\alpha}}_{t|t-1}) + \boldsymbol{\varepsilon}_t, \tag{8.3.6}$$

which has the linear mixed model form (6.2.1) with $\mathbf{y} = \mathbf{y}_t - \mathbf{Z}_t\tilde{\boldsymbol{\alpha}}_{t|t-1}$, $\mathbf{Z} = \mathbf{Z}_t$, $\mathbf{v} = \boldsymbol{\alpha}_t - \tilde{\boldsymbol{\alpha}}_{t|t-1}$, $\mathbf{G} = \mathbf{P}_{t|t-1}$, $\mathbf{X}\boldsymbol{\beta}$ absent and $\mathbf{V} = \mathbf{F}_t$, where

$$\mathbf{F}_t = \mathbf{Z}_t\mathbf{P}_{t|t-1}\mathbf{Z}_t^T + \boldsymbol{\Sigma}_t.$$

Therefore, the BLUP estimator $\tilde{\mathbf{v}} = \mathbf{G}\mathbf{Z}^T\mathbf{V}^{-1}\mathbf{y}$ reduces to

$$\tilde{\boldsymbol{\alpha}}_t = \tilde{\boldsymbol{\alpha}}_{t|t-1} + \mathbf{P}_{t|t-1}\mathbf{Z}_t^T\mathbf{F}_t^{-1}(\mathbf{y}_t - \mathbf{Z}_t\tilde{\boldsymbol{\alpha}}_{t|t-1}). \tag{8.3.7}$$

Further, it follows from (6.2.12) that the covariance matrix of the prediction errors $\tilde{\mathbf{v}} - \mathbf{v}$ is $\mathbf{G} - \mathbf{G}\mathbf{Z}^T\mathbf{V}^{-1}\mathbf{Z}\mathbf{G}$, which in the case of the prediction errors $\tilde{\boldsymbol{\alpha}}_t - \boldsymbol{\alpha}_t$ reduces to

$$\mathbf{P}_t = \mathbf{P}_{t|t-1} - \mathbf{P}_{t|t-1}\mathbf{Z}_t^T\mathbf{F}_t^{-1}\mathbf{Z}_t\mathbf{P}_{t|t-1}. \tag{8.3.8}$$

The general state space model, (8.3.4) and (8.3.5), covers both area level and unit level models.

To illustrate the Kalman filter, consider the area level model

$$\hat{\theta}_{it} = \mathbf{z}_{it}^T\boldsymbol{\beta}_{it} + e_{it} \tag{8.3.9}$$

with the coefficients β_{itj} obeying the random walk model:

$$\beta_{itj} = \beta_{i,t-1,j} + v_{itj}, \quad j = 1, \ldots, p. \tag{8.3.10}$$

The sampling errors, e_{it}, for each area i are assumed to be serially uncorrelated with mean 0 and variance ψ_{it}. Further, the model errors v_{itj} for each area i are uncorrelated over time with mean zero and covariances $E_m(v_{itj}v_{itl}) = \sigma_{vjl}$, and for areas $i \neq h$, $E_m(v_{itj}v_{htj}) = \rho_j\sigma_{vjj}$ and $E_m(v_{itj}v_{htl}) = 0$, $j \neq l$. In this case, $\mathbf{y}_t = \hat{\boldsymbol{\theta}}_t = (\hat{\theta}_{1t}, \ldots, \hat{\theta}_{mt})^T$, $\mathbf{z}_{it}^T\boldsymbol{\alpha}_{it} = \mathbf{z}_{it}^T\boldsymbol{\beta}_{it} = \theta_{it}$, $\mathbf{H} = \mathbf{I}$, $\mathbf{A} = \mathbf{I}$. The BLUP estimator of the realized $\boldsymbol{\theta}_t = (\theta_{1t}, \ldots, \theta_{mt})^T = \mathbf{Z}_t\boldsymbol{\alpha}_t$ is $\tilde{\boldsymbol{\theta}}_t^H = \mathbf{Z}_t\tilde{\boldsymbol{\alpha}}_t$, where $\mathbf{Z}_t$ is block diagonal with the blocks defined by the rows $\mathbf{z}_{it}^T$, $\boldsymbol{\alpha}_t = (\boldsymbol{\beta}_{1t}^T, \ldots, \boldsymbol{\beta}_{mt}^T)^T$ and $\tilde{\boldsymbol{\alpha}}_t$ is obtained from the Kalman filter equation

(8.3.7). After simplification, it can be shown that the BLUP estimator $\tilde{\theta}_{it}^{H}$ is a weighted combination of the direct estimator $\hat{\theta}_{it}$, the regression synthetic estimator $\mathbf{z}_{it}^T\tilde{\boldsymbol{\beta}}_{it|t-1}$, where $\tilde{\boldsymbol{\beta}}_{it|t-1}$ is the BLUP estimator of $\boldsymbol{\beta}_{it} = \boldsymbol{\alpha}_{it}$ at time $t-1$, and "adjustment factors" $\hat{\theta}_{ht} - \mathbf{z}_{ht}^T\tilde{\boldsymbol{\beta}}_{ht|t-1}$ based on areas $h \neq i$ (Pfeffermann and Burck (1990)). The third component vanishes when $\rho_j = 0$ for all j, that is, when the spatial correlations, ρ_j, are absent. In this case, the BLUP estimator $\tilde{\theta}_{it}^{H}$ has the familiar Fay-Herriot form (7.1.3) with the weight attached to the direct estimator decreasing as the sampling variance ψ_{it} increases.

To implement the recursive calculations (8.3.7) and (8.3.8), we need to specify a mean $\tilde{\boldsymbol{\alpha}}_0$ and a covariance matrix $\mathbf{P}_0$ of the initial state vector $\boldsymbol{\alpha}_0$. When the transition equation is non-stationary, as in the case of the random walk model (8.3.10), the values $\tilde{\boldsymbol{\alpha}}_0 = \mathbf{0}$ and $\mathbf{P}_0 = \kappa\mathbf{I}$ could be used, where κ is a large positive constant. This choice corresponds to a diffuse or non-informative prior distribution. For the stationary case, $\tilde{\boldsymbol{\alpha}}_0$ and $\mathbf{P}_0$ are taken as the unconditional mean and covariance matrix of $\boldsymbol{\alpha}_t$. We refer the reader to Harvey (1989, subsection 3.3.4) for a careful discussion of the initial conditions.

The BLUP estimator $\tilde{\boldsymbol{\alpha}}_t$ involves unknown parameters $\boldsymbol{\delta}$ that specify the covariance matrices $\boldsymbol{\Sigma}_t$ and $\boldsymbol{\Gamma}$. The vector $\boldsymbol{\delta}$ may contain also unknown elements of the transition matrix. Assuming normality of the errors $\boldsymbol{\varepsilon}_t$ and $\boldsymbol{\eta}_t$, ML estimators of these parameters can be obtained by expressing the loglikelihood in the *prediction error decomposition* form:

$$\log L = \text{const.} - \frac{1}{2}\sum_{t=1}^{T}\log|\mathbf{F}_t| - \frac{1}{2}\sum_{t=1}^{T}(\mathbf{y}_t - \tilde{\mathbf{y}}_{t|t-1})^T\mathbf{F}_t^{-1}(\mathbf{y}_t - \tilde{\mathbf{y}}_{t|t-1}), \quad (8.3.11)$$

where $\tilde{\mathbf{y}}_{t|t-1} = \mathbf{Z}_t\tilde{\alpha}_{t|t-1}$ is the BLUP of $\mathbf{y}_t$ at time $t-1$ and $\mathbf{y}_t - \tilde{\mathbf{y}}_{t|t-1}$ is the vector of prediction errors. The representation (8.3.11) follows by writing L as

$$L = \prod_{t=1}^{T} p(\mathbf{y}_t|\mathbf{Y}_{t-1}),$$

where $p(\mathbf{y}_t|\mathbf{Y}_{t-1})$ is the conditional distribution of $\mathbf{y}_t$ given $\mathbf{Y}_{t-1} = \{\mathbf{y}_{t-1}, \ldots, \mathbf{y}_1\}$, and noting that $p(\mathbf{y}_t|\mathbf{Y}_{t-1})$ is normal with mean $\tilde{\mathbf{y}}_{t|t-1}$ and covariance matrix $\mathbf{F}_t$. The ML estimator of $\boldsymbol{\delta}$ and the parameters of the initial specification, if any, can be obtained by using the method of scoring with a variable step length (Pfeffermann and Burck (1990)). We solve the following set of equations iteratively:

$$\boldsymbol{\delta}^{(a+1)} = \boldsymbol{\delta}^{(a)} + r_i[\mathcal{I}(\boldsymbol{\delta}^{(a)})]^{-1}\mathbf{s}(\boldsymbol{\delta}^{(a)}).$$

Here, $\boldsymbol{\delta}^{(a)}$ is the value of the estimator at the ath iteration, $\mathbf{s}(\boldsymbol{\delta})$ is the score vector consisting of the elements $\partial L/\partial\delta_i$, $\mathcal{I}(\boldsymbol{\delta})$ is the information matrix and

r_i is a variable step length introduced to ensure that the likelihood is non-decreasing, that is, $L(\boldsymbol{\delta}^{(a+1)}) \geq L(\boldsymbol{\delta}^{(a)})$ at each iteration a. The method of moments can also be used to estimate the model parameters $\boldsymbol{\delta}$ without the normality assumption. Singh, Mantel and Thomas (1994) presented such estimators for special cases, using a moment method analogous to the Fay and Herriot (1979) method for cross-sectional data.

The EBLUP estimator $\hat{\boldsymbol{\alpha}}_t$ is obtained by substituting an estimator $\hat{\boldsymbol{\delta}}$ for $\boldsymbol{\delta}$ in $\tilde{\boldsymbol{\alpha}}_t$. The resulting EBLUP estimator of $\boldsymbol{\theta}_t = \mathbf{Z}_t\boldsymbol{\alpha}_t$ is

$$\hat{\boldsymbol{\theta}}_t^H = \mathbf{Z}_t\hat{\boldsymbol{\alpha}}_t. \tag{8.3.12}$$

A naive covariance matrix of the prediction errors $\hat{\boldsymbol{\alpha}}_t - \boldsymbol{\alpha}_t$ is obtained by substituting $\hat{\boldsymbol{\delta}}$ for $\boldsymbol{\delta}$ in the formula (8.3.8) for $\mathbf{P}_t$, ignoring the variability associated with $\hat{\boldsymbol{\delta}}$. The resulting estimated covariance matrix of $\hat{\boldsymbol{\theta}}_t^H - \boldsymbol{\theta}_t$ is

$$\text{mse}_N(\hat{\boldsymbol{\theta}}_t^H) = \mathbf{Z}_t\hat{\mathbf{P}}_t\mathbf{Z}_t^T, \tag{8.3.13}$$

where $\hat{\mathbf{P}}_t$ is the estimator of $\mathbf{P}_t$. We refer the reader to Harvey (1989, Chapter 3) for a detailed account of the state space model and the Kalman filter.

Methods of estimating the MSE of the EBLUP estimator $\hat{\theta}_{it}^H$ that account for the variability associated with $\hat{\boldsymbol{\delta}}$ have also been studied. Under normality of the errors,

$$\text{MSE}(\hat{\theta}_{it}^H) = \text{MSE}(\tilde{\theta}_{it}^H) + E(\hat{\theta}_{it}^H - \tilde{\theta}_{it}^H)^2, \tag{8.3.14}$$

where

$$\text{MSE}(\tilde{\theta}_{it}^H) = \mathbf{z}_{it}^T\mathbf{P}_t\mathbf{z}_{it}. \tag{8.3.15}$$

Ansley and Kohn (1986) expanded $\hat{\theta}_{it}^H = \tilde{\theta}_{it}^H(\hat{\boldsymbol{\delta}})$ around $\boldsymbol{\delta}$ to approximate the last term of (8.3.14). This leads to the following estimator:

$$\text{est } [E(\hat{\theta}_{it}^H - \tilde{\theta}_{it}^H)^2] \approx [\partial\tilde{\theta}_{it}(\boldsymbol{\delta})/\partial\boldsymbol{\delta}]^T_{\boldsymbol{\delta}=\hat{\boldsymbol{\delta}}}[\mathcal{I}(\hat{\boldsymbol{\delta}})]^{-1}[\partial\tilde{\theta}_{it}(\boldsymbol{\delta})/\partial\boldsymbol{\delta}]_{\boldsymbol{\delta}=\hat{\boldsymbol{\delta}}}, \tag{8.3.16}$$

where $\hat{\boldsymbol{\delta}}$ is the ML estimator of $\boldsymbol{\delta}$ and $\mathcal{I}(\hat{\boldsymbol{\delta}})$ is the corresponding information matrix evaluated at $\boldsymbol{\delta} = \hat{\boldsymbol{\delta}}$. A naive estimator of $\text{MSE}(\tilde{\theta}_{it}^H)$ is given by

$$\text{mse}_N(\tilde{\theta}_{it}^H) = \mathbf{z}_{it}^T\mathbf{P}_t(\hat{\boldsymbol{\delta}})\mathbf{z}_{it}. \tag{8.3.17}$$

Ansley and Kohn used the sum of (8.3.16) and (8.3.17) as the estimator of $\text{MSE}(\hat{\theta}_{it}^H)$. But the bias of $\text{mse}_N(\tilde{\theta}_{it}^H)$ is of the same order as $E(\hat{\theta}_{it}^H - \tilde{\theta}_{it}^H)^2$. As a result, the MSE estimator of Ansley and Kohn is not correct to second-order.

Pfeffermann and Tiller (2001) proposed an estimator of MSE based on the parametric bootstrap. Assuming normality of the errors, their method consists of the following steps:

(1) Generate a large number, B, of bootstrap series $\{\hat{\theta}_{it}^{b}\}, b = 1, \ldots, B$ from the model fitted to the original series, with model parameters $\boldsymbol{\delta} = \hat{\boldsymbol{\delta}}$.

(2) Re-estimate $\boldsymbol{\delta}$ for each of the generated series using the same method as used for estimating δ from the original series $\{\hat{\theta}_{it}\}$. Denote the bootstrap estimators as $\hat{\boldsymbol{\delta}}^{b}, b = 1, \ldots, B$.

(3) Estimate $E(\hat{\theta}_{it}^{H} - \tilde{\theta}_{it}^{H})^2$ as

$$\text{est. } E(\hat{\theta}_{it}^{H} - \tilde{\theta}_{it}^{H})^2 = \frac{1}{B}\sum_{b=1}^{B}[\tilde{\theta}_{it}^{H}(\hat{\boldsymbol{\delta}}^{b}) - \tilde{\theta}_{it}^{H}(\hat{\boldsymbol{\delta}})]^2, \qquad (8.3.18)$$

noting that $\hat{\theta}_{it}^{H} = \tilde{\theta}_{it}^{H}(\hat{\boldsymbol{\delta}})$ and $\tilde{\theta}_{it}^{H} = \tilde{\theta}_{it}^{H}(\boldsymbol{\delta})$.

(4) Correct $\text{mse}_N(\tilde{\theta}_{it}^{H})$ for its bias as follows:

$$\text{mse}_a(\tilde{\theta}_{it}^{H}) = 2\,\text{mse}_N(\tilde{\theta}_{it}^{H}) - \frac{1}{B}\sum_{b=1}^{B}[\mathbf{z}_{it}^{T}\mathbf{P}_t(\hat{\boldsymbol{\delta}}^{b})\mathbf{z}_{it}]. \qquad (8.3.19)$$

(5) The sum of (8.3.18) and (8.3.19) is the bootstrap estimator of $\text{MSE}(\hat{\theta}_{it}^{H})$.

Pfeffermann and Tiller (2001) established the validity of the bootstrap MSE estimator to second-order terms, under certain regularity conditions. The bootstrap MSE estimator is similar to the jackknife MSE estimator studied in Section 9.2, Chapter 9.

The U.S. Bureau of Labor Statistics uses area level state-space models for the prediction of all the major employment and unemployment estimates in the 50 states and the District of Columbia. The CPS direct estimates over time are used to produce the model based estimates which are time indirect in the terminology of Section 1.2 of Chapter 1 (Tiller (1982)). The models used in this application account for seasonal effects, trend, a covariate, and sampling error correlations over time. Pfeffermann, Feder and Signorelli (1998) applied a similar state-space model to labor force data from Australia.

Example 8.3.4. Simulation Study. Singh, Mantel and Thomas (1994) conducted a simulation study of the EBLUP (state-space) estimator at current time T relative to Fay-Herriot (FH), synthetic (SYN) and sample size dependent (SSD) estimators based only on the current cross-sectional data. They generated a pseudo-population from Statistics Canada's biannual farm survey data from Quebec over six time points (June 1988, January 1989, ..., January, 1991). The farm survey used dual frames, a list frame and an area frame, but the pseudo-population was generated from the list frame survey data only. The farm units in the list frame were replicated proportional to their sampling weights to form a pseudo-population of N = 10,362 farm units. The parameter of interest, θ_{iT}, is the total number of cattle and calves for each crop district i (small area) in Quebec for the current period T (January, 1991), $i = 1, \ldots, 12$.

The pseudo-population was stratified into four take-some and one take-all strata using 1986 Agricultural Census count data on cattle and calves as the stratification variable. Independent stratified random samples were generated for each occasion from the pseudo-population, and the estimates SYN, SSD, FH and EBLUP (state space) were computed from each simulated sample r, using the following formulae. SYN is given by

$$\text{SYN}_{iT} = \hat{\beta}_1 + \hat{\beta}_2 z_{iT},$$

where z_{iT} is the ratio synthetic estimate of θ_{iT} for small area i at time T, obtained from the 1986 census counts, θ_i^c and $\theta_\cdot^c = \sum_i \theta_i^c$, and the farm survey direct estimate $\hat{\theta}_{\cdot T}$ of $\theta_{\cdot T} = \sum_i \theta_{iT}$: $z_{iT} = (\hat{\theta}_{\cdot T}/\theta_\cdot^c)\theta_i^c$. Further, $\hat{\beta}_1$ and $\hat{\beta}_2$ are the least square estimates of β_1 and β_2 obtained from the model $\hat{\theta}_{iT} = \beta_1 + \beta_2 z_{iT} + e_{iT}$, where $\hat{\theta}_{iT}$ is the direct estimate of θ_{iT} and $e_{iT} \stackrel{\text{iid}}{\sim} (0, \psi_T)$; $i = 1, \ldots, m$. SSD is a weighted combination of $\hat{\theta}_{iT}$ and SYN: $\phi_{iT}(\text{S1})\hat{\theta}_{iT} + [1-\phi_{iT}(\text{S1})]\text{SYN}_{iT}$, where $\phi_{iT}(\text{S1})$ is given by (4.3.8) with $\delta = 1.0$. FH is given by (7.1.11), Chapter 7, where the model variance, σ_{vT}^2, is estimated by the Fay-Herriot method of moments (see (7.1.12), Chapter 7). EBLUP (state-space) is based on a model of the form (8.3.9) and (8.3.10) with $\hat{\theta}_{it} = \beta_{t1} + \beta_{t2} z_{it} + a_{it}$, $\beta_{tj} = \beta_{t-1,j} + v_{tj}, j = 1, 2$ and $a_{it} = a_{i,t-1} + \epsilon_{it}$. Further, the errors v_{tj} and a_{it} are assumed to be uncorrelated with mean 0 and variances σ_{vj}^2 and σ_a^2, respectively. The parameters σ_{vj}^2 and σ_a^2 were estimated by a method similar to the Fay-Herriot method for cross-sectional data.

Table 8.2 presents the simulated values of absolute relative bias and relative root MSE (or CV) of the estimates averaged over $m = 12$ small areas within specified size groups, denoted by $\overline{\text{ARB}}$ and $\overline{\text{RRMSE}}$. It is clear from Table 8.2 that EBLUP (state-space) leads to significant reduction in $\overline{\text{RRMSE}}$ relative to FH which, in turn, performs better than SSD. In terms of $\overline{\text{ARB}}$, EBLUP (state-space) is better than SYN and FH. As expected, SSD has the smallest $\overline{\text{ARB}}$.

TABLE 8.2 Average Absolute Relative Bias ($\overline{\text{ARB}}$) and Average Relative Root MSE ($\overline{\text{RRMSE}}$) of SYN, SSD, FH and EBLUP (State-Space)

	Estimator			
Size Group	SYN	SSD	FH	EBLUP
		$\overline{\text{ARB}}$ %		
Small	8.7	2.9	5.2	3.9
Medium	14.1	1.7	9.5	8.0
Large	6.3	0.6	4.9	4.2
		$\overline{\text{RRMSE}}$ %		
Small	15.4	19.4	16.1	12.0
Medium	15.9	18.5	15.1	12.7
Large	9.8	14.9	10.0	8.4

SOURCE: Adapted from Table 4 in Singh, Mantel and Thomas (1994).

8.4 Spatial Models

The spatial models discussed in subsection 5.4.4 are similar to the Fay-Herriot model except for spatial correlations among the random small area effects v_i. Therefore, $\mathbf{G} = \sigma_v^2 \mathbf{I}_m$ is changed to $\mathbf{G} = \mathbf{\Gamma}(\boldsymbol{\delta})$, where the covariance matrix $\mathbf{\Gamma}(\boldsymbol{\delta})$ may take the form $\mathbf{\Gamma}(\boldsymbol{\delta}) = \sigma_v^2(\mathbf{I} - \rho\mathbf{Q})^{-1}\mathbf{B}$ as specified in (5.4.17) or $\mathbf{\Gamma}(\boldsymbol{\delta}) = \sigma_v^2(\delta_1\mathbf{I} + \delta_2\mathbf{D})$ or $\mathbf{\Gamma}(\boldsymbol{\delta}) = \sigma_v^2[\delta_1\mathbf{I} + \delta_2\mathbf{D}(\delta_3)]$ as used in the geostatistics literature, where $\mathbf{D}$ and $\mathbf{D}(\delta_3)$ are defined in subsection 5.4.4.

Spatial models are special cases of the general linear mixed model (6.2.1). Therefore, the BLUP estimator of $\theta_i = \mathbf{z}_i^T\boldsymbol{\beta} + b_i v_i$ can be obtained from the general formula (6.2.4), for a specified $\boldsymbol{\delta}$. In practice, $\boldsymbol{\delta}$ is unknown and we need to replace it by an estimator $\hat{\boldsymbol{\delta}}$ to obtain the EBLUP estimator of θ_i. ML or REML estimators of $\boldsymbol{\delta}$ may be obtained from the general results in subsection 6.2.4. Cressie and Chan (1989) discussed ML estimation of $\boldsymbol{\delta}$ for spatial models. Estimators of MSE of the EBLUP estimators that take account of the uncertainty in $\hat{\boldsymbol{\delta}}$ have not been spelled out.

Example 8.4.1. 1980 U.S. Census Undercount. As noted in Example 5.2.2, Chapter 5, Cressie (1989) modelled the state adjustment factor $\theta_i = T_i/C_i$, using the basic area level model, $\hat{\theta}_i = \mathbf{z}_i^T\boldsymbol{\beta} + b_i v_i + e_i$, with $b_i = 1/\sqrt{C_i}$ $(i = 1, \ldots, 51)$. Cressie (1991) extended this model by allowing spatial correlations among the random state effects v_i. The covariance matrix of the $b_i v_i$'s is given by $\mathbf{\Gamma}(\boldsymbol{\delta}) = \sigma_v^2(\mathbf{I} - \rho\mathbf{Q})^{-1}\mathbf{B}$, where $\mathbf{B} = \text{diag}(b_1^2, \ldots, b_m^2)$ and the elements of $\mathbf{Q}$ are given by $q_{ii} = 0, q_{il} = \sqrt{C_l/C_i}$ if the "distance" between the ith and the lth states, d_{il}, is less than or equal to 700 miles, and $q_{il} = 0$ otherwise $(i \neq l)$; see Example 5.4.6. Dual System PES estimates, $\hat{\theta}_i$, were used in the spatial model.

A subset of eight explanatory variables was selected following the variable selection method of Ericksen, Kadane and Tukey (1989), but using weighted least squares with census counts, C_i, as weights. This procedure essentially ignores sampling errors, e_i, and spatial dependence, and fits the model $\hat{\theta}_i = \mathbf{z}_i^T\boldsymbol{\beta} + v_i/\sqrt{C_i}$ by weighted least squares to select the z-variables. The variable selection method is as follows: (1) Consider all subsets of two, three and four variables. (2) Consider all regression equations in which each weighted least squares regression coefficient is at least twice its standard error. (3) Select the regression equation that minimizes the residual mean sum of squares. This method yielded the following variables: z_2 =minority percentage, z_3 =education (percentage of persons age 25 and older who have not graduated from high school) and the constant term $z_1 = 1$.

Using $\mathbf{z} = (z_1, z_2, z_3)^T$ and assuming normality, ML estimates of $\boldsymbol{\beta}$, σ_v^2 and ρ and the resulting EBLUP estimates were calculated. Denote these estimates with no restrictions on $\boldsymbol{\beta}$ and ρ as $\hat{\theta}_i^H(2,2)$. To study the effect of spatial modeling, models with the variable z_3 deliberately omitted (or not) and spatial dependence set equal to zero (or not) were also considered: (1,1): $\beta_3 = 0, \rho = 0$; (1,2):$\beta_3 = 0$; (2,1): $\rho = 0$. Estimate of σ_v^2 for each of the four models was used to summarize the model fit; see Example 7.1.1. The following estimates were obtained: $\hat{\sigma}_v^2(1,1) = 109, \hat{\sigma}_v^2(1,2) = 47.32, \hat{\sigma}_v^2(2,1) = 18.39$ and $\hat{\sigma}_v^2(2,2) = 0$. It is interesting to note that the spatial model (1,2) with omitted variable has a smaller $\hat{\sigma}_v^2$ than the nonspatial model (2,1) with the variable included. This result suggests that spatial models can compensate for omitted variables.

The difference between the "raw" adjustment factor $\hat{\theta}_i$ and no adjustment was decomposed into differences that correspond to the effects of smoothing, spatial modeling, omitting the variable z_3 and adjusting:

$$\begin{aligned}\hat{\theta}_i - 1 &= (\hat{\theta}_i - \hat{\theta}_i^H(1,1)) + (\hat{\theta}_i^H(1,1) - \hat{\theta}_i^H(1,2)) \\ &\quad + (\hat{\theta}_i^H(1,2) - \hat{\theta}_i^H(2,2)) + (\hat{\theta}_i^H(2,2) - 1).\end{aligned}$$

The components of $\hat{\theta}_i - 1$ were calculated for each state i and summarized as weighted sum of squares (WSS) with weights C_i; for example, $\sum_i C_i(\hat{\theta}_i - \hat{\theta}_i^H(1,1))^2$ for the first component $\hat{\theta}_i - \hat{\theta}_i^H(1,1)$. The following WSS-values for the four components were obtained: 27838, 2618, 4248 and 28602. Large WSS-values for the first and the fourth components suggest that the smoothing of raw adjustment factors and the adjustment of census counts are both important. On the other hand, small WSS-values for the second and third components indicate that the omission of an explanatory variable or the spatial correlation has relatively smaller effect on the smoothed adjustment factors.

8.5 Multivariate Nested Error Regression Model

The sample part of the multivariate nested error regression model (5.5.1) can be expressed as a special case of the general mixed model (6.3.1) with block diagonal covariance structure. We have

$$\begin{aligned}&\mathbf{y}_i^T = (\mathbf{y}_{i1}^T, \ldots, \mathbf{y}_{in_i}^T), \quad \mathbf{e}_i^T = (\mathbf{e}_{i1}^T, \ldots, \mathbf{e}_{in_i}^T) \\ &\mathbf{v}_i = \mathbf{v}_i, \quad \boldsymbol{\beta} = \text{vec}(\mathbf{B}), \\ &\mathbf{X}_i^T = [(\mathbf{I}_r \otimes \mathbf{x}_{i1}^T)^T, \ldots, (\mathbf{I}_r \otimes \mathbf{x}_{in_i}^T)^T] \\ &\mathbf{Z}_i = \mathbf{1}_{n_i} \otimes \mathbf{I}_r, \quad i = 1, \ldots, m,\end{aligned}$$

where the operator $\otimes$ denotes the direct product, $\mathbf{1}_{n_i}$ is the n_i-vector of 1s, vec($\mathbf{B}$) is the $pr \times 1$ vector obtained by listing the columns of the $r \times p$ matrix $\mathbf{B}$ one beneath the other beginning with the first column, and n_i is the number of sample multivariate observations $\mathbf{y}_{ij}$. Further,

$$\mathbf{G}_i = \boldsymbol{\Sigma}_v, \quad \mathbf{R}_i = \mathbf{I}_{n_i} \otimes \boldsymbol{\Sigma}_e$$

so that

$$\mathbf{V}_i = (\mathbf{J}_{n_i} \otimes \boldsymbol{\Sigma}_v) + (\mathbf{I}_{n_i} \otimes \boldsymbol{\Sigma}_e),$$

where $\mathbf{J}_{n_i} = \mathbf{1}_{n_i}\mathbf{1}_{n_i}^T$ and $\mathbf{I}_{n_i}$ is the identity matrix of order n_i. The parameters of interest are the small area mean vectors $\bar{\mathbf{Y}}_i \approx \mathbf{B}\bar{\mathbf{X}}_i + \mathbf{v}_i = (\mathbf{I}_r \otimes \bar{\mathbf{X}}_i^T)\boldsymbol{\beta} + \mathbf{v}_i$ if the population size N_i is large. Therefore, $\bar{\mathbf{Y}}_i$ is of the form $\boldsymbol{\mu}_i = \mathbf{L}_i\boldsymbol{\beta} + \mathbf{M}_i\mathbf{v}_i$ with $\mathbf{L}_i = \mathbf{I}_r \otimes \bar{\mathbf{X}}_i^T$ and $\mathbf{M}_i = \mathbf{I}_r$.

Making the above substitutions in the general formula (6.3.19) we get the BLUP estimator $\tilde{\boldsymbol{\mu}}_i^H$ of $\boldsymbol{\mu}_i = \mathbf{B}^T\bar{\mathbf{X}}_i + \mathbf{v}_i$. The EBLUP estimator $\hat{\boldsymbol{\mu}}_i^H$ is obtained from by substituting suitable estimators $\hat{\boldsymbol{\Sigma}}_v$ and $\hat{\boldsymbol{\Sigma}}_e$ for $\boldsymbol{\Sigma}_v$ and $\boldsymbol{\Sigma}_e$.

Fuller and Harter (1987) used a method of moments to estimate $\boldsymbol{\Sigma}_v$ and $\boldsymbol{\Sigma}_e$. We can also obtain ML or REML estimators, using the general results of subsection 6.2.4. MSE estimators may also be obtained from the general results for block diagonal covariance structures given in subsection 6.3.2 (see Datta, Day and Basawa (1999)).

Lohr and Prasad (2001) considered sampling on two occasions with partial sample overlap and developed small area estimators by combining information from the two samples. They used a multivariate nested error model to derive EBLUP estimators of the area means on the second occasion, and associated MSE estimators that are nearly unbiased. Significant gains in efficiency, relative to EBLUP estimators based only on the second occasion sample, may be achieved if the measurements on the matched portion are highly correlated. The method of Lohr and Prasad (2001) may also be used to combine information from separate surveys that overlap in some of the small areas.

Example 8.5.1. Simulation Study. Datta, Day and Basawa (1999) used simulation to compare the efficiencies of EBLUP estimators under bivariate ($r = 2$) and univariate nested error regression models. They estimated the model parameters using the LANDSAT data of Battese, Harter and Fuller (1989), and treated $\hat{\mathbf{B}}$ and the diagonal elements of $\hat{\boldsymbol{\Sigma}}_v$ as the true values, σ_{vii}. The vectors $\mathbf{y}_{ij}$ and $\mathbf{x}_{ij} = (1, x_{ij1}, x_{ij2})^T$ are as described in Example 5.5.1. Using selected values of $r_1 = \sigma_{v11}/\sigma_{e11}$ and $r_2 = \sigma_{v22}/\sigma_{e22}$ and correlation coefficients, ρ_v and ρ_e, associated with $\boldsymbol{\Sigma}_v$ and $\boldsymbol{\Sigma}_e$, respectively, they generated several pairs $(\boldsymbol{\Sigma}_v, \boldsymbol{\Sigma}_e)$. From each parameter setting, 100 data sets were generated under the bivariate nested error regression model, and MSE estimates of EBLUP estimates (univariate and bivariate) for $m = 12$ counties (small areas) were computed from each data set. Based on the average values of MSE estimates over the 100 data sets for each parameter setting, they computed the estimated reduction in MSE of the bivariate method relative to the univariate method for each of the two components (corn and soybeans). Their results suggest that little improvement over the univariate method is achieved if ρ_v and ρ_e have the same sign, while ρ_v and ρ_e with opposite signs and large absolute values lead to large improvement in efficiency. It may be noted that the reduction in MSE does not depend on the choice of $\mathbf{B}$ and $(\sigma_{v11}, \sigma_{v22})$ in the simulation.

8.6 Random Error Variances Linear Model

In subsection 5.5.2 we introduced a simple random effects model with random error variances, σ_{ei}^2. Assuming equal small area sample sizes, $n_i = \bar{n}$, the sample part of this model may be written as

$$
\begin{aligned}
&y_{ij} = \beta + v_i + e_{ij}, \quad j = 1, \ldots, \bar{n}; \quad i = 1, \ldots, m \\
&v_i \overset{\text{iid}}{\sim} (0, \sigma_v^2), \quad e_{ij}|\sigma_{ei}^2 \overset{\text{iid}}{\sim} (0, \sigma_{ei}^2) \text{ for each } i, \\
&\sigma_{ei}^2 \overset{\text{iid}}{\sim} (\sigma_e^2, \delta_e).
\end{aligned} \tag{8.6.1}
$$

Distributional assumptions on v_i, $e_{ij}|\sigma_{ei}^2$ and σ_{ei}^2 are not imposed.

The EBLUP estimator of the small area mean $\mu_i = \mu + v_i$ under (8.6.1) is identical to the EBLUP estimator obtained under the assumption of equal error variances $\sigma_{ei}^2 = \sigma_e^2$ (i.e., $\delta_e = 0$). It is given by

$$\hat{\mu}_i^H = \bar{y}_i + \hat{\gamma}(\bar{y}_i - \bar{y}), \tag{8.6.2}$$

where $\bar{y}_i$ and $\bar{y}$ are the ith area sample mean and the overall sample mean, and $\hat{\gamma} = \hat{\sigma}_v^2/(\hat{\sigma}_v^2 + \hat{\sigma}_e^2/\bar{n})$ with $m(\bar{n}-1)\hat{\sigma}_e^2 = \sum_i \sum_j (y_{ij} - \bar{y}_i)^2$ and $(m-1)(\hat{\sigma}_v^2 + \hat{\sigma}_e^2/\bar{n}) = \sum_i (\bar{y}_i - \bar{y})^2$; see Kleffe and Rao (1992). Note that $\hat{\sigma}_v^2$ and $\hat{\sigma}_e^2$ are the customary ANOVA estimators of σ_v^2 and σ_e^2.

The MSE of $\hat{\mu}_i^H$ under (8.6.1) is different from the MSE under the equal error variance model, although the EBLUP estimators are identical. Kleffe and Rao (1992) derived approximations to $\text{MSE}(\hat{\mu}_i^H)$ and its estimator, $\text{mse}(\hat{\mu}_i^H)$, correct to terms of order $o(m^{-1})$ for large m, assuming normality of the errors v_i and $e_{ij}|\sigma_{ei}^2$ but no distributional assumption on σ_{ei}^2. The MSE estimator is given by

$$
\begin{aligned}
\text{mse}(\hat{\mu}_i^H) = {} & (1 - \hat{\gamma})\left(\hat{\sigma}_v^2 + \frac{\hat{\sigma}_e^2}{m\bar{n}}\right) + \frac{4(m-1)}{m\bar{n}^2 \sum_i (\bar{y}_i - \bar{y})^2} \\
& \times \left[\left\{(1 - \hat{\gamma}) + \frac{1}{\bar{n} - 1}\right\}(\hat{\delta}_e + \hat{\sigma}_e^4) + \frac{\hat{\sigma}_e^4 \hat{\sigma}_v^2}{\sum_i (\bar{y}_i - \bar{y})^2}\right],
\end{aligned} \tag{8.6.3}
$$

where $\hat{\delta}_e = \frac{\bar{n}-1}{m(\bar{n}+1)} \sum_i \hat{\sigma}_{ei}^4 - \hat{\sigma}_e^4$ with $\hat{\sigma}_{ei}^2 = \sum_j (y_{ij} - \bar{y}_i)^2/(\bar{n}-1)$. Derivation of (8.6.3) is highly technical, and the reader is referred to Kleffe and Rao (1992) for details. Extension of the MSE approximation to nested error regression models with random error variances has not been spelled out in the literature.

Example 8.6.1. Simulation Study. Kleffe and Rao (1992) conducted a simulation study to confirm the accuracy of the MSE approximation and the approximate unbiasedness of the estimator of the MSE. To this end, they employed the following parameter values: $\beta = 0$, $\sigma_v^2 = 1$ (without loss of generality), $\sigma_e^2 = 5$, $m = 30$, and two values of $\bar{n}$: 3 and 10. Using these parameter values, 10,000 independent data sets $\{\bar{y}_i,\ \hat{\sigma}_{ei}^2 = \sum_j (y_{ij} - \bar{y}_i)^2/(\bar{n} - 1)\}$ were generated as follows:

Step 1. For each data set, generate $\sigma_{e1}^2, \ldots, \sigma_{e,30}^2$ from a χ^2 distribution with $\sigma_e^2 = 5$ df ($\delta_e = 2\sigma_e^2 = 10$ in this case).

Step 2. For each set, generate v_i from $N(0,1)$ and $\bar{e}_i$ from $N(0, \sigma_{ei}^2/\bar{n})$ and let $\bar{y}_i = \beta + v_i + \bar{e}_i$, $i = 1, \ldots, 30$. Further, noting that $(\bar{n}-1)\hat{\sigma}_{ei}^2/\sigma_{ei}^2$ is a χ^2 variable with $\bar{n}-1$ df under normal errors for a given σ_{ei}^2, and that $\bar{y}_i$ and $\hat{\sigma}_{ei}^2$ are independently distributed, generate a_i from a χ^2 distribution with $\bar{n}-1$ df and let $\hat{\sigma}_{ei}^2 = a_i\sigma_{ei}^2/(\bar{n}-1)$; $i = 1, \ldots, 30$. Generate all the variables independently from the specified distributions.

The EBLUP estimator, $\hat{\mu}_i^H$, has the same MSE for all i because the sample size, $\bar{n}$, is the same for all the m areas. Therefore, there is no loss of generality in confining ourselves to the estimation of MSE($\hat{\mu}_1^H$). From each data set, the EBLUP estimate $\hat{\mu}_1^H$, mse($\hat{\mu}_1^H$), and a naive MSE estimate, $\text{mse}_N(\hat{\mu}_1^H)$, were computed. The naive MSE estimator is given by

$$\text{mse}_N(\hat{\mu}_1^H) = (1 - \hat{\gamma})(\hat{\sigma}_v^2 + \hat{\sigma}_e^2/(m\bar{n})). \tag{8.6.4}$$

Simulation values of the relative bias (RB) of mse($\hat{\mu}_1^H$) and $\text{mse}_N(\hat{\mu}_1^H)$ were then computed from the 10,000 simulated values of $\hat{\mu}_1^H$, mse($\hat{\mu}_1^H$) and $\text{mse}_N(\hat{\mu}_1^H)$. The naive MSE estimator led to serious underestimation for $\bar{n} = 3$: RB=−35%, while the RB of mse($\hat{\mu}_1^H$) remained small even for $\bar{n} = 3$: RB=3.4%. As $\bar{n}$ is increased from 3 to 10, RB=−9% for $\text{mse}_N(\hat{\mu}_1^H)$ compared to RB=0.2% for mse($\hat{\mu}_1^H$).

8.7 Two-fold Nested Error Regression Model

In subsection 5.5.3, we described a two-fold nested error regression model appropriate for two-stage sampling within areas: a sample, s_i, of m_i primary units (clusters) is selected from the i th area and if the j th cluster is sampled then a subsample, s_{ij}, of n_{ij} elements is selected from the j th cluster and the associated values $(y_{ijl}, \mathbf{x}_{ijl}); l = 1, \ldots, n_{ij}$ are observed. We assume that the sample obeys the two-fold model given by (5.5.2) so that

$$y_{ijl} = \mathbf{x}_{ijl}^T\boldsymbol{\beta} + v_i + u_{ij} + e_{ijl}; l = 1, \ldots, n_{ij}; j = 1, \ldots, m_i;\ i = 1, \ldots, m, \tag{8.7.1}$$

where $v_i \overset{\text{iid}}{\sim} (0, \sigma_v^2), u_{ij} \overset{\text{iid}}{\sim} (0, \sigma_u^2)$ and $e_{ijl} = k_{ijl}\tilde{e}_{ijl}$ with $\tilde{e}_{ijl} \overset{\text{iid}}{\sim} (0, \sigma_e^2)$, and known constants k_{ijl}. Further, $\{v_i\}, \{u_{ij}\}$ and $\{\tilde{e}_{ijl}\}$ are mutually independent.

The two-fold model (8.7.1) is a special case of the general linear mixed model (6.3.1) with block diagonal covariance structure. Letting $\text{col}_{1\le i\le t}(\mathbf{a}_i) = (\mathbf{a}_1^T, \ldots, \mathbf{a}_t^T)^T$ and $\text{diag}_{1\le i\le t}(\mathbf{A}_i) = \text{blockdiag}(\mathbf{A}_1, \ldots, \mathbf{A}_t)$, we have

$$\mathbf{y}_i = \text{col}_{1\le j\le m_i}(\mathbf{y}_{ij}),\quad \mathbf{y}_{ij} = \text{col}_{1\le l\le n_{ij}}(y_{ijl}),$$

$$\mathbf{X}_i = \text{col}_{1\le j\le m_i}(\text{col}_{1\le l\le n_{ij}}(\mathbf{x}_{ijl}^T)),\quad \mathbf{Z}_i = (\mathbf{z}_i|\mathbf{Z}_{2i}),$$

$$\mathbf{v}_i = \begin{pmatrix} v_i \\ \mathbf{v}_{2i} \end{pmatrix}, \quad \boldsymbol{\beta} = (\beta_1, \ldots, \beta_p)^T,$$

and $\mathbf{e}_i$ is obtained from $\mathbf{y}_i$ by changing y_{ijl} to e_{ijl}, where

$$\mathbf{z}_i = \mathrm{col}_{1 \le j \le m_i}(\mathrm{col}_{1 \le k \le n_{ij}}(1)) = \mathrm{col}_{1 \le j \le m_i}(\mathbf{z}_{ij}),$$

$$\mathbf{Z}_{2i} = \mathrm{diag}_{1 \le j \le m_i}(\mathbf{z}_{ij}), \quad \mathbf{v}_{2i} = \mathrm{col}_{1 \le j \le m_i}(u_{ij}).$$

Further,

$$\mathbf{G}_i = \begin{bmatrix} \sigma_v^2 & \mathbf{0}^T \\ \mathbf{0} & \sigma_u^2 \mathbf{I}_{m_i} \end{bmatrix}, \quad \mathbf{R}_i = \mathrm{diag}_{1 \le j \le m_i}(\mathbf{R}_{ij})$$

and $\mathbf{R}_{ij} = \mathrm{diag}_{1 \le l \le n_{ij}}(\sigma_e^2 k_{ijl}^2)$, where $\mathbf{I}_b$ is the identity matrix of order b and $\mathbf{0}$ is a vector of zeros. Using an explicit form for $\mathbf{V}_i^{-1}$, the general formula (6.3.4) for $\tilde{\boldsymbol{\beta}}$ reduces to

$$\tilde{\boldsymbol{\beta}} = \mathbf{A}^{-1}\mathbf{b}, \tag{8.7.2}$$

$$\sigma_e^2 \mathbf{A} = \sum_i \sum_j \sum_l a_{ijl} \mathbf{x}_{ijl} \mathbf{x}_{ijl}^T - \sum_i \sum_j a_{ij\cdot} \gamma_{ij} \bar{\mathbf{x}}_{ija} \bar{\mathbf{x}}_{ija}^T - (\sigma_e^2/\sigma_u^2) \sum_i \gamma_i \Big(\sum_j \gamma_{ij} \Big) \bar{\mathbf{x}}_{i\gamma} \bar{\mathbf{x}}_{i\gamma}^T$$

and

$$\sigma_e^2 \mathbf{b} = \sum_i \sum_j \sum_l a_{ijl} \mathbf{x}_{ijl} y_{ijl} - \sum_i \sum_j a_{ij\cdot} \gamma_{ij} \bar{\mathbf{x}}_{ija} \bar{y}_{ija} - (\sigma_e^2/\sigma_u^2) \sum_i \gamma_i \Big(\sum_j \gamma_{ij} \Big) \bar{\mathbf{x}}_{i\gamma} \bar{y}_{i\gamma},$$

where $a_{ijl} = k_{ijl}^{-2}$, $a_{ij\cdot} = \sum_l a_{ijl}$, $\bar{y}_{i\gamma} = \sum_j \gamma_{ij} \bar{y}_{ija} / (\Sigma_j \gamma_{ij})$, $\bar{\mathbf{x}}_{i\gamma} = \sum_j \gamma_{ij} \bar{\mathbf{x}}_{ija} / (\Sigma_j \gamma_{ij})$ with $\bar{y}_{ija} = \sum_l a_{ijl} y_{ijl} / a_{ij\cdot}$, $\bar{\mathbf{x}}_{ija} = \sum_l a_{ijl} \mathbf{x}_{ijl} / a_{ij\cdot}$, and

$$\gamma_{ij} = \sigma_u^2 / (\sigma_u^2 + \sigma_e^2 / a_{ij\cdot}), \gamma_i = \sigma_v^2 / [\sigma_v^2 + \sigma_u^2 / (\Sigma_j \gamma_{ij})); \tag{8.7.3}$$

see Stukel and Rao (1999). Further, the general formula (6.3.3) for the BLUP estimator of $\mathbf{v}_i$ reduces to $\tilde{\mathbf{v}}_i = (\tilde{v}_i, \mathbf{v}_{2i}^T)^T$ with elements

$$\tilde{v}_i = \gamma_i (\bar{y}_{i\gamma} - \bar{\mathbf{x}}_{i\gamma}^T \tilde{\boldsymbol{\beta}}) \tag{8.7.4}$$

and

$$\tilde{u}_{ij} = \gamma_{ij} (\bar{y}_{ija} - \bar{\mathbf{x}}_{ija}^T \tilde{\boldsymbol{\beta}}) - \gamma_i \gamma_{ij} (\bar{y}_{i\gamma} - \bar{\mathbf{x}}_{i\gamma}^T \tilde{\boldsymbol{\beta}}). \tag{8.7.5}$$

Note that $\tilde{\boldsymbol{\beta}}$ and $\tilde{v}_i$ involve the inversion of only a $p \times p$ matrix, where p is the dimension of $\mathbf{x}_{ijl}$.

The BLUP estimator of the small area mean $\bar{Y}_i$ is given by

$$\tilde{\bar{Y}}_i^H = \frac{1}{N_i}\left[\sum_{j\in s_i}\sum_{k\in s_{ij}} y_{ijk} + \sum_{j\in s_i}\left(\sum_{l\in \bar{s}_{ij}} y_{ijl}^*\right) + \sum_{j\in \bar{s}_i}\left(\sum_{l=1}^{N_{ij}} y_{ijl}^{**}\right)\right], \qquad (8.7.6)$$

where $\bar{s}_{ij}$ is the set of nonsampled elements in the jth sampled cluster, $\bar{s}_i$ is the set of nonsampled clusters, N_{ij} is the number of elements in the jth cluster of the ith area ($j = 1, \ldots, M_i$) and $N_i = \sum_j N_{ij}$ is the total number of elements in the ith area. Further,

$$y_{ijl}^* = \mathbf{x}_{ijl}^T\tilde{\boldsymbol{\beta}} + \tilde{v}_i + \tilde{u}_{ij} \qquad (8.7.7)$$

and

$$y_{ijl}^{**} = \mathbf{x}_{ijl}^T\tilde{\boldsymbol{\beta}} + \tilde{v}_i. \qquad (8.7.8)$$

The BLUP property of $\tilde{\bar{Y}}_i^H$ is established by showing that $\mathrm{Cov}(\mathbf{a}^T\mathbf{y}, \tilde{\bar{Y}}_i^H - \bar{Y}_i) = 0$ for every zero function $\mathbf{a}^T\mathbf{y} = \sum_i \mathbf{a}_i^T\mathbf{y}_i$, that is, $E(\mathbf{a}^T\mathbf{y}) = 0$; see Stukel (1991, Chapter 3). The form (8.7.6) for $\tilde{\bar{Y}}_i^H$ shows that $\tilde{\bar{Y}}_i^H$ is readily obtained from $\tilde{\boldsymbol{\beta}}$, $\tilde{v}_i$ and $\tilde{u}_{ij}$ given by (8.7.2), (8.7.4) and (8.7.5), respectively.

The BLUP estimator $\tilde{\bar{Y}}_i^H$ depends on the variance components $\boldsymbol{\delta} = (\sigma_v^2, \sigma_u^2, \sigma_e^2)^T$. Stukel and Rao (1997) extended the simple transformation method to obtain an estimator, $\hat{\boldsymbol{\delta}}_m$, of $\boldsymbol{\delta}$. Fuller and Battese (1973) proposed a similar extension for the special case of $k_{ijl} = 1$ for all (i, j, l). The transformation method is identical to the fitting-of-constants method, but it is computationally less cumbersome and more stable than the latter method, as noted in subsection 7.2.2. One could also use the REML estimator $\hat{\boldsymbol{\delta}}_{\mathrm{RE}}$. Replacing $\boldsymbol{\delta}$ in (8.7.6) by an estimator $\hat{\boldsymbol{\delta}}$, we get the EBLUP estimator $\hat{\bar{Y}}_i^H$.

If the number of primary units, N_i, is large, then $\bar{Y}_i$ may be approximated as $\bar{Y}_i \approx \mu_i = \bar{\mathbf{X}}_i^T\boldsymbol{\beta} + v_i$, as noted in subsection 5.5.3, where $\bar{\mathbf{X}}_i$ is the vector of known population x-means. In this case, the BLUP estimator of μ_i is given by

$$\tilde{\mu}_i^H = \gamma_i[\bar{y}_{i\gamma} + (\bar{\mathbf{X}}_i - \bar{\mathbf{x}}_{i\gamma})^T\tilde{\boldsymbol{\beta}}] + (1-\gamma_i)\bar{\mathbf{X}}_i^T\tilde{\boldsymbol{\beta}}. \qquad (8.7.9)$$

The form (8.7.9) shows that the BLUP estimator of μ_i is a weighted average of the survey regression estimator $\bar{y}_{i\gamma} + (\bar{\mathbf{X}}_i - \bar{\mathbf{x}}_{i\gamma})^T\tilde{\boldsymbol{\beta}}$ and the regression synthetic estimator $\bar{\mathbf{X}}_i^T\tilde{\boldsymbol{\beta}}$. Note that $\tilde{\mu}_i^H$ depends on the population x-values only through the mean $\bar{\mathbf{X}}_i$. The MSE formula (6.3.5) reduces to $\mathrm{MSE}(\tilde{\mu}_i^H) = g_{1i}(\boldsymbol{\delta}) + g_{2i}(\boldsymbol{\delta})$ with

$$g_{1i}(\boldsymbol{\delta}) = \sigma_v^2(1-\gamma_i) \qquad (8.7.10)$$

and

$$g_{2i}(\boldsymbol{\delta}) = (\bar{\mathbf{X}}_i - \gamma_i\bar{\mathbf{x}}_{i\gamma})^T\mathbf{A}^{-1}(\bar{\mathbf{X}}_i - \gamma_i\bar{\mathbf{x}}_{i\gamma}). \qquad (8.7.11)$$

The first term, $g_{1i}(\boldsymbol{\delta})$, is of order $O(1)$ while the second term, $g_{2i}(\boldsymbol{\delta})$, due to estimating $\boldsymbol{\beta}$, is of order $O(m^{-1})$ for large m.

The EBLUP estimator $\hat{\mu}_i^H$ is obtained by replacing $\boldsymbol{\delta}$ by $\hat{\boldsymbol{\delta}}$ in (8.7.9). Under normality of the errors v_i, u_{ij} and e_{ijk}, the estimator of MSE, given by (6.3.11), reduces to

$$\text{mse}(\hat{\mu}_i^H) = g_{1i}(\hat{\boldsymbol{\delta}}) + g_{2i}(\hat{\boldsymbol{\delta}}) + 2g_{3i}(\hat{\boldsymbol{\delta}}), \tag{8.7.12}$$

where

$$\begin{aligned} g_{3i}(\boldsymbol{\delta}) = \sigma_u^{-2}\sigma_v^4(1-\gamma_i)^2 \Bigg[& (1-\gamma_i)\sum_j \gamma_{ij} \\ & \times \bar{V}\left\{(\sigma_v^{-2}\hat{\sigma}_v^2 - \sigma_e^{-2}\hat{\sigma}_e^2)(\Sigma_j\gamma_{ij})^{-1}\sum_j \gamma_{ij}^2(\sigma_u^{-2}\hat{\sigma}_u^2 - \sigma_e^{-2}\hat{\sigma}_e^2)\right\} \\ & + (\Sigma_j\gamma_{ij})^{-1}\left\{\left(\sum_j \gamma_{ij}^3\right)(\Sigma_j\gamma_{ij}) - \left(\sum_j \gamma_{ij}^2\right)^2\right\} \\ & \times \bar{V}(\sigma_u^{-2}\hat{\sigma}_u^2 - \sigma_e^{-2}\hat{\sigma}_e^2)\Bigg] \end{aligned} \tag{8.7.13}$$

and $\bar{V}$ denotes the asymptotic variance (Stukel and Rao (1999)). Formula (8.7.12) is valid for $\hat{\boldsymbol{\delta}}_{\text{RE}}$ and $\hat{\boldsymbol{\delta}}_m$, and the bias of $\text{mse}(\hat{\mu}_i^H)$ is of lower order than m^{-1}. The asymptotic covariance matrix of $\hat{\boldsymbol{\delta}}_m$ is given in Stukel and Rao (1999), while the asymptotic covariance matrix of $\hat{\boldsymbol{\delta}}_{\text{RE}}$ may be obtained from the general results in subsection 6.2.4. Area-specific versions of (8.7.12) may be obtained from (6.3.14)–(6.3.16).

Example 8.7.1. Simulation Study. Stukel and Rao (1999) conducted a simulation study on the relative bias (RB) of $\text{mse}(\hat{\mu}_i^H)$. To this end, they employed a two-fold model without auxiliary variables $\mathbf{x}_{ijl} : y_{ijl} = \beta + v_i + u_{ij} + e_{ijl}$ with $k_{ijl} = 1, m_i = \bar{m}, n_{ij} = \bar{n}$, and normally distributed errors. In this balanced case, fitting-of-constants estimators, $\hat{\delta}_m$, reduces to the simple ANOVA estimators: $\hat{\sigma}_v^2 = \max(0, \tilde{\sigma}_v^2), \hat{\sigma}_u^2 = \max(0, \tilde{\sigma}_u^2)$ and $\hat{\sigma}_e^2 = \tilde{\sigma}_e^2 = [m\bar{m}(\bar{n}-1)]^{-1}\sum_i\sum_j\sum_l(y_{ijl}-\bar{y}_{ij\cdot})^2$, where $\tilde{\sigma}_u^2 = [m(\bar{m}-1)]^{-1}\sum_i\sum_j(\bar{y}_{ij\cdot}-\bar{y}_{i\cdot\cdot})^2 - \tilde{\sigma}_e^2$, $\tilde{\sigma}_v^2 = (m-1)^{-1}\sum_i(\bar{y}_{i\cdot\cdot} - \bar{y}_{\cdot\cdot\cdot})^2 - \tilde{\sigma}_u^2/\bar{m} - \tilde{\sigma}_e^2/(\bar{m}\bar{n}), \bar{y}_{ij\cdot} = \sum_l y_{ijl}/\bar{n}$ and $\bar{y}_{i\cdot\cdot} = \sum_j \bar{y}_{ij\cdot}/\bar{m}$. Asymptotically, $\hat{\boldsymbol{\delta}}$ has the same covariance matrix as $\tilde{\boldsymbol{\delta}}$; the latter may be obtained from Searle, Casella and McCulloch (1992, p. 148).

Simulated values of RB were computed using independent data sets, each of size 2000, generated from the model with $\beta = 0$, $\sigma_e^2 = 300$ and selected values, 0.1, 0.2, 0.5, 1.0, 2.0 of the ratios σ_u^2/σ_e^2 and σ_v^2/σ_e^2, leading to 25 combinations $(\sigma_v^2/\sigma_e^2, \sigma_u^2/\sigma_e^2)$. The RB values were then averaged over the areas i for each combination $(\sigma_v^2/\sigma_e^2, \sigma_u^2/\sigma_e^2)$; note that the theoretical RB of $\text{mse}(\hat{\mu}_i^H)$ does not depend on i under the balanced two-fold model. The average RB values suggest the following properties of $\text{mse}(\hat{\mu}_i^H)$: (1) RB depends on both σ_v^2/σ_e^2 and σ_u^2/σ_e^2. (2) RB is small ($< 10\%$) in a parameter region of interest where $\sigma_v^2/\sigma_e^2 \geq \sigma_u^2/\sigma_e^2$ and $\sigma_v^2/\sigma_e^2 \geq 0.5$. However, RB is quite large (i.e., significant overestimation) when σ_v^2/σ_e^2 is small (≤ 0.2), especially when σ_v^2/σ_e^2 is significantly lower than σ_u^2/σ_e^2. On the other hand, the naive MSE estimator that ignores the variability in $\hat{\boldsymbol{\delta}}$ almost always leads to significant underestimation, except for very few cases of the parameter region under consideration, where the underestimation is mild. Note that the REML estimator $\hat{\boldsymbol{\delta}}_{\text{RE}}$ is not identical to the ANOVA estimator $\hat{\boldsymbol{\delta}}$ in the balanced two-fold case because REML uses a different form of truncation to handle negative values of $\tilde{\sigma}_v^2$ and $\tilde{\sigma}_u^2$ (see Searle et al. (1992), p. 149). It would be interesting to study the properties of RB under REML estimation of $\boldsymbol{\delta}$.

8.8 Two-level Model

In subsection 5.5.4, we introduced a two-level model that effectively integrates the use of unit level and area level covariates into a single model (5.5.8). We assume that the sample obeys the two-level model so that

$$\mathbf{y}_i = \tilde{\mathbf{X}}_i \tilde{\mathbf{Z}}_i \boldsymbol{\alpha} + \tilde{\mathbf{X}}_i \mathbf{v}_i + \mathbf{e}_i; \quad i = 1, \ldots, m \tag{8.8.1}$$

where $(\mathbf{y}_i, \tilde{\mathbf{X}}_i, \mathbf{e}_i)$ correspond to the sample part of the population model (5.5.8); we use $\tilde{\mathbf{X}}_i$ here to denote the sample part of $\mathbf{X}_i^P$ to avoid confusing with $\mathbf{X}_i$ of the general linear mixed model (see below). The model (8.8.1) is a special case of the general linear mixed model (6.3.1) with block diagonal covariance structure. We have

$$\mathbf{y}_i = \mathbf{y}_i = \text{col}_{1 \leq j \leq n_i}(y_{ij}), \quad \mathbf{X}_i = \tilde{\mathbf{X}}_i \tilde{\mathbf{Z}}_i, \quad \mathbf{Z}_i = \tilde{\mathbf{X}}_i,$$

$$\boldsymbol{\beta} = \boldsymbol{\alpha}, \quad \mathbf{v}_i = \mathbf{v}_i, \quad \mathbf{e}_i = \mathbf{e}_i,$$

$$\mathbf{G}_i = \boldsymbol{\Sigma}_v, \quad \mathbf{R}_i = \sigma_e^2 \, \text{diag}_{1 \leq j \leq n_i}(k_{ij}^2),$$

where n_i is the sample size in the ith area ($i = 1, \ldots, m$) and k_{ij}'s are known constants.

If N_i is large, we can express the mean $\bar{Y}_i$ as $\bar{\mathbf{X}}_i^T \mathbf{Z}_i \boldsymbol{\alpha} + \bar{\mathbf{X}}_i^T \mathbf{v}_i$, where $\bar{\mathbf{X}}_i$ is the vector of known population x-means. It now follows that the BLUP estimator, $\tilde{\mu}_i^H$, and its MSE are readily obtained from (6.3.2) and (6.3.5) using $\mathbf{l}_i^T = \bar{\mathbf{X}}_i^T \mathbf{Z}_i$ and $\mathbf{m}_i^T = \bar{\mathbf{X}}_i^T$.

The BLUP estimator, $\tilde{\mu}_i^H$, depends on the $p(p+1)/2$ distinct elements of the $p \times p$ matrix $\boldsymbol{\Sigma}_v$ and the error variance σ_e^2. Moura and Holt (1999) used restricted iterative generalized least squares (RIGLS) to estimate the above parameters, $\boldsymbol{\delta}$. Goldstein (1989) proposed this method for multi-level models. It is equivalent to REML under normality of the errors. The EBLUP estimator, $\hat{\mu}_i^H$, is obtained from the BLUP estimator by replacing $\boldsymbol{\delta}$ by the RIGLS estimator $\hat{\boldsymbol{\delta}}$. An estimator of MSE($\hat{\mu}_i^H$) is obtained from the general formula (6.3.11) and its bias is of lower order than m^{-1}.

When the sampling fractions, $f_i = n_i/N_i$, are not small, the EBLUP estimator of $\bar{Y}_i$ and its MSE estimator are obtained along the lines of subsection 7.2.5. We refer the readers to Moura and Holt (1999) for details.

Example 8.8.1. Simulation Study. Moura and Holt (1999) investigated the properties of alternative estimators, using data from 38,740 households in $m = 140$ enumeration districts (small areas) in one county in Brazil. The data consisted of y = income of the household head, x_1 = number of rooms in the house (1–11+), and x_2 = educational attainment of the household head (ordinal scale of 0–5) for each household (unit) centered around the means $(\overline{x}_{1i}, \overline{x}_{2i})$, as well as one area level covariate, z = number of cars per household, for each area. A two-level model of the form

$$y_{ij} = \beta_{0i} + \beta_{1i}x_{1ij} + \beta_{2i}x_{2ij} + e_{ij}, \tag{8.8.2}$$

$$\beta_{0i} = \alpha_{00} + v_{0i}, \beta_{1i} = \alpha_{10} + \alpha_{11}z_i + v_{1i}, \beta_{2i} = \alpha_{20} + \alpha_{21}z_i + v_{2i} \tag{8.8.3}$$

with diagonal $\mathbf{\Sigma}_v$ was fitted to the data to estimate the elements of $\boldsymbol{\delta}$, namely, $\sigma_{v0}^2, \sigma_{v1}^2, \sigma_{v2}^2$ and σ_e^2, and the regression parameters $\alpha_{00}, \alpha_{10}, \alpha_{11}, \alpha_{20}$ and α_{21}. Note that z_i is not used in modeling the random intercept term, β_{0i}.

To simulate samples from the two-level model, a 10% random sample of (x_1, x_2)-values was selected from each area and treated as fixed; average small area sample size $\bar{n} = 28$. Treating the parameter estimates as true values, a realization, $\boldsymbol{\beta}_i^{(1)}$, of $\boldsymbol{\beta}_i = (\beta_{0i}, \beta_{1i}, \beta_{2i})^T$ was first generated from (8.8.3), assuming normality of the errors $\mathbf{v}_i = (v_{0i}, v_{1i}, v_{2i})^T$. This, in turn, leads to a realization of μ_i, namely, $\mu_i^{(1)} = \beta_{0i}^{(1)} + \beta_{1i}^{(1)}\bar{X}_{1i} + \beta_{2i}^{(1)}\bar{X}_{2i}$, where $\bar{X}_{1i}$ and $\bar{X}_{2i}$ are the known population x-means. At the second stage, y_{ij}-values were generated from $y_{ij} = \beta_{0i}^{(1)} + \beta_{1i}^{(1)}x_{1ij} + \beta_{2i}^{(1)}x_{2ij} + e_{ij}$, where $e_{ij} \overset{\text{iid}}{\sim} N(0, \sigma_e^2)$. This two-stage process resulted in a simulated sample $\{y_{ij}^{(1)}, x_{1ij}, x_{2ij}\}$. It was repeated $R = 5,000$ times, and from each simulated sample $\{y_{ij}^{(r)}, x_{1ij}, x_{2ij}\}$, the EBLUP estimate $\hat{\mu}_i^{H(r)}$ and the corresponding EBLUP estimate, $\hat{\mu}_{i*}^{H(r)}$, under the two-level model without the covariate z_i, were computed. Simulated values of MSE and absolute relative error (ARE) of $\hat{\mu}_i^H$ and $\hat{\mu}_{i*}^H$ were then computed, using

$$\text{MSE}_i = \frac{1}{R}\sum_{r=1}^{R}\left(\text{est}_i^{(r)} - \mu_i^{(r)}\right)^2, \quad \text{ARE}_i = \frac{1}{R}\sum_{r=1}^{R}\left|\text{est}_i^{(r)} - \mu_i^{(r)}\right|/\mu_i^{(r)},$$

where $\text{est}_i^{(r)}$ denotes the value of an estimator for the rth simulated sample from the ith area. Denote the average values of MSE and ARE over i by $\overline{\text{MSE}}$ and $\overline{\text{ARE}}$. Efficiency of $\hat{\mu}_i^H$ relative to $\hat{\mu}_{i*}^H$ was measured by $R_1 = \overline{\text{MSE}}(\hat{\mu}_{i*}^H)/\overline{\text{MSE}}(\hat{\mu}_i^H)$ and $R_2 = \overline{\text{ARE}}(\hat{\mu}_{i*}^H)/\overline{\text{ARE}}(\hat{\mu}_i^H)$: $R_1 = 110.3\%, R_2 = 125.4\%$. The values of R_1 and R_2 suggest that the introduction of the area level covariate, z_i, leads to moderate improvement in efficiency. It may be noted that area level covariates, z_i, related to the random parameters, $\boldsymbol{\beta}_i$, are more difficult to find in practice than unit level covariates $\mathbf{x}_{ij}$ related to y_{ij}.

Chapter 9

Empirical Bayes (EB) Method

9.1 Introduction

The EBLUP method, studied in Chapters 6–8, is applicable to linear mixed models which cover many applications of small area estimation. Normality of random effects and errors is not needed for point estimation, but normality is used for getting accurate MSE estimators. The MSE estimator for the basic area level model remains valid under nonnormality of the random effects, v_i (see subsection 7.1.5), but normality is generally needed.

Linear mixed models are designed for continuous variables, y, but they are not suitable for handling binary or count data. In Section 5.6, Chapter 5, we proposed suitable models for binary and count data; in particular, logistic regression models with random area effects for binary data, and loglinear models with random effects for count data. Empirical Bayes (EB) and hierarchical Bayes (HB) methods are applicable more generally in the sense of handling models for binary and count data as well as normal linear mixed models. In the latter case, EB and EBLUP estimators are identical.

In this chapter, we study the EB method in the context of small area estimation. The EB approach may be summarized as follows: (i) Obtain the posterior density, $f(\boldsymbol{\mu}|\mathbf{y}, \boldsymbol{\lambda})$, of the small area (random) parameters of interest, $\boldsymbol{\mu}$, given the data $\mathbf{y}$, using the conditional density, $f(\mathbf{y}|\boldsymbol{\mu}, \boldsymbol{\lambda}_1)$, of $\mathbf{y}$ given $\boldsymbol{\mu}$ and the density $f(\boldsymbol{\mu}|\boldsymbol{\lambda}_2)$ of $\boldsymbol{\mu}$, where $\boldsymbol{\lambda} = (\boldsymbol{\lambda}_1^T, \boldsymbol{\lambda}_2^T)^T$ denotes the vector of model parameters. (ii) Estimate the model parameters, $\boldsymbol{\lambda}$, from the marginal density, $f(\mathbf{y}|\boldsymbol{\lambda})$, of $\mathbf{y}$. (iii) Use the estimated posterior density, $f(\boldsymbol{\mu}|\mathbf{y}, \hat{\boldsymbol{\lambda}})$, for making inferences about $\boldsymbol{\mu}$, where $\hat{\boldsymbol{\lambda}}$ is an estimator of $\boldsymbol{\lambda}$. The density of $\boldsymbol{\mu}$ is often interpreted as prior density on $\boldsymbol{\mu}$, but it is actually a part of the postulated model on $(\mathbf{y}, \boldsymbol{\mu})$ and it can be validated from the data, unlike subjective priors on model parameters, $\boldsymbol{\lambda}$, used in the HB approach. In this sense the EB approach is essentially frequentist, and EB inferences refer to averaging over the joint distribution of $\mathbf{y}$ and $\boldsymbol{\mu}$. Sometimes, a prior density is chosen for $\boldsymbol{\lambda}$ but used only to derive estimators and associated measures of variability as well as confidence intervals with good frequentist properties (Morris (1983b)).

In the parametric empirical Bayes (PEB) approach, a parametric form, $f(\boldsymbol{\mu}|\boldsymbol{\lambda}_2)$, is assumed for the density of $\boldsymbol{\mu}$. On the other hand, nonparametric empirical Bayes (NPEB) methods do not specify the form of the (prior) distribution of $\boldsymbol{\mu}$. Nonparametric maximum likelihood is used to estimate the (prior) distribution of $\boldsymbol{\mu}$ (Laird (1978)). Semi-nonparametric (SNP) representations of the density of $\boldsymbol{\mu}$ are also used in making EB inferences. For example, Zhang and Davidian (2001) studied linear mixed models with block diagonal covariance structures by approximating the density of the random effects by a SNP representation. Their representation includes normality as a special case and it provides flexibility in capturing nonnormality through a user-chosen tuning parameter. In this chapter, we focus on PEB approach to small area estimation. We refer the reader to Maritz and Lwin (1989) and Carlin and Louis (2000) for excellent accounts of the EB methodology.

The basic area level model (7.1.1) with normal random effects, v_i, is used in Section 9.2 to introduce the EB methodology. A jackknife method of MSE estimation is given in subsection 9.2.2. This method is applicable more generally, as shown in subsequent sections. Inferences based on the estimated posterior density of the small area parameters, θ_i, do not account for the variability in the estimators of model parameters $(\boldsymbol{\beta}, \sigma_v^2)$. Methods that account for the variability are studied in subsection 9.2.3. Confidence interval estimation is addressed in subsection 9.2.4. Section 9.3 provides extensions to linear mixed models with a block diagonal covariance structure. The case of binary data is studied in Section 9.4. Applications to disease mapping using count data are given in Section 9.5. The EB (or EBLUP) estimator, $\hat{\theta}_i^{\text{EB}}$, may not perform well in estimating the histogram of the θ_i's or in ranking them; Section 9.6 proposes constrained EB and other estimators that address this problem. Finally, empirical linear Bayes (ELB) and empirical constrained linear Bayes (ECLB) methods are studied in Sections 9.7 and 9.8, respectively. These methods avoid distributional assumptions.

9.2 Basic Area Level Model

Assuming normality, the basic area level model (7.1.1) may be expressed as a two-stage hierarchical model: (i) $\hat{\theta}_i|\theta_i \overset{\text{ind}}{\sim} N(\theta_i, \psi_i)$, $i = 1, \ldots, m$; (ii) $\theta_i \overset{\text{ind}}{\sim} N(\mathbf{z}_i^T\boldsymbol{\beta}, b_i^2\sigma_v^2)$, $i = 1, \ldots, m$, where $\boldsymbol{\beta}$ is the $p \times 1$ vector of regression parameters. In the Bayesian framework, the model parameters $\boldsymbol{\beta}$ and σ_v^2 are random, and the two-stage hierarchical model is called the conditionally independent hierarchical model (CIHM) because the pairs $(\hat{\theta}_i, \theta_i)$ are independent across areas i, conditionally on $\boldsymbol{\beta}$ and σ_v^2 (Kass and Steffey (1989)).

9.2.1 EB Estimator

The "optimal" estimator of the realized value of θ_i is given by the conditional expectation of θ_i given $\hat{\theta}_i, \boldsymbol{\beta}$ and σ_v^2:

$$E(\theta_i|\hat{\theta}_i, \boldsymbol{\beta}, \sigma_v^2) = \hat{\theta}_i^B = \gamma_i\hat{\theta}_i + (1-\gamma_i)\mathbf{z}_i^T\boldsymbol{\beta}, \tag{9.2.1}$$

where $\gamma_i = b_i^2\sigma_v^2/(b_i^2\sigma_v^2 + \psi_i)$. The result (9.2.1) follows from the posterior (or conditional) distribution of θ_i given $\hat{\theta}_i$, $\boldsymbol{\beta}$ and σ_v^2:

$$\theta_i|\hat{\theta}_i, \boldsymbol{\beta}, \sigma_v^2 \overset{\text{ind}}{\sim} N(\hat{\theta}_i^B, g_{1i}(\sigma_v^2) = \gamma_i\psi_i). \tag{9.2.2}$$

The estimator $\hat{\theta}_i^B = \hat{\theta}_i^B(\boldsymbol{\beta}, \sigma_v^2)$ is the "Bayes" estimator under squared error loss and it is optimal in the sense that its MSE, $\text{MSE}(\hat{\theta}_i^B) = E(\hat{\theta}_i^B - \theta_i)^2$, is smaller than the MSE of any other estimator of θ_i, linear or nonlinear in the $\hat{\theta}_i$'s (see Section 6.2, Chapter 6). It may be more appropriate to name $\hat{\theta}_i^B$ as the best prediction (BP) estimator of θ_i because it is obtained from the conditional distribution (9.2.2) without assuming a prior distribution on the model parameters (Jiang, Lahiri and Wan (2002)).

The Bayes estimator $\hat{\theta}_i^B$ depends on the model parameters $\boldsymbol{\beta}$ and σ_v^2, which are estimated from the marginal distribution: $\hat{\theta}_i \overset{\text{ind}}{\sim} N(\mathbf{z}_i^T\boldsymbol{\beta}, b_i^2\sigma_v^2 + \psi_i)$ using ML or REML. Denoting the estimators as $\hat{\boldsymbol{\beta}}$ and $\hat{\sigma}_v^2$, we obtain the EB (or the empirical BP (EBP)) estimator of θ_i from $\hat{\theta}_i^B$ by substituting $\hat{\boldsymbol{\beta}}$ for $\boldsymbol{\beta}$ and $\hat{\sigma}_v^2$ for σ_v^2:

$$\hat{\theta}_i^{\text{EB}} = \hat{\theta}_i^B(\hat{\boldsymbol{\beta}}, \hat{\sigma}_v^2) = \hat{\gamma}_i\hat{\theta}_i + (1-\hat{\gamma}_i)\mathbf{z}_i^T\hat{\boldsymbol{\beta}}. \tag{9.2.3}$$

The EB estimator, $\hat{\theta}_i^{\text{EB}}$, is identical to the EBLUP estimator $\hat{\theta}_i^H$ given by (7.1.11). Note that $\hat{\theta}_i^{\text{EB}}$ is also the mean of the estimated posterior density, $f(\theta_i|\hat{\boldsymbol{\theta}}, \hat{\boldsymbol{\beta}}, \hat{\sigma}_v^2)$, of θ_i, namely, $N(\hat{\theta}_i^{\text{EB}}, \hat{\gamma}_i\psi_i)$.

For the special case of equal sampling variances $\psi_i = \psi$ and $b_i = 1$ for all i, Morris (1983b) studied the use of an unbiased estimator of $1-\gamma_i = 1-\gamma$ given by

$$1-\gamma^* = \psi(m-p-2)/S, \tag{9.2.4}$$

where $S = \sum_i(\hat{\theta}_i - \mathbf{z}_i^T\hat{\boldsymbol{\beta}}_{\text{LS}})^2$ and $\hat{\boldsymbol{\beta}}_{\text{LS}}$ is the least squares estimator of $\boldsymbol{\beta}$ (see subsection 7.1.2). The resulting EB estimator

$$\hat{\theta}_i^{\text{EB}} = \gamma^*\hat{\theta}_i + (1-\gamma^*)\mathbf{z}_i^T\hat{\boldsymbol{\beta}}_{\text{LS}} \tag{9.2.5}$$

is identical to the James-Stein estimator, studied in subsection 4.4.2. Note that (9.2.4) may be expressed as

$$1-\gamma^* = [(m-p-2)/(m-p)]\psi/(\psi+\tilde{\sigma}_v^2), \tag{9.2.6}$$

where $\tilde{\sigma}_v^2 = S/(m-p) - \psi$ is the unbiased moment estimator of σ_v^2. Morris (1983b) used the REML estimator $\hat{\sigma}_v^2 = \max(0, \tilde{\sigma}_v^2)$ in (9.2.6), instead of the unbiased estimator $\tilde{\sigma}_v^2$, to ensure that $1-\gamma^* < 1$. The multiplying constant $(m-p-2)/(m-p)$ offsets the positive bias introduced by the substitution of a nearly unbiased estimator $\hat{\sigma}_v^2$ into $1-\gamma = \psi/(\psi+\sigma_v^2)$; note that $1-\gamma$ is a convex nonlinear function of σ_v^2 so that $E[\psi/(\psi+\tilde{\sigma}_v^2)] > 1-\gamma$ by Jensen's inequality.

For the case of unequal sampling variances ψ_i and $b_i = 1$, Morris (1983b) used the multiplying constant $(m-p-2)/(m-p)$ in the EB estimator (9.2.3); that is, $1-\hat{\gamma}_i$ is replaced by

$$1-\gamma_i^* = [(m-p-2)/(m-p)]\psi_i/(\psi_i+\hat{\sigma}_v^2), \tag{9.2.7}$$

where $\hat{\sigma}_v^2$ is the REML estimator of σ_v^2. He also proposed an alternative estimator of σ_v^2. It is similar to the Fay-Herriot moment estimator (subsection 7.1.2). It is obtained by solving

$$\sigma_v^2 = \left(\sum_i \alpha_i\right)^{-1}\left[\sum_i \alpha_i \left\{\frac{m}{m-p}(\hat{\theta}_i - \mathbf{z}_i^T\tilde{\boldsymbol{\beta}})^2 - \psi_i\right\}\right] \tag{9.2.8}$$

iteratively for σ_v^2, where $\tilde{\boldsymbol{\beta}} = \tilde{\boldsymbol{\beta}}(\sigma_v^2)$ is the weighted least squares estimator of $\boldsymbol{\beta}$ and $\alpha_i = 1/(\sigma_v^2+\psi_i)$. If the solution, $\tilde{\sigma}_v^2$, is negative, we take $\hat{\sigma}_v^2 = \max(\tilde{\sigma}_v^2, 0)$. If α_i is replaced by α_i^2 in (9.2.8), then the resulting solution is approximately equal to the REML estimator of σ_v^2.

An advantage of EB (or EBP) is that it can be applied to find the EB estimator of any function $\phi_i = h(\theta_i)$; in particular, $\overline{Y}_i = g^{-1}(\theta_i) = h(\theta_i)$. The EB estimator is obtained from the Bayes estimator $\hat{\phi}_i^B = E(\phi_i|\hat{\theta}_i, \boldsymbol{\beta}, \sigma_v^2)$ by substituting $(\hat{\boldsymbol{\beta}}, \hat{\sigma}_v^2)$ for $(\boldsymbol{\beta}, \sigma_v^2)$. The computation of the EB estimator $\hat{\phi}_i^{\text{EB}}$ might require the use of Monte Carlo or numerical integration. For example, $\{\theta_i^{(r)}, r = 1, \ldots, R\}$ can be simulated from the estimated posterior density, namely, $N(\hat{\theta}_i^{\text{EB}}, \hat{\gamma}_i\psi_i)$, to obtain a Monte Carlo approximation:

$$\hat{\phi}_i^{\text{EB}} \approx \frac{1}{R}\sum_{r=1}^{R} h\left(\theta_i^{(r)}\right). \tag{9.2.9}$$

The computation of (9.2.9) can be simplified by rewriting (9.2.9) as

$$\hat{\phi}_i^{\text{EB}} \approx \frac{1}{R}\sum_{r=1}^{R} h\left(\hat{\theta}_i^{\text{EB}} + z_i^{(r)}\sqrt{\hat{\gamma}_i\psi_i}\right), \tag{9.2.10}$$

where $\{z_i^{(r)}, r = 1, \ldots, R\}$ are generated from $N(0,1)$.

The approximation (9.2.10) will be accurate if the number of simulated samples, R, is large. Note that we used $h(\hat{\theta}_i^H) = h(\hat{\theta}_i^{\text{EB}})$ in subsection 7.1.8 and remarked that the estimator $h(\hat{\theta}_i^H)$ does not retain the optimality of $\hat{\theta}_i^H$.

9.2.2 MSE Estimation

The results in subsection 7.1.5 on the estimation of MSE of the EBLUP estimator, $\hat{\theta}_i^H$, are applicable to the EB estimator $\hat{\theta}_i^{\text{EB}}$ because $\hat{\theta}_i^H$ and $\hat{\theta}_i^{\text{EB}}$

are identical under normality. Also, the area-specific estimator (7.1.24) or (7.1.25) may be used as an estimator of the conditional MSE, $\text{MSE}(\hat{\theta}_i^{\text{EB}}) = E[(\hat{\theta}_i^{\text{EB}} - \theta_i)^2|\hat{\theta}_i]$, where the expectation is conditional on the observed $\hat{\theta}_i$ for the ith area (see subsection 7.1.6). As noted in subsection 7.1.5, the MSE estimators are nearly unbiased in the sense of bias lower order than m^{-1}, for large m.

Jiang, Lahiri and Wan (2002) proposed a jackknife method of estimating the MSE of EB estimators. This method is more general than the methods of subsection 7.1.5 for $\hat{\theta}_i^H = \hat{\theta}_i^{\text{EB}}$ (see Sections 9.4 and 9.5 for binary and count data), but we illustrate its use here for estimating $\text{MSE}(\hat{\theta}_i^{\text{EB}})$.

We first note the following decomposition of $\text{MSE}(\hat{\theta}_i^{\text{EB}})$:

$$\begin{aligned}\text{MSE}(\hat{\theta}_i^{\text{EB}}) &= E(\hat{\theta}_i^{\text{EB}} - \hat{\theta}_i^B)^2 + E(\hat{\theta}_i^B - \theta_i)^2 \qquad (9.2.11)\\ &= E(\hat{\theta}_i^{\text{EB}} - \hat{\theta}_i^B)^2 + g_{1i}(\sigma_v^2)\\ &=: M_{2i} + M_{1i}, \qquad (9.2.12)\end{aligned}$$

where the expectation is over the joint distribution of $(\hat{\theta}_i, \theta_i), i = 1, \ldots, m$; see Section 9.9 for a proof of (9.2.11). Note that $\hat{\theta}_i^{\text{EB}}$ depends on all the data, $\hat{\boldsymbol{\theta}} = (\hat{\theta}_1, \ldots, \hat{\theta}_m)^T$, through the estimators $\hat{\boldsymbol{\beta}}$ and $\hat{\sigma}_v^2$.

The jackknife steps for estimating the two terms, M_{2i} and M_{1i}, in (9.2.12) are as follows. Let $\hat{\theta}_i^{\text{EB}} = k_i(\hat{\theta}_i, \hat{\boldsymbol{\beta}}, \hat{\sigma}_v^2)$ be the EB estimator of θ_i expressed as a function of the direct estimator $\hat{\theta}_i$ and the parameter estimators $\hat{\boldsymbol{\beta}}$ and $\hat{\sigma}_v^2$.

Step 1. Calculate the delete-l estimators $\hat{\boldsymbol{\beta}}_{-l}$ and $\hat{\sigma}_{v,-l}^2$ by deleting the lth area data set $(\hat{\theta}_l, \mathbf{z}_l)$ from the full data set $\{(\hat{\theta}_i, \mathbf{z}_i); i = 1, \ldots, m\}$. This calculation is done for each l to get m estimators of $\boldsymbol{\beta}$ and σ_v^2: $\{(\hat{\boldsymbol{\beta}}_{-l}, \hat{\sigma}_{v,-l}^2); l = 1, \ldots, m\}$ which, in turn, provide m estimators of θ_i: $\{\hat{\theta}_{i,-l}^{\text{EB}}; l = 1, \ldots, m\}$, where $\hat{\theta}_{i,-l}^{\text{EB}} = k_i(\hat{\theta}_i, \hat{\boldsymbol{\beta}}_{-l}, \hat{\sigma}_{v,-l}^2)$.

Step 2. Calculate the estimator of M_{2i} as

$$\hat{M}_{2i} = \frac{m-1}{m} \sum_{l=1}^{m} (\hat{\theta}_{i,-l}^{\text{EB}} - \hat{\theta}_i^{\text{EB}})^2. \qquad (9.2.13)$$

Step 3. Calculate the estimator of M_{1i} as

$$\hat{M}_{1i} = g_{1i}(\hat{\sigma}_v^2) - \frac{m-1}{m} \sum_{l=1}^{m} [g_{1i}(\hat{\sigma}_{v,-l}^2) - g_{1i}(\hat{\sigma}_v^2)]. \qquad (9.2.14)$$

The estimator $\hat{M}_{1i}$ corrects the bias of $g_{1i}(\hat{\sigma}_v^2)$.

Step 4. Calculate the jackknife estimator of $\text{MSE}(\hat{\theta}_i^{\text{EB}})$ as

$$\text{mse}_J(\hat{\theta}_i^{\text{EB}}) = \hat{M}_{1i} + \hat{M}_{2i}. \qquad (9.2.15)$$

Note that $\hat{M}_{1i}$ estimates the MSE when the model parameters are known, and $\hat{M}_{2i}$ estimates the additional variability due to estimating the model parameters. The jackknife estimator of MSE, given by (9.2.15), is also nearly unbiased in the sense of having bias of lower order than m^{-1}, for large m. The proof of this result is highly technical and we refer the reader to Jiang, Lahiri and Wan (2002) for details.

The jackknife method is applicable to ML, REML or moment estimators of the model parameters. It is computer-intensive compared to the MSE estimators studied in subsection 7.1.5. In the case of ML or REML, computation of $\{(\hat{\boldsymbol{\beta}}_{-l}, \hat{\sigma}^2_{v,-l}); l = 1, \ldots, m\}$ may be simplified by performing only a single step of the Newton-Raphson algorithm using $(\hat{\boldsymbol{\beta}}, \hat{\sigma}^2_v)$ as the starting values of $\boldsymbol{\beta}$ and σ^2_v. However, properties of the resulting simplified jackknife MSE estimator are not known.

The jackknife method is applicable to the EB estimator of any function $\phi_i = h(\theta_i)$; in particular, $\overline{Y}_i = g^{-1}(\theta_i) = h(\theta_i)$. However, the computation of $\text{mse}_J(\hat{\phi}_i^{\text{EB}}) = \hat{M}_{1i} + \hat{M}_{2i}$ might require repeated Monte Carlo or numerical integration to obtain $\hat{M}_{2i} = (m-1)m^{-1}\sum_l(\hat{\phi}^{\text{EB}}_{i,-l} - \hat{\phi}^{\text{EB}}_i)^2$ and the bias corrected estimator, $\hat{M}_{1i}$, of $E(\hat{\phi}_i^B - \phi_i)^2$.

Example 9.2.1. Visits to Doctor's Office. Jiang, Lahiri, Wan and Wu (2001) applied the jackknife method to data from the U.S. National Health Interview Survey (NHIS). The objective here is to estimate the proportion of individuals who did not visit a doctor's office during the previous twelve months for all the fifty states and the District of Columbia (regarded as small areas). Direct NHIS estimates, $\hat{P}_i$, of the proportions P_i are not reliable for the smaller states. The arcsine transformation was used to stabilize the variances of the direct estimates. We have $\hat{\theta}_i = \arcsin\sqrt{\hat{P}_i}$ and $\hat{\theta}_i \overset{\text{ind}}{\sim} (\theta_i, \psi_i = \hat{D}_i/(4n_i))$, where $\hat{D}_i$ is the estimated design effect (deff) of $\hat{P}_i$ and n_i is the sample size from the ith area. Note that $V(\hat{\theta}_i) \approx D_i/(4n_i)$, where $D_i = V_i[P_i(1-P_i)/n_i]^{-1}$ is the population deff of $\hat{P}_i$ and $V_i = V(\hat{P}_i)$. The estimated deff $\hat{D}_i$ was calculated as $\hat{D}_i = \hat{V}_i[\hat{P}_i(1-\hat{P}_i)/n_i]^{-1}$, where $\hat{V}_i$ is the NCHS variance estimate of $\hat{P}_i$.

For the covariate selection, the largest 15 states with small ψ_i-values were chosen. This permitted the use of standard selection methods for linear regression models because the basic area level model for those states may be treated as $\hat{\theta}_i \approx \mathbf{z}_i^T\boldsymbol{\beta} + v_i$ with $v_i \overset{\text{iid}}{\sim} (0, \sigma^2_v)$. Based on C_p, R^2 and adjusted R^2 criteria used in SAS, the following covariates were selected: $z_1 = 1$, $z_2 =$1995 Bachelor's degree completion for 25+ population and $z_3 =$1995 health insurance coverage. The simple moment estimator $\hat{\sigma}^2_{vs} = \max(\tilde{\sigma}^2_{vs}, 0)$ was used to estimate σ^2_v, where $\tilde{\sigma}^2_{vs}$ is given by (7.1.13) with $m = 51$ and $p = 3$. The resulting weights $\hat{\gamma}_i = \hat{\sigma}^2_v/(\hat{\sigma}^2_v + \psi_i)$ varied from 0.09 (South Dakota) to 0.95 (California). Using the estimated weights $\hat{\gamma}_i$, the EB (EBLUP) estimates $\hat{\theta}_i^{\text{EB}}$ were computed for each of the small areas which in turn provided the estimates of the proportions P_i as $\tilde{P}_i^{\text{EB}} = \sin^2(\hat{\theta}_i^{\text{EB}})$. Note that $\tilde{P}_i^{\text{EB}}$ is not equal to the true EB estimate $\hat{P}_i^{\text{EB}} = E(P_i|\hat{\theta}_i, \hat{\boldsymbol{\beta}}, \hat{\sigma}^2_v)$, but it simplifies the computations.

MSE estimates of the estimated proportions $\tilde{P}_i^{\rm EB}$ were computed from the jackknife estimates $\text{mse}_J(\hat{\theta}_i^{\rm EB})$ given by (9.2.15). Using Taylor linearization, we have $\text{mse}_J(\tilde{P}_i^{\rm EB}) \approx 4\tilde{P}_i^{\rm EB}(1-\tilde{P}_i^{\rm EB})\text{mse}_J(\hat{\theta}_i^{\rm EB}) = s_J^2(\tilde{P}_i^{\rm EB})$. Performance of $\tilde{P}_i^{\rm EB}$ relative to the direct estimate $\hat{P}_i$ was measured by the percent improvement, $PI_i = 100[s(\hat{P}_i) - s_J(\tilde{P}_i^{\rm EB})]/s(\hat{P}_i)$, where $s^2(\hat{P}_i) = \hat{V}_i$, the NCHS variance estimate of $\hat{P}_i$. Values of PI_i indicated that the improvement is quite substantial (30 to 55%) for small states (e.g., South Dakota and Vermont). On the other hand, the improvement is small for large states (e.g., California and Texas), as expected. Note that $\hat{\theta}_i^{\rm EB}$ gives more weight, $\hat{\gamma}_i$, to the direct estimate $\hat{\theta}_i$ for large states (e.g., $\hat{\gamma}_i = 0.95$ for California and $\hat{\gamma}_i = 0.94$ for Texas).

9.2.3 Approximation to Posterior Variance

In subsection 9.2.2, we studied MSE estimation, but alternative measures of variability associated with $\hat{\theta}_i^{\rm EB}$ have also been proposed. Those measures essentially use a HB approach and provide approximations to the posterior variance of θ_i, denoted by $V(\theta_i|\hat{\boldsymbol{\theta}})$, based on a prior distribution on the model parameters $\boldsymbol{\beta}$ and σ_v^2.

If the model parameters $\boldsymbol{\beta}$ and σ_v^2 are given, then the posterior (or conditional) distribution $f(\boldsymbol{\theta}|\hat{\boldsymbol{\theta}})$ is completely known and it provides a basis for inference on $\boldsymbol{\theta} = (\theta_1, \ldots, \theta_m)^T$. In particular, the Bayes estimator $\hat{\theta}_i^B = E(\theta_i|\hat{\theta}_i, \boldsymbol{\beta}, \sigma_v^2)$ is used to estimate the realized value of θ_i, and the posterior variance $V(\theta_i|\hat{\theta}_i, \boldsymbol{\beta}, \sigma_v^2) = g_{1i}(\sigma_v^2) = \gamma_i\psi_i$ is used to measure the variability associated with θ_i. The posterior variance is identical to $\text{MSE}(\hat{\theta}_i^B)$, and $\hat{\theta}_i^B$ is the BP estimator of θ_i. Therefore, the frequentist approach agrees with the Bayesian approach for the basic area level model when the model parameters are known. This agreement also holds for the general linear mixed model, but not necessarily for nonlinear models with random effects.

In practice, the model parameters are not known and the EB approach uses the marginal distribution of the $\hat{\theta}_i$'s, namely, $\hat{\theta}_i \overset{\text{ind}}{\sim} N(\mathbf{z}_i^T\boldsymbol{\beta}, b_i^2\sigma_v^2 + \psi_i)$, to estimate $\boldsymbol{\beta}$ and σ_v^2. A naive EB approach uses the estimated posterior density of θ_i, namely $N(\hat{\theta}_i^{\rm EB}, \hat{\gamma}_i\psi_i)$, to make inferences on θ_i. In particular, θ_i is estimated by $\hat{\theta}_i^{\rm EB}$ and the estimated posterior variance $g_{1i}(\hat{\sigma}_v^2) = \hat{\gamma}_i\psi_i$ is used as a measure of variability. The use of $g_{1i}(\hat{\sigma}_v^2)$, however, leads to severe underestimation of $\text{MSE}(\hat{\theta}_i^{\rm EB})$. Note that the naive EB approach treats $\boldsymbol{\beta}$ and σ_v^2 as fixed, unknown parameters, so no prior distributions are involved. However, if we adopt a HB approach by treating the model parameters $\boldsymbol{\beta}$ and σ_v^2 as random with a prior density $f(\boldsymbol{\beta}, \sigma_v^2)$, then the posterior mean $\hat{\theta}_i^{\rm HB} = E(\theta_i|\hat{\boldsymbol{\theta}})$ is used as an estimator of θ_i and the posterior variance $V(\theta_i|\hat{\boldsymbol{\theta}})$ as a measure of variability associated with $\hat{\theta}_i^{\rm HB}$. We can express $E(\theta_i|\hat{\boldsymbol{\theta}})$ as

$$E(\theta_i|\hat{\boldsymbol{\theta}}) = E_{\boldsymbol{\beta},\sigma_v^2}[E(\theta_i|\hat{\theta}_i, \boldsymbol{\beta}, \sigma_v^2)] = E_{\boldsymbol{\beta},\sigma_v^2}[\hat{\theta}_i^B(\boldsymbol{\beta}, \sigma_v^2)] \tag{9.2.16}$$

and $V(\theta_i|\hat{\boldsymbol{\theta}})$ as

$$\begin{aligned} V(\theta_i|\hat{\boldsymbol{\theta}}) &= E_{\boldsymbol{\beta},\sigma_v^2}[V(\theta_i|\hat{\theta}_i,\boldsymbol{\beta},\sigma_v^2)] + V_{\boldsymbol{\beta},\sigma_v^2}[E(\theta_i|\hat{\theta}_i,\boldsymbol{\beta},\sigma_v^2)] \\ &= E_{\boldsymbol{\beta},\sigma_v^2}[g_{1i}(\sigma_v^2)] + V_{\boldsymbol{\beta},\sigma_v^2}[\hat{\theta}_i^B(\boldsymbol{\beta},\sigma_v^2)], \end{aligned} \tag{9.2.17}$$

where $E_{\boldsymbol{\beta},\sigma_v^2}$ and $V_{\boldsymbol{\beta},\sigma_v^2}$ respectively denote the expectation and the variance with respect to the posterior distribution of $\boldsymbol{\beta}$ and σ_v^2, given $\hat{\boldsymbol{\theta}}$, that is, $f(\boldsymbol{\beta},\sigma_v^2|\hat{\boldsymbol{\theta}})$.

For large m, we can approximate the last term of (9.2.16) by $\hat{\theta}_i^B(\hat{\boldsymbol{\beta}},\hat{\sigma}_v^2) = \hat{\theta}_i^{\text{EB}}$, where $\hat{\boldsymbol{\beta}}$ and $\hat{\sigma}_v^2$ are ML (REML) estimators. More precisely, we have $E(\theta_i|\hat{\boldsymbol{\theta}}) = \hat{\theta}_i^B(\hat{\boldsymbol{\beta}},\hat{\sigma}_v^2)[1 + O(m^{-1})]$ regardless of the prior $f(\boldsymbol{\beta},\sigma_v^2)$. Hence, the EB estimator $\hat{\theta}_i^{\text{EB}}$ tracks $E(\theta_i|\hat{\boldsymbol{\theta}})$ well. However, the naive EB measure of variability, $g_{1i}(\hat{\sigma}_v^2)$, provides only a first-order approximation to the first variance term $E_{\boldsymbol{\beta},\sigma_v^2}[g_{1i}(\sigma_v^2)]$ on the right hand side of (9.2.17). The second variance term, $V_{\boldsymbol{\beta},\sigma_v^2}[\hat{\theta}_i^B(\boldsymbol{\beta},\sigma_v^2)]$, account for the uncertainty about the model parameters, and the naive EB approach ignores this uncertainty. As a result, the naive EB approach can lead to severe underestimation of the true posterior variance, $V(\theta_i|\hat{\boldsymbol{\theta}})$. Note that the naive measure of variability, $g_{1i}(\hat{\sigma}_v^2)$, also underestimates $\text{MSE}(\hat{\theta}_i^{\text{EB}})$.

The HB approach may be used to evaluate the posterior mean, $E(\theta_i|\hat{\boldsymbol{\theta}})$, and the posterior variance, $V(\theta_i|\hat{\boldsymbol{\theta}})$, exactly, for any specified prior on the model parameters $\boldsymbol{\beta}$ and σ_v^2. Moreover, it can handle complex small area models (see Chapter 10). Typically, "improper" priors that reflect lack of information on the model parameters are used in the HB calculations; for example, $f(\boldsymbol{\beta},\sigma_v^2) \propto 1$ may be used for the basic area level model. In the EB context, two methods of approximating the posterior variance, regardless of the prior, have been proposed. The first method, based on bootstrap resampling, imitates the decomposition (9.2.17) of the posterior variance $V(\theta_i|\hat{\boldsymbol{\theta}})$. Thus it attempts to account for the underestimation of $V(\theta_i|\hat{\boldsymbol{\theta}})$. In the bootstrap method, a large number, B, of independent samples $\{\hat{\theta}_i^*(b), b = 1,\ldots,B\}$ are first drawn from the estimated marginal distribution, $N(\mathbf{z}_i^T\hat{\boldsymbol{\beta}}, b_i^2\hat{\sigma}_v^2 + \psi_i), i = 1,\ldots,m$. Estimates $\{\hat{\boldsymbol{\beta}}^*(b), \hat{\sigma}_v^{*2}(b)\}$ and the EB estimates $\hat{\theta}_i^{*\text{EB}}(b)$ are then computed from the bootstrap data $\{\hat{\theta}_i^*(b), \mathbf{z}_i; i = 1,\ldots,m\}$. This leads to the following approximation to the posterior variance:

$$V_{\text{LL}}(\theta_i|\hat{\boldsymbol{\theta}}) = \frac{1}{B}\sum_{b=1}^{B} g_{1i}(\hat{\sigma}_v^{*2}(b)) + \frac{1}{B}\sum_{b=1}^{B}[\hat{\theta}_i^{*\text{EB}}(b) - \hat{\theta}_i^{*\text{EB}}(\cdot)]^2, \tag{9.2.18}$$

where $\hat{\theta}_i^{*\text{EB}}(\cdot) = B^{-1}\sum_b \hat{\theta}_i^{*\text{EB}}(b)$; see Laird and Louis (1987). Note that (9.2.18) has the same form as the decomposition (9.2.17) of the posterior variance. The last term of (9.2.18) is designed to account for the uncertainty about the model parameters. The bootstrap method uses $\hat{\theta}_i^{\text{EB}}$ as the estimator of θ_i and $V_{\text{LL}}(\theta_i|\hat{\boldsymbol{\theta}})$ as a measure of its variability.

Butar and Lahiri (2001) studied the performance of the bootstrap measure, $V_{\text{LL}}(\theta_i|\hat{\boldsymbol{\theta}})$, as an estimator of $\text{MSE}(\hat{\theta}_i^{\text{EB}})$. They obtained an analytical approximation to (9.2.18) by letting $B \to \infty$. Under REML, we have

$$V_{\text{LL}}(\theta_i|\hat{\boldsymbol{\theta}}) \approx g_{1i}(\hat{\sigma}_v^2) + g_{2i}(\hat{\sigma}_v^2) + g_{3i}^*(\hat{\sigma}_v^2, \hat{\theta}_i) =: \tilde{V}_{\text{LL}}(\theta_i|\hat{\boldsymbol{\theta}}), \tag{9.2.19}$$

for large m, where $g_{2i}(\sigma_v^2)$ and $g_{3i}^*(\hat{\sigma}_v^2, \hat{\theta}_i)$ are given by (7.1.8) and (7.1.23), respectively. Comparing (9.2.19) to the nearly unbiased MSE estimator (7.1.25), it follows that $V_{\text{LL}}(\theta_i|\hat{\boldsymbol{\theta}})$ is not nearly unbiased for $\text{MSE}(\hat{\theta}_i^{\text{EB}})$, that is, its bias involves terms of order m^{-1}. In particular,

$$E[\tilde{V}_{\text{LL}}(\theta_i|\hat{\boldsymbol{\theta}})] = \text{MSE}(\hat{\theta}_i) - g_{3i}(\sigma_v^2). \tag{9.2.20}$$

A bias-corrected MSE estimator is therefore given by

$$\text{mse}_{\text{BL}}(\hat{\theta}_i^{\text{EB}}) = \tilde{V}_{\text{LL}}(\theta_i|\hat{\boldsymbol{\theta}}) + g_{3i}(\hat{\sigma}_v^2). \tag{9.2.21}$$

This estimator is nearly unbiased and also area-specific. It is identical to $\text{mse}_2(\hat{\theta}_i^H)$ given by (7.1.25).

In the second method, due to Kass and Steffey (1989), $\hat{\theta}_i^{\text{EB}}$ is taken as the estimator of θ_i, noting that

$$E(\theta_i|\hat{\boldsymbol{\theta}}) = \hat{\theta}_i^{\text{EB}}[1 + O(m^{-1})], \tag{9.2.22}$$

but a positive correction term is added to the estimated posterior variance, $g_{1i}(\hat{\sigma}_v^2)$, to account for the underestimation of $V(\theta_i|\hat{\boldsymbol{\theta}})$. This positive term depends on the information matrix and the partial derivatives of $\hat{\theta}_i^B$ evaluated at the ML (REML) estimates $\hat{\boldsymbol{\beta}}$ and $\hat{\sigma}_v^2$. The following first-order approximation to $V(\theta_i|\hat{\boldsymbol{\theta}})$ is obtained, noting that the information matrix for $\boldsymbol{\beta}$ and σ_v^2 is block diagonal:

$$\begin{aligned} V_{\text{KS}}(\theta_i|\hat{\boldsymbol{\theta}}) &= g_{1i}(\hat{\sigma}_v^2) + (\partial\hat{\theta}_i^B/\partial\boldsymbol{\beta})^T\overline{\mathbf{V}}(\hat{\boldsymbol{\beta}})(\partial\hat{\theta}_i^B/\partial\boldsymbol{\beta})\big|_{\boldsymbol{\beta}=\hat{\boldsymbol{\beta}},\sigma_v^2=\hat{\sigma}_v^2} \\ &\quad + (\partial\hat{\theta}_i^B/\partial\sigma_v^2)^2\overline{V}(\hat{\sigma}_v^2)\big|_{\boldsymbol{\beta}=\hat{\boldsymbol{\beta}},\sigma_v^2=\hat{\sigma}_v^2}, \end{aligned} \tag{9.2.23}$$

where $\overline{\mathbf{V}}(\hat{\boldsymbol{\beta}}) = [\sum_i \mathbf{z}_i\mathbf{z}_i^T/(\psi_i + \sigma_v^2 b_i^2)]^{-1}$ is the asymptotic covariance matrix of $\hat{\boldsymbol{\beta}}$, $\overline{V}(\hat{\sigma}_v^2) = [\mathcal{I}(\sigma_v^2)]^{-1}$ is the asymptotic variance of $\hat{\sigma}_v^2$ with $\mathcal{I}(\sigma_v^2)$ given by (7.1.14),

$$\frac{\partial\hat{\theta}_i^B}{\partial\boldsymbol{\beta}} = (1 - \gamma_i)\mathbf{z}_i \tag{9.2.24}$$

and

$$\frac{\partial\hat{\theta}_i^B}{\partial\sigma_v^2} = \frac{\psi_i b_i^2}{(\psi_i + \sigma_v^2 b_i^2)^2}(\hat{\theta}_i - \mathbf{z}_i^T\boldsymbol{\beta}). \tag{9.2.25}$$

After simplification, (9.2.23) reduces to

$$V_{\text{KS}}(\theta_i|\hat{\boldsymbol{\theta}}) = g_{1i}(\hat{\sigma}_v^2) + g_{2i}(\sigma_v^2) + g_{3i}^*(\sigma_v^2, \hat{\theta}_i), \tag{9.2.26}$$

which is identical to $\tilde{V}_{\mathrm{LL}}(\theta_i|\hat{\boldsymbol{\theta}})$. Therefore, $V_{\mathrm{KS}}(\theta_i|\hat{\boldsymbol{\theta}})$ is also not nearly unbiased for $\mathrm{MSE}(\hat{\theta}_i)$. Kass and Steffey (1989) have given a more accurate approximation to $V(\theta_i|\hat{\boldsymbol{\theta}})$. This approximation ensures that the neglected terms are of lower order than m^{-1}, but it depends on the prior density, $f(\boldsymbol{\beta}, \sigma_v^2)$, unlike the first-order approximation (9.2.23). If the prior needs to be specified, then it may be better to use the HB approach (see Chapter 10) because it is free of asymptotic approximations. Singh, Stukel and Pfeffermann (1998) studied Kass and Steffey's (1989) approximations in the context of small area estimation.

Kass and Steffey's (1989) method is applicable to general functions $\phi_i = h(\theta_i)$, but the calculations might require the use of Monte Carlo or numerical integration. We have

$$E(\phi_i|\hat{\boldsymbol{\theta}}) = \hat{\phi}_i^{\mathrm{EB}}[1 + O(m^{-1})] \tag{9.2.27}$$

and the first-order approximation to $V(\phi_i|\hat{\boldsymbol{\theta}})$ is given by

$$\begin{aligned} V_{\mathrm{KS}}(\phi_i|\hat{\boldsymbol{\theta}}) &= V(\phi_i|\hat{\boldsymbol{\theta}}, \hat{\boldsymbol{\beta}}, \hat{\sigma}_v^2) + (\partial\phi_i^B/\partial\boldsymbol{\beta})^T \overline{\mathbf{V}}(\hat{\boldsymbol{\beta}})(\partial\phi_i^B/\partial\boldsymbol{\beta})|_{\boldsymbol{\beta}=\hat{\boldsymbol{\beta}}, \sigma_v^2=\hat{\sigma}_v^2} \\ &+ [(\partial\phi_i^B/\partial\sigma_v^2)^2 \overline{V}(\hat{\sigma}_v^2)]|_{\boldsymbol{\beta}=\hat{\boldsymbol{\beta}}, \sigma_v^2=\hat{\sigma}_v^2}, \end{aligned} \tag{9.2.28}$$

where $V(\phi_i|\hat{\boldsymbol{\theta}}, \hat{\boldsymbol{\beta}}, \hat{\sigma}_v^2)$ denotes $V(\phi_i|\hat{\boldsymbol{\theta}}, \boldsymbol{\beta}, \sigma_v^2)$ calculated at $\boldsymbol{\beta} = \hat{\boldsymbol{\beta}}$ and $\sigma_v^2 = \hat{\sigma}_v^2$.

Morris (1983a) studied the basic area level model without auxiliary information for the case of equal sampling variances, $\psi_i = \psi$, with known ψ. This model may be expressed as $\hat{\theta}_i|\theta_i \overset{\text{ind}}{\sim} N(\theta_i, \psi)$ and $\theta_i|\mu, \sigma_v^2 \overset{\text{iid}}{\sim} N(\mu, \sigma_v^2), i = 1, \ldots, m$. The Bayes estimator (9.2.1) reduces to

$$\hat{\theta}_i^B = \gamma\hat{\theta}_i + (1-\gamma)\mu, \tag{9.2.29}$$

where $\gamma = \sigma_v^2/(\sigma_v^2 + \psi)$. We first obtain the HB estimator, $\hat{\theta}_i^{\mathrm{HB}}(\sigma_v^2)$, for a given σ_v^2, assuming that $\mu \sim \text{uniform}(-\infty, \infty)$ to reflect the absence of prior information on μ; that is, $f(\mu) = $ constant. It is easy to verify that

$$\mu|\hat{\boldsymbol{\theta}}, \sigma_v^2 \sim N(\hat{\theta}_\cdot, m^{-1}(\sigma_v^2 + \psi)), \tag{9.2.30}$$

where $\hat{\theta}_\cdot = \sum_i \hat{\theta}_i/m$; see Section 9.9. It now follows from (9.2.29) that

$$\begin{aligned} \hat{\theta}_i^{\mathrm{HB}}(\sigma_v^2) &= \gamma\hat{\theta}_i + (1-\gamma)E(\mu|\hat{\boldsymbol{\theta}}, \sigma_v^2) \\ &= \hat{\theta}_i - (1-\gamma)(\hat{\theta}_i - \hat{\theta}_\cdot). \end{aligned} \tag{9.2.31}$$

The HB estimator $\hat{\theta}_i^{\mathrm{HB}}(\sigma_v^2)$ is identical to the BLUP estimator $\tilde{\theta}_i^H$.

The posterior variance of θ_i given $\hat{\boldsymbol{\theta}}$ and σ_v^2 is given by

$$\begin{aligned} V(\theta_i|\hat{\boldsymbol{\theta}}, \sigma_v^2) &= E_\mu[V(\theta_i|\hat{\theta}_i, \mu, \sigma_v^2)] + V_\mu[E(\theta_i|\hat{\theta}_i, \mu, \sigma_v^2)] \\ &= E_\mu[g_1(\sigma_v^2)] + V_\mu(\hat{\theta}_i^B), \end{aligned} \tag{9.2.32}$$

where E_μ and V_μ respectively denote the expectation and the variance with respect to the posterior distribution of μ given $\hat{\boldsymbol{\theta}}$ and σ_v^2, and

$$g_1(\sigma_v^2) = \gamma\psi = \psi - (1-\gamma)\psi. \tag{9.2.33}$$

It now follows from (9.2.29), (9.2.30) and (9.2.32) that

$$V(\theta_i|\hat{\boldsymbol{\theta}}, \sigma_v^2) = g_1(\sigma_v^2) + g_2(\sigma_v^2), \tag{9.2.34}$$

where

$$g_2(\sigma_v^2) = (1-\gamma)^2(\sigma_v^2 + \psi)/m = (1-\gamma)\psi/m. \tag{9.2.35}$$

The conditional posterior variance (9.2.34) is identical to the MSE of BLUP estimator.

To take account of the uncertainty associated with σ_v^2, Morris (1983a) assumed that $\sigma_v^2 \sim$ uniform$[0,\infty)$; that is, $f(\sigma_v^2)$ =constant. The resulting posterior density $f(\sigma_v^2|\hat{\boldsymbol{\theta}})$, however, does not yield closed form expressions for $E[(1-\gamma)|\hat{\boldsymbol{\theta}})]$ and $E[(1-\gamma)^2|\hat{\boldsymbol{\theta}}]$. The latter terms are needed in the evaluation of

$$\hat{\theta}_i^{\text{HB}} = E_{\sigma_v^2}[\hat{\theta}_i^{\text{HB}}(\sigma_v^2)] \tag{9.2.36}$$

and the posterior variance

$$V(\theta_i|\hat{\boldsymbol{\theta}}) = E_{\sigma_v^2}[V(\theta_i|\hat{\boldsymbol{\theta}}, \sigma_v^2)] + V_{\sigma_v^2}[\hat{\theta}_i^{\text{HB}}(\sigma_v^2)], \tag{9.2.37}$$

where $E_{\sigma_v^2}$ and $V_{\sigma_v^2}$ respectively denote the expectation and the variance with respect to $f(\sigma_v^2|\hat{\boldsymbol{\theta}})$. Morris (1983a) evaluated $\hat{\theta}_i^{\text{HB}}$ and $V(\theta_i|\hat{\boldsymbol{\theta}})$ numerically.

Morris (1983a) also obtained closed form approximations to $E[(1-\gamma)|\hat{\boldsymbol{\theta}}]$ and $E[(1-\gamma)^2|\hat{\boldsymbol{\theta}}]$ by expressing them as ratios of definite integrals over the range $[0,1)$ and then replacing $\int_0^1$ by $\int_0^\infty$. This method is equivalent to assuming that $\sigma_v^2 + \psi$ is uniform on $[0,\infty)$. We have

$$E_{\sigma_v^2}[(1-\gamma)|\hat{\boldsymbol{\theta}}] \approx \psi(m-3)/S = 1-\gamma^*, \tag{9.2.38}$$

where $S = \sum_i(\hat{\theta}_i - \hat{\theta}_\cdot)^2$. Hence, it follows from (9.2.31) and (9.2.36) that

$$\hat{\theta}_i^{\text{HB}} \approx \gamma^*\hat{\theta}_i + (1-\gamma^*)\hat{\theta}_\cdot\,, \tag{9.2.39}$$

which is identical to $\hat{\theta}_i^{\text{EB}}$ given by (9.2.5). Turning to the approximation to the posterior variance $V(\theta_i|\hat{\boldsymbol{\theta}})$, we first note that

$$E_{\sigma_v^2}[(1-\gamma)^2|\hat{\boldsymbol{\theta}}] \approx 2\psi^2(m-3)/S^2 = 2(1-\gamma^*)^2/(m-3). \tag{9.2.40}$$

It now follows from (9.2.31), (9.2.33), (9.2.34) and (9.2.37) that $V(\theta_i|\hat{\boldsymbol{\theta}}) \approx V_M(\theta_i|\hat{\boldsymbol{\theta}})$, where

$$V_M(\theta_i|\hat{\boldsymbol{\theta}}) = \psi - \frac{m-1}{m}(1-\gamma^*)\psi + \frac{2(1-\gamma^*)^2}{m-3}(\hat{\theta}_i - \hat{\theta}_\cdot)^2 \tag{9.2.41}$$

$$= \psi\gamma^* + \frac{(1-\gamma^*)\psi}{m} + \frac{2(1-\gamma^*)^2}{m-3}(\hat{\theta}_i - \hat{\theta}_\cdot)^2. \tag{9.2.42}$$

Morris (1983a) obtained (9.2.41). The alternative form (9.2.42) shows that $V_M(\theta_i|\hat{\boldsymbol{\theta}})$ is asymptotically equivalent to $\text{mse}_2(\hat{\theta}_i^H)$ to terms of order m^{-1}, noting that

$$\psi\gamma^* = \psi\hat{\gamma} + \frac{2}{m-1}(1-\hat{\gamma})\psi \tag{9.2.43}$$

and

$$\begin{aligned}\text{mse}_2(\hat{\theta}_i^H) &= \psi\hat{\gamma} + \frac{(1-\hat{\gamma})\psi}{m} + \frac{2(1-\hat{\gamma})\psi}{m} + \frac{2(1-\hat{\gamma})^2}{m}(\hat{\theta}_i - \hat{\theta}_.)^2 \\ &= g_1(\hat{\sigma}_v^2) + g_2(\hat{\sigma}_v^2) + g_3(\hat{\sigma}_v^2) + g_{3i}^*(\hat{\sigma}_v^2, \hat{\theta}_i),\end{aligned} \tag{9.2.44}$$

where $1-\hat{\gamma} = \psi(m-1)/S$ and $\hat{\sigma}_v^2$ is the REML estimator of σ_v^2. This result shows that the frequentist inference, based on the MSE estimator, and the HB inference, based on the approximation to posterior variance, agree to terms of order m^{-1}. Neglected terms in this comparison are of lower order than m^{-1}.

The approximation (9.2.41) to the posterior variance extends to the regression model, $\theta_i = \mathbf{z}_i^T\boldsymbol{\beta} + v_i$, with equal sampling variances $\psi_i = \psi$ (Morris (1983b)). We have

$$V_M(\theta_i|\hat{\boldsymbol{\theta}}) \approx \psi - \frac{m-p}{m}(1-\gamma^*)\psi + \frac{2(1-\gamma^*)^2}{m-p-2}(\hat{\theta}_i - \mathbf{z}_i^T\hat{\boldsymbol{\beta}}_{\text{LS}})^2, \tag{9.2.45}$$

where $1-\gamma^*$ is given by (9.2.6). Note that the approximation to $\hat{\theta}_i^{\text{HB}}$ is identical to the EB estimator given by (9.2.5). Morris (1983b) also proposed an extension to the case of unequal sampling variances ψ_i and $b_i = 1$. It is obtained from (9.2.45) by changing ψ to ψ_i, $1-\gamma^*$ to $1-\gamma_i^*$ given by (9.2.7), $\hat{\boldsymbol{\beta}}_{\text{LS}}$ to the weighted least squares estimator $\hat{\boldsymbol{\beta}}$, and finally multiplying the last term of (9.2.45) by the factor $(\bar{\psi}_w + \hat{\sigma}_v^2)/(\psi_i + \hat{\sigma}_v^2)$, where $\bar{\psi}_w = \sum_i(\hat{\sigma}_v^2 + \psi_i)^{-1}\psi_i / \sum_i(\hat{\sigma}_v^2 + \psi_i)^{-1}$ and $\hat{\sigma}_v^2$ is either the REML estimator or the estimator obtained by solving (9.2.8) iteratively. Note that the adjustment factor $(\bar{\psi}_w + \hat{\sigma}_v^2)/(\psi_i + \hat{\sigma}_v^2)$ reduces to 1 in the case of equal sampling variances, $\psi_i = \psi$. Theoretical justification of this extension remains to be studied.

Example 9.2.2. Simulation Study. Jiang, Lahiri and Wan (2002) reported simulation results on the relative performance of estimators of $\text{MSE}(\hat{\theta}_i^{\text{EB}})$ under the simple model $\hat{\theta}_i|\theta_i \overset{\text{ind}}{\sim} N(\theta_i, \psi)$ and $\theta_i \overset{\text{iid}}{\sim} N(\mu, \sigma_v^2)$, where $\hat{\theta}_i^{\text{EB}}$ is given by (9.2.39), the approximation of Morris (1983a) to $\hat{\theta}_i^{\text{HB}}$. The estimators studied include (i) the Prasad-Rao estimator $\text{mse}_{\text{PR}}(\hat{\theta}_i^{\text{EB}}) = g_1(\hat{\sigma}_v^2) + g_2(\hat{\sigma}_v^2) + 2g_3(\hat{\sigma}_v^2)$, (ii) the area-specific estimator $\text{mse}_2(\hat{\theta}_i^{\text{EB}})$ which is equivalent to $\text{mse}_{\text{BL}}(\hat{\theta}_i^{\text{EB}})$ given by (9.2.21), (iii) the approximation to Laird and Louis bootstrap, $\tilde{V}_{\text{LL}}(\theta_i|\hat{\boldsymbol{\theta}})$, given by (9.2.19), (iv) the jackknife estimator $\text{mse}_J(\hat{\theta}_i^{\text{EB}})$, (v) the Morris estimator, $V_M(\theta_i|\hat{\boldsymbol{\theta}})$, given by (9.2.41), and (vi) the naive estimator $\text{mse}_N(\hat{\theta}_i^{\text{EB}}) = g_1(\hat{\sigma}_v^2) + g_2(\hat{\sigma}_v^2)$ which ignores the variability associated with $\hat{\sigma}_v^2$, where $g_1(\sigma_v^2)$ and $g_2(\sigma_v^2)$ are given by (9.2.33) and (9.2.35) respectively, and $g_3(\sigma_v^2) = 2\psi(1-\gamma)/m$. Average relative bias, $\overline{\text{RB}}$, was used as the criterion for comparison of the estimators, where $\overline{\text{RB}} = m^{-1}\sum_i \text{RB}_i$, $\text{RB}_i = [E(\text{mse}_i) - \text{MSE}_i]/\text{MSE}_i$ and mse_i denotes an estimator of MSE for the ith area and $\text{MSE}_i = \text{MSE}(\hat{\theta}_i^{\text{EB}})$. Note that $\text{mse}_{\text{PR}}, \text{mse}_{\text{BL}}, V_M(\theta_i|\hat{\boldsymbol{\theta}})$ and mse_J are nearly unbiased unlike $\tilde{V}_{\text{LL}}$.

One million samples were simulated from the model by letting $\mu = 0$ (without loss of generality), $\sigma_v^2 = \psi = 1$ and $m = 30, 60$ and 90. Values of $\overline{\text{RB}}$ calculated from the simulated samples are reported in Table 9.1. As expected, the naive estimator and, to a lesser extent, the Laird and Louis estimator $\tilde{V}_{\text{LL}}$ lead to underestimation of MSE. The remaining estimators are nearly unbiased with $\overline{\text{RB}}$ less than 1%. The performance of all the estimators improves as m increases, that is, $\overline{\text{RB}}$ decreases as m increases.

TABLE 9.1 Percent Average Relative Bias ($\overline{\text{RB}}$) of MSE Estimators

MSE Estimators	$m = 30$	$m = 60$	$m = 90$
mse_N	−8.4	−4.8	−3.2
$\tilde{V}_{\text{LL}}(\theta_i\|\hat{\boldsymbol{\theta}})$	−3.8	−2.9	−2.0
mse_J	0.6	0.2	0.1
$V_M(\theta_i\|\hat{\boldsymbol{\theta}})$	0.7	−0.1	0.0
mse_{BL}	0.1	−0.2	−0.1
mse_{PR}	0.7	−0.1	0.0

SOURCE: Adapted from Table 1 in Jiang, Lahiri and Wan (2002).

9.2.4 EB Confidence Intervals

We now turn to EB confidence intervals (CI) on the individual small area parameters θ_i; in particular, say θ_1. An EB interval on θ_1 refers to the joint distribution of $(\hat{\boldsymbol{\theta}}, \boldsymbol{\theta})$. We define a $(1-\alpha)$-level CI on θ_1 as $I_1(\alpha)$ such that

$$P_{\hat{\boldsymbol{\theta}},\boldsymbol{\theta}}[\theta_1 \in I_1(\alpha)] = 1 - \alpha \tag{9.2.46}$$

for all possible values of the model parameters $\boldsymbol{\beta}$ and σ_v^2. Normal-theory EB intervals are given by

$$I_1(\alpha) : \left[\hat{\theta}_1^{\text{EB}} - z_{\alpha/2}s(\hat{\theta}_1^{\text{EB}}), \hat{\theta}_1^{\text{EB}} + z_{\alpha/2}s(\hat{\theta}_1^{\text{EB}})\right], \tag{9.2.47}$$

where $z_{\alpha/2}$ is the upper $\alpha/2$-point of $N(0,1)$ and $s(\hat{\theta}_1^{\text{EB}})$ is an estimate of the standard error of $\hat{\theta}_1$. For example, we may use one of the MSE estimators, studied in subsections 9.2.2 and 9.2.3, as $s^2(\hat{\theta}_1^{\text{EB}})$.

For the special case of equal sampling variances, $\psi_i = \psi$, Morris (1983a) used $V_M(\theta_i|\hat{\boldsymbol{\theta}})$, given by (9.2.41), as $s^2(\hat{\theta}_1^{\text{EB}})$ and conjectured that the CI coverage is at least $1-\alpha$. Note that $V_M(\theta_i|\hat{\boldsymbol{\theta}})$ is the Morris approximation to the posterior variance $V(\theta_i|\hat{\boldsymbol{\theta}})$. Morris (1983a) used the posterior variance $V(\theta_i|\hat{\boldsymbol{\theta}})$ as $s^2(\hat{\theta}_1^{\text{EB}})$ in (9.2.47) for the simple area level model without auxiliary information, and provided some evidence in support of his conjecture. Also, $\hat{\theta}_1^{\text{HB}}$

was used instead of $\hat{\theta}_1^{\text{EB}}$. In particular, he noted that the coverage probability tends to $1-\alpha$ as the number of small areas, m, goes to infinity or as the sampling variance, ψ, tends to zero.

For the simple area level model, Datta, Ghosh, Smith and Lahiri (2002) studied some theoretical properties of the naive EB interval based on $s^2(\hat{\theta}_1^{\text{EB}}) = g_1(\hat{\sigma}_v^2) = \gamma^*\psi$ and $\hat{\theta}_1^{\text{EB}} = \gamma^*\hat{\theta}_1 + (1-\gamma^*)\hat{\theta}_.$:

$$I_{1\text{N}}(\alpha) : [\hat{\theta}_1^{\text{EB}} - z_{\alpha/2}(\gamma^*\psi)^{1/2}, \hat{\theta}_1^{\text{EB}} + z_{\alpha/2}(\gamma^*\psi)^{1/2}], \qquad (9.2.48)$$

where $1-\gamma^* = min[(m-3)\psi/S, (m-3)/(m-1)]$ and $S = \sum_i(\hat{\theta}_i - \hat{\theta}_.)^2 = (m-1)(\tilde{\sigma}_v^2 + \psi)$. Note that $1-\gamma^*$ is equivalent to (9.2.6) with $p=1$ and $\tilde{\sigma}_v^2$ changed to $\hat{\sigma}_v^2 = \max(\tilde{\sigma}_v^2, 0)$. Also, note that the naive MSE estimator $g_1(\hat{\sigma}_v^2)$ ignores the variability associated with $\hat{\mu}$ and $\hat{\sigma}_v^2$. Datta et al. (2002) proved the following asymptotic result (as $m \to \infty$):

$$P_{\hat{\boldsymbol{\theta}},\boldsymbol{\theta}}[\theta_1 \in I_{1\text{N}}(\alpha)]$$
$$\approx 1 - \alpha - \frac{z_{\alpha/2}\phi(z_{\alpha/2})}{m}\left[\frac{3\psi}{\sigma_v^2} + \frac{\left(1+z_{\alpha/2}^2\right)}{2}\left(\frac{\psi}{\sigma_v^2}\right)^2\right], \qquad (9.2.49)$$

where the neglected terms are of order $m^{-3/2}$, and $\phi(t)$ is the density function of $N(0,1)$ variable. The proof of (9.2.49) is very technical, and we refer the reader to Datta et al. (2002) for details.

It follows from (9.2.49) that the coverage probability of the naive EB interval $I_{1\text{N}}(\alpha)$ tends to the desired level $1-\alpha$ as $m \to \infty$ or as $\psi \to 0$. This result confirms Morris (1983a). However, in general, the naive EB interval leads to coverage probabilities smaller than the nominal $1-\alpha$, and the underestimation increases as ψ/σ_v^2 increases. In fact, the approximation (9.2.49) breaks down for large values of ψ/σ_v^2 (or equivalently small values of the ratio σ_v^2/ψ) even for large m. It may be noted that $\hat{\sigma}_v^2$ becomes zero with nonnegligible probability for small σ_v^2/ψ, even if m is large.

Datta et al. (2002) corrected the underestimation of the naive EB interval, $I_{1\text{N}}(\alpha)$, by changing $z_{\alpha/2}$ in (9.2.48) to

$$z_{\alpha/2}^* = z_{\alpha/2}\left[1 + \frac{1}{m}\left\{\frac{1+z_{\alpha/2}^2}{4}\left(\frac{1-\gamma^*}{\gamma^*}\right)^2 + \frac{3}{2}\left(\frac{1-\gamma^*}{\gamma^*}\right)\right\}\right], \qquad (9.2.50)$$

and proved that the resulting interval $I_{1\text{N}}^*$ is more accurate:

$$P_{\hat{\boldsymbol{\theta}},\boldsymbol{\theta}}[\theta_1 \in I_{1\text{N}}^*(\alpha)] \approx 1 - \alpha, \qquad (9.2.51)$$

where the neglected terms are of order $m^{-3/2}$. The bias-corrected interval $I_{1\text{N}}^*(\alpha)$ generally leads to coverage probabilities larger than the nominal $1-\alpha$. However, the overestimation increases as ψ/σ_v^2 increases. This follows by noting that the length of the interval $I_{1\text{N}}^*(\alpha)$ increases with $(1-\gamma^*)/\gamma^*$, which is an estimator of ψ/σ_v^2. Simulation results reported by Datta et al. (2002) confirm the above behavior of the naive and the adjusted EB intervals.

Carlin and Gelfand (1991) also proposed bias-corrected EB intervals. They used a parametric bootstrap method to implement bias correction.

Nandram (1999) obtained an EB interval for a small area mean under a simple random error variances model. This model is similar to (8.6.1) under normality of the effects v_i and the errors e_{ij}, but the assumption $v_i \overset{\text{iid}}{\sim} N(0, \sigma_v^2)$ is replaced by $v_i|\sigma_{ei}^2 \overset{\text{ind}}{\sim} N(0, \tau\sigma_{ei}^2)$. Further, the random error variances σ_{ei}^2 are assumed to be iid inverted gamma (IG) variables (or $\sigma_{ei}^{-2} \overset{\text{iid}}{\sim}$ gamma). Nandram showed that the EB interval is asymptotically correct in the sense that the coverage probability converges to the nominal $1 - \alpha$, as $m \to \infty$. Also, it is optimal in the sense that $E|\ell_{\text{EB}} - \ell_B| \to 0$ as $m \to \infty$, where ℓ_{EB} and ℓ_B are the lengths of the EB and the Bayes intervals, respectively.

Example 9.2.3. Batting Averages. We revisit Example 4.4.1, Chapter 4 on batting averages of major league baseball players to illustrate the use of normal-theory EB intervals based on $(\hat{\theta}_i^{\text{HB}}, V(\theta_i|\hat{\boldsymbol{\theta}}))$ and $(\hat{\theta}_i^{\text{EB}}, V_M(\theta_i|\hat{\boldsymbol{\theta}}))$. Morris (1983a) used the arcsine transformation on the direct estimators, $\hat{P}_i, i = 1, \ldots, 18$. In particular, $\hat{\theta}_i = 0.484 + 0.066\sqrt{45}$ arcsin $(2\hat{P}_i - 1)$, where the batting averages, $\hat{P}_i$, are based on the first 45 times at bat for each player i. The arcsine transformation stabilizes the variance, and the constants 0.484 and 0.066 are chosen so that the mean and standard deviation of $\hat{\theta}_1, \ldots, \hat{\theta}_{18}$ equals the mean, 0.266, and standard deviation, 0.066, of $\hat{P}_1, \ldots, \hat{P}_{18}$, respectively. As a result, we take $\theta_i = 0.484 + 0.066\sqrt{45}$ arcsin $(2P_i - 1)$, where P_i is the batting average during the remainder of the season for player i. We have $\hat{\theta}_i|\theta_i \overset{\text{ind}}{\sim} N(\theta_i, (0.066)^2 = \psi)$ approximately, noting that the variance of $\sqrt{45}$ arcsin $(2\hat{P}_i - 1)$ is approximately equal to 1.

Table 9.2 reports the values of $\hat{\theta}_i, \hat{\theta}_i^{\text{EB}}$ and $\hat{\theta}_i^{\text{HB}}$ (based on uniform priors for μ and σ_v^2) and the true θ_i based on P_i. The EB estimates were calculated from $\gamma^* = 0.208$ and $\hat{\theta}_. = 0.2657$, using $\hat{\theta}_i^{\text{EB}} = \gamma^*\hat{\theta}_i + (1 - \gamma^*)\hat{\theta}_.$. The associated standard errors, $s_{iM}(\hat{\theta}_i^{\text{EB}})$, calculated from (9.2.41) are also reported in Table 9.2. The HB estimates were calculated from $1 - \hat{\gamma}^{\text{HB}} = E[(1 - \gamma)|\hat{\boldsymbol{\theta}}] = 0.675$ and $\hat{\theta}_.$, using $\hat{\theta}_i^{\text{HB}} = \hat{\gamma}^{\text{HB}}\hat{\theta}_i + (1 - \hat{\gamma}^{\text{HB}})\hat{\theta}_.$. The associated standard errors, $s_i(\hat{\theta}_i^{\text{HB}})$, calculated from the posterior variance, $V(\theta_i|\hat{\boldsymbol{\theta}})$ are also reported in Table 9.2. Normal-theory intervals corresponding to $1 - \alpha = 0.95$ are also given in Table 9.2, using $[\hat{\theta}_i^{\text{EB}} - 2s_{iM}(\hat{\theta}_i^{\text{EB}}), \hat{\theta}_i^{\text{EB}} + 2s_{iM}(\hat{\theta}_i^{\text{EB}})]$ and $[\hat{\theta}_i^{\text{HB}} - 2s_i(\hat{\theta}_i^{\text{HB}}), \hat{\theta}_i^{\text{HB}} + 2s_i(\hat{\theta}_i^{\text{HB}})]$.

It is clear from Table 9.2 that the standard errors of EB and HB estimates are much smaller than $\sqrt{\psi} = 0.066$, the standard error of the direct estimates. As a result, the associated intervals are narrower than the intervals $[\hat{\theta}_i - 2\sqrt{\psi}, \hat{\theta}_i + 2\sqrt{\psi}]$ based on the direct estimates. We also note from Table 9.2 that $s_{iM}(\hat{\theta}_i^{\text{EB}})$ is a good approximation to $s_i(\hat{\theta}_i^{\text{HB}})$. The s_{iM}-values reported in Ghosh (1992a) are incorrect and, in fact, consistently larger than $\sqrt{\psi} = 0.066$.

Morris (1983a) also conducted a simulation study to determine the coverage probability associated with the interval $\hat{\theta}_i^{\text{HB}} \pm 2s_i(\hat{\theta}_i^{\text{HB}})$. He used the true θ_i reported in Table 9.2 and generated $\hat{\theta}_i$'s from $N(\theta_i, (0.066)^2), i = 1, \ldots, 18$. Based on 100 simulation runs, the average coverage probability was 0.97 compared to nominal 0.95. The average length of the EB intervals was 35% shorter than the intervals based on the direct estimates. The average coverage probability was computed by averaging the player-specific coverage probabilities, CP_i, over i. Similarly, the average lengths were calculated. Note that the average coverage probability and the average length refer to the conditional distribution of $\hat{\boldsymbol{\theta}}$ given $\boldsymbol{\theta}$.

TABLE 9.2 True θ_i, Direct Estimates $\hat{\theta}_i$, EB and HB Estimates and Associated Standard Errors and Normal Theory 95% Confidence Intervals: Batting Averages

Player i	θ_i	$\hat{\theta}_i$	$\hat{\theta}_i^{\text{EB}}$	$s_{iM}(\hat{\theta}_i^{\text{EB}})$	$\left[\hat{\theta}_i^{\text{EB}} \pm 2s_{iM}(\hat{\theta}_i^{\text{EB}})\right]$	$\hat{\theta}_i^{\text{HB}}$	$s_i(\hat{\theta}_i^{\text{HB}})$	$\left[\hat{\theta}_i^{\text{HB}} \pm 2s_i(\hat{\theta}_i^{\text{HB}})\right]$
1	0.346	0.395	0.293	0.050	[0.193, 0.393]	0.308	0.046	[0.216, 0.400]
2	0.300	0.375	0.288	0.046	[0.196, 0.380]	0.301	0.044	[0.213, 0.389]
3	0.279	0.355	0.284	0.042	[0.200, 0.368]	0.295	0.043	[0.209, 0.381]
4	0.223	0.334	0.280	0.039	[0.203, 0.357]	0.288	0.042	[0.204, 0.372]
5	0.276	0.313	0.275	0.036	[0.203, 0.347]	0.281	0.041	[0.199, 0.363]
6	0.273	0.313	0.275	0.036	[0.203, 0.347]	0.281	0.041	[0.199, 0.363]
7	0.266	0.291	0.271	0.034	[0.203, 0.339]	0.274	0.040	[0.194, 0.354]
8	0.211	0.269	0.266	0.033	[0.200, 0.332]	0.267	0.040	[0.187, 0.347]
9	0.271	0.247	0.262	0.034	[0.195, 0.329]	0.260	0.040	[0.180, 0.340]
10	0.232	0.247	0.262	0.034	[0.195, 0.329]	0.260	0.040	[0.180, 0.340]
11	0.266	0.224	0.257	0.035	[0.186, 0.328]	0.252	0.040	[0.172, 0.332]
12	0.258	0.224	0.257	0.035	[0.186, 0.328]	0.252	0.040	[0.172, 0.332]
13	0.306	0.224	0.257	0.035	[0.186, 0.328]	0.252	0.040	[0.172, 0.332]
14	0.267	0.224	0.257	0.035	[0.186, 0.328]	0.252	0.040	[0.172, 0.332]
15	0.228	0.224	0.257	0.035	[0.186, 0.328]	0.252	0.040	[0.172, 0.332]
16	0.288	0.200	0.252	0.038	[0.176, 0.328]	0.244	0.041	[0.162, 0.326]
17	0.318	0.175	0.247	0.042	[0.162, 0.331]	0.236	0.043	[0.150, 0.322]
18	0.200	0.148	0.241	0.047	[0.146, 0.336]	0.227	0.045	[0.137, 0.317]

SOURCE: Adapted from Table 2 in Morris (1983a).

9.3 Linear Mixed Models

9.3.1 EB Estimation

Assuming normality, the linear mixed model (6.3.1) with block diagonal covariance structure may be expressed as

$$\mathbf{y}_i | \mathbf{v}_i \overset{\text{ind}}{\sim} N(\mathbf{X}_i\boldsymbol{\beta} + \mathbf{Z}_i\mathbf{v}_i, \mathbf{R}_i) \qquad (9.3.1)$$

$$\mathbf{v}_i \overset{\text{ind}}{\sim} N(\mathbf{0}, \mathbf{G}_i); \quad i = 1, \ldots, m,$$

where $\mathbf{G}_i$ and $\mathbf{R}_i$ depend on variance parameters $\boldsymbol{\delta}$. The Bayes or best predictor (BP) estimator of realized $\mu_i = \mathbf{l}_i^T\boldsymbol{\beta} + \mathbf{m}_i^T\mathbf{v}_i$ is given by the conditional expectation of μ_i given $\mathbf{y}_i, \boldsymbol{\beta}$ and $\boldsymbol{\delta}$:

$$\hat{\mu}_i^B = \hat{\mu}_i^B(\boldsymbol{\beta}, \boldsymbol{\delta}) = E(\mu_i|\mathbf{y}_i, \boldsymbol{\beta}, \boldsymbol{\delta}) = \mathbf{l}_i^T\boldsymbol{\beta} + \mathbf{m}_i^T\hat{\mathbf{v}}_i^B, \tag{9.3.2}$$

where

$$\hat{\mathbf{v}}_i^B = E(\mathbf{v}_i|\mathbf{y}_i, \boldsymbol{\beta}, \boldsymbol{\delta}) = \mathbf{G}_i\mathbf{Z}_i^T\mathbf{V}_i^{-1}(\mathbf{y}_i - \mathbf{X}_i\boldsymbol{\beta}) \tag{9.3.3}$$

and $\mathbf{V}_i = \mathbf{R}_i + \mathbf{Z}_i\mathbf{G}_i\mathbf{Z}_i^T$. The result (9.3.2) follows from the posterior (or conditional) distribution of μ_i given $\mathbf{y}_i$:

$$\mu_i|\mathbf{y}_i, \boldsymbol{\beta}, \boldsymbol{\delta} \overset{\text{ind}}{\sim} N(\hat{\mu}_i^B, g_{1i}(\boldsymbol{\delta})), \tag{9.3.4}$$

where $g_{1i}(\boldsymbol{\delta})$ is given by (6.3.6).

The estimator $\hat{\mu}_i^B$ depends on the model parameters $\boldsymbol{\beta}$ and $\boldsymbol{\delta}$ which are estimated from the marginal distribution

$$\mathbf{y}_i \overset{\text{ind}}{\sim} N(\mathbf{X}_i\boldsymbol{\beta}, \mathbf{V}_i), i = 1, \ldots, m \tag{9.3.5}$$

using ML or REML. Denoting the estimators as $\hat{\boldsymbol{\beta}}$ and $\hat{\boldsymbol{\delta}}$, we obtain the EB or the empirical BP (EB) estimator of μ_i from $\hat{\mu}_i^B$ by substituting $\hat{\boldsymbol{\beta}}$ for $\boldsymbol{\beta}$ and $\hat{\boldsymbol{\delta}}$ for $\boldsymbol{\delta}$:

$$\hat{\mu}_i^{\text{EB}} = \hat{\mu}_i^{\text{EB}}(\hat{\boldsymbol{\beta}}, \hat{\boldsymbol{\delta}}) = \mathbf{l}_i^T\hat{\boldsymbol{\beta}} + \mathbf{m}_i^T\hat{\mathbf{v}}_i^B(\hat{\boldsymbol{\beta}}, \hat{\boldsymbol{\delta}}). \tag{9.3.6}$$

The EB estimator $\hat{\mu}_i^{\text{EB}}$ is identical to the EBLUP estimator (6.3.8). Note that $\hat{\mu}_i^{\text{EB}}$ is also the mean of the estimated posterior density, $f(\mu_i|\mathbf{y}_i, \hat{\boldsymbol{\beta}}, \hat{\boldsymbol{\delta}})$, of μ_i, namely, $N(\hat{\mu}_i^{\text{EB}}, g_{1i}(\hat{\boldsymbol{\delta}})$.

9.3.2 MSE Estimation

The results in subsection 6.3.2 on the estimation of the MSE of the EBLUP estimator $\hat{\mu}_i^H$ are applicable to the EB estimator $\hat{\mu}_i^{\text{EB}}$ because $\hat{\mu}_i^H$ and $\hat{\mu}_i^{\text{EB}}$ are identical under normality. Also, the area-specific estimators (6.3.15)–(6.3.18) may be used as estimators of the conditional MSE, $\text{MSE}_c(\hat{\mu}_i^{\text{EB}}) = E[(\hat{\mu}_i^{\text{EB}} - \mu_i)^2|\mathbf{y}_i]$, where the expectation is conditional on the observed $\mathbf{y}_i$ for the ith area. As noted in subsection 6.3.2, the MSE estimators are nearly unbiased in the sense of bias lower than m^{-1}, for large m.

The jackknife MSE estimator, (9.2.12), for the basic area level model, readily extends to the linear mixed model with block diagonal covariance structure. We calculate the delete-l estimators $\hat{\boldsymbol{\beta}}_{-l}$ and $\hat{\boldsymbol{\delta}}_{-l}$ by deleting the lth area data set $(\mathbf{y}_l, \mathbf{X}_l, \mathbf{Z}_l)$ from the full data set. This calculation is done for each l to get m estimators $\{(\hat{\boldsymbol{\beta}}_{-l}, \hat{\boldsymbol{\delta}}_{-l}), l = 1, \ldots, m\}$ which, in turn, provide m estimators of μ_i: $\{\hat{\mu}_{-l}^{\text{EB}}, l = 1, \ldots, m\}$, where $\hat{\mu}_{-l}^{\text{EB}}$ is obtained from $\hat{\mu}_i^B = k_i(\mathbf{y}_i, \boldsymbol{\beta}, \boldsymbol{\delta})$ by changing $\boldsymbol{\beta}$ and $\boldsymbol{\delta}$ to $\hat{\boldsymbol{\beta}}_{-l}$ and $\hat{\boldsymbol{\delta}}_{-l}$, respectively. The jackknife estimator is given by

$$\text{mse}_J\left(\hat{\mu}_i^B\right) = \hat{M}_{1i} + \hat{M}_{2i} \tag{9.3.7}$$

with

$$\hat{M}_{1i} = g_{1i}(\hat{\boldsymbol{\delta}}) - \frac{m-1}{m}\sum_{l=1}^{m}\left[g_{1i}(\hat{\boldsymbol{\delta}}_{-l}) - g_{1i}(\hat{\boldsymbol{\delta}})\right] \tag{9.3.8}$$

and

$$\hat{M}_{2i} = \frac{m-1}{m}\sum_{l=1}^{m}\left(\hat{\mu}_{i,-l}^{\text{EB}} - \hat{\mu}_i^{\text{EB}}\right)^2, \tag{9.3.9}$$

where $g_{1i}(\boldsymbol{\delta})$ is given by (6.3.6). The jackknife estimator of MSE, $\text{mse}_J(\hat{\mu}_i^B)$ is also nearly unbiased in the sense of bias of order m^{-1}, for large m (Jiang, Lahiri and Wan (2001)).

9.3.3 Approximations to the Posterior Variance

As noted in subsection 9.2.3, the naive EB approach uses the estimated posterior density, $f(\mu_i|\mathbf{y}_i, \hat{\boldsymbol{\beta}}, \hat{\boldsymbol{\delta}})$ to make inference on μ_i. In practice, μ_i is estimated by $\hat{\mu}_i^{\text{EB}}$ and the estimated posterior variance $V(\mu_i|\mathbf{y}_i, \hat{\boldsymbol{\beta}}, \hat{\boldsymbol{\delta}}) = g_{1i}(\hat{\boldsymbol{\delta}})$ is used as a measure of variability. But the naive measure, $g_{1i}(\hat{\boldsymbol{\delta}})$ can lead to severe underestimation of the true posterior variance $V(\mu_i|\mathbf{y})$ as well as of $\text{MSE}(\hat{\mu}_i^{\text{EB}})$.

The bootstrap method of Laird and Louis (1987) readily extends to the linear mixed model. Bootstrap data $\{\mathbf{y}_i^*(b), \mathbf{X}_i, \mathbf{Z}_i; i = 1, \ldots, m\}$ are first generated from the estimated marginal distribution, $N(\mathbf{X}_i\hat{\boldsymbol{\beta}}, \hat{\mathbf{V}}_i)$, for $b = 1, \ldots, B$. Estimates $\{\hat{\boldsymbol{\beta}}(b), \hat{\boldsymbol{\delta}}(b)\}$ and the EB estimates $\hat{\mu}_i^{*\text{EB}}(b)$ are then computed from the bootstrap data. The bootstrap approximation to the posterior variance is given by

$$V_{\text{LL}}(\mu_i|\mathbf{y}) = \frac{1}{B}\sum_{b=1}^{B} g_{1i}(\hat{\boldsymbol{\delta}}(b)) + \frac{1}{B}\sum_{b=1}^{B}\left[\hat{\mu}_i^{*\text{EB}}(b) - \hat{\mu}_i^{*\text{EB}}(\cdot)\right]^2, \tag{9.3.10}$$

where $\hat{\mu}_i^{*\text{EB}}(\cdot) = B^{-1}\sum_b \hat{\mu}_i^{*\text{EB}}(b)$.

Butar and Lahiri (2001) showed that $V_{\text{LL}}(\mu_i|\mathbf{y})$ is not approximately unbiased for $\text{MSE}(\hat{\mu}_i^{\text{EB}})$, by deriving an approximation, $\tilde{V}_{\text{LL}}(\mu_i|\mathbf{y})$, as $B \to \infty$. They proposed a bias-corrected MSE estimator

$$\text{mse}_{\text{BL}}(\hat{\mu}_i^{\text{EB}}) = \tilde{V}_{\text{LL}}(\mu_i|\mathbf{y}) + g_{3i}(\hat{\boldsymbol{\delta}}), \tag{9.3.11}$$

where $g_{3i}(\boldsymbol{\delta})$ is given by (6.3.10) and

$$\tilde{V}_{\text{LL}}(\mu_i|\mathbf{y}) = g_{1i}(\hat{\boldsymbol{\delta}}) + g_{2i}(\hat{\boldsymbol{\delta}}) + g_{3i}^*(\hat{\boldsymbol{\delta}}, \mathbf{y}_i), \tag{9.3.12}$$

where $g_{2i}(\boldsymbol{\delta})$ and $g_{3i}^*(\boldsymbol{\delta}, \mathbf{y}_i)$ are given by (6.3.7) and (6.3.14), respectively. It now follows from (9.3.11) and (9.3.12) that $\text{mse}_{\text{BL}}(\hat{\mu}_i^{\text{EB}})$ is identical to the area-specific MSE estimator given by (6.3.16).

Kass and Steffey (1989) obtained a first-order approximation to $V(\mu_i|\mathbf{y})$. After simplification, the first-order approximation, $V_{\text{KS}}(\mu_i|\mathbf{y})$, reduces to $\tilde{V}_{\text{LL}}(\mu_i|\mathbf{y})$ given by (9.3.12). Therefore, $V_{\text{KS}}(\mu_i|\mathbf{y})$ is also not nearly unbiased for $\text{MSE}(\hat{\mu}_i^{\text{EB}})$. A second-order approximation to $V(\mu_i|\mathbf{y})$ ensures that the neglected terms are of lower order than m^{-1}, but it depends on the prior density, $f(\boldsymbol{\beta}, \boldsymbol{\delta})$, unlike the first-order approximation.

9.4 Binary Data

In this section, we study unit level models for binary responses, y_{ij}, that is, $y_{ij} = 1$ or 0. In this case, linear mixed models are not suitable and alternative models have been proposed. If all the covariates $\mathbf{x}_{ij}$ associated with y_{ij} are area-specific, that is, $\mathbf{x}_{ij} = \mathbf{x}_i$, then we can transform the sample area proportions, $\hat{p}_i = \sum_j y_{ij}/n_i = y_i/n_i$, using an arcsine transformation, as in Example 9.2.1, and reduce the model to an area level model. However, the resulting transformed estimators, $\hat{\theta}_i$, may not satisfy the sampling model with zero mean sampling errors if n_i is small. Also, the estimator of the proportion, p_i, obtained from $\hat{\theta}_i^{\text{EB}}$ is not equal to the true EB estimator $\hat{p}_i^{\text{EB}}$.

Unit level modelling avoids the above difficulties, and EB estimators of proportions may be obtained directly for the general case of unit-specific covariates. In subsection 9.4.1, we study the special case of no covariates. A generalized linear mixed model is used in subsection 9.4.2 to handle covariates.

9.4.1 Case of no Covariates

We assume a two-stage model on the sample observations $y_{ij}, j = 1, \ldots, n_i; i = 1, \ldots, m$. In the first stage, we assume that $y_{ij}|p_i \overset{\text{iid}}{\sim} \text{Bernoulli}(p_i), i = 1, \ldots, m$. A model linking the p_i's is assumed in the second-stage; in particular, $p_i \overset{\text{iid}}{\sim} \text{beta}(\alpha, \beta); \alpha > 0, \beta > 0$, where $\text{beta}(\alpha, \beta)$ denotes the beta distribution with parameters α and β:

$$f(p_i|\alpha, \beta) = \frac{\Gamma(\alpha+\beta)}{\Gamma(\alpha)\Gamma(\beta)} p_i^{\alpha-1}(1-p_i)^{\beta-1}; \alpha > 0, \beta > 0 \tag{9.4.1}$$

where $\Gamma(\cdot)$ is the gamma function. We reduce $\mathbf{y}_i = (y_{i1}, \ldots, y_{in_i})^T$ to the sample total $y_i = \sum_j y_{ij}$, noting that y_i is a minimal sufficient statistic for the first-stage model.

We note that $y_i|p_i \overset{\text{ind}}{\sim}$ Binomial (n_i, p_i), that is,

$$f(y_i|p_i) = \binom{n_i}{y_i} p_i^{y_i}(1-p_i)^{n_i-y_i}. \tag{9.4.2}$$

It follows from (9.4.1) and (9.4.2) that $p_i|y_i,\alpha,\beta \overset{\text{ind}}{\sim} \text{beta}(y_i+\alpha, n_i-y_i+\beta)$. Therefore, the Bayes estimator of p_i and the posterior variance of p_i are given by

$$\hat{p}_i^B(\alpha,\beta) = E(p_i|y_i,\alpha,\beta) = (y_i+\alpha)/(n_i+\alpha+\beta) \tag{9.4.3}$$

and

$$V(p_i|y_i,\alpha,\beta) = \frac{(y_i+\alpha)(n_i-y_i+\beta)}{(n_i+\alpha+\beta+1)(n_i+\alpha+\beta)^2}. \tag{9.4.4}$$

Note that the linking distribution, $f(p_i|\alpha,\beta)$, is a "conjugate prior" in the sense that the posterior (or conditional) distribution, $f(p_i|y_i,\alpha,\beta)$, has the same form as the prior distribution.

We obtain estimators of the model parameters from the marginal distribution: $y_i|\alpha,\beta \overset{\text{ind}}{\sim}$ Beta-binomial given by

$$f(y_i|\alpha,\beta) = \binom{n_i}{y_i}\frac{\Gamma(\alpha+y_i)\Gamma(\beta+n_i-y_i)}{\Gamma(\alpha+\beta+n_i)}\frac{\Gamma(\alpha+\beta)}{\Gamma(\alpha)\Gamma(\beta)}. \tag{9.4.5}$$

Maximum likelihood (ML) estimators, $\hat{\alpha}_{\text{ML}}$ and $\hat{\beta}_{\text{ML}}$, may be obtained by maximizing the log likelihood:

$$l(\alpha,\beta) = \text{const} + \sum_{i=1}^m\left[\sum_{h=0}^{y_i-1}\log(\alpha+h) + \sum_{h=0}^{n_i-y_i-1}\log(\beta+h\right.$$
$$\left. - \sum_{h=0}^{n_i-1}\log(\alpha+\beta+h)\right], \tag{9.4.6}$$

where $\sum_{h=0}^{y_i-1}\log(\alpha+h)$ is taken as zero if $y_i=0$ and $\sum_{h=0}^{n_i-y_i-1}\log(\beta+h)$ is taken as zero if $y_i=n_i$. A convenient representation is in terms of the mean $E(y_{ij})=\mu=\alpha/(\alpha+\beta)$ and $\tau=1/(\alpha+\beta)$ which is related to the intraclass correlation $\rho=\text{Corr}(y_{ij},y_{ik})=1/(\alpha+\beta+1)$ for $j\neq k$. Using μ and τ, (9.4.6) takes the form

$$l(\mu,\tau) = \text{const} + \sum_{i=1}^m\left[\sum_{h=0}^{y_i-1}\log(\mu+h\tau) + \sum_{h=0}^{n_i-y_i-1}\log(1-\mu+h\tau)\right.$$
$$\left. - \sum_{h=0}^{n_i-1}\log(1+h\tau)\right]. \tag{9.4.7}$$

Closed-form expressions for $\hat{\alpha}_{\text{ML}}$ and $\hat{\beta}_{\text{ML}}$ (or $\hat{\mu}_{\text{ML}}$ and $\hat{\tau}_{\text{ML}}$) do not exist, but the ML estimates may be obtained by the Newton-Raphson method or some other iterative method. McCulloch and Searle (2001, Section 2.6) have given the asymptotic covariance matrices of $(\hat{\alpha}_{\text{ML}},\hat{\beta}_{\text{ML}})$ and $(\hat{\mu}_{\text{ML}},\hat{\tau}_{\text{ML}})$. Substituting $\hat{\alpha}_{\text{ML}}$ and $\hat{\beta}_{\text{ML}}$ in (9.4.3) and (9.4.4) we get the EB estimator of p_i and the estimated posterior variance.

We can also use simple method-of-moments estimators of α and β. We equate the weighted sample mean $\hat{p} = \sum_i (n_i/n_T)\hat{p}_i$ and the weighted sample variance $s_p^2 = \sum_i (n_i/n_T)(\hat{p}_i - \hat{p})^2$ to their expected values and solve the resulting moment equations for α and β, where $n_T = \sum_i n_i$. This leads to moment estimators, $\hat{\alpha}$ and $\hat{\beta}$, given by

$$\hat{\alpha}/(\hat{\alpha}+\hat{\beta}) = \hat{p} \tag{9.4.8}$$

and

$$\frac{1}{\hat{\alpha}+\hat{\beta}+1} = \frac{n_T s_p^2 - \hat{p}(1-\hat{p})(m-1)}{\hat{p}(1-\hat{p})[n_T - \sum_i n_i^2/n_T - (m-1)]}; \tag{9.4.9}$$

see Kleinman (1973). Note that the moment estimators are not unique, unlike the ML estimators.

We substitute the moment estimators $\hat{\alpha}$ and $\hat{\beta}$ into (9.4.3) to get an EB estimator of p_i as

$$\hat{p}_i^{\text{EB}} = \hat{p}_i^B(\hat{\alpha},\hat{\beta}) = \hat{\gamma}_i\hat{p}_i + (1-\hat{\gamma}_i)\hat{p}, \tag{9.4.10}$$

where $\hat{\gamma}_i = n_i/(n_i+\hat{\alpha}+\hat{\beta})$. Note that $\hat{p}_i^{\text{EB}}$ is a weighted average of the direct estimator $\hat{p}_i$ and the synthetic estimator $\hat{p}$, and more weight is given to $\hat{p}_i$ as the ith area sample size, n_i, increases. It is therefore similar to the Fay-Herriot estimator for the basic area level model, but the weight $\hat{\gamma}_i$ avoids the assumption of known sampling variance of $\hat{p}_i$. The estimator $\hat{p}_i^{\text{EB}}$ is nearly unbiased for p_i in the sense that its bias, $E(\hat{p}_i^{\text{EB}} - p_i)$, is of order m^{-1}, for large m.

A naive EB approach uses $\hat{p}_i^{\text{EB}}$ as the estimator of realized p_i, and its variability is measured by the estimated posterior variance $V(p_i|y_i,\hat{\alpha},\hat{\beta}) = g_{1i}(\hat{\alpha},\hat{\beta},y_i)$, say. However, the estimated posterior variance, $g_{1i}(\hat{\alpha},\hat{\beta},y_i)$, can lead to severe underestimation of MSE($\hat{p}_i^{\text{EB}}$) because it ignores the variability associated with $\hat{\alpha}$ and $\hat{\beta}$. Note that MSE($\hat{p}_i^{\text{EB}}$) $= E[g_{1i}(\alpha,\beta,y_i)] + E(\hat{p}_i^{\text{EB}} - \hat{p}_i^B)^2 = M_{1i} + M_{2i}$, say.

The jackknife method is readily applicable for estimating MSE($\hat{p}_i^{\text{EB}}$). We have $\hat{p}_i^{\text{EB}} = k_i(\hat{p}_i,\hat{\alpha},\hat{\beta}), \hat{p}_{i,-l}^{\text{EB}} = k_i(\hat{p}_i,\hat{\alpha}_{-l},\hat{\beta}_{-l})$ and

$$\hat{M}_{2i} = \frac{m-1}{m}\sum_{l=1}^{m}\left(\hat{p}_{i,-l}^{\text{EB}} - \hat{p}_i^{\text{EB}}\right)^2, \tag{9.4.11}$$

where $\hat{\alpha}_{-l}$ and $\hat{\beta}_{-l}$ are the delete$-l$ moment estimators obtained from $\{(\hat{p}_i,n_i), i \neq l = 1,\ldots,m\}$. Further,

$$\hat{M}_{1i} = g_{1i}(\hat{\alpha},\hat{\beta},y_i) - \frac{m-1}{m}\sum_{l=1}^{m}[g_{1i}(\hat{\alpha}_{-l},\hat{\beta}_{-l},y_i) - g_{1i}(\hat{\alpha},\hat{\beta},y_i)]. \tag{9.4.12}$$

The jackknife estimator is given by

$$\text{mse}_J(\hat{p}_i^{\text{EB}}) = \hat{M}_{1i} + \hat{M}_{2i}. \tag{9.4.13}$$

It is approximately unbiased for $\text{MSE}(\hat{p}_i^{\text{EB}})$ in the sense of bias of lower order than m^{-1}, for large m. Note also that the leading term $g_{1i}(\hat{\alpha}, \hat{\beta}, y_i)$ of $\text{mse}_J(\hat{p}_i^{\text{EB}})$ is area-specific in the sense that it depends on y_i. Our method of estimating M_{1i} differs from the unconditional method used by Jiang, Lahiri and Wan (2001). They evaluated $E[g_{1i}(\alpha, \beta, y_i)] = \tilde{g}_{1i}(\alpha, \beta)$ using the marginal distribution of y_i, and then used it in (9.4.12) to get an estimator of M_{1i}. This estimator is computationally more complex than $\hat{M}_{1i}$ and its leading term, $\tilde{g}_{1i}(\hat{\alpha}, \hat{\beta})$, is not area-specific, unlike $g_{1i}(\hat{\alpha}, \hat{\beta}, y_i)$.

Kass and Steffey's (1989) first-order approximation to the true posterior variance (assuming random α and β) requires the use of ML estimators and the observed information matrix (see Searle, Casella and McCulloch (1992), Chapter 9, p. 347). The Laird and Louis (1987) parametric bootstrap is implemented by drawing repeated bootstrap samples from the estimated marginal distribution $\{f(y_i|\hat{\alpha}, \hat{\beta}); i = 1, \ldots, m\}$. But it appears that both estimators are not approximately unbiased for $\text{MSE}(\hat{p}_i^{\text{EB}})$, as in the case of linear mixed models.

Alternative two-stage models have also been proposed. The first-stage model is not changed, but the second-stage model is changed to either (i) $\text{logit}(p_i) = \log[p_i/(1-p_i)] \overset{\text{iid}}{\sim} N(\mu, \sigma^2)$ or (ii) $\Phi^{-1}(p_i) \overset{\text{iid}}{\sim} N(\mu, \sigma^2)$, where $\Phi(\cdot)$ is the cumulative distribution function (CDF) of a $N(0,1)$ variable. The models (i) and (ii) are called logit-normal and probit-normal models, respectively. Implementation of EB is more complicated for the alternative models because no closed-form expressions for the Bayes estimator and the posterior variance of p_i exist.

For the logit-normal model, the Bayes estimator of p_i can be expressed as a ratio of single-dimensional integrals. Writing p_i as $p_i = h_1(\mu + \sigma z_i)$, where $h_1(a) = e^a/(1+e^a)$ and $z_i \sim N(0,1)$, we obtain $\hat{p}_i^B(\mu, \sigma) = E(p_i|y_i, \mu, \sigma)$ from the conditional distribution of z_i given y_i. We have

$$\hat{p}_i^B(\mu, \sigma) = A(y_i, \mu, \sigma)/B(y_i, \mu, \sigma), \tag{9.4.14}$$

where

$$A(y_i, \mu, \sigma) = E[h_1(\mu + \sigma z)\exp\{h_2(y_i, \mu + \sigma z)\}] \tag{9.4.15}$$

and

$$B(y_i, \mu, \sigma) = E[\exp\{h_2(y_i, \mu + \sigma z)\}], \tag{9.4.16}$$

where $h_2(y_i, a) = a y_i - n_i \log(1 + e^a)$ and the expectation is over $z \sim N(0,1)$; see McCulloch and Searle (2001, p. 67). We can evaluate (9.4.15) and (9.4.16) by simulating samples from $N(0,1)$. Alternatively, numerical integration, as sketched below, can be used.

The log likelihood, $l(\mu, \sigma)$, for the logit-normal model may be written as

$$l(\mu, \sigma) = \text{const} + \sum_{i=1}^{m} \log[B(y_i, \mu, \sigma)], \tag{9.4.17}$$

where $B(y_i, \mu, \sigma)$ is given by (9.4.16). Each of the single-dimensional integrals in (9.4.15)–(9.4.17) is of the form $E[a(z)]$ which may be approximated by a finite sum of the form $\sum_{k=1}^{d} b_k a(z_k)/\sqrt{\pi}$, using Gauss-Hermite quadrature (McCulloch and Searle (2001), Section 10.3). The weights b_k and the evaluation points z_k for a specified value of d can be calculated using mathematical software. For ML estimation of μ and σ, $d = 20$ usually provides good approximations. Derivatives of $l(\mu, \sigma)$, needed in Newton-Raphson type methods for calculating ML estimates, can be approximated in a similar manner.

Using ML estimators $\hat{\mu}$ and $\hat{\sigma}$, we obtain an EB estimator of p_i as $\hat{p}_i^{\mathrm{EB}} = \hat{p}_i^{\mathrm{B}}(\hat{\mu}, \hat{\sigma})$. The posterior variance, $V(p_i|y_i, \mu, \sigma)$, may also be expressed in terms of expectation over $z \sim N(0,1)$, noting that $V(p_i|y_i, \mu, \sigma) = E(p_i^2|y_i, \mu, \sigma) - [\hat{p}_i^B(\mu, \sigma)]^2$. Denoting $V(p_i|y_i, \mu, \sigma) = g_{1i}(\mu, \sigma, y_i)$, the jackknife method can be applied to obtain a nearly unbiased estimator of $\mathrm{MSE}(\hat{p}_i^{\mathrm{EB}})$. We obtain the jackknife estimator, $\mathrm{mse}_J(\hat{p}_i^{\mathrm{EB}})$, from (9.4.13) by substituting $\hat{p}_i^{\mathrm{EB}} = k_i(y_i, \hat{\mu}, \hat{\sigma})$ and $\hat{p}_{i,-l}^{\mathrm{EB}} = k_i(y_i, \hat{\mu}_{-l}, \hat{\sigma}_{-l})$ in (9.4.11) and $g_{1i}(\hat{\mu}, \hat{\sigma}, y_i)$ and $g_{1i}(\hat{\mu}_{-l}, \hat{\sigma}_{-l}, y_i)$ in (9.4.12), where $\hat{\mu}_{-l}$ and $\hat{\sigma}_{-l}$ are the delete$-l$ ML estimators obtained from $\{(y_i, n_i), i \neq l = 1, \ldots, m\}$.

Computation of $\mathrm{mse}_J(\hat{p}_i^{\mathrm{EB}})$ using ML estimators is very cumbersome. However, computations may be simplified by using moment estimators of μ and σ obtained by equating the weighted mean $\hat{p}$ and the weighted variance s_p^2 to their expected values, as in the case of beta-binomial model, and solving the resulting equations for μ and σ. The expected values involve the marginal moments $E(y_{ij}) = E(p_i)$ and $E(y_{ij}y_{ik}) = E(p_i^2)$, $j \neq k$, which can be calculated by numerical or Monte Carlo integration, using $E(p_i) = E[h_1(\mu + \sigma z)]$ and $E(p_i^2) = E[h_1^2(\mu + \sigma z)]$. Jiang (1998) equated $\sum_i y_i = n_T \hat{p}$ and $\sum_i (\sum_{j \neq k} y_{ij} y_{ik}) = \sum_i (y_i^2 - y_i) = \sum_i n_i^2 \hat{p}_i^2 - n_T \hat{p}$ to their expected values to obtain different moment estimators, and established asymptotic consistency as $m \to \infty$. We solve

$$\sum_i y_i = n_T E[h_1(\mu + \sigma z)]; \quad \sum_i (y_i^2 - y_i) = \left[\sum_i n_i(n_i - 1)\right] E[h_1^2(\mu + \sigma z)] \tag{9.4.18}$$

for μ and σ to get the moment estimators. Jiang and Zhang (2001) proposed more efficient moment estimators, using a two-step procedure. In the first step, (9.4.18) is solved for μ and σ, and then used in the second step to produce two "optimal" weighted combinations of $(\sum_i y_i, y_1^2 - y_1, \ldots, y_m^2 - y_m)$. Note that the first step uses $(1, 0, \ldots, 0)$ and $(0, 1, \ldots, 1)$ as the weights. The optimal weights involve the derivatives of $E[h_1(\mu + \sigma z)]$ and $E[h_1^2(\mu + \sigma z)]$ with respect to μ and σ as well as the covariance matrix of $(\sum_i y_i, y_1^2 - y_1, \ldots, y_m^2 - y_m)$, which depend on μ and σ. Replacing μ and σ by the first-step estimators, estimated weighted combinations are obtained. The second-step moment estimators are then obtained by equating the estimated weighted combinations to their expectations, treating the estimated weights as fixed. Simulation studies indicated considerable overall gain in efficiency over the first-step estimators in the case of unequal n_i's. Note that the optimal weights reduce to first-step weights in the balanced case, $n_i = n$. Jiang and Zhang (2001) established the asymptotic consistency of the second-step moment estimators.

Jiang and Lahiri (2001) used $\hat{p}_i^{\text{EB}}$ based on the first-step moment estimators, and obtained a Taylor expansion estimator of MSE($\hat{p}_i^{\text{EB}}$), similar to the second-order MSE estimators of Chapter 7 for the linear mixed model. This MSE estimator is nearly unbiased, as in the case of the jackknife MSE estimator.

9.4.2 Models with Covariates

The logit-normal model of subsection 9.4.1 readily extends to the case of covariates. In the first stage, we assume that $y_{ij}|p_{ij} \stackrel{\text{ind}}{\sim} \text{Bernoulli}(p_{ij})$ for $j = 1, \ldots, N_i; i = 1, \ldots, m$. The p_{ij}'s are linked in the second stage by assuming a logistic regression model with random area effects: logit $(p_{ij}) = \mathbf{x}_{ij}^T\boldsymbol{\beta} + v_i$, where $v_i \stackrel{\text{iid}}{\sim} N(0, \sigma_v^2)$ and $\mathbf{x}_{ij}$ is the vector of fixed covariates. The two-stage model is called a logistic linear mixed model. It belongs to the class of generalized linear mixed models.

The small area parameters are the small area proportions $P_i = \bar{Y}_i = \sum_j y_{ij}/N_i$. As in the case of the basic unit level model, we assume that the model holds for the sample $\{(y_{ij}, \mathbf{x}_{ij}), j \in s_i; i = 1, \ldots, m\}$, where s_i is the sample of size n_i from the ith area.

We express P_i as $P_i = f_i\bar{y}_i + (1 - f_i)\bar{y}_i^*$, where $f_i = n_i/N_i$, $\bar{y}_i$ is the sample mean (proportion) and $\bar{y}_i^* = \sum_{l\in\bar{s}_i} y_{il}/(N_i - n_i)$ is the mean of the nonsampled units $\bar{s}_i$ in the ith area. Now, noting that $E(y_{il}|p_{il}, \mathbf{y}_i, \boldsymbol{\beta}, \sigma_v) = p_{il}$ for $l \in \bar{s}_i$, the Bayes estimator of $\bar{y}_i^*$ is given by $\hat{p}_{i(c)}^B = E(p_{i(c)}|\mathbf{y}_i, \boldsymbol{\beta}, \sigma_v)$, where $p_{i(c)} = \sum_{l\in\bar{s}_i} p_{il}/(N_i - n_i)$ and $\mathbf{y}_i$ is the vector of sample y-values from the ith area. Therefore, the Bayes estimator of P_i may be expressed as

$$\hat{P}_i^B = \hat{P}_i^B(\boldsymbol{\beta}, \sigma_v) = f_i\bar{y}_i + (1 - f_i)\hat{p}_{i(c)}^B. \tag{9.4.19}$$

If the sampling fractionf_i is negligible, we can express $\hat{P}_i^B$ as

$$\hat{P}_i^B \approx \frac{1}{N_i} E\left(\sum_{l=1}^{N_i} p_{il}|\mathbf{y}_i, \boldsymbol{\beta}, \sigma_v\right). \tag{9.4.20}$$

The posterior variance of P_i reduces to

$$\begin{aligned} &V(P_i|\mathbf{y}_i, \boldsymbol{\beta}, \sigma_v) \\ &= (1 - f_i)^2 E(\bar{y}_i^* - \hat{p}_{i(c)}^B)^2 \\ &= N_i^{-2}\left[E\left\{\sum_{l\in\bar{s}_i} p_{il}(1 - p_{il})|\mathbf{y}_i, \boldsymbol{\beta}, \sigma_v)\right\} + V\left\{\sum_{l\in\bar{s}_i} p_{il}|\mathbf{y}_i, \boldsymbol{\beta}, \sigma_v\right\}\right]; \end{aligned} \tag{9.4.21}$$

see Malec et al. (1997). Note that (9.4.21) involves expectations of the form $E(\sum_l p_{il}^2|\mathbf{y}_i, \boldsymbol{\beta}, \sigma_v)$ and $E[(\sum_l p_{il})^2|\mathbf{y}_i, \boldsymbol{\beta}, \sigma_v]$ as well as $E(\sum_l p_{il}|\mathbf{y}_i, \boldsymbol{\beta}, \sigma_v) =$

$(N_i - n_i)\hat{p}_{ic}^B$. No closed-form expressions for these expectations exist. However, we can express the expectations as ratios of single-dimensional integrals, similar to (9.4.14). For example, writing $\sum_l p_{il}^2$ as a function of $z \sim N(0,1)$, we can write

$$E\left(\sum_l p_{il}^2 | \mathbf{y}_i, \boldsymbol{\beta}, \sigma_v\right) = \frac{E\left[\left(\sum_l p_{il}^2\right) \exp\left\{h_i\left(\sum_{j \in s_i} \mathbf{x}_{ij}^T y_{ij}, y_i, \sigma z, \boldsymbol{\beta}\right)\right\}\right]}{E\left[\exp\left\{h_i\left(\sum_{j \in s_i} \mathbf{x}_{ij}^T y_{ij}, y_i, \sigma z, \boldsymbol{\beta}\right)\right\}\right]}, \tag{9.4.22}$$

where

$$h_i\left(\sum_{j \in s_i} \mathbf{x}_{ij}^T y_{ij}, y_i, \sigma z, \boldsymbol{\beta}\right)$$
$$= \left(\sum_{j \in s_i} \mathbf{x}_{ij}^T y_{ij}\right)\boldsymbol{\beta} + (\sigma z) y_i - \sum_{j \in s_i} \log[1 + \exp(\mathbf{x}_{ij}^T \boldsymbol{\beta} + \sigma z)] \tag{9.4.23}$$

and the expectation is over $z \sim N(0,1)$. Note that (9.4.23) reduces to $h_2(y_i, \mu + \sigma z)$ for the logit-normal model without covariates, that is, $\mathbf{x}_{ij}^T \boldsymbol{\beta} = \mu$.

Maximum likelihood estimation of model parameters, $\boldsymbol{\beta}$ and σ_v, for the logistic linear mixed model and other generalized linear mixed models has received considerable attention in recent years. Methods proposed include numerical quadrature, EM algorithm, Markov chain Monte Carlo (MCMC) and stochastic approximation. We refer the readers to McCulloch and Searle, (2001, Section 10.4) for details of the algorithms. Simpler methods, called penalized quasi-likelihood (PQL), based on maximizing the joint distribution of $\mathbf{y} = (\mathbf{y}_1^T, \ldots, \mathbf{y}_m^T)^T$ and $\mathbf{v} = (v_1, \ldots, v_m)^T$ with respect to $\boldsymbol{\beta}$ and $\mathbf{v}$ have also been proposed. For the special case of linear mixed models under normality, PQL leads to "mixed model" equations (6.2.8), Chapter 6 whose solution is identical to the BLUP of $\boldsymbol{\beta}$ and the BLUE of $\mathbf{v}$. However, for the logistic linear mixed model, the PQL estimator of $\boldsymbol{\beta}$ is asymptotically biased (as $m \to \infty$) and hence inconsistent, unlike the ML estimator of $\boldsymbol{\beta}$.

Consistent moment estimators of $\boldsymbol{\beta}$ and σ_v may be obtained by equating $\sum_i \sum_j \mathbf{x}_{ij} y_{ij}$ and $\sum_i (\sum_{j \neq k} y_{ij} y_{ik})$ to their expectations and then solving the equations for $\boldsymbol{\beta}$ and σ_v. The expectations depend on $E(y_{ij}) = E(p_{ij}) = E[(h_1(\mathbf{x}_{ij}^T \boldsymbol{\beta} + \sigma z)]$ and $E(y_{ij} y_{ik}) = E(p_{ij} p_{ik}) = E[h_1(\mathbf{x}_{ij}^T \boldsymbol{\beta} + \sigma z) h_1(\mathbf{x}_{ik}^T \boldsymbol{\beta} + \sigma z)]$. The two-step method of subsection 9.4.1 can be extended to get more efficient moment estimators.

Using either ML or moment estimators $\hat{\boldsymbol{\beta}}$ and $\hat{\sigma}_v$, we obtain an EB estimator of the ith area proportion P_i as $\hat{P}_i^{\text{EB}} = \hat{P}_i^B(\hat{\boldsymbol{\beta}}, \hat{\sigma}_v)$. The estimator $\hat{P}_i^{\text{EB}}$ is nearly unbiased for P_i in the sense that its bias, $E(\hat{P}_i^{\text{EB}} - \hat{P}_i^{\text{B}})$, is of order m^{-1} for large m. A naive EB approach uses $\hat{P}_i^{\text{EB}}$ as the estimator of realized P_i and its variability is measured by the estimated posterior variance $V(P_i | \mathbf{y}_i, \hat{\boldsymbol{\beta}}, \hat{\sigma}_v) = g_{1i}(\hat{\boldsymbol{\beta}}, \hat{\sigma}_v, \mathbf{y}_i)$, using (9.4.21). However, the estimated posterior variance ignores the variability associated with $\hat{\boldsymbol{\beta}}$ and $\hat{\sigma}_v$. The jackknife method can be applied to obtain an approximately unbiased estimator of MSE($\hat{P}_i^{\text{EB}}$). We obtain the jackknife estimator, mse$_J(\hat{P}_i^{\text{EB}})$, from (9.4.13) by substituting $\hat{P}_i^{\text{EB}} = k_i(\mathbf{y}_i, \boldsymbol{\beta}, \sigma_v)$ and $\hat{P}_{i,-l}^{\text{EB}} = k_i(\mathbf{y}_i, \hat{\boldsymbol{\beta}}_{-l}, \hat{\sigma}_{v,-l})$ in (9.4.11) and $g_{1i}(\hat{\boldsymbol{\beta}}, \hat{\sigma}_v, \mathbf{y}_i)$ and $g_{1i}(\hat{\boldsymbol{\beta}}_{-l}, \hat{\sigma}_{v,-l}, \mathbf{y}_i)$ in (9.4.12), where $\hat{\boldsymbol{\beta}}_{-l}$ and $\hat{\sigma}_{v,-l}$ are the delete$-l$ estimators obtained from $\{(y_{ij}, \mathbf{x}_{ij}), j = 1, \ldots, m;\ i \neq l = 1, \ldots, m\}$.

It is clear from the foregoing account that the implementation of EB for the logistic linear mixed model is quite cumbersome computationally. Approximate EB methods that are computationally simpler have been proposed in the literature, but those methods are not asymptotically valid as $m \to \infty$. As a result, the approximate methods might perform poorly in practice, especially for small sample sizes, n_i. We give a brief account of an approximate method proposed by MacGibbon and Tomberlin (1989), based on approximating the posterior distribution $f(\boldsymbol{\beta}, \mathbf{v}|\mathbf{y}, \sigma_v^2)$ by a multivariate normal distribution, assuming a flat prior $f(\boldsymbol{\beta}) =$ constant. We have

$$f(\boldsymbol{\beta}, \mathbf{v}|\mathbf{y}, \sigma_v^2) \propto \left[\prod_{i,j} p_{ij}^{y_{ij}}(1-p_{ij})^{1-y_{ij}}\right] \exp\left(-\sum_i v_i^2/2\sigma_v^2\right), \qquad (9.4.24)$$

where $\text{logit}(p_{ij}) = \mathbf{x}_{ij}^T\boldsymbol{\beta} + v_i$. The posterior (9.4.24) is approximated by a multivariate normal having its mean at the mode of (9.4.24) and covariance matrix equal to the inverse of the information matrix evaluated at the mode. The elements of the information matrix are given by the second derivatives of $\log(f(\boldsymbol{\beta}, \mathbf{v}|\mathbf{y}, \sigma_v^2))$ with respect to the elements of $\boldsymbol{\beta}$ and $\mathbf{v}$, multiplied by -1. The mode $(\boldsymbol{\beta}^*, \mathbf{v}^*)$ is obtained by maximizing $\log(f(\boldsymbol{\beta}, \mathbf{v}|\mathbf{y}, \sigma_v^2))$ with respect to $\boldsymbol{\beta}$ and $\mathbf{v}$, using the Newton-Raphson algorithm, and then substituting the ML estimator of σ_v^2 obtained from the EM algorithm. An EB estimator of P_i is taken as

$$\hat{P}_{i*}^{\text{EB}} = \frac{1}{N_i}\left(n_i\bar{y}_i + \sum_{j \in \bar{s}_i} p_{ij}^*\right), \qquad (9.4.25)$$

where

$$p_{ij}^* = \exp(\mathbf{x}_{ij}^T\boldsymbol{\beta}^* + v_i^*)/[1 + \exp(\mathbf{x}_{ij}^T\boldsymbol{\beta}^* + v_i^*)]. \qquad (9.4.26)$$

Unlike the EB estimator $\hat{P}_i^{\text{EB}} = \hat{P}_i^B(\hat{\boldsymbol{\beta}}, \hat{\sigma}_v)$, the estimator $\hat{P}_{i*}^{\text{EB}}$ is not nearly unbiased for P_i. Its bias is of order n_i^{-1} and hence it may not perform well when n_i is small; even v_i^* is a biased estimator of v_i to the same order.

Farrell, MacGibbon and Tomberlin (1997a) used a bootstrap measure of accuracy of $\hat{P}_{i*}^{\text{EB}}$, similar to the Laird and Louis (1987) parametric bootstrap (also called type III bootstrap) to account for the sampling variability of the estimates of $\boldsymbol{\beta}$ and σ_v^2. Simulation results, based on $m = 20$ and $n_i = 50$, indicated good performance of the proposed EB method. Farrell, MacGibbon and Tomberlin (1997b) relaxed the assumption of normality on the v_i's, using the following linking model: $\text{logit}(p_{ij}) = \mathbf{x}_{ij}^T\boldsymbol{\beta} + v_i$ and $v_i \overset{\text{iid}}{\sim}$ unspecified distribution. They used a nonparametric ML method, proposed by Laird (1978), to obtain an EB estimator of P_i, similar in form to (9.4.25). They also used the type II bootstrap of Laird and Louis (1987) to obtain a bootstrap measure of accuracy.

The EB estimators $\hat{P}_i^{\text{EB}}$ and $\hat{P}_{i*}^{\text{EB}}$ require the knowledge of individual $\mathbf{x}$ values in the population, unlike the EB estimator for the basic unit level model which depends only on the population mean $\bar{\mathbf{X}}_i = N_i^{-1}\sum_j \mathbf{x}_{ij}$. This is a practical drawback of logistic linear mixed models and other nonlinear models because microdata for all individuals in a small area may not be available. Farrell, MacGibbon and Tomberlin (1997a) obtained an approximation to $\hat{P}_{i*}^{\text{EB}}$ depending only on the mean $\bar{\mathbf{X}}_i$ and the cross-product matrix $\sum_j \mathbf{x}_{ij}\mathbf{x}_{ij}^T$ with elements $\sum_j x_{ija}x_{ijb}; a, b = 1, \ldots, p$. It uses a second-order multivariate Taylor expansion of p_{ij}^* around $\bar{\mathbf{x}}_i^*$, the mean vector of $\mathbf{x}_{ij}$'s for $j \in \bar{s}_i$. However, the deviations $\mathbf{x}_{ij} - \bar{\mathbf{x}}_i^*$ in the expansion are of order $O(1)$, and hence terms involving higher powers of the deviations are not necessarily negligible. Farrell et al. (1997a) investigated the accuracy of the second-order approximation for various distributions of $\mathbf{x}_{ij}^T\boldsymbol{\beta} + v_i$, and provided some guidelines.

9.5 Disease Mapping

Mapping of small area mortality (or incidence) rates of diseases such as cancer is a widely used tool in public health research. Such maps permit the analysis of geographical variation in rates of diseases which may be useful in formulating and assessing aetiological hypotheses, resource allocation, and identifying areas of unusually high risk warranting intervention. Examples of disease rates studied in the literature include lip cancer rates in the 56 counties (small areas) of Scotland (Clayton and Kaldor (1987)), incidence of leukemia in 281 census tracts (small areas) of upstate New York (Datta, Ghosh and Waller (2000)), stomach cancer mortality rates in Missouri cities (small areas) for males aged 47–64 years (Tsutakawa, Shoop and Marienfeld (1985)), all cancer mortality rates for white males in health service areas (small areas) of the United States (Nandram, Sedransk and Pickle (1999)), prostate cancer rates in Scottish counties (Langford, Layland, Rasbash and Goldstein (1999)), and infant mortality rates for local health areas (small areas) in British Columbia, Canada (Dean and MacNab (2001)). We refer the reader to the October 2000 issue of *Statistics in Medicine* for recent developments in disease mapping. Typically, administrative data on event counts and related auxiliary variables are used in disease mapping; sampling is not used.

Suppose that the country (or the region) used for disease mapping is divided into m non-overlapping small areas. Let θ_i be the unknown relative risk (RR) in the ith area. A direct (or crude) estimator of θ_i is given by the standardized mortality ratio (SMR), $\hat{\theta}_i = y_i/e_i$, where y_i and e_i respectively denote the observed and expected number of deaths (cases) over a given period ($i =$

$1, \ldots, m$). The e_i's are calculated as

$$e_i = n_i \left(\sum_i y_i / \sum_i n_i \right), \tag{9.5.1}$$

where n_i is the number of person-years at risk in the ith area, and then treated as fixed. Some authors use mortality (event) rates, τ_i, as parameters, instead of relative risks, and a crude estimator of τ_i is then given by $\hat{\tau}_i = y_i/n_i$. The two approaches, however, are equivalent because $\sum_i y_i / \sum_i n_i$ is treated as a constant.

A common assumption in disease mapping is that $y_i|\theta_i \overset{\text{ind}}{\sim} \text{Poisson}(e_i\theta_i)$. Under this assumption, the ML estimator of θ_i is the SMR, $\hat{\theta}_i = y_i/e_i$. However, a map of crude rates $\{\hat{\theta}_i\}$ can badly distort the geographical distribution of disease incidence or mortality because it tends to be dominated by areas of low population, e_i, exhibiting extreme SMR's that are least reliable. Note that $V(\hat{\theta}_i) = \theta_i/e_i$ is large if e_i is small.

EB or HB methods provide reliable estimators of RR by borrowing strength across areas. As a result, maps based on EB or HB estimates are more reliable compared to crude maps. In this section we give a brief account of EB methods based on simple linking models. Various extensions have been proposed in the literature, including models exhibiting spatial correlation and bivariate models.

9.5.1 Poisson-Gamma Model

We first study a two-stage model for count data $\{y_i\}$ similar to the beta-binomial model for binary data. In the first stage, we assume that $y_i \overset{\text{ind}}{\sim} \text{Poisson}(e_i\theta_i)$, $i = 1, \ldots, m$. A "conjugate" model linking the relative risks θ_i is assumed in the second stage: $\theta_i \overset{\text{iid}}{\sim} \text{gamma}(\nu, \alpha)$, where $\text{gamma}(\nu, \alpha)$ denotes the gamma distribution with shape parameter $\nu(>0)$ and scale parameter $\alpha(>0)$. We have

$$f(\theta_i|\alpha, \nu) = \frac{\alpha^\nu}{\Gamma(\nu)} e^{-\alpha\theta_i} \theta_i^{\nu-1} \tag{9.5.2}$$

and

$$E(\theta_i) = \nu/\alpha = \mu, \quad V(\theta_i) = \nu/\alpha^2. \tag{9.5.3}$$

Noting that $\theta_i|y_i, \alpha, \nu \overset{\text{ind}}{\sim} \text{gamma}(y_i + \nu, e_i + \alpha)$, the Bayes estimator of θ_i and the posterior variance of θ_i are obtained from (9.5.3) by changing α to $e_i + \alpha$ and ν to $y_i + \nu$:

$$\hat{\theta}_i^B(\alpha, \nu) = E(\theta_i|y_i, \alpha, \nu) = (y_i + \nu)/(e_i + \alpha) \tag{9.5.4}$$

and

$$V(\theta_i|y_i, \alpha, \nu) = g_{1i}(\alpha, \nu, y_i) = (y_i + \nu)/(e_i + \alpha)^2. \tag{9.5.5}$$

We can obtain ML estimators of α and ν from the marginal distribution, $y_i|\alpha,\nu \overset{\text{iid}}{\sim}$ negative binomial, using the log likelihood

$$l(\alpha,\nu)=\sum_{i=1}^{m}\left[\sum_{h=0}^{y_i-1}\log(\nu+h)+\nu\log(\alpha)-(y_i+\nu)\log(e_i+\alpha)\right]. \qquad (9.5.6)$$

Closed-form expressions for $\hat{\alpha}_{\text{ML}}$ and $\hat{\nu}_{\text{ML}}$ do not exist. Marshall (1991) obtained simple moment estimators by equating the weighted sample mean $\hat{\theta}_{e\cdot}=\frac{1}{m}\sum_l(e_l/e.)\hat{\theta}_l$ and the weighted sample variance $s_e^2=\frac{1}{m}\sum_i(e_i/e.)(\hat{\theta}_i-\hat{\theta}_{e\cdot})^2$ to their expected values and then solving the resulting moment equations for α and ν, where $e.=\sum_i e_i/m$. This leads to moment estimators, $\hat{\alpha}$ and $\hat{\nu}$, given by

$$\frac{\hat{\nu}}{\hat{\alpha}}=\hat{\theta}_{e\cdot} \qquad (9.5.7)$$

and

$$\frac{\hat{\nu}}{\hat{\alpha}^2}=s_e^2-\frac{\hat{\theta}_{e\cdot}}{e.}. \qquad (9.5.8)$$

Lahiri and Maiti (1999) provided more efficient moment estimators. The moment estimators may also be used as starting values for ML iterations.

We substitute the moment estimators $\hat{\alpha}$ and $\hat{\nu}$ into (9.5.4) to get an EB estimator of θ_i as

$$\hat{\theta}_i^{\text{EB}}=\hat{\theta}_i^{B}(\hat{\alpha},\hat{\nu})=\hat{\gamma}_i\hat{\theta}_i+(1-\hat{\gamma}_i)\hat{\theta}_{e\cdot}, \qquad (9.5.9)$$

where $\hat{\gamma}_i=e_i/(e_i+\hat{\alpha})$. Note that $\hat{\theta}_i^{\text{EB}}$ is a weighted average of the direct estimator (SMR) $\hat{\theta}_i$ and the synthetic estimator $\hat{\theta}_{e\cdot}$, and more weight is given to $\hat{\theta}_i$ as the ith area expected deaths, e_i, increases. If $s_e^2<\hat{\theta}_{e\cdot}/e.$ then $\hat{\theta}_i^{\text{EB}}$ is taken as the synthetic estimator $\hat{\theta}_{e\cdot}$. The EB estimator is nearly unbiased for θ_i in the sense that its bias is of order m^{-1}, for large m.

As in the binary case, the jackknife method may be used to obtain a nearly unbiased estimator of $\text{MSE}(\hat{\theta}_i^{\text{EB}})$. We obtain the jackknife estimator, $\text{mse}_J(\hat{\theta}_i^{\text{EB}})$ from (9.4.13) by substituting $\hat{\theta}_i^{\text{EB}}=k_i(y_i,\hat{\alpha},\hat{\nu})$ for $\hat{p}_i^{\text{EB}}$ and $\hat{\theta}_{i,-l}^{\text{EB}}=k_i(y_i,\hat{\alpha}_{-l},\hat{\nu}_{-l})$ for $\hat{p}_{i,l}^{\text{EB}}$ in (9.4.11) and using $g_{1i}(\hat{\alpha},\hat{\nu},y_i)$ and $g_{1i}(\hat{\alpha}_{-l},\hat{\nu}_{-l},y_i)$ in (9.4.12), where $\hat{\alpha}_{-l}$ and $\hat{\nu}_{-l}$ are the delete$-l$ moment estimators obtained from $\{(y_i,e_i), i\neq l=1,\ldots,m\}$. Note that $\text{mse}_J(\hat{\theta}_i^{\text{EB}})$ is area-specific in the sense that it depends on y_i. Lahiri and Maiti (1999) obtained a Taylor expansion estimator of MSE, using a parametric bootstrap to estimate the covariance matrix of $(\hat{\alpha},\hat{\nu})$.

The linking gamma model on the θ_i's can be extended to allow for area level covariates, $\mathbf{z}_i$, such as degree of urbanization of areas, to improve the estimates of relative risk. Clayton and Kaldor (1987) allowed varying scale parameters, α_i, and assumed a loglinear model on $E(\theta_i)=\nu/\alpha_i$: $\log(E(\theta_i))=\mathbf{z}_i^T\boldsymbol{\beta}$. EB estimation for this extension is implemented by changing α to α_i in (9.5.4) and (9.5.5) and using ML or moment estimators of α and $\boldsymbol{\beta}$. Christiansen and Morris (1991) studied the Poisson-gamma regression model in detail, and proposed accurate approximations to the posterior mean and the posterior variance of θ_i. The posterior mean approximation is used as the EB estimator and the posterior variance approximation as a measure of its variability.

9.5.2 Log-normal Models

Log-normal two-stage models have also been proposed. The first-stage model is not changed, but the second-stage linking model is changed to $\xi_i = \log(\theta_i) \stackrel{\text{iid}}{\sim} N(\mu, \sigma^2), i = 1, \ldots, m$. As in the case of logit-normal models, implementation of EB is more complicated for the log-normal model because no closed-form expression for the Bayes estimator, $\hat{\theta}_i^B(\mu, \sigma^2)$, and the posterior variance, $V(\theta_i|y_i, \mu, \sigma^2)$, exist. Clayton and Kaldor (1987) approximated the posterior density, $f(\boldsymbol{\xi}|\mathbf{y}, \mu, \sigma^2)$, by a multivariate normal which gives an explicit approximation to $\hat{\xi}_i^B$, where $\boldsymbol{\xi} = (\xi_1, \ldots, \xi_m)^T$ and $\mathbf{y} = (y_1, \ldots, y_m)^T$. ML estimators of model parameters μ and σ^2 were obtained using the EM algorithm, and then used in the approximate formula for $\hat{\xi}_i^B$ to get EB estimators $\hat{\xi}_i^{\text{EB}}$ of ξ_i and $\hat{\theta}_i^{\text{EB}} = \exp(\hat{\xi}_i^{\text{EB}})$ of θ_i. The EB estimator $\hat{\theta}_i^{\text{EB}}$, however, is not nearly unbiased for θ_i. We can employ numerical integration, as done in subsection 9.4.1, to get nearly unbiased EB estimators, but we omit details here. Moment estimators of μ and σ (Jiang and Zhang (2001)) may be used to simplify the calculation of jackknife estimator of MSE($\hat{\theta}_i^{\text{EB}}$).

The basic log-normal model readily extends to the case of covariates, $\mathbf{z}_i$: $\xi_i = \log(\theta_i) \stackrel{\text{ind}}{\sim} N(\mathbf{z}_i^T\boldsymbol{\beta}, \sigma^2)$. Also, the basic model can be extended to allow spatial correlations; mortality data sets often exhibit significant spatial relationships between the log relative risks, $\xi_i = \log(\theta_i)$. A simple conditional autoregression (CAR)-normal model on $\boldsymbol{\xi}$ assumes that $\boldsymbol{\xi}$ is a multivariate normal specified by

$$E(\xi_i|\xi_l, l \neq i) = \mu + \rho \sum_{l(\neq i)} q_{il}(\xi_l - \mu), \tag{9.5.10}$$

$$V(\xi_i|\xi_l, l \neq i) = \sigma^2, \tag{9.5.11}$$

where ρ is the correlation parameter and $\mathbf{Q} = (q_{il})$ is the "adjacency" matrix of the map given by $q_{il} = 1$ if i and l are adjacent areas and $q_{il} = 0$ otherwise. It follows from Besag (1974) that $\boldsymbol{\xi}$ is multivariate normal with mean $\boldsymbol{\mu} = \mu\mathbf{1}$ and covariance matrix $\boldsymbol{\Sigma} = \sigma^2(\mathbf{I} - \rho\mathbf{Q}^{-1})$, where ρ is bounded above by the inverse of the largest eigenvalue of $\mathbf{Q}$. Clayton and Kaldor (1987) approximated the posterior density, $f(\boldsymbol{\xi}|\mathbf{y}, \mu, \sigma^2, \rho)$, similar to the log-normal case.

The assumption (9.5.11) of a constant conditional variance for the ξ_i's results in the conditional mean (9.5.10) proportional to the sum of the neighboring ξ_i's rather than the mean of the neighboring ξ_i's (local mean). Clayton and Bernardinelli (1992) proposed an alternative joint density of the ξ_i's given by

$$f(\boldsymbol{\xi}) \propto (\sigma^2)^{-m/2} \exp\left[-\sum_{i \neq l}\sum(\xi_i - \xi_l)^2 q_{il}/(2\sigma^2)\right]. \tag{9.5.12}$$

This specification leads to

$$E(\xi_i|\xi_l,\ l \neq i) = \Sigma_l q_{il}\xi_l / \Sigma_l q_{il} \tag{9.5.13}$$

and

$$V(\xi_i|\xi_l,\ l \neq i) = \sigma^2/\Sigma_l q_{il}. \tag{9.5.14}$$

Note that the conditional variance is now inversely proportional to $\Sigma_l q_{il}$, the number of neighbors of area i, and the conditional mean is equal to the mean of the neighboring values ξ_l. In the context of disease mapping, the alternative specification may be more appropriate.

Example 9.5.1. Lip Cancer. Clayton and Kaldor (1987) applied EB estimation to data on observed cases, y_i, and expected cases, e_i, of lip cancer registered during the period 1975–1980 in each of 56 counties (small areas) of Scotland. They reported the SMR, the EB estimate of θ_i based on the Poisson-gamma model (denoted $\hat{\theta}_i^{\text{EB}}(1)$) and the approximate EB estimates of θ_i based on the log-normal model and the CAR-normal model (denoted $\hat{\theta}_i^{\text{EB}}(2)$ and $\hat{\theta}_i^{\text{EB}}(3)$) for each of the 56 counties (all values multiplied by 100). The SMR-values varied between 0 and 652 while the EB estimates showed considerably less variability across counties, as expected: $\hat{\theta}_i^{\text{EB}}(1)$ varied between 31 and 422 (with CV=0.78) and $\hat{\theta}_i^{\text{EB}}(2)$ varied between 34 to 495 (with CV=0.85), suggesting little difference between the two sets of EB estimates. Ranks of EB estimates differed little from the corresponding ranks of the SMRs for most counties, despite less variability exhibited by the EB estimates.

Turning to the CAR-normal model, the adjacency matrix, $\mathbf{Q}$, was specified by listing adjacent counties for each county i. The ML estimate of ρ was 0.174 compared to the upper bound of 0.175, suggesting a high degree of spatial relationship in the data set. Most of the CAR estimates, $\hat{\theta}_i^{\text{EB}}(3)$, differed little from the corresponding estimates $\hat{\theta}_i^{\text{EB}}(1)$ and $\hat{\theta}_i^{\text{EB}}(2)$ based on the independence assumption. Counties with few cases, y_i, and SMRs differing appreciably from adjacent counties are the only counties affected substantially by spatial correlation. For example, county number 24 with $y_{24} = 7$ is adjacent to several low-risk counties, and the CAR estimate $\hat{\theta}_{24}^{\text{EB}}(3) = 83.5$ is substantially smaller than $\hat{\theta}_{24}^{\text{EB}}(1) = 127.7$ and $\hat{\theta}_{24}^{\text{EB}}(2) = 123.6$ based on the independence assumption.

9.5.3 Extensions

Various extensions of the disease mapping models, studied in subsections 9.5.1 and 9.5.2, have been proposed in the recent literature. De Souza

(1992) proposed a two-stage, bivariate logit-normal model to study joint relative risks (or mortality rates), θ_{1i} and θ_{2i}, of two cancer sites (e.g., lung and large bowel cancers), or two groups (e.g., lung cancer in males and females) over several geographical areas. Denote the observed and expected number of deaths at the two sites as (y_{1i}, y_{2i}) and (e_{1i}, e_{2i}) respectively for the ith area $(i = 1, \ldots, m)$. The first stage assumes that $(y_{1i}, y_{2i})|(\theta_{1i}, \theta_{2i}) \stackrel{\text{ind}}{\sim} \text{Poisson}(e_{1i}\theta_{1i}) * \text{Poisson}(e_{2i}\theta_{2i}), i = 1, \ldots, m$, where $*$ denotes that $f(y_{1i}, y_{2i}|\theta_{1i}, \theta_{2i}) = f(y_{1i}|\theta_{1i})f(y_{2i}|\theta_{2i})$. The joint risks $(\theta_{1i}, \theta_{2i})$ are linked in the second stage by assuming that $(\text{logit}(\theta_{1i}), \text{logit}(\theta_{2i})) \stackrel{\text{ind}}{\sim}$ bivariate normal with means μ_1, μ_2, standard deviations σ_1 and σ_2 and correlation ρ, denoted $N(\mu_1, \mu_2, \sigma_1, \sigma_2, \rho)$. Bayes estimators of θ_{1i} and θ_{2i} involve double integrals which may be calculated numerically using Gauss-Hermite quadrature. EB estimators are obtained by substituting ML estimators of model parameters in the Bayes estimators. De Souza (1997) applied the bivariate EB method to two data sets consisting of cancer mortality rates in 115 counties of the state of Missouri during 1972–1981: (i) Lung and large bowel cancers, (ii) lung cancer in males and females. The EB estimates based on the bivariate model lead to improved efficiency for each site (group) compared to the EB estimates based on the univariate logit-normal model, because of significant correlation: $\hat{\rho} = 0.54$ for data set (i) and $\hat{\rho} = 0.76$ for data set (ii). Kass and Steffey's (1989) first-order approximation to the posterior variance was used as a measure of variability of the EB estimates.

Kim, Sun and Tsutakawa (2001) extended the bivariate model by introducing spatial correlations (via CAR) and covariates into the model. They used a HB approach instead of the EB approach. They applied the bivariate spatial model to male and female lung cancer mortality in the State of Missouri, and constructed disease maps of male and female lung cancer mortality rates by age group and time period.

Dean and MacNab (2001) extended the Poisson-gamma model to handle nested data structures, such as a hierarchical health administrative structure consisting of health districts, i, in the first level and local health areas, j, within districts in the second level $(j = 1, \ldots, n_i;\ i = 1, \ldots, m)$. The data consist of incidence or mortality counts, y_{ij}, and the corresponding population at risk counts, n_{ij}. Dean and MacNab (2001) derived EB estimators of the local health area rates, θ_{ij}, using a nested error Poisson-gamma model. The Bayes estimator of θ_{ij} is a weighted combination of the crude local area rate, y_{ij}/n_{ij}, the correspond crude district rate $y_{i\cdot}/n_{i\cdot}$ and the overall rate $y_{\cdot\cdot}/n_{\cdot\cdot}$, where $y_{i\cdot} = \sum_j y_{ij}$ and $y_{\cdot\cdot} = \sum_i y_{i\cdot}$, and $(n_{i\cdot}, n_{\cdot\cdot})$ similarly defined. Dean and MacNab (2001) used the Kass and Steffey (1989) first-order approximation to posterior variance as a measure of variability. They applied the nested error model to infant mortality data from the province of British Columbia, Canada.

9.6 Triple-goal Estimation

We have focused so far on the estimation of area-specific parameters (means, relative risks, etc.), but in some applications the main objective is to produce an ensemble of parameter estimates whose distribution is in some sense close enough to the distribution of area-specific parameters, θ_i. For example, Spjøtvoll and Thomsen (1987) were interested in finding how 100 municipalities in Norway were distributed according to proportions of persons in the labor force. By comparing with the actual distribution in their example, they showed that the EB estimates, $\hat{\theta}_i^{\text{EB}}$, under a simple area level model distort the distribution by over-shrinking towards the synthetic component $\hat{\theta}_.$. In particular, the variability of the EB estimates was smaller than the variability of the θ_i's. On the other hand, the set of direct estimates $\{\hat{\theta}_i\}$ were overdispersed in the sense of variability larger than the variability of the θ_i's.

We are also often interested in the ranks of the θ_i's (e.g., ranks of schools, hospitals or geographical areas) or in identifying domains (areas) with extreme θ_i's. Ideally, it is desirable to construct a set of "triple-goal" estimates that can produce good ranks, a good histogram and good area-specific estimates. However, simultaneous optimization is not feasible, and it is necessary to seek a compromise set that can strike an effective balance between the three goals (Shen and Louis (1998)).

9.6.1 Constrained EB

Consider a two-stage model of the form $\hat{\theta}_i|\theta_i \overset{\text{ind}}{\sim} f(\hat{\theta}_i|\theta_i, \boldsymbol{\lambda}_1)$, and $\theta_i \overset{\text{iid}}{\sim} f(\theta_i|\boldsymbol{\lambda}_2)$, where $\boldsymbol{\lambda} = (\boldsymbol{\lambda}_1, \boldsymbol{\lambda}_2)^T$ is the vector of model parameters. The set of direct estimators $\{\hat{\theta}_i\}$ are generally overdispersed under this model. For example, consider the simple model $\hat{\theta}_i = \theta_i + e_i$ with $e_i \overset{\text{iid}}{\sim} (0, \psi)$ independent of $\theta_i \overset{\text{iid}}{\sim} (\mu, \sigma_v^2), i = 1, \ldots, m$. Noting that $\hat{\theta}_i \overset{\text{iid}}{\sim} (\mu, \psi + \sigma_v^2)$, it immediately follows that

$$E\Big[\frac{1}{m-1}\sum_i(\hat{\theta}_i - \hat{\theta}_.)^2\Big] = \psi + \sigma_v^2 > \sigma_v^2 = E\Big[\frac{1}{m-1}\sum_i(\theta_i - \theta_.)^2\Big], \quad (9.6.1)$$

where $\hat{\theta}_. = \sum_i \hat{\theta}_i/m$ and $\theta_. = \sum_i \theta_i/m$. On the other hand, the set of Bayes estimators $\{\hat{\theta}_i^B\}$ exhibits underdispersion under the two-stage model $\hat{\theta}_i|\theta_i \overset{\text{ind}}{\sim} f(\hat{\theta}_i|\theta_i, \boldsymbol{\lambda}_1)$ and $\theta_i \overset{\text{iid}}{\sim} f(\theta_i|\boldsymbol{\lambda}_2)$, where $\hat{\theta}_i^{\text{B}} = E(\theta_i|\hat{\theta}_i, \boldsymbol{\lambda})$ is the Bayes estimator of θ_i under squared error. We have

$$E\Big[\frac{1}{m-1}\sum_i(\theta_i - \theta_.)^2|\hat{\boldsymbol{\theta}}\Big] = \frac{1}{m-1}\sum_i V(\theta_i - \theta_.|\hat{\boldsymbol{\theta}}) + \frac{1}{m-1}\sum_i\left(\hat{\theta}_i^B - \hat{\theta}_.^B\right)^2 \quad (9.6.2)$$

$$> \frac{1}{m-1}\sum_i(\hat{\theta}_i^B - \hat{\theta}_.^B)^2, \quad (9.6.3)$$

where $\hat{\boldsymbol{\theta}} = (\hat{\theta}_1, \ldots, \hat{\theta}_m)^T$, $\hat{\theta}_.^{\text{B}} = \Sigma_i\hat{\theta}_i^{\text{B}}/m$, and the dependence on $\boldsymbol{\lambda}$ is suppressed for simplicity. It follows from (9.6.3) that $E[\sum_i(\theta_i - \theta_.)^2] > E[\sum_i(\hat{\theta}_i^B -$

$\hat{\theta}_{\cdot}^{B})^2]$. However, note that $\{\hat{\theta}_i^B\}$ match the ensemble mean because $\hat{\theta}_{\cdot}^B = E(\theta_{\cdot}|\hat{\boldsymbol{\theta}})$ which, in turn, implies $E(\hat{\theta}_{\cdot}^B) = E(\theta_{\cdot})$.

We can match the ensemble variance by minimizing the posterior expected squared error loss $E[\sum_i(\theta_i - t_i)^2|\hat{\boldsymbol{\theta}}]$ subject to the constraints

$$t_{\cdot} = \hat{\theta}_{\cdot}^B \tag{9.6.4}$$

$$\frac{1}{m-1}\sum_i (t_i - t_{\cdot})^2 = E\left[\frac{1}{m-1}\sum_i(\theta_i - \theta_{\cdot})^2|\hat{\boldsymbol{\theta}}\right], \tag{9.6.5}$$

where $t_{\cdot} = \Sigma_i t_i/m$. Using Lagrange multipliers, we obtain the constrained Bayes (CB) estimators $\{\hat{\theta}_i^{\text{CB}}\}$ as the solution to the minimization problem, where

$$t_{i,\text{opt}} = \hat{\theta}_i^{\text{CB}} = \hat{\theta}_{\cdot}^B + a(\hat{\boldsymbol{\theta}}, \boldsymbol{\lambda})(\hat{\theta}_i^B - \hat{\theta}_{\cdot}^B) \tag{9.6.6}$$

with

$$a(\hat{\boldsymbol{\theta}}, \boldsymbol{\lambda}) = \left[1 + \frac{(1/m)\sum_i V(\theta_i|\hat{\theta}_i, \boldsymbol{\lambda})}{\{1/(m-1)\}\sum_i(\hat{\theta}_i^B - \hat{\theta}_{\cdot}^B)^2}\right]^{1/2}. \tag{9.6.7}$$

Louis (1984) derived the CB estimator under normality. Ghosh (1992b) obtained (9.6.6) for arbitrary distributions. A proof of (9.6.6) is given in Section 9.9. It follows from (9.6.6) that $\sum_i(\hat{\theta}_i^{\text{CB}} - \hat{\theta}_{\cdot}^{\text{CB}})^2 > \sum_i(\hat{\theta}_i^B - \hat{\theta}_{\cdot}^B)^2$ because the term $a(\hat{\boldsymbol{\theta}}, \boldsymbol{\lambda})$ in (9.6.6) is greater than 1 and $\hat{\theta}_{\cdot}^{\text{CB}} = \hat{\theta}_{\cdot}^B$. That is, the variability of $\{\hat{\theta}_i^{\text{CB}}\}$ is larger than that of $\{\hat{\theta}_i^B\}$. Note that the constraint (9.6.5) implies that $E[\sum_i(\hat{\theta}_i^{\text{CB}} - \hat{\theta}_{\cdot}^{\text{CB}})^2] = E[\sum_i(\theta_i - \theta_{\cdot})^2]$ so that $\{\hat{\theta}_i^{\text{CB}}\}$ matches the variability of $\{\theta_i\}$.

The CB estimator $\hat{\theta}_i^{\text{CB}}$ is a function of the set of posterior variances $\{V(\theta_i|\hat{\boldsymbol{\theta}}, \boldsymbol{\lambda})\}$ and the set of Bayes estimators $\{\hat{\theta}_i^B = E(\theta_i|\hat{\boldsymbol{\theta}}, \boldsymbol{\lambda})\}$. Replacing the model parameters $\boldsymbol{\lambda}$ by suitable estimators $\hat{\boldsymbol{\lambda}}$, we obtain an empirical CB (ECB) estimator $\hat{\theta}_i^{ECB} = \hat{\theta}_i^{\text{CB}}(\hat{\boldsymbol{\lambda}})$.

Example 9.5.1. Simple Model. We illustrate the calculation of $\hat{\theta}_i^{\text{CB}}$ for the simple model $\hat{\theta}_i = \theta_i + e_i$, with $e_i \stackrel{\text{iid}}{\sim} N(0, \psi)$ and independent of $\theta_i \stackrel{\text{iid}}{\sim} N(\mu, \sigma_v^2)$. The Bayes estimator is given by $\hat{\theta}_i^B = \gamma\hat{\theta}_i + (1-\gamma)\mu$, where $\gamma = \sigma_v^2/(\sigma_v^2 + \psi)$. Further, $V(\theta_i|\hat{\theta}_i, \boldsymbol{\lambda}) = \gamma\psi$, $\hat{\theta}_{\cdot}^B = \gamma\hat{\theta}_{\cdot} + (1-\gamma)\mu$ and $\sum_i(\hat{\theta}_i^B - \hat{\theta}_{\cdot}^B)^2 = \gamma^2\sum_i(\hat{\theta}_i - \hat{\theta}_{\cdot})^2$. Hence,

$$\hat{\theta}_i^{\text{CB}} = [\gamma\hat{\theta}_{\cdot} + (1-\gamma)\mu] + \left[1 + \frac{\psi/\gamma}{\{1/(m-1)\}\sum_i(\hat{\theta}_i - \hat{\theta}_{\cdot})^2}\right]^{1/2}\gamma(\hat{\theta}_i - \hat{\theta}_{\cdot}). \tag{9.6.8}$$

Noting that $\hat{\theta}_i \stackrel{\text{iid}}{\sim} N(\mu, \psi + \sigma_v^2)$, it follows that $\hat{\theta}_{\cdot}$ and $(m-1)^{-1}\sum_i(\hat{\theta}_i - \hat{\theta}_{\cdot})^2$ converge in probability to μ and $\psi + \sigma_v^2 = \psi/(1-\gamma)$, respectively, as $m \to \infty$. Hence,

$$\hat{\theta}_i^{\text{CB}} \approx \gamma^{1/2}\hat{\theta}_i + (1 - \gamma^{1/2})\mu. \tag{9.6.9}$$

It follows from (9.6.9) that the weight attached to the direct estimator is larger than the weight used by the Bayes estimator, and the shrinkage toward the synthetic component μ is reduced. Assuming normality, Ghosh (1992) proved that the total MSE, $\sum_i \text{MSE}(\hat{\theta}_i^{\text{CB}}) = \sum_i E(\hat{\theta}_i^{\text{CB}} - \theta_i)^2$, of the CB estimators is smaller than the total MSE of the direct estimators if $m \geq 4$. Hence, the CB estimators perform better than the direct estimators, but are less efficient than the Bayes estimators.

Shen and Louis (1998) studied the performance of CB estimators for exponential families $f(\hat{\theta}_i|\theta, \boldsymbol{\lambda}_1)$ and conjugate $f(\theta_i|\boldsymbol{\lambda}_2)$. They showed that, for large m, CB estimators are always more efficient than the direct estimators in terms of total MSE. Further, the maximum loss in efficiency relative to the Bayes estimators is 24%. Note that the exponential families cover many commonly used distributions, including the binomial, Poisson, normal and gamma distributions.

9.6.2 Histogram

The empirical distribution function based on the CB estimators is given by $F_m^{\text{CB}}(t) = m^{-1}\sum_i I(\hat{\theta}_i^{\text{CB}} \leq t), -\infty < t < \infty$, where $I(\theta_i \leq t) = 1$ if $\theta_i \leq t$ and $I(\theta_i \leq t) = 0$ otherwise. The estimator $F_m^{\text{CB}}(t)$ is generally not consistent for the true distribution of the θ_i's as $m \to \infty$, though the CB approach matches the first and second moments. As a result, $F_m^{\text{CB}}(t)$ can perform poorly as an estimator of $F_m(t) = m^{-1}\sum_i I(\theta_i \leq t)$.

An "optimal" estimator of $F_m(t)$ is obtained by minimizing the posterior expected integrated squared error loss $E[\int\{A(t) - F_m(t)\}^2 dt|\hat{\boldsymbol{\theta}}]$. The optimal $A(\cdot)$ is given by

$$A_{\text{opt}}(t) = \bar{F}_m(t) = \frac{1}{m}\sum_{i=1}^{m} P(\theta_i \leq t|\hat{\theta}_i). \tag{9.6.10}$$

Adding the constraint that $A(\cdot)$ is a discrete distribution with at most m mass points, the optimal estimator $\hat{F}_m(\cdot)$ is discrete with mass $1/m$ at $\hat{U}_l = \bar{F}_m^{-1}(\frac{2l-1}{2m})$, $l = 1, \ldots, m$; see Shen and Louis (1998) for a proof. Note that $\hat{U}_l$ depends on model parameters, $\boldsymbol{\lambda}$, and an EB version of $\hat{U}_l$ is obtained by substituting a suitable estimator $\hat{\boldsymbol{\lambda}}$ for $\boldsymbol{\lambda}$.

9.6.3 Ranks

How good are the ranks based on the $\hat{\theta}_i^B$'s compared to those based on the true (realized but unobservable) values θ_i? In the context of BLUP estimators for the linear mixed model, Portnoy (1982) showed that ranking based on the BLUPs is "optimal" in the sense of maximizing the probability of correctly ranking the θ_is. Also, the ranks based on the $\hat{\theta}_i^B$'s often agree with the ranks based on the direct estimators $\hat{\theta}_i$, and it follows from (9.6.6) that the ranks based on the CB estimators $\hat{\theta}_i^{\text{CB}}$ are always identical with the ranks based on the $\hat{\theta}_i^B$s.

Let $R(i)$ be the rank of the true θ_i, that is, $R(i) = \sum_{l=1}^{m} I(\theta_i \geq \theta_l)$. Then the "optimal" estimator of $R(i)$ that minimizes the expected posterior squared error loss $E[\sum_i (Q(i) - R(i))^2 | \hat{\boldsymbol{\theta}}]$ is given by the Bayes estimator

$$Q_{\text{opt}}(i) = \tilde{R}^B(i) = E[R(i)|\hat{\boldsymbol{\theta}}] = \sum_{l=1}^{m} P(\theta_i \geq \theta_l | \hat{\boldsymbol{\theta}}). \tag{9.6.11}$$

Generally, the estimators $\tilde{R}^B(i)$ are not integers, so we rank the $\tilde{R}^B(i)$ to produce integer ranks $\hat{R}^B(i)$ =rank of $\tilde{R}^B(i)$ in the set $\{\tilde{R}^B(1), \ldots, \tilde{R}^B(m)\}$.

Shen and Louis (1998) proposed $\hat{\theta}_i^{\text{TG}} = \hat{U}_{\hat{R}^B(i)}$ as a compromise triple-goal estimator of the realized θ_i. The set $\{\hat{\theta}_i^{\text{TG}}\}$ is optimal for estimating the distribution $F_m(t)$ as well as the ranks $\{R(i)\}$. Simulation results indicate that the proposed method performs better than the CB method and achieves the three inferential goals.

9.7 Empirical Linear Bayes

EB methods we have studied so far are based on distributional assumptions on $\hat{\theta}_i | \theta_i$ and θ_i. Empirical linear Bayes (ELB) methods avoid distributional assumptions by specifying only the first and second moments but confining to the linear class of estimators, as in the case of EBLUP for the linear mixed models. Maritz and Lwin (1989) provide an excellent account of linear Bayes (LB) methods.

9.7.1 LB Estimation

We assume a two-stage model of the form $\hat{\theta}_i | \theta_i \overset{\text{ind}}{\sim} (\theta_i, \psi_i(\theta_i))$ and $\theta_i \overset{\text{ind}}{\sim} (\mu_i, \sigma_i^2)$, $i = 1, \ldots, m$. Then, we have $\hat{\theta}_i \overset{\text{ind}}{\sim} (\mu_i, \psi_i + \sigma_i^2)$ unconditionally, where $\psi_i = E(\psi_i(\theta_i))$. This result follows by noting that $E(\hat{\theta}_i) = E\{E(\hat{\theta}_i|\theta_i)\} = E(\theta_i) = \mu_i$ and $V(\hat{\theta}_i) = E\{V(\hat{\theta}_i|\theta_i)\} + V\{E(\hat{\theta}_i|\theta_i)\} = E(\psi_i(\theta_i)) + V(\theta_i) = \psi_i + \sigma_i^2$. We consider a linear class of estimators of the realized θ_i of the form $a_i\hat{\theta}_i + b_i$ and minimize the unconditional MSE, $E(a_i\hat{\theta}_i + b_i - \theta_i)^2$ with respect to the constants a_i and b_i. The optimal estimator, called the LB estimator, is given by

$$\hat{\theta}_i^{\text{LB}} = \mu_i + \gamma_i(\hat{\theta}_i - \mu_i) = \gamma_i\hat{\theta}_i + (1 - \gamma_i)\mu_i, \tag{9.7.1}$$

where $\gamma_i = \sigma_i^2/(\psi_i + \sigma_i^2)$; see Griffin and Krutchkoff (1971) and Hartigan (1969). A proof of (9.7.1) is given in Section 9.9. The LB estimator (9.7.1) involves $2m$ parameters (μ_i, σ_i^2). In practice, we need to assume that μ_i and σ_i^2 depend on a fixed set of parameters $\boldsymbol{\lambda}$. The MSE of $\hat{\theta}_i^{\text{LB}}$ is given by

$$\text{MSE}\left(\hat{\theta}_i^{\text{LB}}\right) = E(\hat{\theta}_i^{\text{LB}} - \theta_i)^2 = \gamma_i\psi_i. \tag{9.7.2}$$

We estimate $\boldsymbol{\lambda}$ by the method-of-moments and use the estimator $\hat{\boldsymbol{\lambda}}$ in (9.7.1) to obtain the ELB estimator

$$\hat{\theta}_i^{\text{ELB}} = \hat{\gamma}_i \hat{\theta}_i + (1 - \hat{\gamma}_i)\hat{\mu}_i, \tag{9.7.3}$$

where $\hat{\mu}_i = \mu_i(\hat{\boldsymbol{\lambda}})$ and $\hat{\gamma}_i = \gamma_i(\hat{\boldsymbol{\lambda}})$. A naive estimator of $\text{MSE}(\hat{\theta}_i^{\text{ELB}})$ is obtained as

$$\text{mse}_N(\hat{\theta}_i^{\text{ELB}}) = \hat{\gamma}_i \hat{\psi}_i, \tag{9.7.4}$$

where $\hat{\psi}_i = \psi_i(\hat{\boldsymbol{\lambda}})$. But the naive estimator underestimates the MSE because it ignores the variability associated with $\hat{\boldsymbol{\lambda}}$. It is difficult to find approximately unbiased estimators of MSE without further assumptions. In general, $\text{MSE}(\hat{\theta}_i^{\text{ELB}}) \neq \text{MSE}(\hat{\theta}_i^{\text{LB}}) + E(\hat{\theta}_i^{\text{ELB}} - \hat{\theta}_i^{\text{LB}})^2$ because of the nonzero covariance term $E(\hat{\theta}_i^{\text{LB}} - \theta_i)(\hat{\theta}_i^{\text{ELB}} - \hat{\theta}_i^{\text{LB}})$. As a result, the jackknife method is not applicable here without further assumptions.

To illustrate ELB estimation, suppose that θ_i is the relative risk and $\hat{\theta}_i = y_i/e_i$ is the SMR for the ith area. We assume a two-stage model of the form $E(\hat{\theta}_i|\theta_i) = \theta_i, \psi_i(\theta_i) = V(\hat{\theta}_i|\theta_i) = \theta_i/e_i$ and $E(\theta_i) = \mu_i = \mu, V(\theta_i) = \sigma_i^2 = \sigma^2$. The LB estimator is given by (9.7.1) with $\mu_i = \mu$ and $\gamma_i = \sigma^2/(\sigma^2 + \mu/e_i)$. Note that the conditional first and second moments of y_i are identical to the Poisson moments. We obtain moment estimators of μ and σ^2 by equating the weighted sample mean $\hat{\theta}_{e.} = \frac{1}{m}\sum_i (e_i/e.)\hat{\theta}_i$ and the weighted variance, $s_e^2 = \frac{1}{m}\sum_i (e_i/e.)(\hat{\theta}_i - \hat{\theta}_{e.})^2$ to their expected values, as in the case of the Poisson-gamma model (subsection 9.5.1). This leads to moment estimators

$$\hat{\mu} = \hat{\theta}_{e.}, \quad \hat{\sigma}^2 = s_e^2 - \hat{\theta}_{e.}/e.. \tag{9.7.5}$$

Example 9.7.1. Infant Mortality Rates. Marshall (1991) obtained ELB estimates of infant mortality rates in $m = 167$ census area units (CAUs) of Auckland, New Zealand for the period 1977–1985. For this application, we change $(\theta_i, \hat{\theta}_i, e_i)$ to $(\tau_i, \hat{\tau}_i, n_i)$ and let $E(\tau_i) = \mu, V(\tau_i) = \sigma^2$, where τ_i is the mortality rate, $\hat{\tau}_i = y_i/n_i$ is the crude rate, n_i is the number of person-years at risk and y_i is the number of deaths in the ith area. The n_i for a CAU was taken as nine times its recorded population in the 1981 census. This n_i-value should be a good approximation to the true n_i because 1981 is the midpoint of the study period 1977–1985. "Global" ELB estimates of the τ_is were obtained from (9.7.3), using the moment estimator based on (9.7.5). These estimators shrink the crude rates $\hat{\tau}_i$ toward the overall mean $\hat{\mu} = 2.63$ deaths per thousand. Marshall (1991) also obtained "local" estimates by defining the neighbors of each CAU to be those sharing a common boundary; the smallest neighborhood contained 3 CAUs and the largest 13 CAUs. A local estimate of τ_i was obtained by using local estimates $\hat{\mu}_{(i)}$ and $\hat{\sigma}_{(i)}^2$ of μ and σ^2 for each i. The estimates $\hat{\mu}_{(i)}$ and $\hat{\sigma}_{(i)}^2$ were obtained from (9.7.5) using only the neighborhood areas of i to calculate the weighted sample mean and variance. This heuristic method of local smoothing is similar to smoothing based on spatial modelling of the areas. Marshall (1991) did not report standard errors of the estimates. For a global estimate, the naive standard error based on (9.7.4) should be adequate because $m(=167)$ is large, but it may be inadequate as a measure of variability for local estimates based on 3 to 13 CAUs.

Karunamuni and Zhang (2003) studied LB and ELB estimation of finite population small area means, $\bar{Y}_i = \sum_j y_{ij}/N_i$, for unit level models, without covariates. Suppose that the population model is a two-stage model of the form $y_{ij}|\theta_i \overset{\text{ind}}{\sim} (\theta_i, \mu_2(\theta_i))$ for each i $(j = 1, \ldots, N_i)$ and $\theta_i \overset{\text{iid}}{\sim} (\mu, \sigma_v^2)$, $\sigma_e^2 = E[\mu_2(\theta_i)]$. We assume that the model holds for the sample $\{y_{ij}, j = 1, \ldots, n_i; i = 1, \ldots, m\}$. We consider a linear class of estimators of the realized $\bar{Y}_i$ of the form $a_i\bar{y}_i + b_i$, where $\bar{y}_i$ is the ith area sample mean. Minimizing the unconditional MSE, $E(a_i\bar{y}_i + b_i - \bar{Y}_i)^2$, with respect to the constants a_i and b_i, the "optimal" LB estimator of $\bar{Y}_i$ is obtained as

$$\hat{\bar{Y}}_i^{\text{LB}} = f_i\bar{y}_i + (1 - f_i)[\gamma_i\bar{y}_i + (1 - \gamma_i)\mu], \tag{9.7.6}$$

where $\gamma_i = \sigma_v^2/(\sigma_v^2 + \sigma_e^2/n_i)$ and $f_i = n_i/N_i$. If we replace μ by the BLUE estimator $\tilde{\mu} = \sum_i \gamma_i\bar{y}_i / \sum_i \gamma_i$, we get a first-step ELB estimator similar to the BLUP estimator for the basic unit level model (5.3.4) without covariates and $k_{ij} = 1$. Ghosh and Lahiri (1987) used ANOVA-type estimators of σ_e^2 and σ_v^2:

$$\hat{\sigma}_e^2 = \sum_i \sum_j (y_{ij} - \bar{y}_i)^2/(n_T - m) = s_w^2 \tag{9.7.7}$$

and

$$\hat{\sigma}_v^2 = \max\{0, (s_b^2 - s_w^2)(m - 1)g^{-1}\}, \tag{9.7.8}$$

where $s_b^2 = \sum_i n_i(\bar{y}_i - \bar{y}_.)^2/(m-1)$, with $\bar{y}_. = \Sigma_i n_i\bar{y}_i/n_T$, $g = n_T - \sum_i n_i^2/n_T$ and $n_T = \sum_i n_i$. Replacing σ_e^2 and σ_v^2 by $\hat{\sigma}_e^2$ and $\hat{\sigma}_v^2$ in the first-step estimator, we get an ELB estimator of $\bar{Y}_i$. Noting that the first-step estimator depends only on the ratio $\tau = \sigma_v^2/\sigma_e^2$, an approximately unbiased estimator of τ may be used along the lines of Morris (1983b):

$$\tau^* = \max\left[0, \left\{\frac{(m-1)s_b^2}{(m-3)s_w^2} - 1\right\}(m-1)g^{-1}\right], \tag{9.7.9}$$

instead of $\hat{\tau} = \hat{\sigma}_v^2/\hat{\sigma}_e^2$. The multiplier $(m-1)/(m-3)$ corrects the bias of $\hat{\tau}$. The modified ELB estimator is obtained from the first-step estimator by substituting τ^* for τ, and using $\hat{\mu} = \sum_i \gamma_i^*\bar{y}_i / \sum_i \gamma_i^*$ if $\tau^* \neq 0$ and $\hat{\mu} = \sum_i n_i\bar{y}_i/n_T$ if $\tau^* = 0$.

The MSE of $\hat{\bar{Y}}_i^{\text{LB}}$ is given by

$$\begin{aligned} \text{MSE}(\hat{\bar{Y}}_i^{\text{LB}}) &= E(\hat{\bar{Y}}_i^{\text{LB}} - \bar{Y}_i)^2 \\ &= (1 - f_i)^2 g_{1i}(\sigma_v^2, \sigma_e^2) + \frac{1}{N_i}(1 - f_i)\sigma_e^2 \qquad (9.7.10) \\ &\approx (1 - f_i)^2 g_{1i}(\sigma_v^2, \sigma_e^2), \qquad (9.7.11) \end{aligned}$$

where

$$g_{1i}(\sigma_v^2, \sigma_e^2) = \gamma_i \sigma_e^2 / n_i \tag{9.7.12}$$

and $\gamma_i = \sigma_v^2/(\sigma_v^2 + \sigma_e^2/n_i)$. The last term in (9.7.10) is negligible if N_i is large, leading to the approximation (9.7.11). Note that (9.7.10) has the same form as the MSE of the Bayes estimator of $\bar{Y}_i$ under the basic unit level model with normal effects v_i and normal errors e_{ij} with equal variance σ_e^2. A naive estimator of $\text{MSE}\left(\hat{\bar{Y}}_i^{\text{ELB}}\right)$ is obtained by substituting $(\hat{\sigma}_v^2, \hat{\sigma}_e^2)$ for (σ_v^2, σ_e^2) in (9.7.10), but it underestimates $\text{MSE}\left(\hat{\bar{Y}}_i^{\text{ELB}}\right)$ because it ignores the variability associated with $(\hat{\sigma}_v^2, \hat{\sigma}_e^2)$. Again, it is difficult to find approximately unbiased estimators of MSE without further assumptions.

9.7.2 Posterior Linearity

A different approach, called linear EB, has also been used to estimate the small area means $\bar{Y}_i$ under the two-stage unit level model studied in subsection 9.7.1 (Ghosh and Lahiri (1987)). The basic assumption underlying this approach is the posterior linearity condition (Goldstein (1975)):

$$E(\theta_i|\mathbf{y}_i, \boldsymbol{\lambda}) = a_i\bar{y}_i + b_i, \tag{9.7.13}$$

where $\mathbf{y}_i$ is the $n_i \times 1$ vector of sample observations from the ith area and $\boldsymbol{\lambda} = (\mu, \sigma_v^2, \sigma_e^2)^T$. This condition holds for a variety of distributions on the θ_i's. However, the distribution of the θ_i's becomes a conjugate family if the conditional distribution of y_{ij} given θ_i belongs to the exponential family. For example, $y_{ij}|\theta_i \overset{\text{iid}}{\sim} \text{Bernoulli}(\theta_i)$ and posterior linearity implies that $\theta_i \sim \text{beta}(\alpha, \beta); \alpha > 0, \beta > 0$. The linear EB (LEB) approach assumes posterior linearity and uses the two-stage model on the y_{ij}'s without distributional assumptions. It leads to an estimator of $\bar{Y}_i$ identical to the estimator under the linear Bayes approach, namely, $\hat{\bar{Y}}_i^{\text{LB}}$ given by (9.7.6). The two approaches are therefore similar, but the ELB approach is more appealing as it avoids further assumptions on the two-stage model by confining to the linear class of estimators.

We briefly outline the LEB approach. Using the two-stage model and the posterior linearity condition (9.7.13), it follows from subsection 9.7.1 that the optimal a_i and b_i that minimize $E(\theta_i - a_i\bar{y}_i - b_i)^2$ are given by $a_i^* = \gamma_i$ and $b_i^* = (1-\gamma_i)\mu$, where $\gamma_i = \sigma_v^2/(\sigma_v^2 + \sigma_e^2/n_i)$ as in subsection 9.7.1. Therefore, the Bayes estimator under posterior linearity is given by

$$E(\theta_i|\mathbf{y}_i, \boldsymbol{\lambda}) = \gamma_i\bar{y}_i + (1-\gamma_i)\mu, \tag{9.7.14}$$

which is identical to the LB estimator. Using (9.7.14), we get

$$E(\bar{Y}_i|\mathbf{y}_i, \boldsymbol{\lambda}) = f_i\bar{y}_i + (1-f_i)[\gamma_i\bar{y}_i + (1-\gamma_i)\mu], \tag{9.7.15}$$

noting that for any unit j in the nonsampled set $\bar{s}_i$, $E(y_{ij}|\mathbf{y}_i,\boldsymbol{\lambda}) = E[E(y_{ij}|\theta_i,y_i,\boldsymbol{\lambda})|\mathbf{y}_i,\boldsymbol{\lambda}] = E(\theta_i|\mathbf{y}_i,\boldsymbol{\lambda})$; see Ghosh and Lahiri (1987). It follows from (9.7.15) that $E(\bar{Y}_i|\mathbf{y}_i,\boldsymbol{\lambda})$ is identical to the LB estimator $\hat{\bar{Y}}_i^{\text{LB}}$ given by (9.7.6). Substituting moment estimators $\hat{\boldsymbol{\lambda}}$ in (9.7.15) we get the linear EB estimator, $\hat{\bar{Y}}_i^{\text{LEB}} = E(\bar{Y}_i|\mathbf{y}_i,\hat{\boldsymbol{\lambda}})$, which is identical to the empirical LB estimator $\hat{\bar{Y}}_i^{\text{ELB}}$.

It appears that nothing is gained over the ELB approach by making the posterior linearity assumption and then using the LEB approach. However, it permits the use of the jackknife to obtain an approximately unbiased MSE estimator, unlike the ELB approach, by making use of the orthogonal decomposition

$$\begin{aligned}\text{MSE}\left(\hat{\bar{Y}}_i^{\text{LEB}}\right) &= E\left(\hat{\bar{Y}}_i^{\text{LB}} - \bar{Y}_i\right)^2 + E\left(\hat{\bar{Y}}_i^{\text{LEB}} - \hat{\bar{Y}}_i^{\text{LB}}\right)^2 \\ &= \tilde{g}_{1i}(\sigma_v^2,\sigma_e^2) + E\left(\hat{\bar{Y}}_i^{\text{LEB}} - \hat{\bar{Y}}_i^{\text{LB}}\right)^2 = M_{1i} + M_{2i}, \quad (9.7.16)\end{aligned}$$

where $\tilde{g}_{1i}(\sigma_v^2,\sigma_e^2)$ is given by (9.7.10). We obtain a jackknife estimator of the last term in (9.7.16) as

$$\hat{M}_{2i} = \frac{m-1}{m}\sum_{l=1}^{m}\left(\hat{\bar{Y}}_{i,-l}^{\text{LEB}} - \hat{\bar{Y}}_i^{\text{LEB}}\right)^2, \quad (9.7.17)$$

where $\hat{\bar{Y}}_{i,-l}^{\text{LEB}} = E(\bar{Y}_i|\mathbf{y}_i,\hat{\boldsymbol{\lambda}}_{-l})$ is obtained from $\hat{\bar{Y}}_i^{\text{LEB}}$ by substituting the delete$-l$ estimators $\hat{\boldsymbol{\lambda}}_{-l}$ for $\hat{\boldsymbol{\lambda}}$. The leading term, M_{1i}, is estimated as

$$\hat{M}_{1i} = \tilde{g}_{1i}(\hat{\sigma}_v^2,\hat{\sigma}_e^2) - \frac{m-1}{m}\sum_{l=1}^{m}[\tilde{g}_{1i}(\hat{\sigma}_{v,-l}^2,\hat{\sigma}_{e,-l}^2) - \tilde{g}_{1i}(\hat{\sigma}_v^2,\hat{\sigma}_e^2)]. \quad (9.7.18)$$

The jackknife MSE estimator is given by

$$\text{mse}_J\left(\hat{\bar{Y}}_i^{\text{LEB}}\right) = \hat{M}_{1i} + \hat{M}_{2i}. \quad (9.7.19)$$

Results for the infinite population case are obtained by letting $f_i = 0$.

The above results are applicable to two-stage models without covariates. Raghunathan (1993) proposed two-stage models with area level covariates, by specifying only the first and second moments; see subsection 5.6.4. He obtained a quasi-EB estimator of the small area parameter, θ_i, using the following steps: (1) Based on the mean and variance specifications, obtain the conditional quasi-posterior density of θ_i given the data and model parameters $\boldsymbol{\lambda}$. (2) Evaluate the quasi-Bayes estimator and the conditional quasi-posterior variance by numerical integration, using the density from step (1). (3) Use a generalized EM (GEM) algorithm to get quasi-ML estimator $\hat{\boldsymbol{\lambda}}$ of the model parameters. (4) Replace $\boldsymbol{\lambda}$ by $\hat{\boldsymbol{\lambda}}$ in the quasi-Bayes estimator to obtain the quasi-EB estimator of θ_i. Raghunathan (1993) also proposed a jackknife method of estimating the MSE of the quasi-EB estimator, taking account of the variability of $\hat{\boldsymbol{\lambda}}$. This method is different from the jackknife method of Jiang, Lahiri and Wan (2001), and its asymptotic properties have not been studied.

9.8 Constrained LB

Consider the two-stage model (i) $\hat{\theta}_i|\theta_i \overset{\text{ind}}{\sim} (\theta_i, \psi_i(\theta_i))$ and (ii) $\theta_i \overset{\text{iid}}{\sim} (\mu, \sigma^2)$ and $\psi_i = E(\psi_i(\theta_i))$. We consider a linear class of estimators of the realized θ_i of the form $a_i\hat{\theta}_i + b_i$ and determine the constants a_i and b_i to match the mean μ and the variance σ^2 of θ_i:

$$E(a_i\hat{\theta}_i + b_i) = \mu, \tag{9.8.1}$$

$$E(a_i\hat{\theta}_i + b_i - \mu)^2 = \sigma^2. \tag{9.8.2}$$

Noting that $E(\hat{\theta}_i) = \mu$, we get $b_i = \mu - (1 - a_i)$ from (9.8.1), and (9.8.2) reduces to

$$a_i^2 E(\hat{\theta}_i - \mu)^2 = a_i^2(\sigma^2 + \psi_i) = \sigma^2 \tag{9.8.3}$$

or $a_i = \gamma_i^{1/2}$, where $\gamma_i = \sigma^2/(\sigma^2+\psi_i)$. The resulting estimator is a constrained LB (CLB) estimator:

$$\hat{\theta}_i^{\text{CLB}} = \gamma_i^{1/2}\hat{\theta}_i + (1 - \gamma_i^{1/2})\mu \tag{9.8.4}$$

(Spjøtvoll and Thomsen (1987)). Method of moments may be used to estimate the model parameters. The resulting estimator is an empirical CLB estimator. Note that $\hat{\theta}_i^{\text{LB}}$ attaches a smaller weight, γ_i, to the direct estimator $\hat{\theta}_i$ which leads to over-shrinking toward μ compared to $\hat{\theta}_i^{\text{CLB}}$.

As an example, consider the case of binary data: $\hat{\theta}_i|\theta_i \overset{\text{ind}}{\sim} (\theta_i, \theta_i(1-\theta_i)/n_i)$ and $\theta_i \overset{\text{iid}}{\sim} (\mu, \sigma^2)$, where $\hat{\theta}_i$ is the sample proportion and n_i is the sample size in the ith area. Noting that $E[\theta_i(1-\theta_i)] = \mu(1-\mu) - \sigma^2$, we get

$$\gamma_i = \sigma^2 \Big/ \left[\left(1 - \frac{1}{n_i}\right)\sigma^2 + \frac{\mu(1-\mu)}{n_i}\right]. \tag{9.8.5}$$

Spjøtvoll and Thomsen (1987) used empirical CLB to study the distribution $m = 100$ municipalities in Norway with respect to the proportion of persons not in the labor force. A comparison with the actual distribution in their example showed that the CLB method tracked the actual distribution much better than the LB method.

The above CLB method ignores the simultaneous aspect of the problem by considering each area separately. However, the CLB estimator (9.8.4) is similar to the constrained Bayes (CB) estimator, $\hat{\theta}_i^{\text{CB}}$. In fact, as shown in

(9.6.9), $\hat{\theta}_i^{\text{CB}} \approx \hat{\theta}_i^{\text{CLB}}$ under the simple model $\hat{\theta}_i = \theta_i + e_i$ with $e_i \overset{\text{iid}}{\sim} N(0, \psi)$ independent of $\theta_i \overset{\text{iid}}{\sim} N(\mu, \sigma_v^2)$.

To take account of the simultaneous aspect of the problem, we can use the method of subsection 9.6.1 assuming posterior linearity $E(\theta_i|\hat{\theta}_i) = a_i\hat{\theta}_i + b_i$. We minimize the posterior expected squared error loss under the two-stage model subject to the constraints (9.6.4) and (9.6.5) on the ensemble mean and variance, respectively. The resulting CLB estimator is equal to (9.9.6) with $\hat{\theta}_i^B$ and $V(\theta_i|\hat{\theta}_i, \boldsymbol{\lambda})$ changed to $\hat{\theta}_i^{\text{LB}} = \gamma_i\hat{\theta}_i + (1-\gamma_i)\mu$ and $E[(\theta_i - \hat{\theta}_i^{\text{LB}})^2|\hat{\theta}_i, \boldsymbol{\lambda}]$, respectively. This estimator avoids distributional assumptions. Lahiri (1990) called this method "adjusted" Bayes estimation.

The posterior variance term $E[(\theta_i - \hat{\theta}_i^{\text{LB}})^2|\hat{\theta}_i, \boldsymbol{\lambda}]$ cannot be calculated without additional assumptions (Lahiri (1990) and Ghosh (1992) incorrectly stated that it is equal to $\gamma_i\psi_i$). However, for large m, we can approximate $m^{-1}\Sigma_i E[(\theta_i - \hat{\theta}_i^{\text{LB}})^2|\hat{\theta}_i, \boldsymbol{\lambda}]$ by its expectation $m^{-1}\Sigma_i E[(\theta_i - \hat{\theta}_i^{\text{LB}})^2|\boldsymbol{\lambda}] = m^{-1}\Sigma_i\gamma_i\psi_i$. Using this approximation we get the following CLB estimator:

$$\hat{\theta}_i^{\text{CLB}}(1) = \hat{\theta}_{\cdot}^{\text{LB}} + a^*(\hat{\boldsymbol{\theta}}, \boldsymbol{\lambda})(\hat{\theta}_i^{\text{LB}} - \hat{\theta}_{\cdot}^{\text{LB}}), \tag{9.8.6}$$

where

$$a^*(\hat{\boldsymbol{\theta}}, \boldsymbol{\lambda}) = \left[1 + \frac{(1/m)\Sigma_i\gamma_i\psi_i}{\{1/(m-1)\}\Sigma_i(\hat{\theta}_i^{\text{LB}} - \hat{\theta}_{\cdot}^{\text{LB}})^2}\right]^{1/2}. \tag{9.8.7}$$

In general, $\hat{\theta}_i^{\text{CLB}}(1)$ differs from $\hat{\theta}_i^{\text{CLB}}$ given by (9.8.4), but for the special case of equal ψ_i, we get $\hat{\theta}_i^{\text{CLB}}(1) \approx \hat{\theta}_i^{\text{CLB}}$ for large m, similar to the result (9.6.9).

The CLB estimator (9.8.6) depends on the unknown model parameters $\boldsymbol{\lambda} = (\mu, \sigma^2)^T$. Replacing $\boldsymbol{\lambda}$ by moment estimators $\hat{\boldsymbol{\lambda}}$, we obtain an empirical CLB estimator.

9.9 Proofs

9.9.1 Proof of (9.2.11)

We express the expectation E as $E_{\hat{\boldsymbol{\theta}}}E_{\boldsymbol{\theta}|\hat{\boldsymbol{\theta}}}$ and $(\hat{\theta}_i^{\text{EB}} - \theta_i)^2$ as

$$(\hat{\theta}_i^{\text{EB}} - \hat{\theta}_i^B + \hat{\theta}_i^B - \theta_i)^2 = (\hat{\theta}_i^{\text{EB}} - \hat{\theta}_i^B)^2 + (\hat{\theta}_i^B - \theta_i)^2 + 2(\hat{\theta}_i^{\text{EB}} - \hat{\theta}_i^B)(\hat{\theta}_i^B - \theta_i), \tag{9.9.1}$$

where $E_{\boldsymbol{\theta}|\hat{\boldsymbol{\theta}}}$ is the expectation over the conditional distribution of $\boldsymbol{\theta}$ given $\hat{\boldsymbol{\theta}}$ and $E_{\hat{\boldsymbol{\theta}}}$ is the expectation over the marginal distribution of $\hat{\boldsymbol{\theta}}$. Further,

$$E_{\boldsymbol{\theta}|\hat{\boldsymbol{\theta}}}[(\hat{\theta}_i^{\text{EB}} - \hat{\theta}_i^B)(\hat{\theta}_i^B - \theta_i)] = (\hat{\theta}_i^{\text{EB}} - \hat{\theta}_i^B)[E_{\theta_i|\hat{\theta}_i}(\hat{\theta}_i^B - \theta_i)] = 0. \tag{9.9.2}$$

Now taking the expectation of (9.9.1) and using (9.9.2), we get the decomposition (9.2.11), noting that $E = E_{\hat{\boldsymbol{\theta}}}E_{\boldsymbol{\theta}|\hat{\boldsymbol{\theta}}}$.

9.9.2 Proof of (9.2.30)

We note that $\hat{\theta}_i|\mu, \sigma_v^2 \overset{\text{iid}}{\sim} N(\mu, \sigma_v^2 + \psi), i = 1, \ldots, n$ and $f(\mu) = \text{constant}$. Therefore, by Bayes theorem,

$$
\begin{aligned}
f(\mu|\hat{\boldsymbol{\theta}}, \sigma_v^2) &\propto f(\hat{\boldsymbol{\theta}}|\mu, \sigma_v^2) f(\mu) \\
&\propto \exp\left\{-\frac{1}{2}\sum_i (\hat{\theta}_i - \mu)^2/(\sigma_v^2 + \psi)\right\} \\
&\propto \exp\left\{-\frac{m}{2}\sum_i (\mu - \hat{\theta}_\cdot)^2/(\sigma_v^2 + \psi)\right\},
\end{aligned}
$$

which is the kernel of $N(\hat{\theta}_\cdot, m^{-1}(\sigma_v^2 + \psi))$. Hence, we get (9.2.30).

9.9.3 Proof of (9.6.6)

The posterior expected squared error loss $E[\sum_i (\theta_i - t_i)^2|\hat{\boldsymbol{\theta}}]$ may be written as

$$
E\left[\sum_{i=1}^m (\theta_i - t_i)^2|\hat{\boldsymbol{\theta}}, \boldsymbol{\lambda}\right] = E\left[\sum_{i=1}^m (\theta_i - \hat{\theta}_i^B)^2|\hat{\boldsymbol{\theta}}, \boldsymbol{\lambda}\right] + \sum_{i=1}^m (\hat{\theta}_i^B - t_i)^2. \quad (9.9.3)
$$

Therfore, it is sufficient to minimize the last term of (9.9.3). Using Lagrange multipliers a_1 and a_2, we minimize $\sum_i (\hat{\theta}_i^B - t_i)^2$ subject to constraints $\sum_i t_i = c_1$ and $\sum_i (t_i - t_\cdot)^2 = c_2$ or $\sum_i t_i^2 = c_2 + c_1^2/m$. That is, we minimize the objective function $\phi = \sum_i (\hat{\theta}_i^B - t_i)^2 - a_1(\sum_i t_i - c_1) - a_2(\sum_i t_i^2 - c_2 - c_1^2/m)$ with respect to the t_i's. We obtain

$$
t_{i,\text{opt}} = \frac{1}{1 - a_2}\left(\hat{\theta}_i^B + \frac{a_1}{2}\right). \quad (9.9.4)
$$

Imposing the constraints on (9.9.4), we get $a_1/2 = (1 - a_2)c_1/m - \hat{\theta}_\cdot^B$ and $(1 - a_2)^2 = \sum_i (\hat{\theta}_i^B - \hat{\theta}_\cdot^B)^2/c_2$. Now taking the positive square root of $(1 - a_2)^2$, we get

$$
t_{i,\text{opt}} = \frac{c_1}{m} + \frac{1}{1 - a_2}(\hat{\theta}_i^B - \hat{\theta}_\cdot^B) \quad (9.9.5)
$$

from (9.9.4), where $1 - a_2 = [\sum_i (\hat{\theta}_i^B - \hat{\theta}_\cdot^B)^2/c_2]^{1/2}$. The estimator (9.9.5) is valid for any specified c_1 and c_2. In particular, the constraints in (9.6.4) and (9.6.5) specify $c_1 = m\hat{\theta}_\cdot^B$ and $c_2 = E[\sum_i (\theta_i - \theta_\cdot)^2|\hat{\boldsymbol{\theta}}]$. Now using (9.6.2) and these values of c_1 and c_2 in (9.9.5), we get $t_{i,\text{opt}} = \hat{\theta}_i^{\text{CB}}$ given by (9.6.6). This proof is due to Singh and Folsom (2001a) and Shen and Louis (1998).

9.9.4 Proof of (9.7.1)

We minimize $\phi_i = E(a_i\hat{\theta}_i + b_i - \theta_i)^2$ with respect to a_i and b_i to get the optimal a_i^* and b_i^* from

$$a_i^* \mu_i + b_i^* = \mu_i, \tag{9.9.6}$$

$$E(\theta_i\hat{\theta}_i) = a_i^* E(\hat{\theta}_i^2) + b_i^* \mu_i. \tag{9.9.7}$$

Substituting for b_i^* from (9.9.6) into (9.9.7), we get $\text{Cov}(\hat{\theta}_i, \theta_i)/V(\hat{\theta}_i) = \sigma_i^2/(\psi_i + \sigma_i^2) = \gamma_i$, noting that $E(\theta_i\hat{\theta}_i) = E\{\theta_i E(\hat{\theta}_i|\theta_i)\} = E(\theta_i^2) = \sigma_i^2 + \mu_i^2$ and $V(\theta_i) = \psi_i + \sigma_i^2$. Hence,

$$a_i^*\hat{\theta}_i + b_i^* = \mu_i + \gamma_i(\hat{\theta}_i - \mu_i),$$

which is identical to the LB estimator $\hat{\theta}_i^{\text{LB}}$ given by (9.7.1).

Chapter 10

Hierarchical Bayes (HB) Method

10.1 Introduction

In the hierarchical Bayes (HB) approach, a subjective prior distribution $f(\boldsymbol{\lambda})$ on the model parameters $\boldsymbol{\lambda}$ is specified and the posterior distribution $f(\boldsymbol{\mu}|\mathbf{y})$ of the small area (random) parameters of interest $\boldsymbol{\mu}$, given the data $\mathbf{y}$, is obtained. The two-stage model, $f(\mathbf{y}|\boldsymbol{\mu},\boldsymbol{\lambda}_1)$ and $f(\boldsymbol{\mu}|\boldsymbol{\lambda}_2)$, is combined with the subjective prior on $\boldsymbol{\lambda} = (\boldsymbol{\lambda}_1^T, \boldsymbol{\lambda}_2^T)^T$, using Bayes theorem, to arrive at the posterior $f(\boldsymbol{\mu}|\mathbf{y})$. Inferences are based on $f(\boldsymbol{\mu}|\mathbf{y})$; in particular, a parameter of interest, say $\phi = h(\boldsymbol{\mu})$, is estimated by its posterior mean $\hat{\phi}^{\text{HB}} = E[h(\boldsymbol{\mu})|\mathbf{y}]$, and the posterior variance $V[h(\boldsymbol{\mu})|\mathbf{y}]$ is used as a measure of precision of the estimator, provided they are finite.

The HB approach is straightforward, and HB inferences are clear-cut and "exact", but require the specification of a subjective prior $f(\boldsymbol{\lambda})$ on the model parameters $\boldsymbol{\lambda}$. Priors on $\boldsymbol{\lambda}$ may be informative or "diffuse". Informative priors are based on substantial prior information, such as previous studies judged relevant to the current data set $\mathbf{y}$. However, informative priors are seldom available in real HB applications, particularly those related to public policy. Diffuse (or noninformative or default) priors are designed to reflect lack of information about $\boldsymbol{\lambda}$. The choice of a diffuse prior is not unique, and some diffuse improper priors could lead to improper posteriors. It is therefore essential to make sure that the chosen diffuse prior, $f(\boldsymbol{\lambda})$, leads to a proper posterior $f(\boldsymbol{\mu}|\mathbf{y})$. Also, it is desirable to select a diffuse prior that leads to well-calibrated inferences in the sense of validity under the frequentist framework. In practice, the frequentist bias, $E(\hat{\phi}^{\text{HB}} - \phi)$, of the HB estimator $\hat{\phi}^{\text{HB}}$ and the relative frequentist bias of the posterior variance as an estimator of $\text{MSE}(\hat{\phi}^{\text{HB}})$ should be small. In addition, the frequentist coverage of a HB interval on ϕ should be close to the nominal level (Dawid (1985); Browne and Draper (2000)).

Applying Bayes theorem, we have

$$f(\boldsymbol{\mu}, \boldsymbol{\lambda}|\mathbf{y}) = \frac{f(\mathbf{y}, \boldsymbol{\mu}|\boldsymbol{\lambda})f(\boldsymbol{\lambda})}{f_1(\mathbf{y})}, \tag{10.1.1}$$

where $f_1(\mathbf{y})$ is the marginal density of $\mathbf{y}$:

$$f_1(\mathbf{y}) = \int f(\mathbf{y}, \boldsymbol{\mu}|\boldsymbol{\lambda})f(\boldsymbol{\lambda})d\boldsymbol{\mu}d\boldsymbol{\lambda}. \tag{10.1.2}$$

The desired posterior density $f(\boldsymbol{\mu}|\mathbf{y})$ is obtained from (10.1.1) as

$$f(\boldsymbol{\mu}|\mathbf{y}) = \int f(\boldsymbol{\mu}, \boldsymbol{\lambda}|\mathbf{y})d\boldsymbol{\lambda} \tag{10.1.3}$$

$$= \int f(\boldsymbol{\mu}|\mathbf{y}, \boldsymbol{\lambda})f(\boldsymbol{\lambda}|\mathbf{y})d\boldsymbol{\lambda}. \tag{10.1.4}$$

The form (10.1.4) shows that $f(\boldsymbol{\mu}|\mathbf{y})$ is a mixture of conditional densities $f(\boldsymbol{\mu}|\mathbf{y}, \boldsymbol{\lambda})$; note that $f(\boldsymbol{\mu}|\mathbf{y}, \boldsymbol{\lambda})$ is used for EB inferences. Because of the mixture form (10.1.4), HB is also called Bayes EB or fully Bayes.

It is clear from (10.1.1) and (10.1.3) that the evaluation of $f(\boldsymbol{\mu}|\mathbf{y})$ and associated posterior quantities such as $E[h(\boldsymbol{\mu})|\mathbf{y}]$ involves multi-dimensional integrations. However, it is often possible to perform integration analytically with respect to some of the components of $\boldsymbol{\mu}$ and $\boldsymbol{\lambda}$. If the reduced problem involves one- or two-dimensional integration, then direct numerical integration can be used to calculate the desired posterior quantities. For complex problems, however, it becomes necessary to evaluate high dimensional integrals. Recently developed Markov chain Monte Carlo (MCMC) methods seem to overcome the computational difficulties to a large extent. These methods generate samples from the posterior distribution, and then use the simulated samples to approximate the desired posterior quantities. Software packages BUGS and CODA implement MCMC and convergence diagnostics (see subsection 10.2.5). We refer the reader to the review paper by Brooks (1998) and the book edited by Gilks, Richardson and Spiegelhalter (1996) for details on MCMC methods used in Bayesian computations. Section 10.2 gives a brief account of MCMC methods; in particular, the Gibbs sampling algorithm and its extension, the Metropolis-Hastings (M-H) algorithm. Chib and Greenberg (1995) have provided a detailed, introductory exposition of the M-H algorithm.

10.2 MCMC Methods

10.2.1 Markov Chain

Let $\boldsymbol{\eta} = (\boldsymbol{\mu}^T, \boldsymbol{\lambda}^T)^T$ be the vector of small area parameters $\boldsymbol{\mu}$ and model parameters $\boldsymbol{\lambda}$. In general, it is not feasible to draw independent samples from the joint posterior $f(\boldsymbol{\eta}|\mathbf{y})$ because of the intractable denominator $f_1(\mathbf{y})$. MCMC avoids this difficulty by constructing a Markov chain $\{\boldsymbol{\eta}^{(k)}, k = 0, 1, 2, \ldots\}$ such that the distribution of $\boldsymbol{\eta}^{(k)}$ converges to a unique stationary (or invariant) distribution equal to $f(\boldsymbol{\eta}|\mathbf{y})$, denoted $\pi(\boldsymbol{\eta})$. Thus, after a sufficiently large "burn-in", say, d, we can regard $\boldsymbol{\eta}^{(d+1)}, \ldots, \boldsymbol{\eta}^{(d+D)}$ as D dependent samples from the target distribution $f(\boldsymbol{\eta}|\mathbf{y})$, regardless of the starting point $\boldsymbol{\eta}^{(0)}$.

To construct a Markov chain, we need to specify a one-step transition probability (or kernel) $P(\boldsymbol{\eta}^{(k+1)}|\boldsymbol{\eta}^{(k)})$ which depends only on the current "state" $\boldsymbol{\eta}^{(k)}$ of the chain. That is, the conditional distribution of $\boldsymbol{\eta}^{(k+1)}$ given $\boldsymbol{\eta}^{(0)}, \ldots, \boldsymbol{\eta}^{(k)}$ does not depend on the "history" of the chain $\{\boldsymbol{\eta}^{(0)}, \ldots, \boldsymbol{\eta}^{(k-1)}\}$. The transition kernel must satisfy the stationarity condition:

$$\int \pi(\boldsymbol{\eta}^{(k)}) P(\boldsymbol{\eta}^{(k+1)}|\boldsymbol{\eta}^{(k)}) d\boldsymbol{\eta}^{(k)} = \pi(\boldsymbol{\eta}^{(k+1)}). \tag{10.2.1}$$

Equation (10.2.1) says that if $\boldsymbol{\eta}^{(k)}$ is from $\pi(\cdot)$ then $\boldsymbol{\eta}^{(k+1)}$ will also be from $\pi(\cdot)$. Stationarity is satisfied if the chain is "reversible":

$$\pi(\boldsymbol{\eta}^{(k)}) P(\boldsymbol{\eta}^{(k+1)}|\boldsymbol{\eta}^{(k)}) = \pi(\boldsymbol{\eta}^{(k+1)}) P(\boldsymbol{\eta}^{(k)}|\boldsymbol{\eta}^{(k+1)}). \tag{10.2.2}$$

Verification of (10.2.2) ensures that the stationary distribution of the chain generated by $P(\cdot|\cdot)$ is $\pi(\cdot)$.

It is also necessary to ensure that the distribution of $\boldsymbol{\eta}^{(k)}$ given $\boldsymbol{\eta}^{(0)}$, denoted $P^{(k)}(\boldsymbol{\eta}^{(k)}|\boldsymbol{\eta}^{(0)})$, converges to $\pi(\boldsymbol{\eta}^{(k)})$ regardless of $\boldsymbol{\eta}^{(0)}$. For this, the chain needs to be "irreducible" and "aperiodic". Irreducibility means that from all starting points $\boldsymbol{\eta}^{(0)}$ the chain will eventually reach any nonempty set in the state space with positive probability. Aperiodicity means that the chain is not permitted to oscillate between different sets in a periodic manner. For an irreducible and aperiodic chain, the ergodic theorem also holds:

$$\bar{h}_D = \frac{1}{D} \sum_{k=d+1}^{d+D} h(\boldsymbol{\eta}^{(k)}) \rightarrow_p E[h(\boldsymbol{\eta})|\mathbf{y}] \tag{10.2.3}$$

as $D \rightarrow \infty$, where $\rightarrow_p$ denotes convergence in probability. Thus, for sufficiently large D, we may be able to obtain an estimator, $\bar{h}_D$, of $E[h(\boldsymbol{\eta})|\mathbf{y}]$ with adequate precision. However, it is not easy to find a Monte Carlo standard error of $\bar{h}_D$ because of dependence in the simulated samples $\boldsymbol{\eta}^{(d+1)}, \ldots, \boldsymbol{\eta}^{(d+D)}$.

10.2.2 Gibbs Sampler

To generate the samples $\boldsymbol{\eta}^{(k)}$, we partition $\boldsymbol{\eta}$ into suitable blocks $\boldsymbol{\eta}_1, \ldots, \boldsymbol{\eta}_r$. Some of the blocks might contain only single elements, whereas others contain more than one element. For example, consider the basic area level model with $\boldsymbol{\mu} = (\theta_1, \ldots, \theta_m)^T = \boldsymbol{\theta}$ and $\boldsymbol{\lambda} = (\boldsymbol{\beta}^T, \sigma_v^2)^T$. In this case $\boldsymbol{\eta}$ may be partitioned as $\boldsymbol{\eta}_1 = \boldsymbol{\beta}, \eta_2 = \theta_1, \ldots, \eta_{m+1} = \theta_m, \eta_{m+2} = \sigma_v^2$ $(r = m + 2)$. We need the following set of Gibbs conditional distributions: $f(\boldsymbol{\eta}_1|\boldsymbol{\eta}_2, \ldots, \boldsymbol{\eta}_r, \mathbf{y})$, $f(\boldsymbol{\eta}_2|\boldsymbol{\eta}_1, \boldsymbol{\eta}_3, \ldots, \boldsymbol{\eta}_r, \mathbf{y})$, ..., $f(\boldsymbol{\eta}_r|\boldsymbol{\eta}_1, \ldots, \boldsymbol{\eta}_{r-1}, \mathbf{y})$. The Gibbs sampler uses these conditional distributions to construct a transition kernel, $P(\cdot|\cdot)$, such that the stationary distribution of the resulting Markov chain is $\pi(\boldsymbol{\eta}) = f(\boldsymbol{\eta}|\mathbf{y})$. This result follows from the fact that $f(\boldsymbol{\eta}|\mathbf{y})$ is uniquely determined by the set of Gibbs conditionals.

If a conditional distribution has a standard, closed form, such as normal or inverted gamma, then samples can be generated directly from the conditional distribution. Otherwise, alternative algorithms, such as Metropolis-Hastings (M-H) rejection sampling, may be used to generate samples from the conditional distribution. Some authors suggest using M-H also for the closed form cases. If M-H is used only for the cases without a closed form, then the algorithm is called M-H within Gibbs.

The Gibbs sampling algorithm involves the following steps.

Step 0. Choose a starting point $\boldsymbol{\eta}^{(0)}$ with components $\boldsymbol{\eta}_1^{(0)}, \ldots, \boldsymbol{\eta}_r^{(0)}$; set $k = 0$. For example, one could use REML or moment estimates of model parameters $\boldsymbol{\lambda}$ and EB estimates of $\boldsymbol{\mu}$ as starting values. However, the starting points can be arbitrary.

Step 1. Generate $\boldsymbol{\eta}^{(k+1)} = (\boldsymbol{\eta}_1^{(k+1)}, \ldots, \boldsymbol{\eta}_r^{(k+1)})$ as follows: Draw $\boldsymbol{\eta}_1^{(k+1)}$ from $f(\boldsymbol{\eta}_1|\boldsymbol{\eta}_2^{(k)}, \ldots, \boldsymbol{\eta}_r^{(k)}, \mathbf{y})$; $\boldsymbol{\eta}_2^{(k+1)}$ from $f(\boldsymbol{\eta}_2|\boldsymbol{\eta}_1^{(k+1)}, \boldsymbol{\eta}_3^{(k)}, \ldots, \boldsymbol{\eta}_r^{(k)}, \mathbf{y})$; ...; $\boldsymbol{\eta}_r^{(k+1)}$ from $f(\boldsymbol{\eta}_r|\boldsymbol{\eta}_1^{(k+1)}, \ldots, \boldsymbol{\eta}_{r-1}^{(k+1)}, \mathbf{y})$.

Step 2. Set $k = k + 1$ and go to Step 1.

Steps 1–2 constitute one cycle for each k. The sequence $\{\boldsymbol{\eta}^{(k)}\}$ generated by the Gibbs sampler is a Markov chain with stationary distribution $\pi(\boldsymbol{\eta}) = f(\boldsymbol{\eta}|\mathbf{y})$; see Gelfand and Smith (1990). Note that the one-step transition kernel is the product of the r Gibbs conditional distributions.

10.2.3 M-H Within Gibbs

If all the Gibbs conditionals have closed forms belonging to standard families, then it is straightforward to generate samples from the conditionals. If a conditional does not admit a closed form, then several methods are available for generating samples from the conditional. In particular, we have adaptive rejection sampling for univariate log-concave conditionals (Gilks and Wild (1992)) and more generally M-H (Metropolis et al. (1953), Hastings (1970)) within Gibbs.

Let $f(\boldsymbol{\eta}_i|\boldsymbol{\eta}_{-i}^{(k)}, \mathbf{y})$ denote the Gibbs conditional after completing the first $i-1$ drawings of the $(k+1)$th cycle of Gibbs sampling, where

$$\boldsymbol{\eta}_{-i}^{(k)} = \{\boldsymbol{\eta}_1^{(k+1)}, \ldots, \boldsymbol{\eta}_{i-1}^{(k+1)}, \boldsymbol{\eta}_{i+1}^{(k)}, \ldots, \boldsymbol{\eta}_r^{(k)}\}.$$

The M-H algorithm for generating a sample, $\boldsymbol{\eta}_i^{(k+1)}$, from $f(\boldsymbol{\eta}_i|\boldsymbol{\eta}_{-i}^{(k)}, \mathbf{y})$ involves the following steps.

Step 1. Approximate $f(\boldsymbol{\eta}_i|\boldsymbol{\eta}_{-i}^{(k)}, \mathbf{y})$ by a candidate (or proposal) density $q_i(\boldsymbol{\eta}_i|\boldsymbol{\eta}_i^{(k)}, \boldsymbol{\eta}_{-i}^{(k)})$ that is easy to sample from, such as a normal or a t

distribution. The proposal density may depend on the current values $\{\boldsymbol{\eta}_i^{(k)}, \boldsymbol{\eta}_{-i}^{(k)}\}$.

Step 2. Generate a "candidate" for $\boldsymbol{\eta}_i$, say $\boldsymbol{\eta}_i^*$, from the candidate density, and u from Uniform(0,1).

Step 3. Set $\boldsymbol{\eta}_i^{(k+1)} = \boldsymbol{\eta}_i^*$ if $u \le a(\boldsymbol{\eta}_{-i}^{(k)}, \boldsymbol{\eta}_i^{(k)}, \boldsymbol{\eta}_i^*)$ and $\boldsymbol{\eta}_i^{(k+1)} = \boldsymbol{\eta}_i^{(k)}$ otherwise, where the acceptance probability $a(\boldsymbol{\eta}_{-i}^{(k)}, \boldsymbol{\eta}_i^{(k)}, \boldsymbol{\eta}_i^*)$ is given by

$$a(\boldsymbol{\eta}_{-i}^{(k)}, \boldsymbol{\eta}_i^{(k)}, \boldsymbol{\eta}_i^*) = \min\left\{\frac{f(\boldsymbol{\eta}_i^*|\boldsymbol{\eta}_{-i}^{(k)}, \mathbf{y})q_i(\boldsymbol{\eta}_i^{(k)}|\boldsymbol{\eta}_i^*, \boldsymbol{\eta}_{-i}^{(k)})}{f(\boldsymbol{\eta}_i^{(k)}|\boldsymbol{\eta}_{-i}^{(k)}, \mathbf{y})q_i(\boldsymbol{\eta}_i^*|\boldsymbol{\eta}_i^{(k)}, \boldsymbol{\eta}_{-i}^{(k)})}, 1\right\}. \quad (10.2.4)$$

Note that the acceptance probability (10.2.4) depends on the ratio $f(\boldsymbol{\eta}_i^*|\cdot)/f(\boldsymbol{\eta}_i^{(k)}|\cdot)$, so we need to know $f(\boldsymbol{\eta}_i|\cdot)$ only up to a constant of proportionality, that is, the normalizing constant cancels out in (10.2.4). Also, if the candidate density is symmetric, that is, $q_i(\boldsymbol{\eta}_i^{(k)}|\boldsymbol{\eta}_i^*, \cdot) = q_i(\boldsymbol{\eta}_i^*|\boldsymbol{\eta}_i^{(k)}, \cdot)$, then the acceptance probability reduces to

$$a(\boldsymbol{\eta}_{-i}^{(k)}, \boldsymbol{\eta}_i^{(k)}, \boldsymbol{\eta}_i^*) = \min\left\{\frac{f(\boldsymbol{\eta}_i^*|\boldsymbol{\eta}_{-i}^{(k)}, \mathbf{y})}{f(\boldsymbol{\eta}_i^{(k)}|\boldsymbol{\eta}_{-i}^{(k)}, \mathbf{y})}, 1\right\} \quad (10.2.5)$$

(Metropolis et al. (1953)). The Gibbs sampler is a special case of M-H with $q_i(\boldsymbol{\eta}_i|\boldsymbol{\eta}_i^{(k)}, \boldsymbol{\eta}_{-i}^{(k)}) = f(\boldsymbol{\eta}_i|\boldsymbol{\eta}_{-i}^{(k)}, \mathbf{y})$, and the corresponding acceptance probability equals 1 so that the Gibbs candidate $\boldsymbol{\eta}_i^*$ is automatically accepted.

To improve convergence of the MCMC sampling, several useful variations of M-H have been proposed. We refer the reader to Chen, Shao and Ibrahim (2000, Chapter 2) for details of these algorithms.

10.2.4 Practical Issues

The HB approach based on MCMC has several limitations, so it is important to exercise caution when implementing MCMC. We now give a brief account of some practical issues associated with MCMC.

Choice of Prior

Diffuse priors $f(\boldsymbol{\lambda})$, reflecting lack of information about the model parameters $\boldsymbol{\lambda}$, are commonly used in the HB approach to small area estimation. However, if the diffuse prior is improper, that is, $\int f(\boldsymbol{\lambda})d\boldsymbol{\lambda} = \infty$, then the Gibbs sampler could lead to seemingly reasonable inferences about a nonexistent posterior $f(\boldsymbol{\mu}, \boldsymbol{\lambda}|\mathbf{y})$. This happens when the posterior is improper and yet all the Gibbs conditionals are proper (Natarajan and McCulloch (1995), Hobert and Casella (1996)). To demonstrate this, consider the simple nested error model without covariates: $y_{ij} = \mu + v_i + e_{ij}$ with $v_i \overset{\text{iid}}{\sim} N(0, \sigma_v^2)$ and independent of $e_{ij} \overset{\text{iid}}{\sim} N(0, \sigma_e^2)$. If we choose an improper prior of the form $f(\mu, \sigma_v^2, \sigma_e^2) = f(\mu)f(\sigma_v^2)f(\sigma_e^2)$ with $f(\mu) \propto 1$, $f(\sigma_v^2) \propto 1/\sigma_v^2$ and $f(\sigma_e^2) \propto$

$1/\sigma_e^2$, then the joint posterior of $\mu, \mathbf{v} = (v_1, \ldots, v_m)^T$, σ_v^2 and σ_e^2 is improper (Hill (1965)). On the other hand, all the Gibbs conditionals are proper for this choice of prior. In particular, $\sigma_v^{-2}|$ all others $\sim$ gamma (G) (or $\sigma_v^2|$ all others $\sim$ inverted gamma (IG)), $\sigma_e^{-2}|$ all others $\sim G$, $\mu|$ all others $\sim$ Normal (N) and $v_i|$ all others $\sim N$. To get around this difficulty, the BUGS software (Spiegelhalter et al. (1996)) uses diffuse (or vague) proper priors of the form $\mu \sim N(0, \sigma_0^2)$, $\sigma_v^{-2} \sim G(a_0, a_0)$ and $\sigma_e^{-2} \sim G(a_0, a_0)$ as default priors, where σ_0^2 is chosen very large (say 10,000) and $a_0(> 0)$ very small (say 0.001) to reflect lack of prior information on μ, σ_v^2 and σ_e^2. (Note that $G(a, b)$ denotes a gamma distribution with shape parameter a and scale parameter b and that the variance of $G(a_0, a_0)$ is $1/a_0$ which becomes very large as $a_0 \to 0$.) The posterior resulting from the above prior remains proper as $\sigma_0^2 \to \infty$ (equivalent to choosing $f(\mu) \propto 1$), but it becomes improper as $a_0 \to 0$. Therefore, the posterior is nearly improper (or barely proper) for very small a_0, and this feature can affect the convergence of the Gibbs sampler. An alternative choice, $\sigma_v^2 \sim \text{uniform}(0, 1/a_0)$ and $\sigma_e^2 \sim \text{uniform}(0, 1/a_0)$, avoids this difficulty in the sense that the posterior remains proper as $a_0 \to 0$.

We have noted in Section 10.1 that it is desirable to choose diffuse priors that lead to well-calibrated inferences. Browne and Draper (2001) compared frequentist performances of posterior quantities under the simple nested error model and $G(a_0, a_0)$ or uniform$(0, 1/a_0)$ priors on the variance parameters. In particular, for σ_v^2 they examined the bias of the posterior mean, median and mode and the Bayesian interval coverage in repeated sampling. All the Gibbs conditionals have closed forms here, so Gibbs sampling was used to generate samples from the posterior. They found that the posterior quantities are generally insensitive to the specific choice of a_0 around 0.001 (default setting used in BUGS with gamma prior). In terms of bias, the posterior median performed better than the posterior mean (HB estimator) for the gamma prior, while the posterior mode performed much better than the posterior mean for the uniform prior. Bayesian intervals for uniform or gamma priors did not attain nominal coverage when the number of small areas, m, and/or the variance ratio $\tau = \sigma_v^2/\sigma_e^2$ are small. Browne and Draper (2001) did not study the frequentist behaviour of the posterior quantities associated with the small area means $\mu_i = \mu + v_i$. Datta, Ghosh and Kim (2001) developed noninformative priors for the simple nested error model, called probability matching priors, and focused on the variance ratio τ. These priors ensure that the frequentist coverage of Bayesian intervals for τ approaches the nominal level asymptotically, as $m \to \infty$. In Section 10.3, we consider priors for the basic area level model that make the posterior variance of a small area mean, μ_i, approximately unbiased for the MSE of EB/HB estimators, as $m \to \infty$.

Single Run Versus Multiple Runs

Subsections 10.2.2 and 10.2.3 discussed the generation of one single long run $\{\boldsymbol{\eta}^{(d+1)}, \ldots, \boldsymbol{\eta}^{(d+D)}\}$. One single long run may provide reliable Monte Carlo estimates of posterior quantities by choosing D sufficiently large, but it may leave a significant portion of the space generated by the posterior, $f(\boldsymbol{\eta}|\mathbf{y})$, unexplored. To avoid this problem, Gelman and Rubin (1992) and others recommended the use of multiple parallel runs with different starting points, leading to parallel samples. Good starting values are required to generate multiple runs, and it may not be easy to find such values. On the other hand, for generating one single run, REML estimates of model parameters, $\boldsymbol{\lambda}$, and EB estimates of small area parameters, $\boldsymbol{\mu}$, may work quite well as starting points. For multiple runs, Gelman and Rubin (1992) recommended generating starting points from an "overdispersed" distribution compared to the target distribution $\pi(\boldsymbol{\eta}|\mathbf{y})$. They proposed some methods for generating overdispersed starting values, but general methods that lead to good multiple starting values are not available. Note that the normalizing factor of $\pi(\boldsymbol{\eta}|\mathbf{y})$ is not known in a closed form.

Multiple runs can be wasteful because initial burn-in periods are discarded from each run, although this may not be a serious limitation if parallel processors are used to generate the parallel samples. Gelfand and Smith (1990) used many short runs, each consisting of $(d+1)$ $\boldsymbol{\eta}$-values, and kept only the last observation from each run. This method provides independent samples $\boldsymbol{\eta}^{(d+1)}(1), \ldots, \boldsymbol{\eta}^{(d+1)}(L)$, where $\boldsymbol{\eta}^{(d+1)}(l)$ is the last $\boldsymbol{\eta}$-value of the lth run $(l = 1, \ldots, L)$, but it discards most of the generated values. Gelman and Rubin (1992) used small L (say 10) and generated $2d$ values for each run and kept the last d values from each run, leading to L independent sets of $\boldsymbol{\eta}$-values. This approach facilitates convergence diagnostics (see below). Note that it is not necessary to generate independent samples for getting Monte Carlo estimates of posterior quantities because of the ergodic theorem for Markov chains. However, independent samples facilitate the calculation of Monte Carlo standard errors.

Proponents of one single long run argue that comparing chains can never "prove" convergence. On the other hand, multiple runs proponents assert that comparing seemingly converged runs might disclose real differences if the runs have not yet attained stationarity.

Burn-in Length

The length of burn-in, d, depends on the starting point $\boldsymbol{\eta}^{(0)}$ and the convergence rate of $P^{(k)}(\cdot)$ to the stationary distribution $\pi(\cdot)$. Convergence rates have been studied in the literature (Roberts and Rosenthal (1998)), but it is not easy to use these rates to determine d. Note that the convergence rates of different MCMC algorithms may vary significantly and also depend on the target distribution $\pi(\cdot)$.

Convergence diagnostics, based on the MCMC output $\{\boldsymbol{\eta}^{(k)}\}$, are often used in practice to determine the burn-in length d. At least 13 convergence diagnostics have been proposed in the literature (see Cowles and Carlin (1994)). A diagnostic measure based on multiple runs and classical ANOVA is currently popular (Gelman and Rubin (1992)). Suppose $h(\boldsymbol{\eta})$ denotes a scalar summary of $\boldsymbol{\eta}$; for example $h(\boldsymbol{\eta}) = \mu_i$, the mean of the ith small area. Denote the values of $h(\boldsymbol{\eta})$ for the lth run as $h_{l,d+1}, \ldots, h_{l,2d}$; $l = 1, \ldots, L$. Calculate the between-run variance $B = d\sum_{l=1}^{L}(\bar{h}_{l\cdot} - \bar{h}_{\cdot\cdot})^2/(L-1)$ and the within-run variance $W = \sum_{l=1}^{L} \sum_{k=d+1}^{2d} (h_{l,k} - \bar{h}_{l \cdot})^2 / [L(d-1)]$, where $\bar{h}_{l\cdot} = \sum_k h_{lk}/d$ and $\bar{h}_{\cdot\cdot} = \sum_l \bar{h}_{l\cdot}/L$. Using the two variance components B and W, we calculate an estimate of the variance of $h(\boldsymbol{\eta})$ in the target distribution as

$$\hat{V} = \frac{d-1}{d} W + \frac{1}{d} B. \tag{10.2.6}$$

Under stationarity, $\hat{V}$ is unbiased but it is an overestimate if the L points $h_{1,d+1}, \ldots, h_{L,d+1}$ are overdispersed. Using the latter property, we calculate the estimated "potential scale reduction" factor

$$\hat{R} = \hat{V}/W. \tag{10.2.7}$$

If stationarity is attained, $\hat{R}$ will be close to 1. Otherwise, $\hat{R}$ will be significantly larger than 1, suggesting a larger burn-in length d. It is necessary to calculate $\hat{R}$ for all scalar summaries of interest; for example, all small area means μ_i.

Unfortunately, most of the proposed diagnostics have shortcomings. Cowles and Carlin (1994) compared the performance of 13 convergence diagnostics, including $\hat{R}$, in two simple models and concluded that "all of the methods can fail to detect the sorts of convergence failure that they were designed to identify".

Methods of generating independent samples from the exact stationary distribution of a Markov chain, that is, $\pi(\boldsymbol{\eta}) = f(\boldsymbol{\eta}|\mathbf{y})$, have been proposed recently. These methods eliminate the difficulties noted above, but currently the algorithms are not easy to implement. We refer the reader to Casella, Lavine and Robert (2001) for a brief introduction to the proposed methods, called perfect (or exact) samplings.

10.2.5 Posterior Quantities

The MCMC output $\{\boldsymbol{\eta}^{(k)}, k = d+1, \ldots, d+D\}$ from a single long run may be employed to compute posterior quantities of interest, noting that $\{\boldsymbol{\mu}^{(k)}\}$ is a sample from the marginal posterior $f(\boldsymbol{\mu}|\mathbf{y})$. In particular, the posterior mean (or the HB estimator) of $\phi = h(\boldsymbol{\mu})$ is estimated by the "ergodic average"

$$\hat{\phi}^{\text{HB}} = \frac{1}{D} \sum_{k=d+1}^{d+D} \phi^{(k)} = \phi^{(\cdot)}, \tag{10.2.8}$$

where $\phi^{(k)} = h(\boldsymbol{\mu}^{(k)})$. Similarly the posterior variance of ϕ is estimated by

$$\hat{V}(\phi|\mathbf{y}) = \frac{1}{D-1} \sum_{k=d+1}^{d+D} (\phi^{(k)} - \phi^{(\cdot)})^2. \tag{10.2.9}$$

By the ergodic theorem for Markov chains, $\hat{\phi}^{\text{HB}}$ converges to $E(\phi|\mathbf{y})$ and $\hat{V}(\phi|\mathbf{y})$ to $V(\phi|\mathbf{y})$ as $D \to \infty$. However, Monte Carlo standard errors of $\hat{\phi}^{\text{HB}}$

and $\hat{V}(\phi|\mathbf{y})$ are not easy to find if the simulated values $\phi^{(d+1)}, \ldots, \phi^{(d+D)}$ are dependent.

If the $\boldsymbol{\eta}^{(k)}$ are iid vectors, as in the multiple runs method of Gelfand and Smith (1990), then estimates of ϕ with reduced standard errors may be obtained if the mathematical form for the conditional expectation of ϕ given the data $\mathbf{y}$ and the model parameters $\boldsymbol{\lambda}$ is known. Suppose that the form of $E(\phi|\mathbf{y}, \boldsymbol{\lambda})$ is known. Then an improved estimate of $E(\phi|\mathbf{y})$ is given by

$$\tilde{\phi}^{\text{HB}} = \frac{1}{D} \sum_{k=d+1}^{d+D} E(\phi|\mathbf{y}, \boldsymbol{\lambda}^{(k)}), \tag{10.2.10}$$

where $\{\boldsymbol{\lambda}^{(k)}\}$ is the iid sample from the marginal posterior $f(\boldsymbol{\lambda}|\mathbf{y})$. This result readily follows by noting that

$$\begin{aligned} V(\phi|\mathbf{y}) &= E[V(\phi|\mathbf{y}, \boldsymbol{\lambda})|\mathbf{y}] + V[E(\phi|\mathbf{y}, \boldsymbol{\lambda})|\mathbf{y}] \\ &\geq V[E(\phi|\mathbf{y}, \boldsymbol{\lambda})|\mathbf{y}], \end{aligned}$$

and that the Monte Carlo variances of $\hat{\phi}^{\text{HB}}$ and $\tilde{\phi}^{\text{HB}}$ are given by $D^{-1}V(\phi|\mathbf{y})$ and $D^{-1}V[E(\phi|\mathbf{y}, \boldsymbol{\lambda})|\mathbf{y}]$, respectively. The improved estimator $\tilde{\phi}^{\text{HB}}$ is called a Rao-Blackwell estimator (Gelfand and Smith (1991)) because of the similarity of the result to the well-known Rao-Blackwell theorem (see, e.g., Casella and Berger (1990), p. 316). An estimate of posterior variance associated with $\tilde{\phi}^{\text{HB}}$ is given by

$$\tilde{V}(\phi|\mathbf{y}) = \frac{1}{D} \sum_{k=d+1}^{d+D} V(\phi|\mathbf{y}, \boldsymbol{\lambda}^{(k)}) + \frac{1}{D-1} \sum_{k=d+1}^{d+D} \left[E(\phi|\mathbf{y}, \boldsymbol{\lambda}^{(k)}) - \tilde{\phi}^{\text{HB}}\right]^2. \tag{10.2.11}$$

For the case of a single run with dependent samples, (10.2.11) may not hold. However, if the D samples are "thinned" by selecting say every 5th or 10th $\boldsymbol{\eta}^{(k)}$, then the reduced MCMC output may be approximately iid and the result (10.2.11) may hold for the thinned sample. Thinning of the chain is sometimes used to save storage space and computational time, especially when the consecutive samples $\boldsymbol{\eta}^{(k)}$ are highly correlated, necessitating a very long run, that is, very large D.

We can also estimate $(1 - 2\alpha)$-level posterior (or credible) intervals for ϕ from the MCMC output $\{\boldsymbol{\eta}^{(k)}\}$. For example, by setting ϕ_α equal to the α-level quantile of $\{\phi^{(k)}\}$ and $\phi_{1-\alpha}$ to $(1-\alpha)$th quantile, we obtain a "central" or "equal-tailed" posterior interval $[\phi_\alpha, \phi_{1-\alpha}]$ for ϕ such that $P(\phi_\alpha \leq \phi \leq \phi_{1-\alpha}|\mathbf{y}) = 1 - 2\alpha$. An alternative "interval", called the highest posterior density (HPD) "interval", is given by $A = \{\phi : f(\phi|y) \geq c\}$, where c is chosen such that $P(\phi \in A|\mathbf{y}) = 1 - 2\alpha$; the set A may not be an interval. The HPD interval is the shortest length interval for a given credible probability, $1 - 2\alpha$, provided $f(\phi|\mathbf{y})$ is unimodal (see Casella and Berger (1990), p. 430). Generally, the central posterior interval is preferred to the HPD interval because it is invariant to one-to-one transformations of ϕ, and is easier to compute. Also, the end points ϕ_α and $\phi_{1-\alpha}$ can be interpreted as the posterior α and $1-\alpha$ quantiles.

A computer program, called Bayesian inference Using Gibbs Sampling (BUGS), is widely used to implement MCMC and to calculate posterior quantities from the MCMC output (Spiegelhalter et al. (1997)). BUGS can handle many of the small area models studied in this chapter. It uses inverse gamma priors on variance parameters and normal priors on regression parameters. BUGS runs are monitored using a menu-driven set of S-Plus functions, called Convergence Diagnosis and Output Analysis (CODA). Convergence diagnostics and statistical and graphical output analysis of the simulated values from BUGS may be performed using CODA. The BUGS software package is freely available and can be downloaded from http://www.mrc-bsu.cam.ac.uk/bugs/. Currently, there are BUGS version 0.5, version 0.6 for UNIX and DOS systems, and WinBUGS for Windows. BUGS version 0.6 includes Metropolis-Hastings within Gibbs steps to handle non-log-concave conditionals not admitting closed forms; version 0.5 handles only log-concave conditionals using adaptive rejection sampling (Gilks and Wild (1992)). CODA is supplied with BUGS. We refer the reader to the BUGS manual for examples of typical applications. Spiegelhalter et al. (1996) presented a case study of MCMC methods, using BUGS.

10.2.6 Model Determination

MCMC methods are also used for model determination. In particular, methods based on Bayes factors, posterior predictive densities and cross-validation predictive densities are employed for model determination. These approaches require the specification of priors on the model parameters associated with the competing models. Typically, noninformative priors are employed with the methods based on posterior predictive and cross-validation predictive densities. Because of the dependence on priors, some authors have suggested a hybrid strategy in which prior-free, frequentist methods are used in the model exploration phase, such as those given in Chapters 6–9, and Bayesian diffuse-prior MCMC methods for inference from the selected model (see, e.g., Browne and Draper (2001)). We refer the reader to Gelfand (1996) for an excellent account of model determination using MCMC methods.

Bayes Factors

Suppose that M_1 and M_2 denote two competing models with associated parameters (random effects and model parameters) $\boldsymbol{\eta}_1$ and $\boldsymbol{\eta}_2$, respectively. We denote the marginal densities of the "observables" $\mathbf{y}$ under M_1 and M_2 by $f(\mathbf{y}|M_1)$ and $f(\mathbf{y}|M_2)$, respectively. Note that

$$f(\mathbf{y}|M_i) = \int f(\mathbf{y}|\boldsymbol{\eta}_i, M_i) f(\boldsymbol{\eta}_i|M_i) d\boldsymbol{\eta}_i, \tag{10.2.12}$$

where $f(\boldsymbol{\eta}_i|M_i)$ is the density of $\boldsymbol{\eta}_i$ under model $M_i (i = 1, 2)$. We denote the actual observations as $\mathbf{y}_{\text{obs}}$. The Bayes factor (BF) is defined as

$$\text{BF}_{12} = \frac{f(\mathbf{y}_{\text{obs}}|M_1)}{f(\mathbf{y}_{\text{obs}}|M_2)}, \tag{10.2.13}$$

where $\mathbf{y}_{\text{obs}}$ denotes the actual observations. It provides relative weight of evidence for M_1 compared to M_2. BF_{12} may be interpreted as the ratio of the posterior odds for M_1 versus M_2 to prior odds for M_1 versus M_2, noting that the posterior odds is $f(M_1|\mathbf{y})/f(M_2|\mathbf{y})$ and the prior odds is $f(M_1)/f(M_2)$, where

$$f(M_i|\mathbf{y}) = f(\mathbf{y}|M_i)f(M_i)/f(\mathbf{y}), \tag{10.2.14}$$

$f(M_i)$ is the prior on M_i and

$$f(\mathbf{y}) = f(\mathbf{y}|M_1)f(M_1) + f(\mathbf{y}|M_2)f(M_2). \tag{10.2.15}$$

Kass and Raftery (1995) classified the evidence against M_2 as follows: (a) BF_{12} between 1 and 3: not worth more than a bare mention; (b) BF_{12} between 3 and 20: positive; (c) BF_{12} between 20 and 50: strong; (d) $\text{BF}_{12} > 50$: very strong. BF is appealing for model selection, but $f(\mathbf{y}|M_i)$ is necessarily improper if the prior on the model parameters $\boldsymbol{\lambda}_i$ is improper, even if the posterior $f(\boldsymbol{\eta}_i|\mathbf{y}, M_i)$ is proper. A proper prior on $\boldsymbol{\lambda}_i$ $(i = 1, 2)$ is needed to implement BF. However, the use of BF with diffuse proper priors usually gives bad answers (Berger and Pericchi (2001)). Gelfand (1996) noted several limitations of BF, including the above mentioned impropriety of $f(\mathbf{y}|M_i)$, and concluded that "use of the Bayes factor often seems inappropriate in real applications". We refer the reader to Kass and Raftery (1995) for methods of calculating Bayes factors.

Posterior Predictive Densities

The posterior predictive density, $f(\mathbf{y}|\mathbf{y}_{\text{obs}})$, is defined as the predictive density of new independent observations, $\mathbf{y}$, under the model, given $\mathbf{y}_{\text{obs}}$. We have

$$f(\mathbf{y}|\mathbf{y}_{\text{obs}}) = \int f(\mathbf{y}|\boldsymbol{\eta})f(\boldsymbol{\eta}|\mathbf{y}_{\text{obs}})d\boldsymbol{\eta}, \tag{10.2.16}$$

and the marginal posterior predictive density of an element y_r of $\mathbf{y}$ is given by

$$f(y_r|\mathbf{y}_{\text{obs}}) = \int f(y_r|\boldsymbol{\eta})f(\boldsymbol{\eta}|\mathbf{y}_{\text{obs}})d\boldsymbol{\eta}. \tag{10.2.17}$$

Using the MCMC output $\{\boldsymbol{\eta}^{(k)}; k = d+1, \ldots, d+D\}$, we can draw a sample $\{\mathbf{y}^{(k)}\}$ from $f(\mathbf{y}|\mathbf{y}_{\text{obs}})$ as follows: For each $\boldsymbol{\eta}^{(k)}$, draw $\mathbf{y}^{(k)}$ from $f(\mathbf{y}|\boldsymbol{\eta}^{(k)})$. Further, $\{y_r^{(k)}\}$ constitutes a sample from the marginal density $f(y_r|\mathbf{y}_{\text{obs}})$, where $y_r^{(k)}$ is the rth element of $\mathbf{y}^{(k)}$.

To check the overall fit of a proposed model to data $\mathbf{y}_{\text{obs}}$, Gelman and Meng (1996) proposed the criterion of posterior predictive p-value, noting that the hypothetical replications $\{\mathbf{y}^{(k)}\}$ should look similar to $\mathbf{y}_{\text{obs}}$ if the assumed model is reasonably accurate. Let $T(\mathbf{y}, \boldsymbol{\eta})$ denote a measure of discrepancy between $\mathbf{y}$ and $\boldsymbol{\eta}$. The posterior predictive p-value is then defined as

$$p = P\{T(\mathbf{y}, \boldsymbol{\eta}) \geq T(\mathbf{y}_{\text{obs}}, \boldsymbol{\eta}) | \mathbf{y}_{\text{obs}}\}. \qquad (10.2.18)$$

The MCMC output $\{\boldsymbol{\eta}^{(k)}\}$ may be used to approximate p by

$$\hat{p} = \frac{1}{D} \sum_{k=d+1}^{d+D} I\{T(\mathbf{y}^{(k)}, \boldsymbol{\eta}^{(k)}) \geq T(\mathbf{y}_{\text{obs}}, \boldsymbol{\eta}^{(k)})\}, \qquad (10.2.19)$$

where $I(\cdot)$ is the indicator function taking the value 1 when its argument is true and 0 otherwise.

A limitation of the posterior predictive p-value is that it makes "double use" of the data $\mathbf{y}_{\text{obs}}$, first to generate $\{\mathbf{y}^{(k)}\}$ from $f(\mathbf{y}|\mathbf{y}_{\text{obs}})$ and then the compute $\hat{p}$ given by (10.2.19). This double use of the data can induce unnatural behaviour, as demonstrated by Bayarri and Berger (2000). They proposed two alternative p-measures, named the partial posterior predictive p-value and the conditional predictive p-value, that attempt to avoid double use of the data. These measures, however, are more difficult to implement than the posterior predictive p-value, especially for small area models.

If the model fits $\mathbf{y}_{\text{obs}}$, then the two values $T(\mathbf{y}^{(k)}, \boldsymbol{\eta}^{(k)})$ and $T(\mathbf{y}_{\text{obs}}, \boldsymbol{\eta}^{(k)})$ should tend to be similar for most k, and $\hat{p}$ close to 0.5. Extreme $\hat{p}$ values (closer to 0 or 1) suggest poor fit. It is informative to plot the realized values $T(\mathbf{y}_{\text{obs}}, \boldsymbol{\eta}^{(k)})$ versus the predictive values $T(\mathbf{y}^{(k)}, \boldsymbol{\eta}^{(k)})$. If the model is a good fit, then about half the points in the scatter plot would fall above the 45° line and the remaining half below the line. Brooks, Catchpole and Morgan (2000) studied competing animal survival models, and used the above scatter plots for four competing models, called discrepancy plots, to select a model.

Another measure of fit is given by the posterior expected predictive deviance, $E\{\Delta(\mathbf{y}, \mathbf{y}_{\text{obs}}) | \mathbf{y}_{\text{obs}}\}$, where $\Delta(\mathbf{y}, \mathbf{y}_{\text{obs}})$ is a measure of discrepancy between $\mathbf{y}_{\text{obs}}$ and $\mathbf{y}$. For count data $\{y_r\}$, we can use a chi-squared measure given by

$$\Delta(\mathbf{y}, \mathbf{y}_{\text{obs}}) = \sum_r (y_{r,\text{obs}} - y_r)^2 / (y_r + 0.5). \qquad (10.2.20)$$

It is a general measure of agreement. Nandram, Sedransk and Pickle (1999) studied (10.2.20) and some other measures in the context of disease mapping (see Section 10.10). Using the predictions $\mathbf{y}^{(k)}$, the posterior expected predictive deviance may be estimated as

$$\hat{E}\{\Delta(\mathbf{y}, \mathbf{y}_{\text{obs}}) | \mathbf{y}_{\text{obs}}\} = \frac{1}{D} \sum_{k=d+1}^{d+D} \Delta(\mathbf{y}^{(k)}, \mathbf{y}_{\text{obs}}). \qquad (10.2.21)$$

Note that the deviance measure (10.2.21) also makes double use of the data.

The deviance measure is useful for comparing the relative fits of candidate models. The model with the smallest deviance value may then be used to check its overall fit to the data $\mathbf{y}_{\text{obs}}$, using the posterior predictive p-value.

Cross-validation Predictive Densities

The cross-validation predictive density of the element y_r is given by

$$f(y_r|\mathbf{y}_{(r)}) = \int f(y_r|\boldsymbol{\eta}, \mathbf{y}_{(r)}) f(\boldsymbol{\eta}|\mathbf{y}_{(r)}) d\boldsymbol{\eta}, \tag{10.2.22}$$

where $\mathbf{y}_{(r)}$ denotes all elements of $\mathbf{y}$ except y_r. This density suggests what values of y_r are likely when the model is fitted to $\mathbf{y}_{(r)}$. The actual observation $y_{r,\text{obs}}$ may be compared to the hypothetical values y_r in a variety of ways to see whether $y_{r,\text{obs}}$, for each r, supports the model. In most applications, $f(y_r|\boldsymbol{\eta}, \mathbf{y}_{(r)}) = f(y_r|\boldsymbol{\eta})$, that is, y_r and $\mathbf{y}_{(r)}$ are conditionally independent given $\boldsymbol{\eta}$. Also, $f(\boldsymbol{\eta}|\mathbf{y}_{(r)})$ is usually proper if $f(\boldsymbol{\eta}|\mathbf{y})$ is proper. As a result, $f(y_r|\mathbf{y}_{(r)})$ will remain proper, unlike $f(\mathbf{y})$ used in the computation of the Bayes Factor (BF). If $f(\mathbf{y})$ is proper, then the set $\{f(y_r|\mathbf{y}_{(r)})\}$ is equivalent to $f(\mathbf{y})$ in the sense that $f(\mathbf{y})$ is uniquely determined from $\{f(y_r|\mathbf{y}_{(r)})\}$ and vice versa.

The product of the densities $f(y_r|\mathbf{y}_{(r)})$ is often used as a substitute for $f(\mathbf{y})$ to avoid an improper marginal $f(\mathbf{y})$. This leads to a pseudo-Bayes factor

$$\text{PBF}_{12} = \prod_r \frac{f(y_{r,\text{obs}}|\mathbf{y}_{(r),\text{obs}}, M_1)}{f(y_{r,\text{obs}}|\mathbf{y}_{(r),\text{obs}}, M_2)}, \tag{10.2.23}$$

which is often used as a substitute for BF. However, PBF cannot be interpreted as the ratio of posterior odds to prior odds, unlike BF.

Global summary measures, such as the p-value (10.2.18) and the posterior expected predictive deviance (10.2.21), are useful for checking the overall fit of a model, but not helpful for discovering the reasons for poor global performance. The univariate density $f(y_r|\mathbf{y}_{(r),\text{obs}})$ can be used through a "checking" function $c(y_r, y_{r,\text{obs}})$ to see whether $y_{r,\text{obs}}$, for each r, supports the model. Gelfand (1996) proposed a variety of checking functions and calculated their expectations over $f(y_r|\mathbf{y}_{(r),\text{obs}})$. In particular, choosing

$$c_1(y_r, y_{r,\text{obs}}) = \frac{1}{2\epsilon} I[y_{r,\text{obs}} - \epsilon \le y_r \le y_{r,\text{obs}} + \epsilon], \tag{10.2.24}$$

then taking its expectation and letting $\epsilon \to 0$, we obtain the conditional predictive ordinate (CPO):

$$\text{CPO}_r = f(y_{r,\text{obs}}|\mathbf{y}_{(r),\text{obs}}). \tag{10.2.25}$$

Models with larger CPO-values provide better fit to the observed data. A plot of CPO_r versus r for each model is useful for comparing the models visually. We can easily see from a CPO plot which models are better than the others, which models are indistinguishable, which points, $y_{r,\text{obs}}$, are poorly fit under all the competing models, and so on. If two or more models are indistinguishable and provide good fits, then we should choose the simplest among the models.

We approximate CPO_r from the MCMC output $\{\boldsymbol{\eta}^{(k)}\}$ as follows:

$$\widehat{\text{CPO}}_r = \hat{f}(y_{r,\text{obs}}|\mathbf{y}_{(r),\text{obs}}) = \left[\frac{1}{D}\sum_{k=d+1}^{d+D}\frac{1}{f(y_{r,\text{obs}}|\mathbf{y}_{(r),\text{obs}},\boldsymbol{\eta}^{(k)})}\right]^{-1}. \quad (10.2.26)$$

That is, $\widehat{\text{CPO}}_r$ is the harmonic mean of the D ordinates $f(y_{r,\text{obs}}|\mathbf{y}_{(r),\text{obs}},\boldsymbol{\eta}^{(k)})$. For several small area models, y_r and $\mathbf{y}_{(r)}$ are conditionally independent given $\boldsymbol{\eta}$, that is, $f(y_{r,\text{obs}}|\mathbf{y}_{(r),\text{obs}},\boldsymbol{\eta}^{(k)}) = f(y_{r,\text{obs}}|\boldsymbol{\eta}^{(k)})$ in (10.2.26). A proof of (10.2.26) is given in Section 10.14.

A checking function of a "residual" form is given by

$$c_2(y_r, y_{r,\text{obs}}) = y_{r,\text{obs}} - y_r \quad (10.2.27)$$

with expectation $d_{2r} = y_{r,\text{obs}} - E(y_r|\mathbf{y}_{(r),\text{obs}})$. We can use a standardized form

$$d_{2r}^* = \frac{d_{2r}}{\sqrt{V(y_r|\mathbf{y}_{(r),\text{obs}})}} \quad (10.2.28)$$

and plot d_{2r}^* versus r, similar to standard residual analysis, where $V(y_r|\mathbf{y}_{(r),\text{obs}})$ is the variance of y_r with respect to $f(y_r|\mathbf{y}_{(r),\text{obs}})$. Another choice of checking function is $c_{3r} = I(-\infty < y_r \le y_{r,\text{obs}})$ with expectation

$$d_{3r} = P(y_r \le y_{r,\text{obs}}|\mathbf{y}_{(r),\text{obs}}). \quad (10.2.29)$$

We can use d_{3r} to measure how unlikely y_r is under $f(y_r|\mathbf{y}_{(r),\text{obs}})$. If y_r is discrete, then we can use $P(y_r = y_{r,\text{obs}}|\mathbf{y}_{(r),\text{obs}})$. Yet another choice is given by $c_4(y_r, \mathbf{y}_{(r),\text{obs}}) = I(y_r \in B_r)$, where

$$B_r = \{y_r : f(y_r|\mathbf{y}_{(r),\text{obs}}) \le f(y_{r,\text{obs}}|\mathbf{y}_{(r),\text{obs}})\}. \quad (10.2.30)$$

Its expectation is equal to

$$d_{4r} = P(B_r|\mathbf{y}_{(r),\text{obs}}). \quad (10.2.31)$$

We can use d_{4r} to see how likely $f(y_r|\mathbf{y}_{(r),\text{obs}})$ will be smaller than the corresponding conditional predictive ordinate, CPO_r.

To estimate the measures d_{2r}, d_{3r} and d_{4r}, we need to evaluate expectations of the form $E[a(y_r)|\mathbf{y}_{(r),\text{obs}}]$ for specified functions $a(y_r)$. Suppose that $f(y_r|\mathbf{y}_{(r),\text{obs}},\boldsymbol{\eta}) = f(y_r|\boldsymbol{\eta})$. Then we can estimate the expectations as

$$\hat{E}[a(y_r)|\mathbf{y}_{(r),\text{obs}}] = \hat{f}(y_r|\mathbf{y}_{(r),\text{obs}})\frac{1}{D}\sum_{k=d+1}^{d+D}\frac{b_r(\boldsymbol{\eta}^{(k)})}{f(y_{r,\text{obs}}|\boldsymbol{\eta}^{(k)})}, \quad (10.2.32)$$

where $b_r(\boldsymbol{\eta})$ is the conditional expectation of $a(y_r)$ over y_r given $\boldsymbol{\eta}$. A proof of (10.2.32) is given by Section 10.14. If a closed-form expression for $b_r(\boldsymbol{\eta})$ is not available, then we have to draw a sample from $f(y_r|\mathbf{y}_{(r)})$ and estimate $E[a(y_r)|\mathbf{y}_{(r),\text{obs}}]$ directly. Gelfand (1996) has given a method of drawing such a sample without rerunning the MCMC sampler, using only $\mathbf{y}_{(r),\text{obs}}$.

10.3 Basic Area Level Model

In this section we apply the HB approach to the basic area level model (7.1.1), assuming a prior distribution on the model parameters $(\boldsymbol{\beta}, \sigma_v^2)$. We first consider the case of known σ_v^2 and assume a "flat" prior on $\boldsymbol{\beta}$ given by $f(\boldsymbol{\beta}) \propto 1$, and rewrite (7.1.1) as a HB model:

$$\begin{aligned} &\text{(i)} \quad \hat{\theta}_i|\theta_i, \boldsymbol{\beta}, \sigma_v^2 \overset{\text{ind}}{\sim} N(\theta_i, \psi_i), \quad i = 1, \ldots, m \\ &\text{(ii)} \quad \theta_i|\boldsymbol{\beta}, \sigma_v^2 \overset{\text{ind}}{\sim} N(\mathbf{z}_i^T\boldsymbol{\beta}, b_i^2\sigma_v^2), \quad i = 1, \ldots, m \\ &\text{(iii)} \quad f(\boldsymbol{\beta}) \propto 1. \end{aligned} \tag{10.3.1}$$

We then extend the results to the case of unknown σ_v^2 by replacing (iii) in (10.3.1) by

$$\text{(iii)}' \quad f(\boldsymbol{\beta}, \sigma_v^2) = f(\boldsymbol{\beta})f(\sigma_v^2) \propto f(\sigma_v^2), \tag{10.3.2}$$

where $f(\sigma_v^2)$ is a prior on σ_v^2.

10.3.1 Known σ_v^2

Straightforward calculations show that the posterior distribution of θ_i given $\hat{\boldsymbol{\theta}} = (\hat{\theta}_1, \ldots, \hat{\theta}_m)^T$ and σ_v^2, under the HB model (10.3.1), is normal with mean equal to the BLUP estimator $\tilde{\theta}_i^H$ and variance equal to $M_{1i}(\sigma_v^2)$ given by (7.1.6). That is, the HB estimator of θ_i is

$$\tilde{\theta}_i^{\text{HB}}(\sigma_v^2) = E(\theta_i|\hat{\boldsymbol{\theta}}, \sigma_v^2) = \tilde{\theta}_i^H, \tag{10.3.3}$$

and the posterior variance of θ_i is

$$V(\theta_i|\hat{\boldsymbol{\theta}}, \sigma_v^2) = M_{1i}(\sigma_v^2) = \text{MSE}(\tilde{\theta}_i^H). \tag{10.3.4}$$

Hence, when σ_v^2 is assumed to be known and $f(\boldsymbol{\beta}) \propto 1$, the HB and BLUP approaches under normality lead to identical point estimates and measures of variability.

10.3.2 Unknown σ_v^2: Numerical Integration

In practice, σ_v^2 is unknown and it is necessary to take account of the uncertainty about σ_v^2 by assuming a prior, $f(\sigma_v^2)$, on σ_v^2. The HB model is given by (i) and (ii) of (10.3.1) and (iii)$'$ given by (10.3.2). We obtain the HB estimator of θ_i as

$$\hat{\theta}_i^{\text{HB}} = E(\theta_i|\hat{\boldsymbol{\theta}}) = E_{\sigma_v^2}[\tilde{\theta}_i^{\text{HB}}(\sigma_v^2)], \tag{10.3.5}$$

where $E_{\sigma_v^2}$ denotes the expectation with respect to the posterior distribution of σ_v^2, $f(\sigma_v^2|\hat{\boldsymbol{\theta}})$. The posterior variance of θ_i is given by

$$V(\theta_i|\hat{\boldsymbol{\theta}}) = E_{\sigma_v^2}[M_{1i}(\sigma_v^2)] + V_{\sigma_v^2}[\tilde{\theta}_i^{\text{HB}}(\sigma_v^2)], \tag{10.3.6}$$

where $V_{\sigma_v^2}$ denotes the variance with respect to $f(\sigma_v^2|\hat{\boldsymbol{\theta}})$. It follows from (10.3.5) and (10.3.6) that the evaluation of $\hat{\theta}_i^{\text{HB}}$ and $V(\theta_i|\hat{\boldsymbol{\theta}})$ involves only single dimensional integrations.

The posterior $f(\sigma_v^2|\hat{\boldsymbol{\theta}})$ may be obtained from the restricted likelihood function $L_R(\sigma_v^2)$ as

$$f(\sigma_v^2|\hat{\boldsymbol{\theta}}) \propto L_R(\sigma_v^2) f(\sigma_v^2), \tag{10.3.7}$$

where

$$\log[L_R(\sigma_v^2)] = \text{const} - \frac{1}{2}\sum_{i=1}^{m} \log(\sigma_v^2 b_i^2 + \psi_i) - \frac{1}{2}\log\left|\sum_{i=1}^{m}(\sigma_v^2 b_i^2 + \psi_i)^{-1}\mathbf{z}_i\mathbf{z}_i^T\right|$$
$$- \frac{1}{2}\sum_{i=1}^{m}[\hat{\theta}_i - \mathbf{z}_i^T\tilde{\boldsymbol{\beta}}(\sigma_v^2)]^2/(\sigma_v^2 b_i^2 + \psi_i), \tag{10.3.8}$$

with $\mathbf{z}_i$ is the $p \times 1$ vector of covariates, and $\tilde{\boldsymbol{\beta}}(\sigma_v^2) = \tilde{\boldsymbol{\beta}}$ is the weighted least squares estimator of $\boldsymbol{\beta}$ (see the equation below (7.1.4)); see Harville (1977). Under a "flat" prior $f(\sigma_v^2) \propto 1$, the posterior $f(\sigma_v^2|\hat{\boldsymbol{\theta}})$ is proper, provided $m > p + 2$, and proportional to $L_R(\sigma_v^2)$; note that $f(\sigma_v^2)$ is improper. More generally, for any improper prior $f(\sigma_v^2) \propto h(\sigma_v^2)$, the posterior $f(\sigma_v^2|\hat{\boldsymbol{\theta}})$ will be proper if $h(\sigma_v^2)$ is a bounded function of σ_v^2 and $m > p + 2$. Note that $h(\sigma_v^2) = 1$ for the flat prior. Morris (1983a) and Ghosh (1992) used the flat prior $f(\sigma_v^2) \propto 1$.

The posterior mean of σ_v^2 under the flat prior $f(\sigma_v^2) \propto 1$ may be expressed as

$$E(\sigma_v^2|\hat{\boldsymbol{\theta}}) = \hat{\sigma}_{v\text{HB}}^2 = \int_0^\infty \sigma_v^2 L_R(\sigma_v^2) d\sigma_v^2 \Big/ \int_0^\infty L_R(\sigma_v^2) d\sigma_v^2. \tag{10.3.9}$$

This estimator is always unique and positive, unlike the posterior mode or the REML estimator of σ_v^2. Also, the HB estimator $\hat{\theta}_i^{\text{HB}}$ attaches a positive weight $\hat{\gamma}_i^{\text{HB}}$ to the direct estimator $\hat{\theta}_i$, where $\hat{\gamma}_i^{\text{HB}} = E[\gamma_i(\sigma_v^2)|\hat{\boldsymbol{\theta}}]$ is obtained from (10.3.9) by changing σ_v^2 to $\gamma_i(\sigma_v^2) = \gamma_i$. On the other hand, a REML estimator $\hat{\sigma}_{v\text{RE}}^2 = 0$ will give a zero weight to $\hat{\theta}_i$ in the EBLUP (or EB) for all the small areas regardless of the area sample sizes. Because of the latter difficulty, the HB estimator $\hat{\theta}_i^{\text{HB}}$ may be more appealing (Bell (1999)).

The frequentist bias of $\hat{\sigma}_{v\text{HB}}^2$ can be substantial when σ_v^2 is small (Browne and Draper (2001)). However, $\hat{\theta}_i^{\text{HB}}$ may not be affected by the bias of $\hat{\sigma}_{v\text{HB}}^2$ because the weighted average $a_i\hat{\theta}_i + (1 - a_i)\mathbf{z}_i^T\hat{\boldsymbol{\beta}}$ is approximately unbiased for θ_i for any choice of the weight $a_i (0 \le a_i \le 1)$, provided the linking model (ii) of (10.3.1) is correctly specified.

The posterior variance (10.3.6) is used as a measure of uncertainty associated with $\hat{\theta}_i^{\text{HB}}$. It is computed similar to (10.3.9). As noted in Section 10.1, it is desirable to select a "matching" improper prior that leads to well-calibrated inferences. In particular, the posterior variance should be second order unbiased for the MSE, that is, $E[V(\theta_i|\hat{\boldsymbol{\theta}})] - \text{MSE}(\hat{\theta}_i^{\text{HB}}) = o(m^{-1})$. Such a dual justification is very appealing to survey practitioners. Datta, Rao and Smith (2002) showed that the matching prior for the special case of $b_i = 1$ is given by

$$f_i(\sigma_v^2) \propto (\sigma_v^2 + \psi_i)^2 \sum_{l=1}^{m} (\sigma_v^2 + \psi_l)^{-2}. \quad (10.3.10)$$

The proof of (10.3.10) is quite technical and we refer the reader to Datta et al. (2002) for details. The matching prior (10.3.10) is a bounded function of σ_v^2 so that the posterior is proper provided $m > p + 2$. Also, the prior (10.3.10) depends collectively on the sampling variances for all the areas as well as on the individual sampling variance, ψ_i, of the ith area. Strictly speaking, a prior on the common parameter σ_v^2 should not vary with area i.

For the balanced case, $\psi_i = \psi$, the matching prior (10.3.10) reduces to the flat prior $f(\sigma_v^2) \propto 1$ and the resulting HB inferences have dual justification. However, the use of a flat prior when the sampling variances vary significantly across areas could lead to posterior variances not tracking the MSE.

Example 10.3.1. Poverty Counts. In Example 7.1.2, we considered EBLUP estimation of school-age children in poverty in the United States at the county and state levels. Bell (1999) used the state model to calculate ML, REML and HB estimates of σ_v^2 for five years, 1989–1993. ML and REML estimates are both zero for the first four years while $\hat{\sigma}_{v\text{HB}}^2$, based on the flat prior $f(\sigma_v^2) \propto 1$, varied from 1.6 to 3.4. The resulting EBLUP estimates of poverty rates attach zero weight to the direct estimates regardless of the Current Population Survey (CPS) state sample sizes n_i (number of households). Also, the leading term, $g_{1i}(\hat{\sigma}_v^2) = \hat{\gamma}_i\psi_i$, of the MSE estimator (7.1.22) becomes zero when the estimate of σ_v^2 is zero. As a result, the contribution to (7.1.22) comes entirely from the terms $g_{2i}(\hat{\sigma}_v^2)$ and $g_{3i}(\hat{\sigma}_v^2)$ that account for the estimation of $\boldsymbol{\beta}$ and σ_v^2, respectively.

TABLE 10.1 MSE Estimates and Posterior Variance for Four States

State	n_i	$\hat{\theta}_i$	ψ_i	$\text{mse}_{\text{N,ML}}$	mse_{ML}	$\text{mse}_{\text{N,RE}}$	mse_{RE}	$V(\theta_i\|\hat{\boldsymbol{\theta}})$
				1992				
CA	4,927	20.9	1.9	1.3	3.6	1.3	2.8	1.4
NC	2,400	23.0	5.5	0.6	2.0	0.6	1.2	2.0
IN	670	11.8	9.3	0.3	1.4	0.3	0.6	1.7
MS	796	29.6	12.4	2.8	3.8	2.8	3.0	4.0
				1993				
CA	4,639	23.8	2.3	1.5	3.2	1.6	2.2	1.7
NC	2,278	17.0	4.5	1.0	2.4	1.7	2.2	2.0
IN	650	10.3	8.5	0.8	1.9	1.8	2.2	3.0
MS	747	30.5	13.6	3.2	4.3	4.2	4.5	5.1

SOURCE: Adapted from Tables 2 and 3 in Bell (1999).

Table 10.1 reports results for four states in increasing order of the sampling variance ψ_i: California (CA), North Carolina (NC), Indiana (IN) and

Mississippi (MS). This table shows the sample sizes n_i, poverty rates $\hat{\theta}_i$ as a percentage, sampling variances ψ_i, MSE estimates mse_{ML} and mse_{RE} based on ML and RE using (7.1.22) and (7.1.26), and the posterior variances $V(\theta_i|\hat{\boldsymbol{\theta}})$. Naive MSE estimates, based on the formula $g_{1i}(\hat{\sigma}_v^2)+g_{2i}(\hat{\sigma}_v^2)$, are also included ($\text{mse}_{N,\text{ML}}$ and $\text{mse}_{N,\text{RE}}$). Results for 1992 (with $\hat{\sigma}_{v\text{ML}}^2 = \hat{\sigma}_{v\text{RE}}^2 = 0, \hat{\sigma}_{v\text{HB}}^2 = 1.6$) are compared in Table 10.1 to those for 1993 (with $\hat{\sigma}_{v\text{ML}}^2 = 0.4, \hat{\sigma}_{v\text{RE}}^2 = 1.7, \hat{\sigma}_{v\text{HB}}^2 = 3.4$).

Comparing $\text{mse}_{N,\text{ML}}$ to mse_{ML} and $\text{mse}_{N,\text{RE}}$ to mse_{RE} given in Table 10.1, we note that the naive MSE estimates lead to significant underestimation. Here σ_v^2 is not estimated precisely and this fact is reflected in the contribution from $2g_{3i}(\hat{\sigma}_v^2)$. For 1992, the leading term $g_{1i}(\hat{\sigma}_v^2)$ is zero and the contribution to MSE estimates comes entirely from $g_{2i}(\hat{\sigma}_v^2)$ and $2g_{3i}(\hat{\sigma}_v^2)$. The MSE estimates for NC and IN turned out to be smaller than the corresponding MSE estimates for CA, despite the smaller sample sizes and larger sampling variances compared to CA. The reason for this occurrence becomes evident by examining $g_{2i}(\hat{\sigma}_v^2)$ and $2g_{3i}(\hat{\sigma}_v^2)$ which reduce to $g_{2i}(\hat{\sigma}_v^2) = \mathbf{z}_i^T(\sum_i \mathbf{z}_i\mathbf{z}_i^T/\psi_i)^{-1}\mathbf{z}_i$ and $2g_{3i}(\hat{\sigma}_v^2) = 4/(\psi_i \sum_i \psi_i^{-2})$ for ML and REML when $\hat{\sigma}_v^2 = 0$. The term $2g_{3i}(\hat{\sigma}_v^2)$ for CA is larger because the sampling variance $\psi_{\text{CA}} = 1.9$, which appears in the denominator, is much smaller than $\psi_{\text{NC}} = 5.5$ and $\psi_{IN} = 9.3$. The naive MSE estimator is also smaller for NC and IN compared to CA, and in this case only the g_{2i}-term contributes. It appears from the form of $g_{2i}(\hat{\sigma}_v^2)$ that the covariates $\mathbf{z}_i$ for CA contribute to this increase.

Turning to the HB values for 1992, we see that the leading term of the posterior variance $V(\theta_i|\hat{\boldsymbol{\theta}})$ is $g_{1i}(\hat{\sigma}_{v\text{HB}}^2) = \hat{\gamma}_i^{\text{HB}}\psi_i$. Here $\hat{\sigma}_{v\text{HB}}^2 = 1.6$ and the leading term dominates the posterior variance. As a result, the posterior variance is the smallest for CA, although the value for NC is slightly larger compared to IN, despite the larger sample size and smaller sampling variance (5.5 vs 9.3). For 1993 with nonzero ML and REML estimates of σ_v^2, mse_{RE} exhibits a trend similar to $V(\theta_i|\hat{\boldsymbol{\theta}})$, but mse_{ML} values are similar to 1992 values due to a small $\hat{\sigma}_{v\text{ML}}^2(=0.4)$ compared to $\hat{\sigma}_{v\text{RE}}^2(=1.7)$ and $\hat{\sigma}_{v\text{HB}}^2(=3.4)$.

The occurrences of zero estimates of σ_v^2 in the frequentist approach is problematic (Hulting and Harville (1991)), but it is not clear if the HB approach based on a flat prior on σ_v^2 leads to well-calibrated inferences. We have already noted that the matching prior is different from the flat prior if the ψ_i-values vary significantly, as in the case of state CPS variance estimates with $\max(\psi_i)/\min(\psi_i)$ as large as 20.

10.3.3 Unknown σ_v^2: Gibbs Sampling

In this subsection we apply Gibbs sampling to the basic area level model, given by (i) and (ii) of (10.3.1), assuming the prior (10.3.2) on $\boldsymbol{\beta}$ and σ_v^2 with $\sigma_v^{-2} \sim G(a,b)$, $a>0, b>0$, that is, a gamma distribution with shape parameter a and scale parameter b. Note that σ_v^2 is distributed as inverted gamma $IG(a,b)$, with $f(\sigma_v^2) \propto \exp(-b/\sigma_v^2)(1/\sigma_v^2)^{a+1}$. The positive constants a and b are set to be very small (BUGS uses $a=b=0.001$ as the default setting). It is easy to verify that the Gibbs conditionals are given by

$$\text{(i)}\ \ [\boldsymbol{\beta}|\boldsymbol{\theta},\sigma_v^2,\hat{\boldsymbol{\theta}}] \sim N_p\left[\boldsymbol{\beta}^*,\sigma_v^2\left(\sum_i \tilde{\mathbf{z}}_i\tilde{\mathbf{z}}_i^T\right)^{-1}\right] \tag{10.3.11}$$

$$\text{(ii)}\ \ [\theta_i|\boldsymbol{\beta},\sigma_v^2,\hat{\boldsymbol{\theta}}] \sim N\left[\hat{\theta}_i^B(\boldsymbol{\beta},\sigma_v^2),\gamma_i\psi_i\right],\ i=1,\ldots,m \tag{10.3.12}$$

$$\text{(iii)}\ \ [\sigma_v^{-2}|\boldsymbol{\beta},\boldsymbol{\theta},\hat{\boldsymbol{\theta}}] \sim G\left[\frac{m}{2}+a,\frac{1}{2}\sum_i(\tilde{\theta}_i-\tilde{\mathbf{z}}_i^T\boldsymbol{\beta})^2+b\right], \tag{10.3.13}$$

where $\tilde{\theta}_i = \theta_i/b_i$, $\tilde{\mathbf{z}}_i = \mathbf{z}_i/b_i$, $\boldsymbol{\beta}^* = (\sum_i \tilde{\mathbf{z}}_i\tilde{\mathbf{z}}_i^T)^{-1}(\sum_i \tilde{\mathbf{z}}_i\tilde{\theta}_i)$, $N_p(\cdot)$ denotes a p-variate normal, and $\hat{\theta}_i^B(\boldsymbol{\beta},\sigma_v^2) = \gamma_i\hat{\theta}_i + (1-\gamma_i)\mathbf{z}_i^T\boldsymbol{\beta}$ is the Bayes estimator of θ_i. A proof of (10.3.11)–(10.3.13) is sketched in Section 10.14. Note that all the Gibbs conditionals have closed forms and hence the MCMC samples can be generated directly from the conditionals (i)–(iii).

Denote the MCMC samples as $\{(\boldsymbol{\beta}^{(k)},\boldsymbol{\theta}^{(k)},\sigma_v^{2(k)}), k=d+1,\ldots,d+D\}$. Using (10.3.12), we can obtain Rao-Blackwell estimators of the posterior mean and the posterior variance of θ_i as

$$\hat{\theta}_i^{\text{HB}} = \frac{1}{D}\sum_{k=d+1}^{d+D} \hat{\theta}_i^B(\boldsymbol{\beta}^{(k)},\sigma_v^{2(k)}) = \hat{\theta}_i^B(\cdot,\cdot) \tag{10.3.14}$$

and

$$\hat{V}(\theta_i|\hat{\boldsymbol{\theta}}) = \frac{1}{D}\sum_{k=d+1}^{d+D} g_{1i}(\sigma_v^{2(k)}) + \frac{1}{D-1}\sum_{k=d+1}^{d+D}\left[\hat{\theta}_i^B(\boldsymbol{\beta}^{(k)},\sigma_v^{2(k)}) - \hat{\theta}_i^B(\cdot,\cdot)\right]^2. \tag{10.3.15}$$

More efficient estimators may be obtained by exploiting the closed-form results of subsection 10.3.1 for known σ_v^2. We have

$$\hat{\theta}_i^{\text{HB}} = \frac{1}{D}\sum_{k=d+1}^{d+D} \tilde{\theta}_i^H(\sigma_v^{2(k)}) = \tilde{\theta}_i^H(\cdot) \tag{10.3.16}$$

and

$$\hat{V}(\theta_i|\hat{\boldsymbol{\theta}}) = \frac{1}{D}\sum_{k=d+1}^{d+D}\left[g_{1i}(\sigma_v^{2(k)}) + g_{2i}(\sigma_v^{2(k)})\right] + \frac{1}{D-1}\sum_{k=d+1}^{d+D}\left[\tilde{\theta}_i^H(\sigma_v^{2(k)}) - \tilde{\theta}_i^H(\cdot)\right]^2. \tag{10.3.17}$$

Based on the Rao-Blackwell estimator $\hat{\theta}_i^{\text{HB}}$, an estimator of the total $Y_i = g^{-1}(\theta_i)$ is given by $g^{-1}(\hat{\theta}_i^{\text{HB}})$, but it is not equal to the desired posterior mean

$E(Y_i|\hat{\boldsymbol{\theta}})$. However, the marginal MCMC samples $\{Y_i^{(k)} = g^{-1}(\theta_i^{(k)})\}$ can be used directly to estimate the posterior mean of Y_i as

$$\hat{Y}_i^{\text{HB}} = \frac{1}{D}\sum_{k=d+1}^{d+D} Y_i^{(k)} = Y_i^{(\cdot)}. \tag{10.3.18}$$

Similarly, the posterior variance of Y_i is estimated as

$$\hat{V}(Y_i|\hat{\boldsymbol{\theta}}) = \frac{1}{D-1}\sum_{k=d+1}^{d+D}\left(Y_i^{(k)} - Y_i^{(\cdot)}\right)^2. \tag{10.3.19}$$

If L independent runs are generated, instead of a single long run, then the posterior mean is estimated as

$$\hat{Y}_i^{\text{HB}} = \frac{1}{Ld}\sum_{l=1}^{L}\sum_{k=d+1}^{2d} Y_i^{(lk)} = \frac{1}{L}\sum_{l=1}^{L} Y_i^{(l\cdot)} = Y_i^{(\cdot\cdot)}, \tag{10.3.20}$$

where $Y_i^{(lk)}$ is the kth retained value in the lth run of length $2d$ with the first d burn-in iterations deleted. The posterior variance is estimated from (10.2.6):

$$\hat{V}(Y_i|\hat{\boldsymbol{\theta}}) = \frac{d-1}{d}W_i + \frac{1}{d}B_i, \tag{10.3.21}$$

where $B_i = d\sum_{l=1}^{L}(Y_i^{(l\cdot)} - Y_i^{(\cdot\cdot)})^2/(L-1)$ is the between-run variance and $W_i = \sum_{l=1}^{L}\sum_{k=d+1}^{2d}(Y_i^{(lk)} - Y_i^{(l\cdot)})^2/[L(d-1)]$ is the within-run variance.

Example 10.3.2. Canadian Unemployment Rates. Example 5.4.5 mentioned the use of time series and cross-sectional models to estimate monthly unemployment rates for $m = 62$ CMAs and CAs (small areas) in Canada, using time series and cross-sectional data. We study this application in Example 10.8.1. Here we consider only the cross-sectional data for the "current" month June 1999 to illustrate the Gibbs sampling HB method for the basic area level model.

To obtain smoothed estimates of the sampling variances ψ_i, You, Rao and Gambino (2001) first computed the average CV (over time), $\overline{\text{CV}}_i$, and then treated $(\overline{\text{CV}}_i)\hat{\theta}_i$ as $\psi_i^{1/2}$, where $\hat{\theta}_i$ is the Labor Force Survey (LFS) unemployment rate for the ith area. Employment Insurance (EI) beneficiary rates were used as auxiliary variables, z_i, in the linking model $\theta_i = \beta_0 + \beta_1 z_i + v_i; i = 1, \ldots, m$, where $m = 62$.

Gibbs sampling was implemented using $L = 10$ parallel runs, each of length $2d = 2000$. The first $d = 1000$ "burn-in" iterations of each run were deleted. The convergence of the Gibbs sampler was monitored using the method of Gelman and Rubin (1992); see subsection 10.2.4. The Gibbs sampler converged very well in terms of the estimated potential scale reduction factor $\hat{R}$ given by (10.2.7).

To check the overall fit of the model, the posterior predictive p-value, p was approximated from each run of the MCMC output $\{\boldsymbol{\eta}^{(lk)}\}, l = 1, \ldots, 10$, using the formula (10.2.18) with the measure of discrepancy given by $T(\hat{\boldsymbol{\theta}}, \boldsymbol{\eta}) = \sum_{i=1}^{62}(\hat{\theta}_i - \theta_i)^2/\psi_i$. The average of the $L = 10$ p-values, $\hat{p} = 0.59$, indicates a good fit of the model to the current cross-sectional data. Note that the hypothetical replication $\hat{\theta}_i^{(lk)}$, used in (10.2.18), is generated from $N(\theta_i^{(lk)}, \psi_i)$ for each $\theta_i^{(lk)}$.

Rao-Blackwell estimators (10.3.16) and (10.3.17) were used to calculate the estimated posterior mean, $\hat{\theta}_i^{\text{HB}}$, and the estimated posterior variance, $\hat{V}(\theta_i|\hat{\boldsymbol{\theta}})$, for each area i. Figure 10.1 displays the CVs of the direct estimates $\hat{\theta}_i$ and the HB estimates $\hat{\theta}_i^{\text{HB}}$; the CV of $\hat{\theta}_i^{\text{HB}}$ is taken as $[\hat{V}(\theta_i|\hat{\boldsymbol{\theta}})]^{1/2}/\hat{\theta}_i^{\text{HB}}$. It is clear from Figure 10.1 that the HB estimates lead to significant reduction in CV, especially for the areas with smaller population sizes.

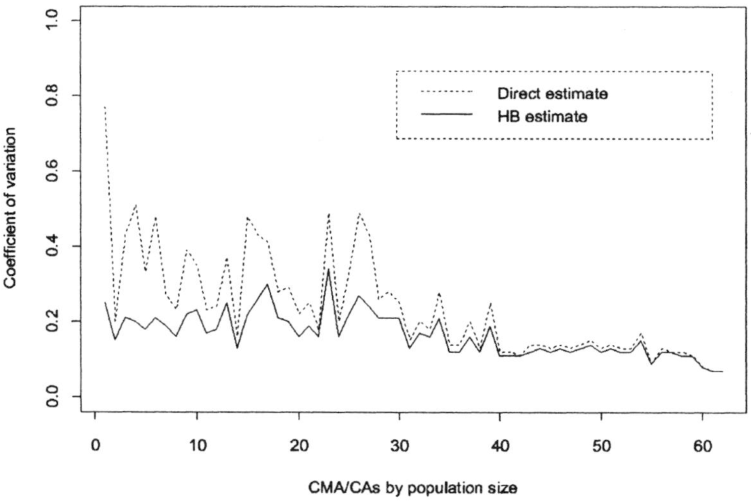

Figure 10.1 Coefficient of Variation (CV) of Direct and HB Estimates. SOURCE: Adapted from Figure 3 in You, Rao and Gambino (2001).

10.4 Unmatched Sampling and Linking Area Level Models

As noted in Section 5.2, the assumption of zero design-bias, that is, $E_p(e_i|\theta_i) = 0$, in the sampling model $\hat{\theta}_i = \theta_i + e_i$ may not be valid if the sample size n_i in the ith area is small and $\theta_i = g(Y_i)$ is a nonlinear function of the total Y_i. A more realistic sampling model is given by

$$\hat{Y}_i = Y_i + e_i^*, \; i = 1, \ldots, m, \tag{10.4.1}$$

with $E_p(e_i^*|Y_i) = 0$, where $\hat{Y}_i$ is a design-unbiased (p-unbiased) estimator of Y_i. A GREG estimator, $\hat{Y}_{\text{GR}}$, of the form (2.4.7) may also be used since it is approximately p-unbiased if the overall sample size is large. The sampling model (10.4.1) is not matched with the linking model $\theta_i = \mathbf{z}_i^T \boldsymbol{\beta} + b_i v_i$, where $v_i \overset{\text{iid}}{\sim} N(0, \sigma_v^2)$. As a result, the two models cannot be combined to produce a basic area level model.

The HB approach readily extends to unmatched sampling and linking models (You and Rao (2002b)). We write the sampling model (10.4.1) under normality as

$$\hat{Y}_i \overset{\text{ind}}{\sim} N(Y_i, \phi_i), \tag{10.4.2}$$

where the sampling variance ϕ_i is known or a known function of Y_i. We first consider the case of known ϕ_i. We combine (10.4.2) with the linking model (ii) or (10.3.1) and the prior (10.3.2) with $\sigma_v^{-2} \sim G(a, b)$. The resulting Gibbs conditionals $[\boldsymbol{\beta}|\mathbf{Y}, \sigma_v^2, \hat{\mathbf{Y}}]$ and $[\sigma_v^{-2}|\boldsymbol{\beta}, \mathbf{Y}, \hat{\mathbf{Y}}]$ are identical to (10.3.11) and (10.3.13), respectively, where $\mathbf{Y} = (Y_1, \ldots, Y_m)^T$ and $\hat{\mathbf{Y}} = (\hat{Y}_1, \ldots, \hat{Y}_m)^T$. However, the Gibbs conditional $[Y_i|\boldsymbol{\beta}, \sigma_v^2, \hat{\mathbf{Y}}]$ does not admit a closed form, unlike (10.3.12). It is easy to verify that

$$f(Y_i|\boldsymbol{\beta}, \sigma_v^2, \hat{\mathbf{Y}}) \propto h(Y_i|\boldsymbol{\beta}, \sigma_v^2) k(Y_i), \tag{10.4.3}$$

where $k(Y_i)$ is the normal density $N(\hat{Y}_i, \phi_i)$, and

$$h(Y_i|\boldsymbol{\beta}, \sigma_v^2) \propto g'(Y_i) \exp\left\{ -\frac{(\tilde{\theta}_i - \tilde{\mathbf{z}}_i^T \boldsymbol{\beta})^2}{2\sigma_v^2} \right\} \tag{10.4.4}$$

is the density of Y_i given $\boldsymbol{\beta}$ and σ_v^2 and $g'(Y_i) = \partial g(Y_i)/\partial Y_i$, where $\tilde{\theta}_i = \hat{\theta}_i / b_i$ and $\tilde{\mathbf{z}}_i = \mathbf{z}_i / b_i$. Using $k(Y_i)$ as the "candidate" density to implement M-H within Gibbs, the acceptance probability (10.2.5) reduces to

$$a(\boldsymbol{\beta}^{(k)}, \sigma_v^{2(k)}, Y_i^{(k)}, Y_i^*) = \min\left\{ \frac{h(Y_i^*|\boldsymbol{\beta}^{(k)}, \sigma_v^{2(k)})}{h(Y_i^{(k)}|\boldsymbol{\beta}^{(k)}, \sigma_v^{2(k)})}, 1 \right\}, \tag{10.4.5}$$

where the candidate Y_i^* is drawn from $N(\hat{Y}_i, \phi_i)$, and $\{Y_i^{(k)}, \boldsymbol{\beta}^{(k)}, \sigma_v^{2(k)}\}$ are the values of Y_i, $\boldsymbol{\beta}$ and σ_v^2 after the kth cycle. The candidate Y_i^* is accepted with probability $a(\boldsymbol{\beta}^{(k)}, \sigma_v^{2(k)}, Y_i^{(k)}, Y_i^*)$; that is, $Y_i^{(k+1)} = Y_i^*$ if Y_i^* is accepted, and $Y_i^{(k+1)} = Y_i^{(k)}$ otherwise. Repeating this updating procedure for $i = 1, \ldots, m$ we get $\mathbf{Y}^{(k+1)}$, noting that the Gibbs conditional of Y_i does not depend on Y_l, $l \neq i$. Given $\mathbf{Y}^{(k+1)}$, we then generate $\boldsymbol{\beta}^{(k+1)}$ from $[\boldsymbol{\beta}|\mathbf{Y}^{(k+1)}, \sigma_v^{2(k)}, \hat{\mathbf{Y}}]$ and $\sigma_v^{2(k+1)}$ from $[\sigma_v^2|\mathbf{Y}^{(k+1)}, \boldsymbol{\beta}^{(k+1)}, \hat{\mathbf{Y}}]$ to complete the $(k+1)$th cycle $\{\mathbf{Y}^{(k+1)}, \boldsymbol{\beta}^{(k+1)}, \sigma_v^{2(k+1)}\}$.

If ϕ_i is a known function of Y_i, we use $h(Y_i|\boldsymbol{\beta}^{(k)}, \sigma_v^{2(k)})$ to draw the candidate Y_i^*, noting that $Y_i = g^{-1}(\theta_i)$ and $\theta_i|\boldsymbol{\beta}, \sigma_v^2 \sim N(z_i^T \boldsymbol{\beta}, b_i^2 \sigma_v^2)$. In this case, the acceptance probability is given by

$$a(Y_i^{(k)}, Y_i^*) = \min\left\{\frac{k(Y_i^*)}{k(Y_i^{(k)})}, 1\right\}. \tag{10.4.6}$$

Example 10.2.2. Canadian Census Undercoverage. In Example 7.1.3, we provided a brief account of an application of the basic area level model to estimate domain undercount in the 1991 Canadian census, using the EBLUP method. You and Rao (2002b) applied the unmatched sampling and linking models to 1991 Canadian census data, using the HB method, and estimated the number of missing persons, $M_i(=Y_i)$ and the undercoverage rate $U_i = M_i/(M_i+C_i)$ for each province i ($i = 1,\ldots,10$), where C_i is the census count.

The sampling variances ϕ_i were estimated through a generalized variance function model of the form $V(\hat{Y}_i) \propto C_i^\gamma$ and then treated as known in the sampling model (10.4.1). The linking model is given by (ii) of (10.3.1) with $\theta_i = \log\{M_i/(M_i + C_i)\}$, $\mathbf{z}_i^T\boldsymbol{\beta} = \beta_0 + \beta_1 \log(C_i)$ and $b_i = 1$. The prior on the model parameters $(\beta_0, \beta_1, \sigma_v^2)$ is given by (10.3.2) with $\sigma_v^{-2} \sim G(0.001, 0.001)$ to reflect lack of prior information. You and Rao (2002b) simulated $L = 8$ parallel M-H within Gibbs runs independently, each of length $2d = 10,000$. The first $d = 5,000$ "burn-in" iterations of each run were deleted. Further, to reduce the autocorrelation within runs, they selected every 10th iteration of the remaining 5,000 iterations of each run, leading to 500 iterations for each run. Thus, the total number of MCMC runs retained is $Ld = 4,000$. The convergence of the M-H within Gibbs sampler was monitored using the method of Gelman and Rubin (1992) described in subsection 10.2.4. The MCMC sampler converged very well in terms of the potential scale reduction factor $\hat{R}$ given by (10.2.7); values of $\hat{R}$ for the 10 provinces were close to 1. We denote the marginal MCMC sample as $\{M_i^{(lk)}, k = d+1,\ldots,2d; l = 1,\ldots,L\}$.

To check the overall fit of the model, the posterior predictive p-value was estimated from each run of the MCMC output $\{\boldsymbol{\eta}^{(lk)}\}, l = 1,\ldots,8$ using the formula (10.2.18) with $T(\hat{\mathbf{Y}},\boldsymbol{\eta}) = \sum_i(\hat{Y}_i - Y_i)^2/\phi_i$. The average of the $L = 8$ p-values, $\hat{p} = 0.38$, indicates a good fit of the model. Note that the hypothetical replications $\hat{Y}_i^{(lk)}$, used in (10.2.18), are generated from $N(Y_i^{(lk)}, \phi_i)$, for each $Y_i^{(lk)}$.

To assess model fit at the individual province level, You and Rao (2002b) computed two statistics proposed by Daniels and Gatsonis (1999). The first statistic, given by

$$p_i = P(\hat{Y}_i < \hat{Y}_{i,\text{obs}}|\hat{\mathbf{Y}}_{\text{obs}}), \tag{10.4.7}$$

provides information on the degree of consistent overestimation or underestimation of $\hat{Y}_{i,\text{obs}}$. This statistic is similar to the cross-validation statistic (10.2.29) but uses the full predictive density $f(Y_i|\hat{\mathbf{Y}}_{\text{obs}})$. As a result, it is computationally simpler than (10.2.29). The p_i-values, computed from the hypothetical replications $\hat{Y}_i^{(lk)}$, ranged from 0.28 to 0.87 with a mean of 0.51

and a median of 0.49, indicating no consistent overestimation or underestimation. The second statistic is similar to the cross-validation standardized residual (10.2.28) but uses the full predictive density. It is given by

$$d_i = [E(\hat{Y}_i|\hat{\mathbf{Y}}_{\text{obs}}) - \hat{Y}_{i,\text{obs}}]/\sqrt{V(\hat{Y}_i|\hat{\mathbf{Y}}_{\text{obs}})}, \qquad (10.4.8)$$

where the expectation and variance are with respect to the full predictive density. The estimated standardized residuals, d_i, ranged from -1.13 (New Brunswick) to 0.57 (Prince Edward Island), indicating adequate model fit for individual provinces.

The posterior mean and the posterior variance of $M_i(= Y_i)$ were estimated using (10.3.20) and (10.3.21) with $L = 8$ and $d = 500$. Table 10.2 reports the direct estimates, $\hat{M}_i$, the HB estimates $\hat{M}_i^{\text{HB}}$ and the associated CVs. For the direct estimate, the CV is based on the design-based variance estimate, while the CV for $\hat{M}_i^{\text{HB}}$ is based on the estimated posterior variance. Table 10.2 also reports the estimates of undercoverage rates U_i, denoted $\hat{U}_i$ and $\hat{U}_i^{\text{HB}}$, and the associated CVs, where $\hat{U}_i$ is the direct estimate of U_i. The HB estimate $\hat{U}_i^{\text{HB}}$ and the associated posterior variance are obtained from (10.3.20) and (10.3.21) by changing $Y_i^{(lk)}$ to $U_i^{(lk)} = M_i^{(lk)}/(M_i^{(lk)} + C_i)$.

It is clear from Table 10.2 that the HB estimates $(\hat{M}_i^{\text{HB}}, \hat{U}_i^{\text{HB}})$ perform better than the direct estimates $(\hat{M}_i, \hat{U}_i)$ in terms of CV, except the estimate for New Brunswick. For Ontario, Quebec and British Columbia, the CVs of HB and direct estimates are nearly equal due to larger sample sizes in these provinces.

Note that the sampling model (10.4.1) assumes that the direct estimate $\hat{Y}_i$ is design-unbiased. In the context of census undercoverage estimation, this assumption may be restrictive because the estimate may be subject to nonsampling bias (Zaslavsky (1993)). This potential bias was ignored due to lack of a reliable estimate of bias.

TABLE 10.2 1991 Canadian Census Undercount Estimates and Associated CVs

Prov	$\hat{M}_i$	CV($\hat{M}_i$)	$\hat{M}_i^{\text{HB}}$	CV($\hat{M}_i^{\text{HB}}$)	$\hat{U}_i$(%)	CV($\hat{U}_i$)	$\hat{U}_i^{\text{HB}}$(%)	CV($\hat{U}_i^{\text{HB}}$)
Nfld	11566	0.16	10782	0.14	1.99	0.16	1.86	0.13
PEI	1220	0.30	1486	0.19	0.93	0.30	1.13	0.19
NS	17329	0.20	17412	0.14	1.89	0.20	1.90	0.14
NB	24280	0.14	18948	0.17	3.25	0.13	2.55	0.17
Que	184473	0.08	189599	0.08	2.58	0.08	2.65	0.08
Ont	381104	0.08	368424	0.08	3.64	0.08	3.52	0.08
Man	20691	0.21	21504	0.14	1.86	0.20	1.93	0.14
Sask	18106	0.19	18822	0.14	1.80	0.18	1.87	0.13
Alta	51825	0.15	55392	0.12	2.01	0.14	2.13	0.12
BC	92236	0.10	89929	0.09	2.73	0.10	2.67	0.09

SOURCE: Adapted from Table 2 in You and Rao (2002b).

10.5 Basic Unit Level Model

In this section, we apply the HB approach to the basic unit level model (7.2.1) with equal error variances (that is, $k_{ij} = 1$), assuming a prior distribution on the model parameters $(\boldsymbol{\beta}, \sigma_v^2, \sigma_e^2)$. We first consider the case of known σ_v^2 and σ_e^2, and assume a "flat" prior on $\boldsymbol{\beta}$: $f(\boldsymbol{\beta}) \propto 1$. We rewrite (7.2.1) as a HB model:

$$\text{(i)} \quad y_{ij}|\boldsymbol{\beta}, v_i, \sigma_e^2 \overset{\text{ind}}{\sim} N(\mathbf{x}_{ij}^T\boldsymbol{\beta} + v_i, \sigma_e^2), \quad j = 1, \ldots, n_i; \quad i = 1, \ldots, m$$

$$\text{(ii)} \quad v_i|\sigma_v^2 \overset{\text{iid}}{\sim} N(0, \sigma_v^2), \quad i = 1, \ldots, m$$

$$\text{(iii)} \quad f(\boldsymbol{\beta}) \propto 1. \tag{10.5.1}$$

We then extend the results to the case of unknown σ_v^2 and σ_e^2 by replacing (iii) in (10.5.1) by

$$\text{(iii)}' \quad f(\boldsymbol{\beta}, \sigma_v^2, \sigma_e^2) = f(\boldsymbol{\beta})f(\sigma_v^2)f(\sigma_e^2) \propto f(\sigma_v^2)f(\sigma_e^2), \tag{10.5.2}$$

where $f(\sigma_v^2)$ and $f(\sigma_e^2)$ are the priors on σ_v^2 and σ_e^2. For simplicity, we take $\mu_i = \overline{\mathbf{X}}_i^T\boldsymbol{\beta} + v_i$ as the ith small area mean, assuming the population size, N_i, is large.

10.5.1 Known σ_v^2 and σ_e^2

When σ_v^2 and σ_e^2 are assumed to be known, the HB and BLUP approaches under normality lead to identical point estimates and measures of variability, assuming a flat prior on $\boldsymbol{\beta}$. This result, in fact, is valid for general linear mixed models with known variance parameters. The HB estimator of μ_i is given by

$$\tilde{\mu}_i^{\text{HB}}(\sigma_v^2, \sigma_e^2) = E(\mu_i|\mathbf{y}, \sigma_v^2, \sigma_e^2) = \tilde{\mu}_i^H, \tag{10.5.3}$$

where $\mathbf{y}$ is the vector of sample observations and $\tilde{\mu}_i^H$ is the BLUP estimator given by (7.2.5). Similarly, the posterior variance of μ_i is

$$V(\mu_i|\sigma_v^2, \sigma_e^2, \mathbf{y}) = M_{1i}(\sigma_v^2, \sigma_e^2) = \text{MSE}(\tilde{\mu}_i^H), \tag{10.5.4}$$

where $M_{1i}(\sigma_v^2, \sigma_e^2)$ is given by (7.2.10).

10.5.2 Unknown σ_v^2 and σ_e^2: Numerical Integration

In practice, σ_v^2 and σ_e^2 are unknown and it is necessary to take account of the uncertainty about σ_v^2 and σ_e^2 by assuming a prior on σ_v^2 and σ_e^2. The HB model is given by (i) and (ii) of (10.5.1) and (iii)′ given by (10.5.2). We obtain the HB estimator of μ_i and the posterior variance of μ_i as

$$\hat{\mu}_i^{\text{HB}} = E(\mu_i|\mathbf{y}) = E_{\sigma_v^2, \sigma_e^2}\left[\tilde{\mu}_i^{\text{HB}}(\sigma_v^2, \sigma_e^2)\right] \tag{10.5.5}$$

and

$$V(\mu_i|\mathbf{y}) = E_{\sigma_v^2,\sigma_e^2}\left[M_{1i}(\sigma_v^2,\sigma_e^2)\right] + V_{\sigma_v^2,\sigma_e^2}\left[\tilde{\mu}_i^{\text{HB}}(\sigma_v^2,\sigma_e^2)\right], \tag{10.5.6}$$

where $E_{\sigma_v^2,\sigma_e^2}$ and $V_{\sigma_v^2,\sigma_e^2}$, respectively denote the expectation and variance with respect to the posterior distribution $f(\sigma_v^2,\sigma_e^2|\mathbf{y})$.

As in Section 10.3, the posterior $f(\sigma_v^2,\sigma_e^2|\mathbf{y})$ may be obtained from the restricted likelihood function $L_R(\sigma_v^2,\sigma_e^2)$ as

$$f(\sigma_v^2,\sigma_e^2|\mathbf{y}) \propto L_R(\sigma_v^2,\sigma_e^2)f(\sigma_v^2)f(\sigma_e^2). \tag{10.5.7}$$

Under flat priors $f(\sigma_v^2) \propto 1$ and $f(\sigma_e^2) \propto 1$, the posterior $f(\sigma_v^2,\sigma_e^2|\mathbf{y})$ is proper (subject to a mild sample size restriction) and proportional to $L_R(\sigma_v^2,\sigma_e^2)$. Evaluation of the posterior mean (10.5.5) and the posterior variance (10.5.6), using $f(\sigma_v^2,\sigma_e^2|\mathbf{y}) \propto L_R(\sigma_v^2,\sigma_e^2)$, involves two-dimensional integrations.

If we assume a diffuse gamma prior on σ_e^{-2}, $G(a_e,b_e)$ with $a_e \geq 0$ and $b_e > 0$, then it is possible to integrate out σ_e^2 with respect to $f(\sigma_e^2|\tau_v,\mathbf{y})$, where $\tau_v = \sigma_v^2/\sigma_e^2$. The evaluation of (10.5.5) and (10.5.6) is now reduced to single-dimensional integration with respect to the posterior of τ_v, $f(\tau_v|\mathbf{y})$. Datta and Ghosh (1991) expressed $f(\tau_v|\mathbf{y})$ as $f(\tau_v|\mathbf{y}) \propto h(\tau_v)$ and obtained an explicit expression for $h(\tau_v)$, assuming a gamma prior on τ_v^{-1} : $G(a_v,b_v)$ with $a_v \geq 0$ and $b_v > 0$; note that a_v is the shape parameter and b_v is the scale parameter. Datta and Ghosh (1991) applied the numerical integration method to the data on county crop areas (Example 7.2.1) studied by Battese, Harter and Fuller (1988). They calculated the HB estimate of mean hectares of soybeans and associated standard error (square root of posterior variance) for each of the $m = 12$ counties in north-central Iowa, using flat priors on $\boldsymbol{\beta}$ and gamma priors on σ_e^{-2} and τ_v^{-1} with $a_e = a_v = 0$ and $b_e = b_v = 0.005$ to reflect lack of prior information. Datta and Ghosh (1991) actually studied the more complex case of finite population means $\overline{Y}_i$ instead of the means μ_i, but the sampling fractions n_i/N_i are negligible in Example 7.2.1.

10.5.3 Unknown σ_v^2 and σ_e^2: Gibbs Sampling

In this section we apply Gibbs sampling to the basic unit level model given by (i) and (ii) of (10.5.1), assuming the prior (10.5.2) on $\boldsymbol{\beta},\sigma_v^2$ and σ_e^2 with $\sigma_v^{-2} \sim G(a_v,b_v), a_v \geq 0, b_v > 0$ and $\sigma_e^{-2} \sim G(a_e,b_e), a_e \geq 0, b_e > 0$. It is easy to verify that the Gibbs conditionals are given by

$$\text{(i)}\quad [\boldsymbol{\beta}|\mathbf{v},\sigma_v^2,\sigma_e^2,\mathbf{y}] \sim N_p\left[\left(\sum_i\sum_j \mathbf{x}_{ij}\mathbf{x}_{ij}^T\right)^{-1}\sum_i\sum_j \mathbf{x}_{ij}(y_{ij}-v_i),\ \sigma_e^2\left(\sum_i\sum_j \mathbf{x}_{ij}\mathbf{x}_{ij}^T\right)^{-1}\right] \tag{10.5.8}$$

(ii) For $i = 1,\ldots,m$,

$$[v_i|\boldsymbol{\beta},\sigma_v^2,\sigma_e^2,\mathbf{y}] \sim N\left[\gamma_i(\bar{y}_i - \bar{\mathbf{x}}_i^T\boldsymbol{\beta}), g_{1i}(\sigma_v^2,\sigma_e^2) = \gamma_i\sigma_e^2/n_i\right] \tag{10.5.9}$$

(iii) $[\sigma_e^{-2}|\boldsymbol{\beta},\mathbf{v},\sigma_v^2,\mathbf{y}] \sim$

$$G\left[\frac{n}{2}+a_e, \frac{1}{2}\sum_i\sum_j(y_{ij}-\mathbf{x}_{ij}^T\boldsymbol{\beta}-v_i)^2+b_e\right] \qquad (10.5.10)$$

(iv) $$[\sigma_v^{-2}|\boldsymbol{\beta},\mathbf{v},\sigma_e^2,\mathbf{y}] \sim G\left(\frac{m}{2}+a_v, \frac{1}{2}\sum_i v_i^2+b_v\right), \qquad (10.5.11)$$

where $n = \Sigma_i n_i$, $\mathbf{v} = (v_1,\ldots,v_m)^T$, $\bar{y}_i = \sum_j y_{ij}/n_i$, $\bar{\mathbf{x}}_i = \sum_j \mathbf{x}_{ij}/n_i$, and $\gamma_i = \sigma_v^2/(\sigma_v^2+\sigma_e^2/n_i)$. The proof of (10.5.8)–(10.5.11) follows along the lines of the proof of (10.3.11)–(10.3.13) given in subsection 10.14.3. Note that all the Gibbs conditionals have closed forms and hence the MCMC samples can be generated directly from the conditionals (i) to (iv). BUGS can be used to generate samples from the above conditionals using either default inverted gamma priors on σ_v^2 and σ_e^2 (with $a_e = b_e = 0.001$ and $a_v = b_v = 0.001$) or with specified values of a_e, b_e, a_v and b_v. CODA can be used to perform convergence diagnostics.

Denote the MCMC samples from a single large run by $\{\boldsymbol{\beta}^{(k)},\mathbf{v}^{(k)},\sigma_v^{2(k)},\sigma_e^{2(k)}, k=d+1,\ldots,d+D\}$. The marginal MCMC samples $\{\boldsymbol{\beta}^{(k)},\mathbf{v}^{(k)}\}$ can be used directly to estimate the posterior mean of μ_i as

$$\hat{\mu}_i^{\text{HB}} = \frac{1}{D}\sum_{k=d+1}^{d+D}\mu_i^{(k)} = \mu_i^{(\cdot)}, \qquad (10.5.12)$$

where $\mu_i^{(k)} = \bar{\mathbf{X}}_i^T\boldsymbol{\beta}^{(k)}+v_i^{(k)}$. Similarly, the posterior variance of μ_i is estimated as

$$V(\mu_i|\mathbf{y}) = \frac{1}{D-1}\sum_{k=d+1}^{d+D}\left(\mu_i^{(k)}-\mu_i^{(\cdot)}\right)^2. \qquad (10.5.13)$$

Alternatively, Rao-Blackwell estimators of the posterior mean and the posterior variance of μ_i may be used along the lines of (10.3.16) and (10.3.17):

$$\mu_i^{\text{HB}} = \frac{1}{D}\sum_{k=d+1}^{d+D}\tilde{\mu}_i^{\text{HB}}(\sigma_v^{2(k)},\sigma_e^{2(k)}) = \tilde{\mu}_i^{\text{HB}}(\cdot,\cdot) \qquad (10.5.14)$$

and

$$V(\mu_i|\mathbf{y}) = \frac{1}{D}\sum_{k=d+1}^{d+D}\left[g_{1i}(\sigma_v^{2(k)},\sigma_e^{2(k)})+g_{2i}(\sigma_v^{2(k)},\sigma_e^{2(k)})\right]$$
$$+\frac{1}{D-1}\sum_{k=d+1}^{d+D}\left[\tilde{\mu}_i^{\text{HB}}(\sigma_v^{2(k)},\sigma_e^{2(k)})-\tilde{\mu}_i^{\text{HB}}(\cdot,\cdot)\right]^2. \qquad (10.5.15)$$

Example 10.5.1. County Crop Areas. The Gibbs sampling method was applied to the data on area under corn for each of $m = 12$ counties in north-central Iowa, excluding the second sample segment in Hardin county (see

Example 7.2.1). We fitted the nested error linear regression model, $y_{ij} = \beta_0 + \beta_1 x_{1ij} + \beta_2 x_{2ij} + v_i + e_{ij}$, for corn ($y_{ij}$) using the LANDSAT satellite readings of corn (x_{1ij}) and soybeans (x_{2ij}); see Example 7.2.1 for details. We used the BUGS program with default flat prior on $\boldsymbol{\beta}$ and gamma priors on σ_v^{-2} and σ_e^{-2} to generate samples from the joint posterior distribution of $(\boldsymbol{\beta}, \mathbf{v}, \sigma_v^2, \sigma_e^2)$. We generated a single long run of length $D = 5000$ after discarding the first $d = 5000$ "burn-in" iterations. We used CODA to implement convergence diagnostics and statistical and output analysis of the simulated samples.

To check the overall fit of the model, the posterior predictive value, p, was estimated, using the formula (10.2.19) with measure of discrepancy given by $T(\mathbf{y}, \boldsymbol{\eta}) = \sum_i \sum_j (y_{ij} - \beta_0 - \beta_1 x_{1ij} - \beta_2 x_{2ij})^2/(\sigma_v^2 + \sigma_e^2)$. The estimated p-value, $p = 0.51$, indicates a good fit of the model. We also fitted the simpler model $y_{ij} = \beta_0 + \beta_1 x_{1ij} + v_i + e_{ij}$, using only the corn ($x_{1ij}$) satellite reading as an auxiliary variable. The resulting p-value, $\hat{p} = 0.497$, indicates that the simpler model also fits the data well. We also compared the CPO plots for the two models using the formula (10.2.26). The CPO plots indicate that the full model with x_1 and x_2 is slightly better than the simpler model with x_1 only.

TABLE 10.3 EBLUP and HB Estimates and Associated Standard Errors: County Corn Areas

		Estimate		Standard Error	
County	n_i	EBLUP	HB	EBLUP	HB
Cerro Gordo	1	122.2	122.2	9.6	8.9
Hamilton	1	126.3	126.1	9.5	8.7
Worth	1	106.2	108.1	9.3	9.8
Humboldt	2	108.0	109.5	8.1	8.1
Franklin	3	145.0	142.8	6.5	7.3
Pocahontas	3	112.6	111.2	6.6	6.5
Winnebago	3	112.4	113.8	6.6	6.5
Wright	3	122.1	122.0	6.7	6.2
Webster	4	115.8	114.5	5.8	6.1
Hancock	5	124.3	124.8	5.3	5.2
Kossuth	5	106.3	108.4	5.2	6.3
Hardin	5	143.6	142.2	5.7	6.0

Table 10.3 gives the EBLUP estimates and associated standard errors (taken from Battese, Harter and Fuller (1988)), and the HB estimates and associated standard errors. It is clear from Table 10.3 that the EBLUP and HB estimates are similar. Standard errors are also similar except for Kossuth county (with $n_i = 5$) where the HB standard error (square root of posterior variance) is about 20% larger than the corresponding EBLUP standard error; in fact, the HB standard error is slightly larger than the corresponding sample regression (SR) standard error reported in Table 7.3. Reasons for this exception are not clear.

10.5.4 Pseudo-HB Estimation

In subsection 7.2.7 of Chapter 7 we obtained a pseudo-EBLUP estimator, $\hat{\mu}_{iw}^{H}$, of the small area mean μ_i that makes use of survey weights w_{ij}, unlike the EBLUP estimator $\hat{\mu}_i^H$. Also, $\hat{\mu}_{iw}^{H}$ satisfies the benchmarking property without any adjustment, unlike $\hat{\mu}_i^H$. In this subsection, we obtain a pseudo-HB estimator, $\hat{\mu}_{iw}^{\text{HB}}$, that is analogous to the pseudo-EBLUP estimator $\hat{\mu}_{iw}^{H}$. The estimator $\hat{\mu}_{iw}^{\text{HB}}$ also makes use of the survey weights and satisfies the benchmarking property without any adjustment.

We make use of the unit level MCMC samples $\{\sigma_v^{2(k)}, \sigma_e^{2(k)}\}$ generated from (i) and (ii) of (10.5.1) and the prior (10.5.2) with $\sigma_v^{-2} \sim G(a_v, b_v)$ and $\sigma_e^{-2} \sim G(a_e, b_e)$ but replace $\boldsymbol{\beta}^{(k)}$ by $\boldsymbol{\beta}_w^{(k)}$ that makes use of the design-weighted estimator $\tilde{\boldsymbol{\beta}}_w(\sigma_v^2, \sigma_e^2) = \tilde{\boldsymbol{\beta}}_w$ given by (7.2.52). Now, assuming a flat prior $f(\boldsymbol{\beta}) \propto 1$ and noting that $\tilde{\boldsymbol{\beta}}_w|\boldsymbol{\beta}, \sigma_v^2, \sigma_e^2 \sim N_p(\boldsymbol{\beta}, \Phi_w)$ we get the posterior distribution $[\boldsymbol{\beta}|\tilde{\boldsymbol{\beta}}_w, \sigma_v^2, \sigma_e^2] \sim N(\tilde{\boldsymbol{\beta}}_w, \Phi_w)$, where $\Phi_w = \Phi_w(\sigma_v^2, \sigma_e^2)$ is given by (7.2.59). We calculate $\tilde{\boldsymbol{\beta}}_w^{(k)} = \tilde{\boldsymbol{\beta}}_w(\sigma_v^{2(k)}, \sigma_e^{2(k)})$ and $\Phi_w(\sigma_v^{2(k)}, \sigma_e^{2(k)})$ using $\sigma_v^{2(k)}$ and $\sigma_e^{2(k)}$ and then generate $\boldsymbol{\beta}_w^{(k)}$ from $N_p(\tilde{\boldsymbol{\beta}}_w^{(k)}, \Phi_w^{(k)})$.

Under the survey-weighted model (7.2.48), the conditional posterior mean $E(\mu_i|\bar{y}_{iw}, \boldsymbol{\beta}, \sigma_v^2, \sigma_e^2)$ is identical to the BLUP estimator $\tilde{\mu}_{iw}^H = \tilde{\mu}_{iw}^H(\boldsymbol{\beta}, \sigma_v^2, \sigma_e^2)$ given by (7.2.49), where $\bar{y}_{iw} = \sum_j w_{ij} y_{ij} / \sum_j w_{ij} = \sum_j \tilde{w}_{ij} y_{ij}$. Similarly, the conditional posterior variance $V(\mu_i|\bar{y}_{iw}, \boldsymbol{\beta}, \sigma_v^2, \sigma_e^2)$ is equal to $g_{1iw}(\sigma_v^2, \sigma_e^2)$ given by (7.2.56). Now using the MCMC samples $\{\boldsymbol{\beta}^{(k)}, \sigma_v^{2(k)}, \sigma_e^{2(k)}; k = d + 1, \ldots, d + D\}$, we get a pseudo-HB estimator of μ_i as

$$\hat{\mu}_{iw}^{\text{HB}} = \frac{1}{D} \sum_{k=d+1}^{d+D} \tilde{\mu}_{iw}\left(\boldsymbol{\beta}^{(k)}, \sigma_v^{2(k)}, \sigma_e^{2(k)}\right)$$

$$= \overline{\mathbf{X}}_i^T \boldsymbol{\beta}_w^{(\cdot)} + \tilde{v}_{iw}^{(\cdot)}, \tag{10.5.16}$$

where $\boldsymbol{\beta}_w^{(\cdot)} = \sum_k \boldsymbol{\beta}_w^{(k)}/D$, $\tilde{v}_{iw}^{(\cdot)}$ is the average of $\tilde{v}_{iw}(\boldsymbol{\beta}^{(k)}, \sigma_v^{2(k)}, \sigma_e^{2(k)})$ over k, and $\tilde{v}_{iw}(\boldsymbol{\beta}, \sigma_v^2, \sigma_e^2)$ is given by (7.2.50). The pseudo-HB estimator (10.5.16) is design-consistent as n_i increases.

It is easy to verify that $\sum_i N_i \hat{\mu}_{iw}^{\text{HB}}$ benchmarks to the direct survey regression estimator $\hat{Y}_w + (\mathbf{X} - \hat{\mathbf{X}}_w)^T \boldsymbol{\beta}_w^{(\cdot)}$, where $\hat{Y}_w = \sum_i \sum_j \tilde{w}_{ij} y_{ij}$ and $\hat{X}_w = \sum_i \sum_j \tilde{w}_{ij} x_{ij}$ are the direct estimators of the overall totals Y, and $\mathbf{X}$, respectively. Note that the direct survey regression estimator here differs from the estimator in subsection 7.2.7. The latter uses $\hat{\boldsymbol{\beta}}_w = \tilde{\boldsymbol{\beta}}_w(\hat{\sigma}_v^2, \hat{\sigma}_e^2)$ but the difference between $\hat{\boldsymbol{\beta}}_w$ and $\boldsymbol{\beta}_w^{(\cdot)}$ should be very small.

A pseudo-posterior variance of μ_i is obtained as

$$\hat{V}^{\text{PHB}}(\mu_i) = \frac{1}{D} \sum_{k=d+1}^{d+D} g_{1iw}(\sigma_v^{2(k)}, \sigma_e^{2(k)})$$

$$+ \frac{1}{D-1} \sum_{k=d+1}^{d+D} \left[\tilde{\mu}_{iw}^H(\boldsymbol{\beta}^{(k)}, \sigma_v^{2(k)}, \sigma_e^{2(k)}) - \hat{\mu}_{iw}^{\text{HB}}\right]^2. \tag{10.5.17}$$

The last term in (10.5.17) accounts for the uncertainty associated with $\boldsymbol{\beta}$, σ_v^2 and σ_e^2. We refer the reader to You and Rao (2003) for details of the pseudo-HB method.

Example 10.5.2. County Corn Areas. In Example 7.2.3, we applied the pseudo-EBLUP method to county corn area data from Battese et al. (1988), assuming simple random sampling within areas, that is, $w_{ij} = w_i = N_i/n_i$. We generated MCMC samples $\{\boldsymbol{\beta}_w^{(k)}, \sigma_v^{2(k)}, \sigma_e^{2(k)}\}$ from this data set, using diffuse priors on $\boldsymbol{\beta}$, σ_v^2 and σ_e^2. Table 10.4 compares the results from the pseudo-HB method to those from the pseudo-EBLUP method (Table 7.6). It is clear from Table 10.4 that the pseudo-HB and the pseudo-EBLUP estimates are very similar. Standard errors are also similar except for Kossuth county (with $n_i = 5$): the pseudo-HB standard error is significantly larger than the corresponding pseudo-EBLUP standard error, similar to the EBLUP and HB standard errors reported in Table 10.3.

TABLE 10.4 Pseudo-HB and Pseudo-EBLUP Estimates and Associated Standard Errors: County Corn Areas

County	n_i	Estimate		Standard Error	
		Pseudo-HB	Pseudo-EBLUP	Pseudo-HB	Pseudo-EBLUP
Cerro Gordo	1	120.6	120.5	9.3	9.9
Hamilton	1	125.2	125.2	9.4	9.7
Worth	1	107.5	106.4	10.2	9.6
Humboldt	2	108.4	107.4	8.2	8.3
Franklin	3	142.5	143.7	7.4	6.6
Pocahontas	3	110.6	111.5	7.0	6.6
Winnebago	3	113.2	112.1	7.0	6.6
Wright	3	121.1	121.3	6.4	6.8
Webster	4	114.2	115.0	6.4	5.8
Hancock	5	124.8	124.5	5.4	5.4
Kossuth	5	108.0	106.6	6.6	5.3
Hardin	5	142.3	143.5	6.1	5.8

SOURCE: Adapted from Table 1 in You and Rao (2003).

We assumed that the basic unit level model holds for the sample $\{y_{ij}, \mathbf{x}_{ij}, j \in s_i;\ i = 1, \ldots, m\}$ in developing the pseudo-HB estimator (10.5.16) and the pseudo-posterior variance (10.5.17). If the model holds for the sample, then the HB estimator (without weights) is optimal in the sense of providing the smallest posterior variance, but it is not design-consistent. As noted in subsection 7.2.7, survey practitioners prefer design-consistent estimators as a form of insurance, and the sample size, n_i, could be moderately large for some of the areas under consideration in which case design-consistency becomes relevant.

Singh and Folsom (2001b) developed methods based on survey-weighted estimating functions (EFs) that account for sample selection bias. The EFs for v_i and $\boldsymbol{\beta}$ are given by

$$\hat{\phi}_i = \sum_{j=1}^{n_i} w_{ij}(y_{ij} - \mathbf{x}_{ij}^T\boldsymbol{\beta} - v_i), \quad i = 1, \ldots, m \tag{10.5.18}$$

and

$$\hat{\boldsymbol{\phi}} = \sum_{i=1}^{m}\sum_{j=1}^{n_i} w_{ij}\mathbf{x}_{ij}(y_{ij} - \mathbf{x}_{ij}^T\boldsymbol{\beta} - v_i), \tag{10.5.19}$$

respectively. We take the expectation of $\hat{\phi}_i$ with respect to the design first (denoted E_p) and then with respect to the population model (denoted E_m) to get $E_mE_p(\hat{\phi}_i) = E_m[\sum_{j=1}^{N_i}(y_{ij} - \mathbf{x}_{ij}^T\boldsymbol{\beta} - v_i)] = E_m[\sum_{j=1}^{N_i} e_{ij}] = 0$ regardless of sample selection bias, noting that $y_{ij} - \mathbf{x}_{ij}^T\boldsymbol{\beta} - v_i = e_{ij}$ and $E_m(e_{ij}) = 0$ for the population model. Similarly, $E_mE_p(\hat{\boldsymbol{\phi}}) = 0$. Turning to the variance of $\hat{\phi}_i$, we have $V(\hat{\phi}_i) = E_mV_p(\hat{\phi}_i) + V_mE_p(\hat{\phi}_i) \approx E_mV_p(\hat{\phi}_i)$ if n_i is much smaller than N_i, where V_p and V_m denote the variances with respect to the design and the population model, respectively. This follows by noting that $E_mV_p(\hat{\phi}_i)$ is of order N_i^2 compared to $V_mE_p(\hat{\phi}_i) = V_m(\sum_{j=1}^{N_i} e_{ij}) = N_i\sigma_e^2$, which is of order N_i. Further, $E_mV_p(\hat{\phi}_i)$ is approximated by $V_p(\hat{\phi}_i)$ which is assumed to be known for given $\boldsymbol{\beta}$ and v_i, as in the Fay-Herriot method for the basic area level model. Similarly, the design-based covariance matrix $\mathbf{V}_p(\hat{\boldsymbol{\phi}})$ and the covariance vector $\text{Cov}_p(\hat{\boldsymbol{\phi}}, \hat{\phi}_i)$ are assumed to be known, for given $\boldsymbol{\beta}$ and v_i's. Further, it is assumed that $\hat{\phi}_i$ is normal with mean zero and variance $V_p(\hat{\phi}_i)$ and $\hat{\boldsymbol{\phi}}$ is normal with mean vector $\mathbf{0}$ and covariance matrix $\mathbf{V}_p(\hat{\boldsymbol{\phi}})$. Normality of $\hat{\boldsymbol{\phi}}$ is reasonable because $\hat{\boldsymbol{\phi}}$ is based on the overall sample, but the normality of $\hat{\phi}_i$ may not hold because it is based on the ith area sample only. Singh and Folsom (2001b) suggested collapsing the $\hat{\phi}_i$'s to improve the normal approximation, and for this purpose they proposed using "EF-collapsing posteriors" with similar values of model-based HB estimates $\hat{v}_i^{\text{HB}}$.

Based on the above assumptions and possible collapsing of the EFs for the random effects v_i, Singh and Folsom (2001b) developed an EF-based normal likelihood which is a function of $\boldsymbol{\beta}$, $\mathbf{v} = (v_1, \ldots, v_m)^T$ and weighted data, say $\mathbf{t} = \mathbf{t}(\mathbf{y}, \mathbf{w})$. Using this likelihood and the flat prior $f(\boldsymbol{\beta}) \propto 1$, the Gibbs conditionals $[\boldsymbol{\beta}|\mathbf{v}, \mathbf{t}]$ and $[v_i|v_{l(l\neq i)}, \boldsymbol{\beta}, \mathbf{t}]$ are obtained. Similarly, assuming $\sigma_v^{-2} \sim G(a, b)$, the Gibbs conditional $[\sigma_v^2|\mathbf{v}]$ is obtained from $f(\mathbf{v}|\sigma_v^2)$, noting that $v_i \overset{\text{iid}}{\sim} N(0, \sigma_v^2)$. The MCMC samples $\{\boldsymbol{\beta}^{(k)}, \mathbf{v}^{(k)}, \sigma_v^{2(k)}\}$ are generated from the above Gibbs conditionals which, in turn, lead to survey weighted pseudo-HB estimates of the μ_i's and associated pseudo-posterior variances. To generate $\{\boldsymbol{\beta}^{(k+1)}, \mathbf{v}^{(k+1)}, \sigma_v^{2(k+1)}\}$ from the previous values $\{\boldsymbol{\beta}^{(k)}, \mathbf{v}^{(k)}, \sigma_v^{2(k)}\}$, the design-based variances and covariances are computed at $\boldsymbol{\beta}^{(k)}$ and $\mathbf{v}^{(k)}$ so the they are known at the $(k+1)$th cycle.

The proposed method looks promising as it takes account of possible sample selection bias. Moreover, it extends to generalized linear mixed population models such as a logistic mixed model: $y_{ij}|p_{ij} \stackrel{\text{ind}}{\sim} \text{Bernoulli}(p_{ij})$ and $\text{logit}(p_{ij}) = \mathbf{x}_{ij}^T\boldsymbol{\beta} + v_i$. However, its scope is limited by the assumption of known design variances and covariances, similar to the Fay-Herriot method for the basic area level model, and by the somewhat ad hoc choice of EF-collapsing "partners". Also, in situations where only some of the areas are selected in the sample, we have to assume that the random area effects, v_i, are not subject to sample selection bias.

10.6 General ANOVA Model

In this section, we apply the HB approach to the general ANOVA model (6.2.15), assuming a prior distribution on the model parameters $(\boldsymbol{\beta}, \boldsymbol{\delta})$, where $\boldsymbol{\delta} = (\sigma_0^2, \ldots, \sigma_r^2)^T$ is the vector of variance components. In particular, we assume that $f(\boldsymbol{\beta}, \boldsymbol{\delta}) = f(\boldsymbol{\beta})\prod_{i=0}^r f(\sigma_i^2)$ with $f(\boldsymbol{\beta}) \propto 1$, $f(\sigma_i^2) \propto (\sigma_i^2)^{-(a_i+1)}, i = 1, \ldots, r$ and $f(\sigma_0^2) \propto (\sigma_0^2)^{-(b+1)}$ for specified values a_i and b. Letting $\sigma_0^2 = \sigma_e^2$, the HB model may be written as

$$\text{(i)}\quad \mathbf{y}|\mathbf{v}, \sigma_e^2, \boldsymbol{\beta}, \sim N\left(\mathbf{X}\boldsymbol{\beta} + \sum_{i=1}^r \mathbf{Z}_i\mathbf{v}_i, \sigma_e^2\mathbf{I}\right)$$

$$\text{(ii)}\quad \mathbf{v}_i|\sigma_1^2, \ldots, \sigma_r^2 \stackrel{\text{ind}}{\sim} N_{h_i}(\mathbf{0}, \sigma_i^2\mathbf{I}_{h_i});\quad i = 1, \ldots, r$$

$$\text{(iii)}\quad f(\boldsymbol{\beta}) \propto 1,\ f(\sigma_i^2) \propto (\sigma_i^2)^{-(a_i+1)},\ f(\sigma_e^2) \propto (\sigma_e^2)^{-(b+1)}. \tag{10.6.1}$$

Under model (10.6.1), Hobert and Casella (1996) derived Gibbs conditionals and showed that the Gibbs conditionals are all proper if $2a_i > -h_i$ for all i and $2b > -n$. These conditions are satisfied for diffuse priors with small values of a_i and b. However, propriety of the conditionals does not imply propriety of the joint posterior $f(\boldsymbol{\beta}, \boldsymbol{\delta}, \mathbf{v}|\mathbf{y})$. In fact, many values of the vector $(a_1, \ldots, a_r, b)$ simultaneously yield proper conditionals and an improper joint posterior. It is therefore important to verify that the chosen improper joint prior yields a proper joint posterior before proceeding with Gibbs sampling. Hobert and Casella (1996) derived conditions on the constants $(a_1, \ldots, a_r, b)$ that ensure propriety of the joint posterior. Theorem 10.6.1 gives these conditions.

Theorem 10.6.1. Let $t = \text{rank}(\mathbf{P_X Z}) = \text{rank}(\mathbf{Z}^T\mathbf{P_X Z}) \le h$, where $h = \sum_i h_i$, $\mathbf{Z} = (\mathbf{Z}_1, \ldots, \mathbf{Z}_r)$ and $\mathbf{P_X} = \mathbf{I}_n - \mathbf{X}(\mathbf{X}^T\mathbf{X})^{-1}\mathbf{X}^T$. Consider the following two cases: (1) $t = h$ or $r = 1$; (2) $t < h$ and $r > 1$. For case 1, the following conditions are necessary and sufficient for the propriety of the joint posterior: (a) $a_i < 0$, (b) $h_i > h - t - 2a_i$, (c) $n + 2\sum_i a_i + 2b - p > 0$. For case 2, conditions (a), (b) and (c) are sufficient for the property of the joint posterior while necessary conditions result when (b) is replaced by (b′) $h_i > -2a_i$.

The proof of Theorem 10.6.1 is very technical and we refer the reader to Hobert and Casella (1996) for details of the proof. If we choose $a_i = -1$ for all

i and $b = -1$, we get uniform (flat) priors on the variance components, that is, $f(\sigma_i^2) \propto 1$ and $f(\sigma_e^2) \propto 1$. This choice typically yields a proper joint posterior in practical applications, but not always. For example, for the balanced one-fold random effects model, $y_{ij} = \mu + v_i + e_{ij}, j = 1, \ldots, \bar{n}, i = 1, \ldots, m$, a flat prior on $\sigma_v^2 (= \sigma_1^2)$ violates condition (b) if $m = 3$, noting that $r = 1$, $h_i = h = 3$, $t = \text{rank}(\mathbf{Z}_i^T \mathbf{P}_\mathbf{X} \mathbf{Z}) = \text{rank}(\mathbf{I}_3 - \mathbf{1}_3 \mathbf{1}_3^T/3) = 2$, and $a_i = -1$. As a result, the joint posterior is improper.

PROC MIXED in SAS, Version 8.0, implements the HB method for the general ANOVA model. It uses flat priors on the variance components $\sigma_e^2, \sigma_1^2, \ldots, \sigma_r^2$ and the regression parameters $\boldsymbol{\beta}$. PRIOR option in SAS generates MCMC samples from the joint posterior of variance components and regression parameters, while RANDOM option generates samples from the marginal posterior of random effects $\mathbf{v}_1, \ldots, \mathbf{v}_r$.

The ANOVA model (10.6.1) covers the one-fold and two-fold nested error regression models as special cases, but not the two-level model (8.8.1) which is a special case of the general linear mixed model with a block diagonal covariance structure; see (6.3.1). In Section 10.7, we apply the HB method to the two-level model (8.8.1).

10.7 Two-level Models

In Section 8.8 of Chapter 8, we studied EBLUP estimation for the two-level model (8.8.1). In this section we study three different HB models, including the HB version of (8.8.1), assuming priors on the model parameters.

Model 1. The HB version of (8.8.1) with $k_{ij} = 1$ may be written as

(i) $y_{ij} | \boldsymbol{\beta}_i, \sigma_e^2 \overset{\text{ind}}{\sim} N(\mathbf{x}_{ij}^T \boldsymbol{\beta}_i, \sigma_e^2)$

(ii) $\boldsymbol{\beta}_i | \boldsymbol{\alpha}, \Sigma_v \overset{\text{ind}}{\sim} N_p(\mathbf{Z}_i \boldsymbol{\alpha}, \boldsymbol{\Sigma}_v)$

(iii) $f(\boldsymbol{\alpha}, \sigma_e^2, \boldsymbol{\Sigma}_v) = f(\boldsymbol{\alpha}) f(\sigma_e^2) f(\boldsymbol{\Sigma}_v)$, where

$$\boldsymbol{\alpha} \sim N_q(\mathbf{0}, \mathbf{D}), \sigma_e^{-2} \sim G(a, b), \boldsymbol{\Sigma}_v^{-1} \sim W_p(d, \boldsymbol{\Delta}), \tag{10.7.1}$$

and $W_p(d, \boldsymbol{\Delta})$ denotes a Wishart distribution with df d and scale matrix $\boldsymbol{\Delta}$:

$$f(\boldsymbol{\Sigma}_v^{-1}) \propto |\boldsymbol{\Sigma}_v^{-1}|^{(d-p-1)/2} \exp\left\{ -\frac{1}{2} \text{tr}(\boldsymbol{\Delta} \boldsymbol{\Sigma}_v^{-1}) \right\}, \quad d \geq p. \tag{10.7.2}$$

The constants $a \geq 0, b > 0$, d and the elements of $\mathbf{D}$ and $\boldsymbol{\Delta}$ are chosen to reflect lack of prior information on the model parameters. In particular, using a diagonal matrix $\mathbf{D}$ with very large diagonal elements, say 10^4, is roughly equivalent to a flat prior $f(\boldsymbol{\alpha}) \propto 1$. Similarly, $d = p, a = 0.001, b = 0.001$ and a scale matrix $\boldsymbol{\Delta}$ with diagonal elements equal to 1 and off diagonals equal to 0.001 may be chosen. Daniels and Kass (1999) studied alternative priors for $\boldsymbol{\Sigma}_v^{-1}$ and discuss the limitations of Wishart prior when the number of small areas, m, is not large.

Model 2. By relaxing the assumption of constant error variance, σ_e^2, we obtain a HB two-level model with unequal error variances:

(i) $y_{ij}|\boldsymbol{\beta}_i, \sigma_{ei}^2 \overset{\text{ind}}{\sim} N(\mathbf{x}_{ij}^T\boldsymbol{\beta}_i, \sigma_{ei}^2)$

(ii) Same as in (ii) Model 1.

(iii) Marginal priors on $\boldsymbol{\alpha}$ and $\boldsymbol{\Sigma}_v^{-1}$ same as in (iii) of Model 1

$$\text{and } \sigma_{ei}^{-2} \overset{\text{ind}}{\sim} G(a_i, b_i), \ i = 1, \ldots, m. \tag{10.7.3}$$

The constants a_i and b_i are chosen to reflect lack of prior information: $a_i = 0.001, b_i = 0.001$.

Model 3. In Section 8.6, we studied a simple random effects model $y_{ij} = \beta + v_i + e_{ij}$ with random error variance σ_{ei}^2. We also noted that the HB approach may be used to handle extensions to nested error regression models with random error variances. Here we consider a HB version of a two-level model with random error variances:

(i) Same as in (i) Model 2.

(ii) Same as in (ii) Model 2.

(iii) $\sigma_{ei}^{-2} \overset{\text{ind}}{\sim} G(\eta, \lambda)$

(iv) Marginal priors on $\boldsymbol{\alpha}$ and $\boldsymbol{\Sigma}_v^{-1}$ same as in (iii) of Model 2,

$$\text{and } \eta \text{ and } \lambda \text{ uniform over a large interval } (0, 10^4]. \tag{10.7.4}$$

Note that (iii) of (10.7.4) is a component of the two-level model, and priors on model parameters are introduced only in (iv) of (10.7.4). The priors on η and λ reflect vague prior knowledge on the model parameters $\eta > 0, \lambda > 0$.

You and Rao (2000c) showed that all the Gibbs conditionals for models 1-3 have closed forms. They also obtained Rao-Blackwell HB estimators of small area means $\mu_i = \overline{\mathbf{X}}_i^T\boldsymbol{\beta}_i$ and posterior variance $V(\mu_i|\mathbf{y}) = \overline{\mathbf{X}}_i^T V(\boldsymbol{\beta}_i|\mathbf{y})\overline{\mathbf{X}}_i$, where $\overline{\mathbf{X}}_i$ is the vector of population x-means for the ith area.

Example 10.7.1. Household Income. In Example 8.8.1 a two-level model with equal error variances σ_e^2 was applied to data from 38,704 households in $m = 140$ enumeration districts (small areas) in one county in Brazil. The two-level model is given by (8.8.2) and (8.8.3), where y_{ij} is the jth household's income in the ith small area, (x_{1ij}, x_{2ij}) are two unit level covariates: number of rooms in the (i,j)th household and educational attainment of head of the (i,j)th household centered around the means $(\overline{x}_{1i}, \overline{x}_{2i})$. The area level covariate z_i, the number of cars per household in the ith area, is related to the random slopes $\boldsymbol{\beta}_i$, using (8.8.3). You and Rao (2000) used a subset of $m = 10$ small areas with $n_i = 28$ households, obtained by simple random sampling in each area, to illustrate model selection and HB estimation.

The Gibbs sampler for the three models was implemented using the BUGS program aided by CODA for convergence diagnostics. Using priors as specified above, the Gibbs sampler for each model was first run for a "burn-in" of $d = 2000$ iterations, then $D = 5000$ more iterations were run and kept for model selection and HB estimation.

For model selection, You and Rao (2000) calculated CPO-values for the three models, using (10.2.26). In particular, for Model 1,

$$\widehat{\mathrm{CPO}}_{ij} = \left[\frac{1}{D} \sum_{k=d+1}^{d+D} \frac{1}{f(y_{ij}|\boldsymbol{\beta}_i^{(k)}, \sigma_e^{2(k)})} \right]^{-1}, \tag{10.7.5}$$

where $\{\boldsymbol{\beta}_i^{(k)}, \sigma_e^{2(k)}, \boldsymbol{\alpha}^{(k)}, \boldsymbol{\Sigma}_v^{(k)};\ k=d+1, \ldots, d+D\}$ denote the MCMC samples. For Models 2 and 3, $\sigma_e^{2(k)}$ in (10.7.5) is replaced by $\sigma_{ei}^{2(k)}$. A CPO plot for the three models is given in Figure 10.2. It shows that a majority of CPO-values for Model 2 were significantly larger than those for Models 1 and 3, thus indicating Model 2 as the best fitting model among the three models.

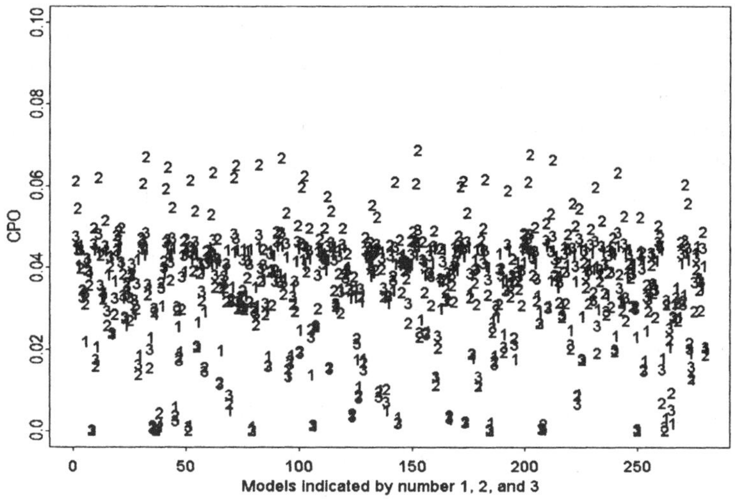

Figure 10.2 CPO Comparison Plot for Models 1, 2 and 3. SOURCE: Adapted from Figure 1 in You and Rao (2000c).

You and Rao (2000) calculated Rao-Blackwell HB estimates of small area means $\mu_i = \beta_0 + \beta_1 \overline{X}_{1i} + \beta_2 \overline{X}_{2i}$ under Model 2 and demonstrated the benefits of Rao-Blackwellization in terms of simulation standard errors.

Example 10.7.2. Korean Unemployment. Chung, Lee and Kim (2001) applied Models 1 and 2 to adjust direct unemployment estimates, y_{ij}, associated with the jth month survey data in the ith small area (here j refers to May, July and December 2000). Auxiliary variables $\mathbf{x}_{ij}$ were obtained from the Economically Active Population Survey (EAPS), census and administrative records. HB analysis for these data was conducted using WinBUGS program (Spiegelhalter et al. (2000)).

A $\widehat{\text{CPO}}_{ij}$-plot for the two models showed that $\widehat{CPO}$-values for Model 2 are significantly larger in every small area than those for Model 1. Chung et al. (2001) also calculated standardized residuals, d_{ik}, similar to (10.4.8), and two other measures of fit: Negative cross-validatory log likelihood $-\sum_i \sum_j \log\{\hat{f}(y_{ij,\text{obs}}|\mathbf{y}_{(ij),\text{obs}})\}$ and posterior mean deviance $-2\sum_i \sum_j [D^{-1} \sum_k \log\{f(y_{ij,\text{obs}}|\boldsymbol{\eta}^{(k)})\}]$, where $\hat{f}(y_{ij,\text{obs}}|\mathbf{y}_{(ij),\text{obs}})$ is obtained from (10.2.26). These measures are computable directly in WinBUGS. Model 2 yielded a negative cross-validatory log-likelihood of 121.5 (compared to 188.7 for Model 1) and a posterior mean deviance of 243.0 (compared to 377.3 for Model 1), thus supporting Model 2 relative to Model 1. Using Model 2, Chung et al. (2001) calculated the posterior means and variances of the small area means $\mu_{ij} = \mathbf{x}_{ij}^T\boldsymbol{\beta}_i$.

10.8 Time Series and Cross-sectional Models

In subsection 8.3.1 we studied EBLUP estimation from time series and cross-sectional data using the sampling model (5.4.6) and the linking model (5.4.7): (i) $\hat{\theta}_{it} = \theta_{it} + e_{it}$; (ii) $\theta_{it} = \mathbf{z}_{it}^T\boldsymbol{\beta} + v_i + u_{it}$, where $\{u_{it}\}$ follows either an AR(1) model $u_{it} = \rho u_{i,t-1} + \epsilon_{it}, |\rho| < 1$, or a random walk model $u_{it} = u_{i,t-1} + \epsilon_{it}$. In this section, we give a brief account of HB estimation under this model, assuming a prior distribution on the model parameters. The HB version of the model with random walk effects u_{it} may be expressed as

(i) $\hat{\boldsymbol{\theta}}_i|\boldsymbol{\theta}_i \overset{\text{ind}}{\sim} N_T(\boldsymbol{\theta}_i, \boldsymbol{\Psi}_i)$, where $\boldsymbol{\Psi}_i$ is the known sampling covariance matrix of $\hat{\boldsymbol{\theta}}_i = (\hat{\theta}_{i1}, \ldots, \hat{\theta}_{iT})^T$.

(ii) $\theta_{it} = \mathbf{Z}_{it}^T\boldsymbol{\beta} + v_i + u_{it}$

(iii) $u_{it}|u_{i,t-1}, \sigma^2 \overset{\text{ind}}{\sim} N(u_{i,t-1}, \sigma^2), v_i \overset{\text{iid}}{\sim} N(0, \sigma_v^2)$, $\{v_i\}$ and $\{u_{it}\}$ are mutually independent.

(iv) $f(\boldsymbol{\beta}, \sigma_v^2, \sigma^2) = f(\boldsymbol{\beta})f(\sigma_v^2)f(\sigma^2)$ with

$$f(\boldsymbol{\beta}) \propto 1, \sigma_v^{-2} \sim G(a_1, b_1), \sigma^{-2} \sim G(a_2, b_2). \tag{10.8.1}$$

For the AR(1) case with known ρ, we replace $u_{i,t-1}$ in (iii) of (10.8.1) by $\rho u_{i,t-1}$.

Datta, Lahiri, Maiti and Lu (1999) and You, Rao and Gambino (2001) obtained Gibbs conditionals under (10.8.1) and showed that all the conditionals have closed forms. You et al. (2000) also obtained Rao-Blackwell estimators of the posterior mean $E(\theta_{iT}|\hat{\boldsymbol{\theta}})$ and the posterior variance $V(\theta_{iT}|\hat{\boldsymbol{\theta}})$ for the current time T. We refer the reader to the above papers for technical details.

Example 10.8.1. Canadian Unemployment Rates. You et al. (2001) used the HB model (10.8.1) to estimate unemployment rates, θ_{iT}, for m = 62 Census Agglomerations (CAs) in Canada, using Canadian Labour Force

Survey (LFS) direct estimates $\hat{\theta}_{it}$ and auxiliary data $\mathbf{z}_{it}$; CA's are treated as small areas. They used data $\{\hat{\theta}_{it}, \mathbf{z}_{it}\}$ for $T = 6$ consecutive months, January 1999 to June 1999, and the parameters of interest are the true unemployment rates, θ_{iT}, in June 1999 for each of the small areas. The choice $T = 6$ was motivated by the fact that the correlation between the estimates $\{\hat{\theta}_{it}\}$ and $\{\hat{\theta}_{is}\}$ $(s \neq t)$ is weak after a lag of six months because of the LFS sample rotation based on a six month cycle; each month one-sixth of the sample is replaced.

To obtain a smoothed estimate of the sampling covariance matrix $\mathbf{\Psi}_i$ used in the model (10.8.1), You et al. (2001) first computed the average CV $(\overline{\text{CV}}_i)$ for each CA i over time and the average lag correlations, $\bar{r}_a$, over time and all CA's. A smoothed estimate of $\mathbf{\Psi}_i$ was then obtained using those smoothed CVs and lag correlations: the tth diagonal element, ψ_{itt}, of $\mathbf{\Psi}_i$ (that is, sampling variance of $\hat{\theta}_{it}$) equals $[(\overline{\text{CV}}_i)\hat{\theta}_{it}]^2$ and the (t,s)th element, ψ_{its}, of $\mathbf{\Psi}_i$ (that is, sampling covariance of $\hat{\theta}_{it}$ and $\hat{\theta}_{is}$) equals $\bar{r}_a^2[\psi_{itt}\psi_{iss}]^{1/2}$ with $a = |t-s|$. The smoothed estimate of $\mathbf{\Psi}_i$ was treated as the true $\mathbf{\Psi}_i$.

You et al. (2001) used a divergence measure proposed by Laud and Ibrahim (1995) to compare the relative fits of the random walk model (10.8.1) and the corresponding AR(1) model with $\rho = 0.75$ and $\rho = 0.50$. This measure is given by $d(\hat{\boldsymbol{\theta}}_*, \hat{\boldsymbol{\theta}}_{\text{obs}}) = E[(mT)^{-1}||\hat{\boldsymbol{\theta}}_* - \hat{\boldsymbol{\theta}}_{\text{obs}}||^2|\hat{\boldsymbol{\theta}}_{\text{obs}}]$, where $\hat{\boldsymbol{\theta}}_{\text{obs}}$ is the (mt)-vector of estimates $\hat{\theta}_{it}$ and the expectation is with respect to the posterior predictive distribution $f(\hat{\boldsymbol{\theta}}_*|\hat{\boldsymbol{\theta}}_{\text{obs}})$ of a new observation $\hat{\boldsymbol{\theta}}_*$. Models yielding smaller values of this measure are preferred.

The Gibbs output with $L = 10$ parallel runs was used to generate samples $\{\boldsymbol{\theta}^{(lk)};\ l = 1, \ldots, 10\}$ from the posterior distribution, $f(\boldsymbol{\theta}|\hat{\boldsymbol{\theta}}_{\text{obs}})$, where $\boldsymbol{\theta}$ is the (mT)-vector of small area parameters θ_{it}. For each $\boldsymbol{\theta}^{(lk)}$, a new observation $\hat{\theta}_*^{(lk)}$ was then generated from $f(\hat{\boldsymbol{\theta}}|\boldsymbol{\theta}^{(lk)})$. The new observations $\{\hat{\boldsymbol{\theta}}_*^{(lk)}\}$ represent simulated samples from $f(\hat{\boldsymbol{\theta}}_*|\hat{\boldsymbol{\theta}}_{\text{obs}})$. The measure $d\ (\hat{\boldsymbol{\theta}}_*, \hat{\boldsymbol{\theta}}_{\text{obs}})$ was approximated by using these observations from the posterior predictive distribution. The following values of $d(\hat{\boldsymbol{\theta}}_*, \hat{\boldsymbol{\theta}}_{\text{obs}})$ were obtained: (i) 13.36 for the random walk model, (ii) 14.62 for the AR(1) model with $\rho = 0.75$, (iii) 14.52 for the AR(1) model with $\rho = 0.5$. Based on these values, the random walk model was selected.

To check the overall fit of the random walk model, the simulated values $\{\boldsymbol{\theta}^{(lk)}, \hat{\boldsymbol{\theta}}_*^{(lk)};\ l = 1, \ldots, 10\}$ were employed to approximate the posterior predictive p value from each run, l, using the formula (10.2.19) with measure of discrepancy given by $T(\hat{\boldsymbol{\theta}}, \boldsymbol{\eta}) = \Sigma_{i=1}^{62}(\hat{\boldsymbol{\theta}}_i - \boldsymbol{\theta}_i)^T\mathbf{\Psi}_i^{-1}(\hat{\boldsymbol{\theta}}_i - \boldsymbol{\theta}_i)$. The average of the $L = 10$ p-values, $\hat{p} = 0.615$, indicated a good fit of the random walk model to the time series and cross-sectional data.

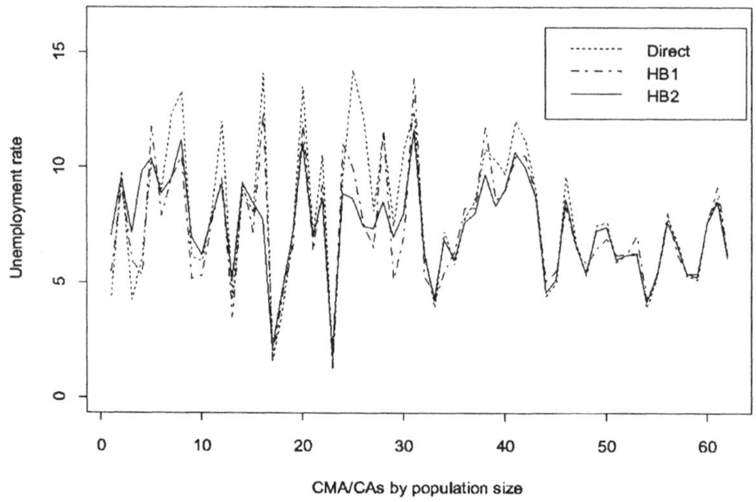

Figure 10.3 Direct, Cross-sectional HB (HB2) and Cross-Sectional and Time Series HB (HB1) Estimates. SOURCE: Adapted from Figure 2 in You, Rao and Gambino (2001).

Rao-Blackwell method was used to calculate the posterior mean, $\hat{\theta}_{i6}^{\mathrm{HB}}$, and the posterior variance, $V(\theta_{i6}|\hat{\boldsymbol{\theta}})$, for each area i. Figure 10.3 displays the HB estimates under model (10.8.1), denoted HB1, the HB estimates under the Fay-Herriot model using only the current cross-sectional data, denoted HB2, and the direct LFS estimates, denoted DIRECT. It is clear from Figure 10.3 that the HB2 estimates tend to be smoother than HB1, whereas HB1 leads to moderate smoothing of the direct estimates. For the CAs with larger population sizes and therefore larger sample sizes, DIRECT and HB1 are very close to each other, whereas DIRECT differs substantially from HB1 for some smaller CAs.

Figure 10.4 displays the CVs of HB1 and HB2; note that we have already compared the CVs of HB2 and DIRECT in Figure 10.1. It is clear from Figure 10.4 that HB1 leads to significant reduction in CV relative to HB2, especially for the smaller CAs.

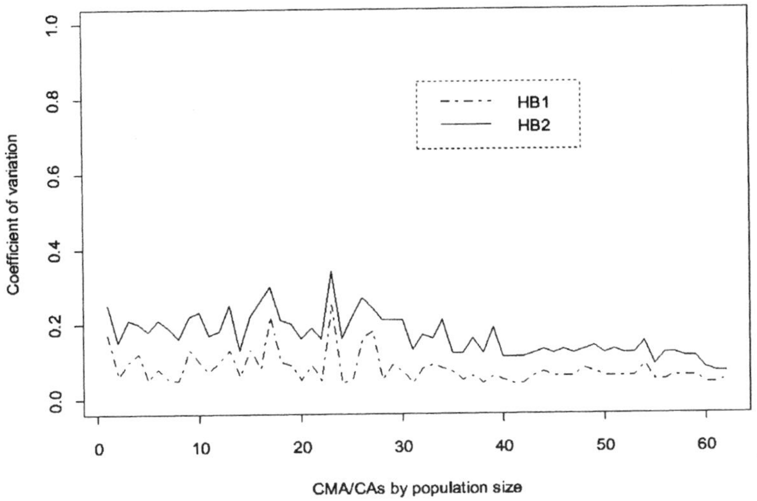

Figure 10.4 Coefficient of Variation of Cross-sectional HB (HB2) and Cross-Sectional and Time Series HB (HB1) Estimates. SOURCE: Adapted from Figure 3 in You, Rao and Gambino (2001).

Example 10.8.2. U.S. Unemployment Rates. Datta, Lahiri, Maiti and Lu (1999) studied an extension of the random walk model (10.8.1) and applied their model to estimate monthly unemployment rate for forty nine U.S. states and the District of Columbia ($m = 50$), using CPS estimates as $\hat{\theta}_{it}$ and Unemployment Insurance (UI) claims rate as z_{it}; New York state was excluded from the study because of very unreliable UI data. They considered the data for the period January 1985 - December 1988 ($T = 48$) to calculate HB estimates, $\hat{\theta}_{iT}$, for the current period T. The HB version of the Datta et al. (1999) model may be written as

(i) $\hat{\boldsymbol{\theta}}_i|\boldsymbol{\theta}_i \overset{\text{ind}}{\sim} N_T(\boldsymbol{\theta}_i, \boldsymbol{\Psi}_i)$,

(ii) $\theta_{it} = \beta_{0i} + \beta_{1i}z_{it} + u_{it} + \sum_{u=1}^{12} a_{itu}\gamma_{iu} + \sum_{k=1}^{4} b_{itk}\lambda_{ik}$,

where $\gamma_{i,12} = -\sum_{u=1}^{11}\gamma_{iu}$ and $\lambda_{i4} = -\sum_{k=1}^{3}\lambda_{ik}$

(iii) $\beta_{0i} \overset{\text{iid}}{\sim} N(\beta_0, \sigma_0^2), \beta_{1i} \overset{\text{iid}}{\sim} N(\beta_1, \sigma_1^2), \boldsymbol{\gamma}_i \sim N_{11}(\boldsymbol{\gamma}, \boldsymbol{\Sigma}_\gamma)$,

$\boldsymbol{\lambda}_i \sim N_3(\boldsymbol{\lambda}, \boldsymbol{\Sigma}_\lambda), u_{it}|u_{i,t-1}, \sigma^2 \sim N(u_{i,t-1}, \sigma^2)$

(iv) Independent gamma priors on $\sigma_0^{-2}, \sigma_1^{-2}$ and σ^{-2}, independent Wishart priors on $\boldsymbol{\Sigma}_\gamma^{-1}$ and $\boldsymbol{\Sigma}_\lambda^{-1}$, and flat priors on β_0 and β_1. (10.8.2)

The last two terms in (ii) of (10.8.2) account for seasonal variation in monthly unemployment rates, where $a_{itu} = 1$ if $t = u, a_{itu} = 0$ otherwise; $a_{it,12} = 1$ if $t = 12, 24, 36, 48, a_{it,12} = 0$ otherwise, and $\boldsymbol{\gamma}_i = (\gamma_{i1}, \ldots, \gamma_{i,11})^T$ represents random month effects. Similarly, $b_{itk} = 1$ if $12(k-1) < t \leq 12k, b_{itk} = 0$ otherwise, and $\boldsymbol{\lambda}_i = (\lambda_{i1}, \lambda_{i2}, \lambda_{i3})^T$ represent random year effects. The random effects $\beta_{0i}, \beta_{1i}, \boldsymbol{\gamma}_i, \boldsymbol{\lambda}_i$ and u_{it} are assumed to be mutually independent.

Datta et al. (1999) used $L = 10$ parallel runs to draw MCMC samples $\{\boldsymbol{\eta}^{(k)}\}$ from the joint posterior distribution of the random effects $\{\beta_{0i}, \beta_{1i}, u_{it}, \boldsymbol{\gamma}_i, \boldsymbol{\lambda}_i\}$ and the model parameters. To check the overall fit of the model (10.8.2), they estimated the posterior predictive value using the discrepancy measure $T(\hat{\boldsymbol{\theta}}, \boldsymbol{\eta})$ of Example 10.8.1. The estimated p-value, $\hat{p} = 0.614$, indicates a good fit of the model to the CPS time series and cross-sectional data. If the seasonal effects are deleted from the model (10.8.2), then $\hat{p} = 0.758$ which suggests that the inclusion of seasonal effects is important because of the long time series ($T = 48$) used in the study. Note that in Example 10.8.1 the time series is short ($T = 6$).

Datta et al. (1999) also calculated the divergence measure proposed by Laud and Ibrahim (1999):

$$d(\hat{\boldsymbol{\theta}}_*, \hat{\boldsymbol{\theta}}_{\text{obs}}) = \frac{1}{nD_0} \sum_{k=1}^{D_0} ||\hat{\boldsymbol{\theta}}_*^{(k)} - \hat{\boldsymbol{\theta}}_{\text{obs}}||^2, \tag{10.8.3}$$

where $n = mT$, D_0 is the total number of MCMC samples, $\hat{\boldsymbol{\theta}}_{\text{obs}}$ is the vector of observed direct estimates with blocks $\hat{\boldsymbol{\theta}}_{i,\text{obs}}$, $\hat{\boldsymbol{\theta}}_*^{(k)}$ is the vector of hypothetical replications with blocks $\hat{\boldsymbol{\theta}}_{i*}^{(k)}$, and $\hat{\boldsymbol{\theta}}_{i*}^{(k)}$ is drawn from $N_T(\boldsymbol{\theta}_i^{(k)}, \boldsymbol{\Psi}_i)$. For the CPS data, $d(\hat{\boldsymbol{\theta}}_*, \hat{\boldsymbol{\theta}}_{\text{obs}}) = 0.0007$ using model (10.8.2) while it increased to $d_{\text{FH}}(\hat{\boldsymbol{\theta}}_*, \hat{\boldsymbol{\theta}}_{\text{obs}}) = 1.63$ using separate Fay-Herriot models. If the seasonal effects γ_{iu} and λ_{ik} are deleted from the model (10.8.2), then $d(\hat{\boldsymbol{\theta}}_*, \hat{\boldsymbol{\theta}}_{\text{obs}}) = 0.0008$ which is close to the value 0.0007 under the full model.

Datta et al. (1999) calculated the HB estimates, $\hat{\theta}_{iT}^{\text{HB}}$, and the posterior variances, $V(\theta_{iT}|\hat{\boldsymbol{\theta}})$, for the current occasion T, undermodel (10.8.2), and then compared the values to the CPS values, $\hat{\theta}_{iT}$ and ψ_{iT}, and the HB values for the Fay-Herriot model fitted only to the current month cross-sectional data. The HB standard errors, $\sqrt{V(\theta_{iT}|\hat{\boldsymbol{\theta}})}$, under the model (10.8.2) were significantly smaller than the corresponding CPS standard errors, $\sqrt{\psi_{iT}}$, uniformly over time and across all states. However, HB was more effective for states with fewer sample households ("indirect-use" states). The reduction in standard error under the Fay-Herriot model for the current period relative to CPS standard error was sizeable for indirect use states, but it is less than 10% (and as small as 0.1% in some months) for "direct-use" states with larger sample sizes.

10.9 Multivariate Models

10.9.1 Area Level Model

In Section 8.1 we considered EBLUP estimation under the multivariate Fay-Herriot model (5.4.3). A HB version of (5.4.3) is given by

(i) $\hat{\boldsymbol{\theta}}_i|\boldsymbol{\theta}_i \overset{\text{ind}}{\sim} N_r(\boldsymbol{\theta}_i, \boldsymbol{\Psi}_i)$, where

$\hat{\boldsymbol{\theta}}_i = (\hat{\theta}_{i1}, \ldots, \hat{\theta}_{ir})^T$ is the vector of direct estimators and $\boldsymbol{\Psi}_i$ is known.

(ii) $\boldsymbol{\theta}_i|\boldsymbol{\beta}, \boldsymbol{\Sigma}_v \sim N_r(\mathbf{Z}_i\boldsymbol{\beta}, \boldsymbol{\Sigma}_v)$

(iii) $f(\boldsymbol{\beta}, \boldsymbol{\Sigma}_v) = f(\boldsymbol{\beta})f(\boldsymbol{\Sigma}_v)$ with $f(\boldsymbol{\beta}) \propto 1, f(\boldsymbol{\Sigma}_v) \propto 1.$ (10.9.1)

Datta, Ghosh, Nangia and Natarajan (1996) derived the Gibbs conditionals, $[\boldsymbol{\beta}|\boldsymbol{\theta}, \boldsymbol{\Sigma}_v, \hat{\boldsymbol{\theta}}]$, $[\boldsymbol{\theta}_i|\boldsymbol{\beta}, \boldsymbol{\Sigma}_v, \hat{\boldsymbol{\theta}}]$ and $[\boldsymbol{\Sigma}_v|\boldsymbol{\beta}, \boldsymbol{\theta}, \hat{\boldsymbol{\theta}}]$, under (10.9.1) and showed that all the conditionals have closed forms. They also obtained Rao-Blackwell estimators of the posterior means $E(\boldsymbol{\theta}_i|\hat{\boldsymbol{\theta}})$ and the posterior variances $V(\boldsymbol{\theta}_i|\hat{\boldsymbol{\theta}})$.

Example 10.9.1. Median Income. Datta et al. (1996) used the multivariate Fay-Herriot model (10.9.1) to estimate the median incomes of four-person families in U.S. states (small areas). Here $\boldsymbol{\theta}_i = (\theta_{i1}, \theta_{i2}, \theta_{i3})^T$ with θ_{i1}, θ_{i2} and θ_{i3} denoting the true median income of four-, three- and five-person families in state i, and θ_{i1}'s are the parameters of interest. The adjusted census median income, z_{ij1}, and base-year census median income, z_{ij2}, for the three groups $j = 1, 2, 3$ were used as explanatory variables. The matrix $\mathbf{Z}_i$ consists of three rows $(\mathbf{z}_{i1}^T, \mathbf{0}_3^T, \mathbf{0}_3^T)$, $(\mathbf{0}_3^T, \mathbf{z}_{i2}^T, \mathbf{0}_3^T)$ and $(\mathbf{0}_3^T, \mathbf{0}_3^T, \mathbf{z}_{i3}^T)$, where $\mathbf{z}_{ij}^T = (1, z_{ij1}, z_{ij2})$ and $\mathbf{0}_3^T = (0, 0, 0)$. Direct survey estimates $\hat{\boldsymbol{\theta}}_i$ of $\boldsymbol{\theta}_i$ and associated sampling covariance matrices $\boldsymbol{\Psi}_i$ were obtained from the 1979 Current Population Survey (CPS). HB estimates of θ_{i1} were obtained using univariate, bivariate and trivariate Fay-Herriot models, denoted as HB^1, HB^2 and HB^3, respectively. Two cases for the bivariate model were studied: $\boldsymbol{\theta}_i = (\theta_{i1}, \theta_{i2})^T$ with the corresponding $\mathbf{Z}_i$ and $\boldsymbol{\theta}_i = (\theta_{i1}, \theta_{i3})^T$ with the corresponding $\mathbf{Z}_i$. Denote the HB estimates for the two cases as HB^{2a} and HB^{2b}, respectively. In terms of posterior variances, HB^{2b} obtained from the bivariate model that uses only the median incomes of four- and five-person families performed better than the other estimates. HB estimates based on the multivariate models performed better than HB^1 based on the univariate model.

Datta et al. (1996) also conducted an external evaluation by treating the census estimates for 1979, available from the 1980 census data, as true values. Table 10.5 reports the absolute relative error averaged over the states ($\overline{\text{ARE}}$), the direct estimates $\hat{\theta}_{i1}$ and the HB estimates HB^3, HB^{2a}, HB^{2b} and HB^1. It is clear from Table 10.4 that the HB estimates performed similarly in terms of $\overline{\text{ARE}}$, and outperformed the direct estimates.

TABLE 10.5 Average Absolute Relative Error (ARE%): Median Income of Four-person Families

Direct	HB^1	HB^{2a}	HB^{2b}	HB^3
4.98	2.07	2.04	2.06	2.02

SOURCE: Adapted from Table 11.2 in Datta et al. (1996).

10.9.2 Unit Level Model

In Section 8.5, we studied EBLUP estimation under the multivariate nested error regression model (5.5.1). Datta, Day and Maiti (1998) studied HB estimation for this model. A HB version of the model may be expressed as

$$\begin{aligned}
&\text{(i) } \mathbf{y}_{ij}|\mathbf{B},\boldsymbol{\Sigma}_e \overset{\text{ind}}{\sim} N_r(\mathbf{B}\mathbf{x}_{ij}+\mathbf{v}_i,\boldsymbol{\Sigma}_e), \quad j=1,\ldots,n_i; i=1,\ldots,m\\
&\text{(ii) } \mathbf{v}_i|\boldsymbol{\Sigma}_v \overset{\text{iid}}{\sim} N_r(\mathbf{0},\boldsymbol{\Sigma}_v) \text{ and } \{\mathbf{v}_i\}, \{\mathbf{e}_{ij}\} \text{ are mutually independent.}\\
&\text{(iii) Flat prior on } \mathbf{B} \text{ and independent Wishart priors on } \boldsymbol{\Sigma}_v^{-1} \text{ and } \boldsymbol{\Sigma}_e^{-1}.
\end{aligned} \tag{10.9.2}$$

Datta et al. (1998) derived Gibbs conditionals under (10.9.2) and showed that all the conditionals have closed forms. They also considered improper priors of the form $f(\mathbf{B},\boldsymbol{\Sigma}_v,\boldsymbol{\Sigma}_e) \propto |\boldsymbol{\Sigma}_v|^{-a_v/2}|\boldsymbol{\Sigma}_e|^{-a_e/2}$ under which Gibbs conditionals are proper but the joint posterior may be improper. They obtained necessary and sufficient conditions on a_v and a_e that ensure property of the joint posterior. Datta et al. (1998) also obtained Rao-Blackwell estimators of the posterior means $E(\overline{\mathbf{Y}}_i|\mathbf{y})$ and the posterior variance matrix $V(\overline{\mathbf{Y}}_i|\mathbf{y})$, where $\overline{\mathbf{Y}}_i$ is the vector of finite population means for the ith area.

The multivariate model was applied to the county corn and soybeans data, y_{ij1}, y_{ij2} of Battese et al. (1988). In terms of posterior variances, the HB estimates under the multivariate model performed better than the HB estimates under the univariate model for corn as well as soybeans. We refer the reader to Datta et al. (1998) for details of MCMC implementation and model fit.

10.10 Disease Mapping Models

In Section 9.5 of Chapter 9, we studied EBLUP estimation for three models useful in disease mapping applications: Poisson-gamma, log-normal and CAR-normal models. In this section we study HB estimation for these models, assuming priors on the model parameters. We also consider extensions to two-level models.

10.10.1 Poisson-gamma Model

Using the notation in Section 9.5, let θ_i, y_i and e_i respectively denote the relative risk (RR), observed and expected number of cases (deaths) over a given period in the ith area ($i = 1, \ldots, m$). A HB version of the Poisson-gamma model, given in subsection 9.5.1, may be written as

$$
\begin{aligned}
&\text{(i)} \quad y_i|\theta_i \overset{\text{ind}}{\sim} \text{Poisson}(e_i\theta_i)\\
&\text{(ii)} \quad \theta_i|\alpha,\nu \overset{\text{iid}}{\sim} G(\nu,\alpha)\\
&\text{(iii)} \quad f(\alpha,\nu) \propto f(\alpha)f(\nu)\\
&\qquad\quad \text{with } f(\alpha) \propto 1/\alpha;\ \nu \sim G(a=1/2, b),\ b>0;
\end{aligned}
\tag{10.10.1}
$$

see Datta, Ghosh and Waller (2000). The joint posterior $f(\boldsymbol{\theta},\alpha,\nu|\mathbf{y})$ is proper if at least one y_i is greater than zero. It is easy to verify that the Gibbs conditionals are given by

$$
\begin{aligned}
&\text{(i)} \quad [\theta_i|\alpha,\nu,\mathbf{y}] \overset{\text{ind}}{\sim} G(y_i+\nu, e_i+\alpha)\\
&\text{(ii)} \quad [\alpha|\boldsymbol{\theta},\nu,\mathbf{y}] \sim G\left(m\nu, \sum_i \theta_i\right)\\
&\text{(iii)} \quad f(\nu|\boldsymbol{\theta},\alpha,\mathbf{y}) \propto \left(\prod \theta_i^{\nu-1}\right)\exp(-b\nu)\alpha^{\nu m}/\Gamma^m(\nu).
\end{aligned}
\tag{10.10.2}
$$

MCMC samples can be generated directly from (i) and (ii) of (10.10.2), but we need to use the Metropolis-Hastings (M-H) algorithm to generate samples from (iii) of (10.10.2). Using the MCMC samples $\{\theta_1^{(k)},\ldots,\theta_m^{(k)},\nu^{(k)},\alpha^{(k)}; k = d+1,\ldots,d+D\}$, posterior quantities of interest may be computed; in particular, the posterior mean $E(\theta_i|\mathbf{y})$ and posterior variance $V(\theta_i|\mathbf{y})$ for each area $i=1,\ldots,m$.

10.10.2 Log-normal Model

A HB version of the basic log-normal model, given in subsection 9.5.2, may be written as

$$
\begin{aligned}
&\text{(i)} \quad y_i|\theta_i \overset{\text{ind}}{\sim} \text{Poisson}(e_i\theta_i)\\
&\text{(ii)} \quad \xi_i = \log(\theta_i)|\mu,\sigma^2 \overset{\text{iid}}{\sim} N(\mu,\sigma^2)\\
&\text{(iii)} \quad f(\mu,\sigma^2) \propto f(\mu)f(\sigma^2) \quad \text{with}\\
&\qquad\quad f(\mu) \propto 1;\ \sigma^{-2} \sim G(a,b); a \ge 0,\ b>0.
\end{aligned}
\tag{10.10.3}
$$

The joint posterior $f(\boldsymbol{\theta},\mu,\sigma^2|\mathbf{y})$ is proper if at least one y_i is greater than zero. It is easy to verify that the Gibbs conditionals are given by

$$\text{(i)}\quad f(\theta_i|\mu,\sigma^2,\mathbf{y}] \propto \theta_i^{y_i-1}\exp\left[-e_i\theta_i - \frac{1}{2\sigma^2}(\xi_i-\mu)^2\right]$$

$$\text{(ii)}\quad [\mu|\boldsymbol{\theta},\sigma^2,\mathbf{y}] \sim N\left(\frac{1}{m}\sum_i \xi_i, \frac{\sigma^2}{m}\right)$$

$$\text{(iii)}\quad [\sigma^2|\boldsymbol{\theta},\mu,\mathbf{y}] \sim G\left(\frac{m}{2}+a, \frac{1}{2}\sum_i(\xi_i-\mu)^2+b\right); \tag{10.10.4}$$

see Maiti (1998). MCMC samples can be generated directly from (ii) and (iii) of (10.10.4), but we need to use the M-H algorithm to generate samples from (i) of (10.10.4). We can express (i) as $f(\theta_i|\mu,\sigma^2,\mathbf{y}) \propto k(\theta_i)h(\theta_i|\mu,\sigma^2)$, where $k(\theta_i) = \exp(-e_i\theta_i)\theta_i^{y_i}$ and $h(\theta_i|\mu,\sigma^2) \propto g'(\theta_i)\exp\{-(\xi_i-\mu)^2/2\sigma^2\}$ with $g'(\theta_i) = \partial g(\theta_i)/\partial\theta_i$ and $g(\theta_i) = \log(\theta_i)$. We can use $h(\theta_i|\mu,\sigma^2)$ to draw the candidate, θ_i^*, noting that $\theta_i = g^{-1}(\xi_i)$ and $\xi_i|\mu,\sigma^2 \sim N(\mu,\sigma^2)$. The acceptance probability used in the M-H algorithm is then given by $a(\theta^{(k)},\theta_i^*) = \min\{k(\theta_i^*)/k(\theta_i^{(k)}),1\}$.

As noted in subsection 9.5.2, the basic log-normal model with Poisson counts, y_i, readily extends to the case of covariates $\mathbf{z}_i$ by changing (ii) of (10.10.3) to $\xi_i|\boldsymbol{\beta},\sigma^2 \sim N(\mathbf{z}_i^T\boldsymbol{\beta},\sigma^2)$. Further, we change (iii) of (10.10.3) to $f(\boldsymbol{\beta},\sigma^2) \propto f(\boldsymbol{\beta})f(\sigma^2)$ with $f(\boldsymbol{\beta}) \propto 1$ and $\sigma^{-2} \sim G(a,b)$. Also, the basic model can be extended to allow spatial covariates. A HB version of the spatial CAR-normal model is given by

$$\text{(i)}\quad y_i|e_i \sim \text{Poisson}(e_i\theta_i)$$

$$\text{(ii)}\quad \xi_i|\xi_{j(j\neq i)},\rho,\sigma^2 \sim N\left[\mu+\rho\sum_l q_{il}(\xi_l-\mu),\sigma^2\right]$$

$$\text{(iii)}\quad f(\mu,\sigma^2,\rho) \propto f(\mu)f(\sigma^2)f(\rho) \quad \text{with}$$

$$f(\mu)\propto 1;\quad \sigma^{-2}\sim G(a,b); a\geq 0, b>0;\quad \rho\sim U(0,\rho_0), \tag{10.10.5}$$

where ρ_0 denotes the maximum value of ρ in the CAR model, and $Q = (q_{il})$ is the "adjacency" matrix of the map with $q_{il} = q_{li}$, $q_{il} = 1$ if i and l are adjacent areas and $q_{il} = 0$ otherwise.

Maiti (1998) obtained the Gibbs conditionals. In particular, $[\mu|\boldsymbol{\theta},\sigma^2,\rho,\mathbf{y}] \sim$ normal, $[\sigma^{-2}|\boldsymbol{\theta},\mu,\rho,\mathbf{y}] \sim$ gamma, $[\rho|\boldsymbol{\theta},\mu,\sigma^2,\mathbf{y}] \sim$ truncated normal and $[\theta_i|\theta_{j(j\neq i)},\mu,\sigma^2,\rho,\mathbf{y}]$ does not admit a closed form in the sense that the conditional is known only up to a multiplicative constant. MCMC samples can be generated directly from the first three conditionals, but we need to use the M-H algorithm to generate samples from the conditionals $[\theta_i|\theta_{j(j\neq i)},\mu,\sigma^2,\rho,\mathbf{y}]$, $i = 1,\ldots,m$.

Example 10.10.1. Lip Cancer. In Example 9.5.1 we applied EB estimation to lip cancer counts, y_i, in each of 56 counties of Scotland. Maiti (1998) applied

HB estimator to the same data using the log-normal and the CAR-normal models. The HB estimates $E(\theta_i|\mathbf{y})$ of lip cancer incidence are very similar for the two models, but the standard errors, $\sqrt{V(\theta_i|\mathbf{y})}$, are smaller for the CAR-normal model as it exploits the spatial structure of the data.

Ghosh, Natarajan, Waller and Kim (1999) proposed a spatial log-normal model that allows covariates $\mathbf{z}_i$. It is given by

(i) $y_i|e_i \overset{\text{ind}}{\sim} \text{Poisson}(e_i\theta_i)$,

(ii) $\xi_i = \mathbf{z}_i^T\boldsymbol{\beta} + u_i + v_i$ where $\mathbf{z}_i^T\boldsymbol{\beta}$ does not include an intercept term,

$v_i \overset{\text{iid}}{\sim} N(0, \sigma_v^2)$ and the u_i's have joint density (9.5.12):

$f(\mathbf{u}) \propto (\sigma_u^2)^{-m/2} \exp\left[-\sum\sum_{i\neq l}(u_i - u_l)^2 q_{il}/(2\sigma^2)\right]$, where $q_{il} \geq 0$ for all $1 \leq i \neq l \leq m$.

(iii) $\boldsymbol{\beta}$, σ_u^2 and σ_v^2 are mutually independent with $f(\boldsymbol{\beta}) \propto 1$,

$$\sigma_u^{-2} \sim G(a_u, b_u) \text{ and } \sigma_v^{-2} \sim G(a_v, b_v). \tag{10.10.6}$$

Ghosh et al. (1999) showed that the Gibbs conditionals except $[\theta_i|\ \theta_{l(l\neq i)}, \boldsymbol{\beta}, \boldsymbol{\mu}, \sigma_u^2, \sigma_v^2, \mathbf{y}]$, admit closed forms. They also established conditions for the propriety of the joint posterior; in particular, we need $b_u > 0, b_v > 0$.

Example 10.10.3. Leukemia Incidence. Ghosh et al. (1999) applied the HB method, based on the model (10.10.6), to leukemia incidence estimation for $m = 281$ census tracts (small areas) in an eight-county region of upstate New York. Here $q_{il} = 1$ if i and l are neighbors and $q_{il} = 0$ otherwise, and $\mathbf{z}_i$ is a scalar ($p = 1$) variable z_i denoting the inverse distance of the centroid of the ith census tract from the nearest hazardous waste site containing trichloroethylene (TCE), a common contaminant of ground water. We refer the reader to Ghosh et al. (1999) for further details.

10.10.3 Two-level Models

Let y_{ij} and n_{ij} respectively denote the number of cases (deaths) and the population at risk in the jth age class in the ith area ($j = 1, \ldots, J$; $i = 1, \ldots, m$). Using the data $\{y_{ij}, n_{ij}\}$ it is of interest to estimate the age-specific mortality rates τ_{ij} and the age-adjusted rates $\sum_j a_j\tau_{ij}$ where the a_j's are specified constant. The basic assumption is

$$y_{ij}|\tau_{ij} \overset{\text{ind}}{\sim} \text{Poisson}(n_{ij}\tau_{ij}). \tag{10.10.7}$$

Nandram, Sedransk and Pickle (1999) studied HB estimation under different linking models:

$$\log(\tau_{ij}) = \mathbf{z}_{ij}^T\boldsymbol{\beta} + v_i,\ \ v_i|\sigma_v^2 \overset{\text{iid}}{\sim} N(0, \sigma_v^2) \tag{10.10.8}$$

$$\log(\tau_{ij}) = \mathbf{z}_j^T\boldsymbol{\beta}_i,\ \ \boldsymbol{\beta}_i|\boldsymbol{\beta}, \boldsymbol{\Delta} \overset{\text{iid}}{\sim} N_p(\boldsymbol{\beta}, \boldsymbol{\Delta}) \tag{10.10.9}$$

$$\log(\tau_{ij}) = \mathbf{z}_j^T\boldsymbol{\beta}_i + \delta_j,\ \ \boldsymbol{\beta}_i|\boldsymbol{\beta}, \boldsymbol{\Delta} \overset{\text{iid}}{\sim} N_p(\boldsymbol{\beta}, \boldsymbol{\Delta}), \delta_j \overset{\text{iid}}{\sim} N(0, \sigma^2), \tag{10.10.10}$$

where $\mathbf{z}_j$ is a $p \times 1$ vector of covariates and δ_j is an "offset" corresponding to age class j. Nandram et al. (1999) assumed the flat prior $f(\boldsymbol{\beta}) \propto 1$ and proper diffuse (that is, proper with very large variance) priors for σ_v^2, $\boldsymbol{\Delta}$ and σ^2. For model selection, they used the posterior expected predictive deviance, the posterior predictive value and measures based on the cross-validation predictive densities (see subsection 10.2.6).

Example 10.10.3. Mortality Rates. Nandram et al. (1999) applied the HB method to estimate age-specific and age-adjusted mortality rates for U.S. Health Service Areas (HSAs). They studied one of the disease categories, all cancer for white males, presented in the 1996 Atlas of United States Mortality. The number of HSAs (small areas), m, is 798 and the number of age categories, J, is 10: 0–4, 5–14, ..., 75–84, 85 and higher, coded as 0.25, 1, ..., 9. The vector of auxiliary variables is given by $\mathbf{z}_j = [1, j-1, (j-1)^2, (j-1)^3, \max\{0, ((d-1)-\text{knot})^3\}]^T$ for $j \geq 2$ and $\mathbf{z}_1 = [1, 0.25, (0.25)^2, (0.25)^3, \max\{0, (0.25-\text{knot})^3\}]^T$, where the value of the "knot" was obtained by maximizing the likelihood based on marginal deaths, $y_j = \sum_i y_{ij}$, and population at risk, $n_j = \sum_i n_{ij}$, where $y_j | n_j, \tau_j \overset{\text{ind}}{\sim} \text{Poisson}(n_j \tau_j)$ with $\log(\tau_j) = \mathbf{z}_j^T \boldsymbol{\beta}$. The auxiliary vector $\mathbf{z}_j$ was used in the Atlas model based on a normal approximation to $\log(r_{ij})$ with mean $\log(\tau_{ij})$ and matching linking model given by (10.10.9), where $r_{ij} = y_{ij}/n_{ij}$ is the crude rate. Nandram et al. (1999) used unmatched sampling and linking models, based on the Poisson sampling model (10.10.7) and the linking models (10.10.8)–(10.10.10). We denote these models as models 1,2 and 3 respectively.

Nandram et al. (1999) used the MCMC samples generated from the three models to calculate the values of the posterior expected predictive deviance $E\{\Delta(\mathbf{y}; \mathbf{y}_{\text{obs}}) | \mathbf{y}_{\text{obs}}\}$ using the chi-square measure $\Delta(\mathbf{y}, \mathbf{y}_{\text{obs}}) = \sum_i \sum_j (y_{ij} - y_{ij,\text{obs}})^2 / (y_{ij} + 0.5)$. They also calculated the posterior predictive p-values, using $T(\mathbf{y}, \boldsymbol{\tau}) = \sum_i \sum_j (y_{ij} - n_{ij}\tau_{ij})^2 / (n_{ij}\tau_{ij})$, the standardized cross-validation residuals

$$d_{2,ij}^* = \frac{r_{ij,\text{obs}} - E(r_{ij} | \mathbf{y}_{(ij),\text{obs}})}{\sqrt{V(r_{ij} | \mathbf{y}_{(ij),\text{obs}})}}, \tag{10.10.11}$$

where $\mathbf{y}_{(ij),\text{obs}}$ denotes all elements of $\mathbf{y}_{\text{obs}}$ except $y_{ij,\text{obs}}$; see (10.2.28). The residuals $d_{2,ij}^*$ were summarized by counting (a) the number of (i, j) such that $|d_{2,ij}^*| \geq 3$, called "outliers", and (b) the number of HSAs, i, such that $|d_{2,ij}^*| \geq 3$ for at least one j, called "# of HSAs".

TABLE 10.6 Comparison of Models 1, 2 and 3: Mortality Rates

Model	EPD	p-value	Outliers	# of HSAs
1	22,307	0.00	284	190
2	16,920	0.00	136	93
3	16,270	0.32	59	54

SOURCE: Adapted from Table 1 in Nandram et al. (1999).

Table 10.6 reports the posterior expected predictive deviance (EPD), the p-value, outliers and number of HSAs for the three models 1, 2 and 3. It is clear from Table 10.6 that model 1 performs poorly with respect to all the four measures. Overall, model 3 with the random age coefficient, δ_j, provides the best fit to the data, although models 2 and 3 are similar with respect to EPD. Based on model 3, Nandram et al. (1999) produced cancer maps of HB estimates of age-specific mortality rates for each age group j. Note that the HB estimate of the mortality rate τ_{ij} is given by the posterior mean $E(\tau_{ij}|\mathbf{y})$. The maps revealed that the mortality rates, for all age classes, are highest among the Appalachian mountain range (Mississippi to West Virginia) and the Ohio River Valley (Illinois to Ohio). Also, the highest rates formed more concentrated clusters in the middle and older age groups (e.g., 45–54), whereas the youngest and oldest age groups exhibited more scattered patterns.

Nandram, Sedranksk and Pickle (2000) used models and methods similar to those in Nandram et al. (1999) to estimate age-specific and age-adjusted mortality rates for chronic obstructive pulmonary disease for white males in HSAs.

10.11 Binary Data

In Section 9.4 of Chapter 9, we studied EB models for binary responses y_{ij}. In this section we study HB estimation for these models, assuming priors on the model parameters. We also consider extensions to logistic linear mixed models.

10.11.1 Beta-binomial Model

A HB-version of the beta-binomial model, studied in subsection 9.4.1, may be written as

$$\begin{aligned}
&\text{(i)}\quad y_i|p_i \overset{\text{ind}}{\sim} \text{binomial}(n_i, p_i)\\
&\text{(ii)}\quad p_i|\alpha,\beta \overset{\text{iid}}{\sim} \text{beta}(\alpha,\beta); \alpha > 0, \beta > 0,\\
&\text{(iii)}\quad \alpha \text{ and } \beta \text{ mutually independent with}\\
&\qquad \alpha \sim G(a_1, b_1); a_1 > 0, b_1 > 0\\
&\qquad \beta \sim G(a_2, b_2); a_2 > 0, b_2 > 0. \qquad (10.11.1)
\end{aligned}$$

The Gibbs conditional $[p_i|p_{l(l\neq i)}, \alpha, \beta, \mathbf{y}]$ is beta$(y_i+\alpha, n_i-y_i+\beta)$ under model (10.11.1), but the remaining conditionals $[\alpha|\mathbf{p}, \beta, \mathbf{y}]$ and $[\beta|\mathbf{p}, \alpha, \mathbf{y}]$ do not admit closed forms. He and Sun (2000) showed that the latter two conditionals are log-concave if $a_1 \geq 1$ and $a_2 \geq 1$. Using this result, they used adaptive rejection sampling to generate samples from these conditionals. BUGS version 0.5 handles log-concave conditionals using adaptive rejection sampling (subsection 10.2.5).

Example 10.11.1. Turkey Hunting. He and Sun (1998) applied the beta-binomial model (10.11.1) to data collected from a mail survey on turkey hunting conducted by Missouri Department of Conservation. The purpose of this survey was to estimate hunter success rates, p_i, and other parameters of interest for $m = 114$ counties in Missouri. Questionnaires were mailed to a random sample of 5151 permit buyers after the 1994 spring turkey hunting season; the response rate after these mailings was 69%. A hunter was allowed at most 14 one-day trips, and each respondent reported the number of trips, whether a trip was success or not, and the county where hunted for each trip. From this data, a direct estimate of p_i is computed as $\hat{p}_i = y_i/n_i$, where n_i is the total number of trips in county i and y_i is the number of successful trips. The direct estimate is not reliable for counties with small n_i; for example $n_i = 11$ and $y_i = 1$ for county 11. The direct state estimate of success rate, $\hat{p} = 10.1\%$, is reliable because of the large overall sample size; the average of HB estimates, $\hat{p}_i^H$, equals 10.2%. For the counties with small n_i, the HB estimates shrink towards $\hat{p}$. For example, $\hat{p}_i = 9.1\%$ and $\hat{p}_i^{\text{HB}} = 10.8\%$ for county 11.

You and Reiss (1999) extended the beta-binomial HB model (10.11.1) to a two-fold beta-binomial HB model. They applied the two-fold model to estimate the proportions, p_{ij}, within groups, j, in each area i. They also derived Rao-Blackwell estimators of the posterior means and posterior variances. Using data from Statistics Canada's 1997 Homeowner Repair and Renovation Survey, they obtained HB estimates of the response rates, p_{ij}, within response homogeneity groups (RHG's), j, in each province i. The number of RHG's varied widely across provinces; for example, 153 RHG's in Ontario compared to only 8 RHG's in Prince Edward Island. Sample sizes, n_{ij}, varied from 1 to 201. As a result, direct estimates of the p_{ij}'s are unreliable for the RHG's with small sample sizes. HB standard errors were substantially smaller than the direct standard errors, especially for the RHG's with small n_{ij}. For example, the sample size is only 4 for the RHG 17 in Newfoundland, and the direct standard error is 0.22 compared to HB standard error equal to 0.03.

10.11.2 Logit-normal Model

As noted in subsection 9.4.2, the logit-normal model readily allows covariates, unlike the beta-binomial model. We first consider the case of area-specific covariates, $\mathbf{z}_i$, and then study the case of unit-level covariates, $\mathbf{x}_{ij}$.

(i) Area level covariates

A HB version of the logit-normal model with area level covariates may be expressed as

$$\begin{aligned}
&\text{(i)}\quad y_i|p_i \overset{\text{ind}}{\sim} \text{binomial}(n_i, p_i)\\
&\text{(ii)}\quad \xi_i = \text{logit}(p_i) = \mathbf{z}_i^T\boldsymbol{\beta} + v_i \text{ with } v_i \overset{\text{iid}}{\sim} N(0, \sigma_v^2)\\
&\text{(iii)}\quad \boldsymbol{\beta} \text{ and } \sigma_v^2 \text{ are mutually independent with } f(\boldsymbol{\beta}) \propto 1\\
&\qquad \text{and } \sigma_v^{-2} \sim G(a, b), a \geq 0, b > 0. \qquad (10.11.2)
\end{aligned}$$

It is easy to verify that the Gibbs conditionals corresponding to the HB model (10.11.2) are given by

$$\text{(i)}\quad [\boldsymbol{\beta}|\mathbf{p},\sigma_v^2,\mathbf{y}] \sim N_p\left[\boldsymbol{\beta}^*,\sigma_v^2(\sum_i \mathbf{z}_i\mathbf{z}_i^T)^{-1}\right]$$

$$\text{(ii)}\quad [\sigma_v^2|\boldsymbol{\beta},\mathbf{p},\mathbf{y}] \sim G\left(\frac{m}{2}+a,\frac{1}{2}\sum_i(\xi_i-\mathbf{z}_i^T\boldsymbol{\beta})^2+b\right).$$

$$\text{(iii)}\quad f(p_i|\boldsymbol{\beta},\sigma_v^2,\mathbf{y}) \propto h(p_i|\boldsymbol{\beta},\sigma_v^2)k(p_i), \tag{10.11.3}$$

where $\boldsymbol{\beta}^* = (\sum_i \mathbf{z}_i^T\mathbf{z}_i)^{-1}(\sum_i \mathbf{z}_i^T\xi_i)$, $k(p_i) = p_i^{y_i}(1-p_i)^{n_i-y_i}$ and

$$h(p_i|\boldsymbol{\beta},\sigma_v^2) \propto g'(p_i)\exp\left\{-\frac{(\xi_i-\mathbf{z}_i^T\boldsymbol{\beta})^2}{2\sigma_v^2}\right\}, \tag{10.11.4}$$

where $g'(p_i) = \partial g(p_i)/\partial p_i$ with $g(p_i) = \text{logit}(p_i)$. It is clear from (10.11.3) that the conditionals (i) and (ii) admit closed forms while (iii) has form similar to (10.4.3). Therefore, we can use $h(p_i|\boldsymbol{\beta},\sigma_v^2)$ to draw the candidate p_i^*, noting that $p_i = g^{-1}(\xi_i)$ and $\xi_i|\boldsymbol{\beta},\sigma_v^2 \sim N(\mathbf{z}_i^T\boldsymbol{\beta},\sigma_v^2)$. In this case, the acceptance probability used in the M-H algorithm is given by $a(p_i^{(k)},p_i^*) = \min\{k(p_i^*)/k(p_i^{(k)}),1\}$.

The HB estimate of p_i and the posterior variance of p_i are obtained directly from the MCMC samples $\{p_i^{(k)},\ldots,p_m^{(k)},\boldsymbol{\beta}^{(k)},\sigma_v^{2(k)};k=d+1,\ldots,d+D\}$ generated from the joint posterior $f(p_1,\ldots,p_m,\boldsymbol{\beta},\sigma_v^2|\mathbf{y})$. We have

$$\hat{p}_i^{\text{HB}} \approx \frac{1}{D}\sum_{k=d+1}^{d+D} p_i^{(k)} = p_i^{(\cdot)} \tag{10.11.5}$$

and

$$V(p_i|\hat{p}) \approx \frac{1}{D-1}\sum_{k=d+1}^{d+D}\left(p_i^{(k)}-p_i^{(\cdot)}\right)^2. \tag{10.11.6}$$

(ii) Unit level covariates

In subsection 9.4.2 we studied EB estimation for a logistic linear mixed model with unit level covariates $\mathbf{x}_{ij}$, assuming that the model holds for the sample $\{(y_{ij},\mathbf{x}_{ij}), j\in s_i; i=1,\ldots,m\}$. A HB version of this model may be expressed as

$$\text{(i)}\quad y_{ij}|p_{ij} \overset{\text{ind}}{\sim} \text{Bernoulli}(p_{ij})$$

$$\text{(ii)}\quad \xi_{ij} = \text{logit}(p_{ij}) = \mathbf{x}_{ij}^T\boldsymbol{\beta}+v_i \quad\text{with}\quad v_i \overset{\text{iid}}{\sim} N(0,\sigma_v^2)$$

$$\text{(iii)}\quad \boldsymbol{\beta} \text{ and } \sigma_v^2 \text{ are mutually independent with}$$

$$f(\boldsymbol{\beta}) \propto 1 \quad\text{and}\quad \sigma_v^{-2} \sim G(a,b), a\ge 0, b>0. \tag{10.11.7}$$

An alternative prior on $\boldsymbol{\beta}$ and σ_v^2 is the flat prior $f(\boldsymbol{\beta}, \sigma_v^2) \propto 1$.

Let $\{v_1^{(k)}, \ldots, v_m^{(k)}, \boldsymbol{\beta}^{(k)}, \sigma_v^{2(k)}; k = d+1, \ldots, d+D\}$ denote the MCMC samples generated from the joint posterior $f(v_1, \ldots, v_m, \boldsymbol{\beta}, \sigma_v^2|\mathbf{y})$. Then the HB estimate of the finite population proportion P_i is obtained as

$$\hat{P}_i^{\text{HB}} \approx \frac{1}{N_i}\left[\sum_{j\in s_i} y_{ij} + \frac{1}{D}\sum_{k=d+1}^{d+D}\sum_{l\in\bar{s}_i} p_{il}^{(k)}\right], \tag{10.11.8}$$

where $p_{il}^{(k)} = \exp(\mathbf{x}_{il}^T\boldsymbol{\beta}^{(k)})/[1 + \mathbf{x}_{il}^T\boldsymbol{\beta}^{(k)}]$ and $\bar{s}_i$ is the set of nonsampled units in the ith area. Similarly, the posterior variance $V(P_i|\mathbf{y})$ is obtained as

$$\begin{aligned} V(P_i|\mathbf{y}) \approx {} & N_i^{-2}\frac{1}{D}\sum_{k=d+1}^{d+D}\left[\sum_{l\in\bar{s}_i} p_{il}^{(k)}(1-p_{il}^{(k)}) + \left(\sum_{l\in\bar{s}_i} p_{il}^{(k)}\right)^2\right] \\ & -N_i^{-2}\left[\frac{1}{D}\sum_{k=d+1}^{d+D} p_{il}^{(k)}\right]^2; \end{aligned} \tag{10.11.9}$$

see (9.4.21). The total $Y_i = N_iP_i$ is estimated by $N_i\hat{P}_i^{\text{HB}}$ and its posterior variance is given by $N_i^2V(P_i|\mathbf{y})$. Note that the $\mathbf{x}_{il}$-values for $l \in \bar{s}_i$ are needed to implement (10.11.8) and (10.11.9).

The Gibbs conditionals corresponding to the HB model (10.11.7) are given by

(i) $f(\beta_1|\beta_2, \ldots, \beta_p, \mathbf{v}, \mathbf{y}) \propto \prod_{i,j} p_{ij}^{y_{ij}}(1-p_{ij})^{1-y_{ij}}$,

(ii) $f(\beta_u|\beta_1, \ldots, \beta_{u-1}, \beta_{u+1}, \ldots, \beta_p, \mathbf{v}, \mathbf{y}) \propto \prod_{i,j} p_{ij}^{y_{ij}}(1-p_{ij})^{1-y_{ij}}$,

(iii) $f(\delta_i|\boldsymbol{\beta}, \sigma_v^2, \mathbf{y}) \propto \prod_{j} p_{ij}^{y_{ij}}(1-p_{ij})^{1-y_{ij}} \exp[-v_i^2/(2\sigma_v^2)]$,

(iv) $\sigma_v^{-2} \sim G\left(\frac{m}{2} + a, \frac{1}{2}\sum_i v_i^2 + b\right)$. (10.11.10)

It is clear from (10.11.10) that (i), (ii) and (iii) do not admit closed forms. Farrell (2000) used the griddy-Gibbs sampler (Ritter and Tanner (1992)) to generate MCMC samples from (i)–(iii). We refer the reader to Farrell (2000) for details of the griddy-Gibbs sampler with regard to the HB model (10.11.10). Alternatively, M-H within Gibbs may be used to generate MCMC samples from (i)–(iii) of (10.11.10).

Example 10.11.2. Simulation Study. Farrell (2000) conducted a simulation study on the frequentist performance of the HB method and the approximate EB method of Farrell et al. (1997); see subsection 9.4.2 for the approximate EB method. He treated a public use microdata sample of the

1950 United States Census as the population and calculated the true local area female participation rates, P_i, for $m = 20$ local areas selected with probability proportional to size and without replacement from $M = 52$ local areas in the population. Auxiliary variables, $\mathbf{x}_{ij}$, related to p_{ij}, were selected by a stepwise logistic regression procedure. Both unit (or individual) level and area level covariates were selected: age, marital status and whether the individual had children (unit level variables); average age, proportions of individuals in various marital status categories, and proportion of individuals having children (area level variables).

Treating the $m = 20$ sampled areas as strata, $R = 500$ independent stratified random samples with $n_i = 50$ individuals in each stratum were drawn. The HB estimates $\hat{P}_i^{\text{HB}}(r)$ and the approximate EB estimates $\hat{P}_i^{\text{EB}}(r)$ were then calculated from each simulation run $r(=1, \ldots, 500)$. Using these estimates, the absolute differences

$$\text{AD}_i^{\text{HB}} = \frac{1}{R}\sum_{r=1}^{R} |\hat{P}_i^{\text{HB}}(r) - P_i|; \text{AD}_i^{\text{EB}} = \frac{1}{R}\sum_{r=1}^{R} |\hat{P}_i^{\text{EB}}(r) - P_i|$$

and the mean absolute differences $\overline{\text{AD}}^{\text{HB}} = m^{-1}\sum_i \text{AD}_i^{\text{HB}}$ and $\overline{\text{AD}}^{\text{EB}} = m^{-1}\sum_i \text{AD}_i^{\text{EB}}$ were obtained. In terms of mean absolute difference, HB performed better than EB: $\overline{\text{AD}}^{\text{HB}} = 0.0031$ compared to $\overline{\text{AD}}^{\text{EB}} = 0.0056$. Moreover, the absolute difference AD_i^{HB} was smaller than AD_i^{EB} for sixteen of the twenty local areas. However, these results should be interpreted with caution because Farrell (2000) used the gamma prior on σ_v^{-2} with both $a = 0$ and $b = 0$ which results in an improper joint posterior.

10.11.3 Logistic Linear Mixed Models

A logistic linear mixed HB model is given by

$$\text{(i)}\ \ y_{ij}|p_{ij} \overset{\text{ind}}{\sim} \text{Bernoulli}(p_{ij})$$

$$\text{(ii)}\ \ \xi_{ij} = \text{logit}(p_{ij}) = \mathbf{x}_{ij}^T\boldsymbol{\beta} + \mathbf{z}_{ij}^T\mathbf{v}_i \text{ with } \mathbf{v}_i \overset{\text{iid}}{\sim} N_q(\mathbf{0}, \boldsymbol{\Sigma}_v)$$

$$\text{(iii)}\ \ f(\boldsymbol{\beta}, \boldsymbol{\Sigma}_v) \propto f(\boldsymbol{\Sigma}_v). \tag{10.11.11}$$

Zeger and Karim (1999) proposed the Jeffrey's prior

$$f(\boldsymbol{\Sigma}_v) \propto |\boldsymbol{\Sigma}_v|^{-(q+1)/2}. \tag{10.11.12}$$

Unfortunately, the choice (10.11.12) leads to an improper joint posterior distribution even though all the Gibbs conditionals are proper (Natarjan and Kass (2000)). A simple choice that avoids this difficulty is the flat prior $f(\boldsymbol{\Sigma}_v) \propto 1$. An alternative prior is obtained by assuming a Wishart distribution on $\boldsymbol{\Sigma}_v^{-1}$ with parameters reflecting lack of information on $\boldsymbol{\Sigma}_v$. Natarajan and Kass (2000) proposed different priors on $\boldsymbol{\Sigma}_v$ that ensure propriety of the joint posterior.

In the context of small area estimation, Malec, Sedransk, Moriarity and LeClaire (1997) studied a two-level HB model with class-specific covariates, $\mathbf{x}_j$, when the population is divided into classes j. Suppose there are M areas (say, counties) in the population and y_{ijl} denotes the binary response variable associated with the lth individual in class j and area i ($l = 1, \ldots, N_{ij}$). The two-level population HB model of Malec et al. (1997) may be expressed as

$$\begin{aligned}
&\text{(i)} \quad y_{ijl}|p_{ij} \overset{\text{ind}}{\sim} \text{Bernoulli}(p_{ij})\\
&\text{(ii)} \quad \xi_{ij} = \text{logit}(p_{ij}) = \mathbf{x}_j^T\boldsymbol{\beta}_i\\
&\qquad \boldsymbol{\beta}_i = \mathbf{Z}_i\boldsymbol{\alpha} + \mathbf{v}_i; \quad \mathbf{v}_i \overset{\text{iid}}{\sim} N_q(\mathbf{0}, \boldsymbol{\Sigma}_v)\\
&\text{(iii)} \quad f(\boldsymbol{\alpha}, \boldsymbol{\Sigma}_v) \propto 1, \qquad (10.11.13)
\end{aligned}$$

where $\mathbf{Z}_i$ is a $p \times q$ matrix of area level covariates. We assume that the model (10.11.13) holds for the sample $\{(y_{ijk}, \mathbf{x}_j, \mathbf{Z}_i); k \in s_{ij}, i \in s\}$, where s is the set of sample areas and s_{ij} is the sample of n_{ij} individuals in class j and area $i \in s$.

Let $\{\boldsymbol{\beta}_i^{(k)} : i \in s, \boldsymbol{\alpha}^{(k)}, \boldsymbol{\Sigma}_v^{(k)}; k = d{+}1, \ldots, d{+}D\}$ denote the MCMC samples generated from the joint posterior $f(\boldsymbol{\beta}_i : i \in s, \boldsymbol{\alpha}, \boldsymbol{\Sigma}_v|\mathbf{y})$. Given $\boldsymbol{\alpha}^{(k)}, \boldsymbol{\Sigma}_v^{(k)}$, we generate $\boldsymbol{\beta}_i^{(k)} : i \notin s$ from $N(\mathbf{Z}_i\boldsymbol{\alpha}^{(k)}, \boldsymbol{\Sigma}_v^{(k)})$ assuming $\mathbf{Z}_i$ for $i \notin s$ is observed. Using these values of $\boldsymbol{\beta}_i$, we calculate $p_{ij}^{(k)} = \exp(\mathbf{x}_j^{(k)}\boldsymbol{\beta}_i^{(k)})/[1+\exp(\mathbf{x}_j^{(k)}\boldsymbol{\beta}_i^{(k)})]$ for all areas $i = 1, \ldots, M$.

Suppose that we are interested in a finite population proportion corresponding to a collection of areas, say I, and a collection of classes, say J; that is,

$$\begin{aligned}
P(I, J) &= \sum_{i\in I}\sum_{j\in J}\sum_{l=1}^{N_{ij}} y_{ij} \Big/ \sum_{i\in I}\sum_{j\in J} N_{ij},\\
&=: Y(I, J)/N(I, J). \qquad (10.11.14)
\end{aligned}$$

Then the HB estimate of the proportion $P(I, J)$ is $\hat{P}^{\text{HB}}(I, J) = \frac{\hat{Y}^{\text{HB}}(I,J)}{N(I,J)}$, where $\hat{Y}^{\text{HB}}(I, J)$ is obtained as

$$\hat{Y}^{\text{HB}}(I, J) = \sum_{i\in I}\sum_{j\in J}\sum_{k\in s_{ij}} y_{ijk} + \frac{1}{D}\sum_{k=d+1}^{d+D}\left[\sum_{i\in I}\sum_{j\in J}(N_{ij} - n_{ij})p_{ij}^{(k)}\right]. \qquad (10.11.15)$$

Similarly, the posterior variance $V[P(I, J)|\mathbf{y}] = V[Y(I, J)|\mathbf{y}]/[N(I, J)]^2$, where $V[Y(I,J)|\mathbf{y}]$ is obtained as

$$V[Y(I,J)|\mathbf{y}] \approx \frac{1}{D}\sum_{k=d+1}^{d+D}\left[\sum_{i\in I}\sum_{j\in J}(N_{ij}-n_{ij})p_{ij}^{(k)}(1-p_{ij}^{(k)}) + \left\{\sum_{i\in I}\sum_{j\in J}(N_{ij}-n_{ij})p_{ij}^{(k)}\right\}^2\right] - \left[\frac{1}{D}\sum_{k=d+1}^{d+D}\sum_{i\in I}\sum_{j\in J}(N_{ij}-n_{ij})p_{ij}^{(k)}\right]^2. \quad (10.11.16)$$

Note that $n_{ij}=0$ if an area $i\in I$ is not sampled.

Example 10.11.3. Visits to Physicians. Malec et al. (1999) applied the two-level model (10.11.13) to estimate health-related proportions from the National Health Interview Survey (NHIS) data for the 50 states and the District of Columbia and for specified subpopulations within the 51 areas. The NHIS sample consisted of 200 primary sampling units (psu's) and households sampled within each selected psu. Each psu is either a county or a group of contiguous counties, and the total sample consisted of approximated 50,000 households and 120,000 individuals.

Selection of variables. Individual-level auxiliary variables included demographic variables such as race, age and sex and social-economic variables such as highest education level attained. County-level covariates, such as mortality rates, counts of hospitals and hospital beds and number of physicians (by field of specialization), were also available.

The population was partitioned into classes defined by the cross-classification of race, sex and age (in 5-year groups). Reliable estimates of the counts, N_{ij}, were available for each county i and class j. The vector of covariates, $\mathbf{x}_j$, used in the model (10.11.13), is assumed to be the same for all individuals in class j. A particular binary variable y, denoting the presence/absence of at least one doctor visit within the past year, was used to illustrate the model fitting and HB estimation of proportions $P(I,J)$.

The auxiliary variables, $\mathbf{x}_j$, and the area level covariates, $\mathbf{Z}_i$, used in the linking model, were selected in two steps using PROC LOGISTIC in SAS. In the first step, the variation in the $\boldsymbol{\beta}_i$'s was ignored by setting $\boldsymbol{\beta}_i=\boldsymbol{\beta}$ and the elements of $\mathbf{x}$ were selected from a list of candidate variables and their interactions, using the model $\text{logit}(p_j)=\mathbf{x}_j^T\boldsymbol{\beta}$ and $y_{ijl}\overset{\text{ind}}{\sim}\text{Bernoulli}(p_j)$. The following variables and interactions were selected by the SAS procedures: $\mathbf{x}_j=(1,x_{0,j},x_{15,j},x_{25,j},x_{55,j},a_jx_{15,j},a_jx_{25,j},b_j)^T$, where a_j and b_j are (0,1) variables with $a_j=1$ if class j corresponds to males, $b_j=1$ if class j corresponds to whites, and $x_{t,j}=\max(0,\ \text{age}\ j-t)$, $t=0,15,20,25$ with age j denoting the midpoint of the age group in class j; for example, age $j=42.5$ if class j corresponds to age group $[40,45]$, and $x_{15,j}=\max(0,42.5-15)=27.5$.

The variables $\mathbf{x}_j$ selected in the first step were then used in the combined model $\text{logit}(p_{ij}) = \mathbf{x}_j^T \mathbf{Z}_i \boldsymbol{\alpha}$ and $y_{ijl} \overset{\text{ind}}{\sim} \text{Bernoulli}(p_{ij})$ to select the covariates $\mathbf{Z}_i$, using the SAS procedure. Note that the combined model is obtained by setting $\boldsymbol{\Sigma}_v = 0$. For the particular binary variable y, the choice $\mathbf{Z}_i = \mathbf{I}$ captured between county variation quite well, that is, county level covariates are not needed. But for other binary variables, county level variables may be needed. Based on the above two-step selection of variables, the final linking model is given by

$$\begin{aligned}\text{logit}(p_{ij}) = \beta_{1i} + \beta_{2i}x_{0,j} + \beta_{3i}x_{15,j} + \beta_{4i}x_{25,j} + \beta_{5i}x_{55,j} \\ + \beta_{6i}a_j x_{15,j} + \beta_{7i}a_j x_{25,j} + \beta_{8i}b_j, \qquad (10.11.17)\end{aligned}$$

where $\boldsymbol{\beta}_i = (\beta_{1i}, \ldots, \beta_{8i})^T = \boldsymbol{\alpha} + \mathbf{v}_i$.

Model fit. To determine the model fit, two kinds of cross-validation were conducted. In the first kind, the sample individuals were randomly divided into five groups, while in the second kind the sample counties were randomly divided into groups. Let s_{ih} be the set of individuals in the hth group in county i, and $\bar{y}_{ih}$ be the corresponding sample mean. Denote the expectation and variance of $\bar{y}_{ih}$ with respect to the full predictive density by $E_2(\bar{y}_{ih})$ and $V_2(\bar{y}_{ih}) = E_2[E_2(\bar{y}_{ih}) - \bar{y}_{ih}]^2$, respectively. Malec et al. (1998) compared $D_{ih}^2 = [E_2(\bar{y}_{ih}) - \bar{y}_{ih}]^2$ to $V_2(\bar{y}_{ih})$ using the following summary measure:

$$C = \left[\sum_i \sum_{h=1}^{5} V_2(\bar{y}_{ih}) / \sum_i \sum_{h=1}^{5} D_{ih}^2 \right]^{1/2}. \qquad (10.11.18)$$

If the assumed model provides an adequate fit, then $|C - 1|$ should be reasonably small. The values of C were calculated for both types of cross-validation and for each of several subpopulations. These values clearly indicated adequacy of the model in the sense of $C-$values close to 1.

The measure C^2 may also be regarded as the ratio of the average posterior variance to the average MSE of HB estimator, treating $\bar{y}_{ih}$ as the "true" value. Under this interpretation, the posterior variance is consistent with the MSE because the $C-$values are close to 1.

Comparison of estimators. The HB method was compared to the EB method and synthetic estimation. For the EB method, the posterior mean and posterior variance were obtained from (10.11.15) and (10.11.16) by setting $\boldsymbol{\alpha}$ and $\boldsymbol{\Sigma}_v$ equal to the corresponding ML estimates and generating the MCMC samples. Similarly, a synthetic estimate (posterior mean) and associated posterior variance were obtained from (10.11.15) and (10.11.16) by setting $\boldsymbol{\beta}_i = \mathbf{Z}_i\boldsymbol{\alpha}$ and $\boldsymbol{\Sigma}_v = 0$ and generating the MCMC samples. As expected, the EB estimates were close to the corresponding HB estimates, and the EB standard errors were considerably smaller than the corresponding HB standard errors. Synthetic standard errors were also much smaller than the corresponding HB standard errors, while the relative differences in the point estimates were small.

Malec et al. (1999) also conducted an external evaluation in a separate study. For this purpose, they used a binary variable, health-related partial work limitation, which was included in the 1990 U.S. Census of Population and Housing Long Form. Treating the small area census proportions for this variable as "true" values, they compared the estimates corresponding to alternative methods and models to the true values.

Folsom, Shah and Vaish (1999) used a different logistic linear mixed HB model to produce HB estimates of small area prevalence rates for states and age groups for up to 20 drug use related binary outcomes, using data from pooled National Household Survey on Drug Abuse (NHSDA) surveys. Let y_{aijl} denote the value of a binary variable, y, for the l th individual in age group $a(=1,\ldots,4)$ belonging to the j th cluster of the i th state, and let $y_{aijl}|p_{aijl} \overset{\text{ind}}{\sim}$ Bernoulli(p_{aijl}). A linking model of the logistic linear mixed model type was assumed:

$$\text{logit}(p_{aijl}) = \mathbf{x}_{aijl}^T\boldsymbol{\beta}_a + v_{ai} + u_{aij}, \qquad (10.11.19)$$

where $\mathbf{x}_{aijl}$ denote a $p_a \times 1$ vector of auxiliary variables associated with age group a and $\boldsymbol{\beta}_a$ is the associated regression parameters. Further, the vectors $\mathbf{v}_i = (v_{1i},\ldots,v_{4i})^T$ and $\mathbf{u}_{ij} = (u_{1ij},\ldots,u_{4ij})^T$ are assumed to be mutually independent with $\mathbf{v}_i \sim N_4(\mathbf{0},\boldsymbol{\Sigma}_v)$ and $\mu_{ij} \sim N_4(\mathbf{0},\boldsymbol{\Sigma}_u)$. The model parameters $\boldsymbol{\beta}, \boldsymbol{\Sigma}_u$ and $\boldsymbol{\Sigma}_v$ are assumed to obey the following prior:

$$f(\boldsymbol{\beta},\boldsymbol{\Sigma}_u^{-1},\boldsymbol{\Sigma}_v^{-1}) \propto f(\boldsymbol{\Sigma}_u^{-1})f(\boldsymbol{\Sigma}_v^{-1}), \qquad (10.11.20)$$

where $f(\boldsymbol{\Sigma}_u^{-1})$ and $f(\boldsymbol{\Sigma}_v^{-1})$ are proper Wishart densities.

The population model is assumed to hold for the sample (that is, absence of sample selection bias), but survey weights, w_{aijl}, were introduced to obtain pseudo-HB estimates and pseudo-HB standard errors, similar to subsection 10.5.4. We refer the reader to Folsom et al. (1999) for details on MCMC sampling, selection of covariates and validation studies.

10.12 Exponential Family Models

In subsection 5.6.3 of Chapter 5 we assumed that the sample statistics y_{ij}, $(j = 1,\ldots,n_i; i = 1,\ldots,m)$, given the θ_{ij}'s, are independently distributed with probability density function belonging to the exponential family with canonical parameters θ_{ij} and known scale parameters $\phi_{ij}(>0)$:

$$f(y_{ij}|\theta_{ij}) = \exp\left[\frac{1}{\phi_{ij}}(\theta_{ij}y_{ij} - a(\theta_{ij})) + b(y_{ij},\phi_{ij})\right], \qquad (10.12.1)$$

where $a(\cdot)$ and $b(\cdot)$ are known functions. For example, $\theta_{ij} = \text{logit}(p_{ij}), \phi_{ij} = 1$ if $y_{ij} \sim \text{binomial}(n_{ij}, p_{ij})$. The linking model on the θ_{ij}'s is given by

$$\theta_{ij} = \mathbf{x}_{ij}^T\boldsymbol{\beta} + v_i + u_{ij}, \qquad (10.12.2)$$

where v_i and u_{ij} are mutually independent with $v_i \overset{\text{iid}}{\sim} N(0, \sigma_v^2)$ and $u_{ij} \overset{\text{iid}}{\sim} N(0, \sigma_u^2)$, and $\boldsymbol{\beta}$ is a $p \times 1$ vector of covariates without the intercept term. For the HB version of the model, we make an additional assumption that the model parameters $\boldsymbol{\beta}, \sigma_v^2$ and σ_u^2 are mutually independent with $f(\boldsymbol{\beta}) \propto 1$, $\sigma_v^{-2} \sim G(a_v, b_v)$ and $\sigma_u^{-2} \sim G(a_u, b_u)$. The objective here is to make inferences on the small area parameters; in particular, evaluate the posterior quantities $E(\theta_{ij}|\mathbf{y})$, $V(\theta_{ij}|\mathbf{y})$ and $\text{Cov}(\theta_{ij}, \theta_{lk}|\mathbf{y})$ for $(ij) \neq (lk)$. For example, $\theta_{ij} = \text{logit}(p_{ij})$ and p_{ij} denotes the proportion associated with a binary variable in the jth age-sex group in the ith region.

Ghosh, Natarajan, Stroud and Carlin (1998) gave sufficient conditions for the propriety of the joint posterior $f(\boldsymbol{\theta}|\mathbf{y})$. In particular, if y_{ij} is either binomial or Poisson, the conditions are: $b_v > 0$, $b_u > 0$, $\sum_i n_i - p + a_u > 0$, $m + a_v > 0$ and $\sum_j y_{ij} > 0$ for each i. The posterior is not identifiable (that is, improper) if an intercept term is included in the linking model (10.12.2).

The Gibbs conditionals are easy to derive. In particular, $[\boldsymbol{\beta}|\boldsymbol{\theta}, \mathbf{v}, \sigma_v^2, \sigma_u^2, \mathbf{y}]$ is p-variate normal, $[v_i|\boldsymbol{\theta}, \boldsymbol{\beta}, \sigma_v^2, \sigma_u^2, \mathbf{y}] \overset{\text{ind}}{\sim}$ normal, $[\sigma_v^{-2}|\boldsymbol{\theta}, \boldsymbol{\beta}, \mathbf{v}, \sigma_u^2, \mathbf{y}] \sim$ gamma, and $[\sigma_u^{-2}|\boldsymbol{\theta}, \boldsymbol{\beta}, \mathbf{v}, \sigma_v^2, \mathbf{y}] \sim$ gamma, but $[\theta_{ij}|\boldsymbol{\beta}, \mathbf{v}, \sigma_v^2, \sigma_u^2, \mathbf{y}]$ does not admit a closed-form density function. However, $\log[f(\theta_{ij}|\boldsymbol{\beta}, \mathbf{v}, \sigma_v^2, \sigma_u^2, \mathbf{y})]$ is a concave function of θ_{ij}, and therefore one can use the adaptive rejection sampling scheme of Gilks and Wild (1992) to generate samples.

Ghosh et al. (1998) generalized the model given by (10.12.1) and (10.12.2) to handle multicategory data sets. They applied this model to a data set from Canada on exposures to health hazards. Sample respondents in 15 regions of Canada were asked whether they experienced any negative impact of health hazards in the work place. Responses were classified into four categories: 1 = yes, 2 = no, 3 = not exposed, and 4 = not applicable or not stated. Here it was desired to make inferences on the category proportions within each age-sex class j in each region i.

10.13 Constrained HB

In subsection 9.6.1 of Chapter 9 we obtained the constrained Bayes (CB) estimator $\hat{\theta}_i^{CB}$, given by (9.6.6), that matches the variability of the small area parameters θ_i. This estimator, $\hat{\theta}_i^{CB}(\boldsymbol{\lambda})$, depends on the unknown model parameters $\boldsymbol{\lambda}$ of the two-stage model $\hat{\theta}_i|\theta_i \overset{\text{ind}}{\sim} f(\hat{\theta}_i|\theta_i, \boldsymbol{\lambda}_1)$ and $\theta_i \overset{\text{iid}}{\sim} f(\theta_i|\boldsymbol{\lambda}_2)$. Replacing $\boldsymbol{\lambda}$ by a suitable estimator $\hat{\boldsymbol{\lambda}}$, we obtained an empirical CB (ECB) estimator $\hat{\theta}_i^{\text{ECB}}(\hat{\boldsymbol{\lambda}})$

We can use a constrained HB (CHB) estimator instead of the ECB estimator. The CHB estimator, $\hat{\theta}_i^{\text{CHB}}$, is obtained by minimizing the posterior squared error $E[\sum_i(\theta_i - t_i)^2|\hat{\boldsymbol{\theta}}]$ subject to $t_{\cdot} = \hat{\theta}_{\cdot}^{\text{HB}}$ and $(m-1)^{-1}\sum_i(t_i - t_{\cdot})^2 = E[(m-1)^{-1}\sum_i(\theta_i - \theta_{\cdot})^2|\hat{\boldsymbol{\theta}}]$. We obtain

$$t_{i(opt)} = \hat{\theta}_i^{\text{CHB}} = \hat{\theta}_{\cdot}^{\text{HB}} + a(\hat{\boldsymbol{\theta}})(\hat{\theta}_i^{\text{HB}} - \hat{\theta}_{\cdot}^{\text{HB}}), \qquad (10.13.1)$$

where $\hat{\theta}_i^{\text{HB}}$ is the HB estimator, $\hat{\theta}_\cdot^{\text{HB}} = \sum_i \hat{\theta}_i^{\text{HB}}/m$ and

$$a(\hat{\boldsymbol{\theta}}) = \left[1 + \frac{\sum_i V(\theta_i - \theta_\cdot|\hat{\boldsymbol{\theta}})}{\sum_i (\hat{\theta}_i^{\text{HB}} - \hat{\theta}_\cdot^{\text{HB}})^2}\right]^{1/2}, \qquad (10.13.2)$$

where $\theta_\cdot = \sum_i \theta_i/m$. The proof (10.13.1) follows along the lines of subsection 9.9.3.

The estimator $\hat{\theta}_i^{\text{CHB}}$ depends on the HB estimators $\hat{\theta}_i^{\text{HB}}$ and the posterior variances $V(\theta_i - \theta_\cdot|\hat{\boldsymbol{\theta}})$ which can be evaluated using MCMC methods. The estimator $\hat{\theta}_i^{\text{CHB}}$ usually employs less shrinking toward the overall average compared to $\hat{\theta}_i^{\text{ECB}}$ (Ghosh and Maiti (1999)).

An advantage of the CHB approach is that it readily provides a measure of uncertainty associated with $\hat{\theta}_i^{\text{CHB}}$. Similar to the posterior variance $V(\theta_i|\hat{\boldsymbol{\theta}})$ associated with $\hat{\theta}_i^{\text{HB}}$, we use the posterior MSE, $E[(\theta_i - \hat{\theta}_i^{\text{CHB}})^2|\hat{\boldsymbol{\theta}}]$, as the measure of uncertainty associated with $\hat{\theta}_i^{\text{CHB}}$. We have

$$\begin{aligned} E[(\theta_i - \hat{\theta}_i^{\text{CHB}})^2|\hat{\boldsymbol{\theta}}] &= E[(\theta_i - \hat{\theta}_i^{\text{HB}})^2|\hat{\boldsymbol{\theta}}] + (\hat{\theta}_i^{\text{HB}} - \hat{\theta}_i^{\text{CHB}})^2 \\ &= V(\theta_i|\hat{\boldsymbol{\theta}}) + (\hat{\theta}_i^{\text{HB}} - \hat{\theta}_i^{\text{CHB}})^2. \end{aligned} \qquad (10.13.3)$$

It is clear from (10.13.3) that the posterior MSE is readily obtained from the posterior variance $V(\theta_i|\hat{\boldsymbol{\theta}})$ and the estimators $\hat{\theta}_i^{\text{HB}}$ and $\hat{\theta}_i^{\text{CHB}}$. On the other hand, it appears difficult to obtain a nearly unbiased estimator of MSE of the ECB estimator $\hat{\theta}_i^{\text{ECB}}$. The jackknife method used in Chapter 9 is not readily applicable to $\hat{\theta}_i^{\text{ECB}}(\hat{\boldsymbol{\lambda}})$.

10.14 Proofs

10.14.1 Proof of (10.2.26)

We express $f(y_r|\mathbf{y}_{(r)})$ as

$$\begin{aligned} f(y_r|\mathbf{y}_{(r)}) = \frac{f(\mathbf{y})}{f(\mathbf{y}_{(r)})} &= \frac{1}{\int \frac{f(\mathbf{y}_{(r)},\boldsymbol{\eta})}{f(\mathbf{y},\boldsymbol{\eta})} f(\boldsymbol{\eta}|\mathbf{y})d\boldsymbol{\eta}} \\ &= \frac{1}{\int \frac{1}{f(y_r|\mathbf{y}_{(r)},\boldsymbol{\eta})} f(\boldsymbol{\eta}|\mathbf{y})d\boldsymbol{\eta}}, \end{aligned} \qquad (10.14.1)$$

noting that $f(\mathbf{y}) = f(\mathbf{y},\boldsymbol{\eta})/f(\boldsymbol{\eta}|\mathbf{y})$. The denominator of (10.14.1) is the expectation of $1/f(y_r|\mathbf{y}_{(r)},\boldsymbol{\eta})$ with respect to $f(\boldsymbol{\eta}|\mathbf{y})$. Hence, we can estimate (10.14.1) by (10.2.26) using the MCMC output $\{\boldsymbol{\eta}^{(k)}\}$.

10.14.2 Proof of (10.2.32)

We write $E[a(y_r)|\mathbf{y}_{(r),\text{obs}}]$ as

$$E[a(y_r)|\mathbf{y}_{(r),\text{obs}}] = E_1E_2[a(y_r)] = E_1[b_r(\boldsymbol{\eta})],$$

where E_2 is the expectation over y_r given $\boldsymbol{\eta}$ and E_1 is the expectation over $\boldsymbol{\eta}$ given $\mathbf{y}_{(r),\text{obs}}$. Note that we have assumed conditional independence of y_r and $\mathbf{y}_{(r)}$ given $\boldsymbol{\eta}$. Therefore,

$$\begin{aligned} E_1[b_r(\boldsymbol{\eta})] &= \int b_r(\boldsymbol{\eta}) f(\boldsymbol{\eta}|\mathbf{y}_{(r),\text{obs}})d\boldsymbol{\eta} \\ &= f(y_{r,\text{obs}}|\mathbf{y}_{(r),\text{obs}}) \int \frac{b_r(\boldsymbol{\eta})}{f(y_{r,\text{obs}}|\boldsymbol{\eta})} f(\boldsymbol{\eta}|\mathbf{y}_{\text{obs}})d\boldsymbol{\eta}, \qquad (10.14.2) \end{aligned}$$

noting that

(i) $$f(\boldsymbol{\eta}|\mathbf{y}_{(r)}) = \frac{f(\mathbf{y}_{(r)}|\boldsymbol{\eta})f(\boldsymbol{\eta})}{f(\mathbf{y}_{(r)})},$$

(ii) $f(\boldsymbol{\eta}) = f(\mathbf{y})f(\boldsymbol{\eta}|\mathbf{y})/f(\mathbf{y}|\boldsymbol{\eta})$, (iii) $f(\mathbf{y})/f(\mathbf{y}_{(r)}) = f(\mathbf{y}|\mathbf{y}_{(r)})$ and (iv) $f(\mathbf{y}|\boldsymbol{\eta}) = f(\mathbf{y}_{(r)}, y_r|\boldsymbol{\eta}) = f(\mathbf{y}_{(r)}|\boldsymbol{\eta})f(y_r|\boldsymbol{\eta})$. The integral in (10.14.2) is the expectation of $b_r(\boldsymbol{\eta})/f(y_{r,\text{obs}}|\boldsymbol{\eta})$ with respect to $f(\boldsymbol{\eta}|\mathbf{y}_{\text{obs}})$. Hence, (10.14.2) may be estimated by (10.2.32) using the MCMC output $\{\boldsymbol{\eta}^{(k)}\}$.

10.14.3 Proof of (10.3.11)–(10.3.13)

The Gibbs conditional $[\boldsymbol{\beta}|\boldsymbol{\theta}, \sigma_v^2, \hat{\boldsymbol{\theta}}]$ may be expressed as

$$\begin{aligned} f(\boldsymbol{\beta}|\boldsymbol{\theta}, \sigma_v^2, \hat{\boldsymbol{\theta}}) &= \frac{f(\boldsymbol{\beta}, \boldsymbol{\theta}, \sigma_v^2|\hat{\boldsymbol{\theta}})}{\int f(\boldsymbol{\beta}, \boldsymbol{\theta}, \sigma_v^2|\hat{\boldsymbol{\theta}})d\boldsymbol{\beta}} \\ &\propto f(\boldsymbol{\beta}, \boldsymbol{\theta}, \sigma_v^2|\hat{\boldsymbol{\theta}}) \qquad (10.14.3) \end{aligned}$$

because the denominator of (10.14.3) is constant with respect to $\boldsymbol{\beta}$. Retaining terms involving only $\boldsymbol{\beta}$ in $f(\boldsymbol{\beta}, \boldsymbol{\theta}, \sigma_v^2|\hat{\boldsymbol{\theta}})$ and letting $\tilde{\mathbf{z}}_i = \mathbf{z}_i/b_i$ and $\tilde{\theta}_i = \theta_i/b_i$, we get

$$\begin{aligned} &\log[f(\boldsymbol{\beta}|\boldsymbol{\theta}, \sigma_v^2, \hat{\boldsymbol{\theta}})] \\ &= \text{const} - \frac{1}{2\sigma_v^2}\boldsymbol{\beta}^T\Big(\sum_i \tilde{\mathbf{z}}_i\tilde{\mathbf{z}}_i^T\Big)\boldsymbol{\beta} - 2\sum \tilde{\theta}_i\tilde{\mathbf{z}}_i^T\boldsymbol{\beta} \\ &= \text{const} - \frac{1}{2\sigma_v^2}\Big[(\boldsymbol{\beta}-\boldsymbol{\beta}^*)^T\Big(\sum_i \tilde{\mathbf{z}}_i\tilde{\mathbf{z}}_i^T\Big)^{-1}(\boldsymbol{\beta}-\boldsymbol{\beta}^*)\Big], \qquad (10.14.4) \end{aligned}$$

where $\boldsymbol{\beta}^* = (\sum_i \tilde{\mathbf{z}}_i \tilde{\mathbf{z}}_i^T)^{-1}(\sum_i \tilde{\mathbf{z}}_i \tilde{\theta}_i)$. It follows from (10.14.4) that $[\boldsymbol{\beta}|\boldsymbol{\theta}, \sigma_v^2, \hat{\boldsymbol{\theta}}]$ is $N_p[\boldsymbol{\beta}^*, \sigma_v^2(\sum_i \tilde{\mathbf{z}}_i \tilde{\mathbf{z}}_i^T)^{-1}]$. Similarly,

$$\begin{aligned}\log[f(\theta_i|\boldsymbol{\beta}, \sigma_v^2, \hat{\boldsymbol{\theta}})] &= \text{const} - \frac{1}{2}\left[\frac{\theta_i^2}{\gamma_i \psi_i} - 2\frac{\theta_i \hat{\theta}_i}{\psi_i} - 2\frac{\theta_i \mathbf{z}_i^T \boldsymbol{\beta}}{\sigma_v^2 b_i^2}\right] \\ &= \text{const} - \frac{1}{2\gamma_i \psi_i}[\theta_i - \hat{\theta}_i^B(\boldsymbol{\beta}, \sigma_v^2)]^2. \qquad (10.14.5)\end{aligned}$$

It now follows from (10.14.5) that $[\theta_i|\boldsymbol{\beta}, \sigma_v^2, \hat{\boldsymbol{\theta}}]$ is $N[\hat{\theta}_i^B(\boldsymbol{\beta}, \sigma_v^2), \gamma_i \psi_i]$. Similarly,

$$\begin{aligned}\log[f(\sigma_v^2|\boldsymbol{\beta}, \boldsymbol{\theta}, \hat{\boldsymbol{\theta}})] &= \text{const} - \frac{1}{2\sigma_v^2}\sum_i (\tilde{\theta}_i - \tilde{\mathbf{z}}_i^T \boldsymbol{\beta})^2 \\ &\quad - \frac{b}{\sigma_v^2} + \log\left[\frac{1}{(\sigma_v^2)^{m/2+a-1}}\right]. \qquad (10.14.6)\end{aligned}$$

It now follows from (10.14.6) that $[\sigma_v^{-2}|\boldsymbol{\beta}, \boldsymbol{\theta}, \hat{\boldsymbol{\theta}}]$ is $G\left[\frac{m}{2} + a,\ \frac{1}{2}\sum_i (\tilde{\theta}_i - \tilde{\mathbf{z}}_i^T \boldsymbol{\beta})^2 + b\right]$.

References

1. Anderson, T.W. (1973), Asymptotically Efficient Estimation of Covariance Matrices with Linear Covariance Structure, *Annals of Statistics*, **1**, 135–141.
2. Anderson, T.W., and Hsiao, C. (1981), Formulation and Estimation of Dynamic Models Using Panel Data, *Journal of Econometrics* **18**, 67–82.
3. Ansley, C.F., and Kohn, R. (1986), Prediction Mean Squared Error for State Space Models with Estimated Parameters, *Biometrika*, **73**, 467–473.
4. Aragon, Y. (1984), Random Variance Linear Models: Estimation, *Computational Statistics Quarterly*, **1**, 295–309.
5. Arora, V., and Lahiri, P. (1997), On the Superiority of the Bayes Method over the BLUP in Small Area Estimation Problems, *Statistica Sinica*, **7**, 1053–1063.
6. Banerjee, M., and Frees, E.W. 1997), Influence Diagnostics for Linear Longitudinal Models, *Journal of the American Statistical Association*, **92**, 999–1005.
7. Battese, G.E., Harter, R.M., and Fuller, W.A. (1988), An Error Component Model for Prediction of County Crop Areas Using Survey and Satelite Date, *Journal of the American Statistical Association*, **83**, 28–36.
8. Bayarri, M.J., and Berger, J. (2000), P Values for Composite Null Models, *Journal of the American Statistical Association*, **95**, 1127–1142.
9. Beckman, R.J., Nachtsheim, C.J., and Cook, R.D. (1987), Diagnostics for Mixed-Model Analysis of Variance, *Technometrics*, 1987, 413–426.
10. Béland, Y., Bailie, L., Catlin, G., and Singh, M.P. (2000), An Improved Health Survey Program at Statistics Canada, *Proceedings of the Section on Survey Research Methods*, American Statistical Association, Washington, DC, pp. 671–676.
11. Bell, W.R. (1999), Accounting for Uncertainty About Variances in Small Area Estimation, *Bulletin of the International Statistical Institute*,
12. Berger, J.O., and Pericchi, L.R. (2001), Objective Bayesian Methods for Model Selection: Introduction and Comparison, in P. Lahiri (ed.), *Model Selection*, Lecture Notes - Monograph Series, Volume 38, Beachwood, OH: Institute of Mathematical Statistics.
13. Besag, J.E. (1974), Spatial Interaction and the Statistical Analysis of Lattice Systems (with Discussion), *Journal of the Royal Statistical Society, Series B*, **35**, 192-236.
14. Bilodeau, M., and Srivastava, M.S. (1988), Estimation of the MSE Matrix of the Stein Estimator, *Canadian Journal of Statistics*, **16**, 153–159.
15. Bogue, D.J. (1950), A Technique for Making Extensive Postcensal Estimates, *Journal of the American Statistical Association*, **45**, 149–163.

16. Bogue, D.J., and Duncan, B.D. (1959), *A Composite Method of Estimating Post Censal Population of Small Areas by Age, Sex and Colour*, Vital Statistics – Special Report 47, No. 6, National Office of Vital Statistics, Washington, DC.
17. Bousfield, M.V. (1994), Estimation of Fertility Rates for Small Areas in Postcensal Years, *Proceedings of Social Statistics Section*, Washington, DC: American Statistical Association, pp. 106–111.
18. Brackstone, G.J. (1987), Small Area Data: Policy Issues and Technical Challenges, in R. Platek, J.N.K. Rao, C.E. Särndal and M.P. Singh (Eds.), *Small Area Statistics*, New York: Wiley, pp. 3–20.
19. Brandwein, A.C., and Strawderman, W.E. (1990), Stein-Estimation: The Spherically Symmetric Case, *Statistical Science*, **5**, 356–369.
20. Breslow, N., and Clayton, D. (1993), Approximate Inference in Generalized Linear Mixed Models, *Journal of the American Statistical Association*, **88**, 9–25.
21. Brewer, K.R.W. (1963), Ratio Estimation and Finite Populations: Some Results Deducible from the Assumption of an Underlying Stochastic Process, *Australian Journal of Statistics*, **5**, 5–13.
22. Brooks, S.P. (1988), Markov Chain Monte Carlo Method and Its Applications, *Statistician*, **47**, 69–100.
23. Brooks, S.P., Catchpole, E.A., and Morgan, B.J. (2000), Bayesian Animal Survival Estimation, *Statistical Science*, **15**, 357–376.
24. Browne, W.J., and Draper, D. (2001), A Comparison of Bayesian and Likelihood-based Methods for Fitting Multilevel Models, Technical Report, Institute for Education, London.
25. Butar, F.B., and Lahiri, P. (2001), On Measures of Uncertainty of Empirical Bayes Small-Area Estimators, Technical Report, Department of Statistics, University of Nebraska, Lincoln.
26. Carlin, B., and Gelfand, A. (1991), A Sample Reuse Method for Accurate Parametric Empirical Bayes Intervals, *Journal of the Royal Statistical Society, Series B*, **53**, 189–200.
27. Carlin, B.P., and Louis, T.A. (2000), *Bayes and Empirical Bayes Methods for Data Analysis (2nd ed.)*, London: Chapman and Hall.
28. Casady, R.J., and Valliant, R. (1993), Conditional Properties of Post-stratified Estimators Under Normal Theory, *Survey Methodology*, **19**, 183–192.
29. Casella, G., and Berger, R.L. (1990), *Statistical Inference*, Belmonte, CA: Wadsworth & Brooks/Cole.
30. Casella, G., and Hwang, J.T. (1982), Limit Expressions for the Risk of James-Stein Estimators, *Canadian Journal of Statistics*, **10**, 305–309.
31. Casella, G., Lavine, M., and Robert, C.P. (2001), Explaining the Perfect Sampler, *American Statistician*, **55**, 299–305.
32. Chambers, R.L., and Feeney, G.A. (1977), Log Linear Models for Small Area Estimation, Unpublished paper, Australian Bureau of Statistics.
33. Chattopadhyay, M., Lahiri, P., Larsen, M., and Reimnitz, J. (1999), Composite Estimation of Drug Prevalences for Sub-State Areas, *Survey Methodology*, **25**, 81–86.

34. Chaudhuri, A. (1994), Small Domain Statistics: A Review, *Statistica Neerlandica*, **48**, 215–236.
35. Chen, M-H., Shao, Q-M., and Ibrahim, J.G. (2000), *Monte Carlo Methods in Bayesian Computation*, New York: Springer-Verlag.
36. Cheng, Y.S., Lee, K-O., and Kim, B.C. (2001), Adjustment of Unemployment Estimates Based on Small Area Estimation in Korea, Technical Report, Department of Mathematics, KAIST, Taejun, Korea.
37. Chib, S., and Greenberg, E. (1995), Understanding the Metropolis-Hastings Algorithm, *American Statistician*, **49**, 327–335.
38. Christiansen, C.L., and Morris, C.N. (1997), Hierarchical Poisson Regression Modeling, *Journal of the American Statistical Association*, **92**, 618–632.
39. Christiansen, C.L., Pearson, L.M., and Johnson, W. (1992), Case-Deletion Diagnostics for Mixed Models, *Technometrics*, **34**, 38–45.
40. Choudhry, G.H., and Rao, J.N.K. (1989), Small Area Estimation Using Models That Combine Time Series and Cross-Sectional Data, in A.C. Singh and P. Whitridge (Eds.), *Proceedings of the Statistics Canada Symposium on Analysis of Data in Time*, Ottawa: Statistics Canada, pp. 67–74.
41. Chung, Y.S., Lee, K-O., and Kim, B.C. (2001), Adjustment of Unemployment Estimates Based on Small Area Estimation in Korea, Technical Report, Department of Mathematics, KAIST, Taejun, Korea.
42. Clayton, D., and Bernardinelli, L. (1992), Bayesian Methods for Mapping Disease Risk, in P. Elliot, J. Cuzick, D. English and R. Stern (Eds.), *Geographical and Environmental Epidemiology: Methods for Small-Area Studies*, London: Oxford University Press.
43. Clayton, D., and Kaldor, J. (1987), Empirical Bayes Estimates of Age-Standardized Relative Risks for Use in Disease Mapping, *Biometrics*, **43**, 671–681.
44. Cochran, W.G. (1977), *Sampling Techniques*, 3rd ed., New York: Wiley.
45. Cook, R.D. (1977), Detection of Influential Observations in Linear Regression, *Technometrics*, **19**, 15–18.
46. Cook, R.D. (1986), Assessment of Local Influence, *Journal of the Royal Statistical Society, Series B*, **48**, 133–155.
47. Cowles, M.K., and Carlin, B.P. (1996), Markov Chain Monte Carlo Convergence Diagnostics: A Comparative Review, *Journal of the American Statistical Association*, **91**, 883–904.
48. Cressie, N. (1989), Empirical Bayes Estimation of Undercount in the Decennial Census, *Journal of the American Statistical Association*, **84**, 1033–1044.
49. Cressie, N. (1991), Small-Area Prediction of Undercount Using the General Linear Model, *Proceedings of Statistics Symposium 90: Measurement and Improvement of Data Quality*, Ottawa: Statistics Canada, pp. 93–105.
50. Cressie, N. (1992), REML Estimation in Empirical Bayes Smoothing of Census Undercount, *Survey Methodology*, **18**, 75–94.
51. Cressie, N., and Chan, N.H. (1989), Spatial Modelling of Regional Variables, *Journal of the American Statistical Association*, **84**, 393–401.
52. Daniels, M.J., and Gatsonis, C. (1999), Hierarchical Generalized Linear Models in the Analysis of Variations in Health Care Utilization, *Journal of the American Statistical Association*, **94**, 29–42.

53. Daniels, M.J., and Kass, R.E. (1999), Nonconjugate Bayesian Estimation of Covariance Matrices and Its Use in Hierarchical Models, *Journal of the American Statistical Association*, **94**, 1254–1263.
54. Das, K., Jiang, J., and Rao, J.N.K. (2001), Mean Squared Error of Empirical Predictor, Technical Report, School of Mathematics and Statistics, Carleton University, Ottawa, Canada.
55. Datta, G.S., Day, B., and Basawa, I. (1999), Empirical Best Linear Unbiased and Empirical Bayes Prediction in Multivariate Small Area Estimation, *Journal of Statistical Planning and Inference*, **75**, 169–179.
56. Datta, B., Day, B., and Maiti, T. (1998), Multivariate Bayesian Small Area Estimation: An Application to Survey and Satellite Data, *Sankhyā, Series A*, **60**, 1–19.
57. Datta, G.S., Fay, R.E., and Ghosh, M. (1991), Hierarchical and Empirical Bayes Multivariate Analysis in Small Area Estimation, in *Proceedings of Bureau of the Census 1991 Annual Research Conference*, U.S. Bureau of the Census, Washington, DC, pp. 63–79.
58. Datta, G.S., Ghosh, M., Nangia, N., and Natarajan, K. (1996), Estimation of Median Income of Four-Person Families: A Bayesian Approach, in W.A. Berry, K.M. Chaloner, and J.K. Geweke (Eds.), *Bayesian Analysis in Statistics and Econometrics*, New York: Wiley, pp. 129-140.
59. Datta, G.S., Ghosh, M., and Waller, L.A. (2000), Hierarchical and Empirical Bayes Methods for Environmental Risk Assessment, in P.K. Sen and C.R. Rao (Eds.), *Handbook of Statistics*, Volume 18, Amsterdam: Elsevier Science B.V., pp. 223–245.
60. Datta, G.S. and Lahiri, P. (2000), A Unified Measure of Uncertainty of Estimated Best Linear Unbiased Predictors in Small Area Estimation Problems, *Statistica Sinica*, **10**, 613–627.
61. Datta, G.S., Lahiri, P., and Maiti, T. (2002), Empirical Bayes Estimation of Median Income of Four-Person Families by State Using Time Series and Cross-Sectional Data, *Journal of Statistical Planning and Inference*, **102**, 83-97.
62. Datta, G.S., Lahiri, P., Maiti, T., and Lu, K.L. (1999), Hierarchical Bayes Estimation of Unemployment Rates for the U.S. States, *Journal of the American Statistical Association*, **94**, 1074–1082.
63. Datta, G.S. and Ghosh, M. (1991), Bayesian Prediction in Linear Models: Applications to Small Area Estimation, *Annals of Statistics*, **19**, 1748–1770.
64. Datta, G.S., Ghosh, M., Huang, E.T., Isaki, C.T., Schultz, L.K., and Tsay, J.H. (1992), Hierarchical and Empirical Bayes Methods for Adjustment of Census Undercount: The 1988 Missouri Dress Rehearsal Data, *Survey Methodology*, **18**, 95–108.
65. Datta, G.S., Ghosh, M., and Kim, Y-H. (2001), Probability Matching Priors for One-Way Unbalanced Random Effect Models, Technical Report, Department of Statistics, University of Georgia, Athens.
66. Datta, G.S., Ghosh, M., Smith, D.D., and Lahiri, P. (2002), On the Asymptotic Theory of Conditional and Unconditional Coverage Probabilities of Empirical Bayes Confidence Intervals, *Scandinavian Journal of Statistics*, **29**, 139–152.

67. Datta, G.S., Kubakawa, T., and Rao, J.N.K. (2002), Estimation of MSE in Small Area Estimation, Technical Report, Department of Statistics, University of Georgia, Athens.
68. Datta, G.S., Rao, J.N.K., and Smith, D.D. (2002), On Measures of Uncertainty of Small Area Estimators in the Fay-Herriot Model, Technical Report, University of Georgia, Athens.
69. Dawid, A.P. (1985), Calibration-Based Empirical Probability, *Annals of Statistics*, **13**, 1251–1274.
70. Dean, C.B., and MacNab, Y.C. (2001), Modeling of Rates Over a Hierarchical Health Administrative Structure, *Canadian Journal of Statistics*, **29**, 405–419.
71. Dempster, A.P., Laird, N.M., and Rubin, D.B. (1977), Maximum Likelihood from Incomplete Data Via the EM Algorithm, *Journal of the Royal Statistical Society, Series B*, **39**, 1–38.
72. Dempster, A.P., and Ryan, L.M. (1985), Weighted Normal Plots, *Journal of the American Statistical Association*, **80**, 845–850.
73. Deming, W.E., and Stephan, F.F. (1940), On a Least Squares Adjustment of a Sample Frequency Table when the Expected Marginal Totals Are Known, *Annals of Mathematical Statistics*, **11**, 427–444.
74. DeSouza, C.M. (1992), An Appropriate Bivariate Bayesian Method for Analysing Small Frequencies, *Biometrics*, **48**, 1113–1130.
75. Deville, J.C., and Särndal, C.E. (1992), Calibration Estimation in Survey Sampling, *Journal of the American Statistical Association*, **87**, 376–382.
76. Dick, P. (1995), Modelling Net Undercoverage in the 1991 Canadian Census, *Survey Methodology*, **21**, 45–54.
77. Ding, Y., and Fienberg, S.E. (1994), Dual System Estimation of Census Undercount in the Presence of Matching Error, *Survey Methodology*, **20**, 149–158.
78. Draper, N.R., and Smith, H. (1981), *Applied Regression Analysis* (2nd ed.), New York: Wiley.
79. Drew, D., Singh, M.P., and Choudhry, G.H. (1982), Evaluation of Small Area Estimation Techniques for the Canadian Labour Force Survey, *Survey Methodology*, **8**, 17–47.
80. Efron, B. (1975), Biased Versus Unbiased Estimation, *Advances in Mathematics*, **16**, 259–277.
81. Efron, B., and Morris, C.E. (1972a), Limiting the Risk of Bayes and Empirical Bayes Estimators, Part II: The Empirical Bayes Case, *Journal of the American Statistical Association*, **67**, 130–139.
82. Efron, B., and Morris, C.E. (1972b), Empirical Bayes on Vector Observations: An Extension of Stein's Method, *Biometrika*, **59**, 335–347.
83. Efron, B., and Morris, C.E. (1973), Stein's Estimation Rule and Its Competitors - An Empirical Bayes Approach, *Journal of the American Statistical Association*, **68**, 117–130.
84. Efron, B., and Morris, C.E. (1975), Data Analysis Using Stein's Estimate and Its Generalizations, *Journal of the American Statistical Association*, **70**, 311–319.
85. Ericksen, E.P. (1974), A Regression Method for Estimating Population Changes of Local Areas, *Journal of the American Statistical Association*, **69**, 867–875.

86. Ericksen, E.P., and Kadane, J.B. (1985), Estimating the Population in Census Year: 1980 and Beyond (with discussion), *Journal of the American Statistical Association*, **80**, 98–131.

87. Ericksen, E.P., Kadane, J.G., and Tukey, J.W. (1989), Adjusting the 1981 Census of Population and Housing, *Journal of the American Statistical Association*, **84**, 927–944.

88. Estevao, V., Hidiroglou, M.A., and Särndal, C.E. (1995), Methodological Principles for a Generalized Estimation Systems at Statistics Canada, *Journal of Official Statistics*, **11**, 181–204.

89. Falorsi, P.D., Falorsi, S., and Russo, A. (1994), Empirical Comparison of Small Area Estimation Methods for the Italian Labour Force Survey, *Survey Methodology*, **20**, 171–176.

90. Farrell, P.J. (2000), Bayesian Inference for Small Area Proportions, *Sankhyā, Series B*, **62**, 402–416.

91. Farrell, P.J., MacGibbon, B., and Tomberlin, T.J. (1997a), Empirical Bayes Estimators of Small Area Proportions in Multistage Designs, *Statistica Sinica*, **7**, 1065–1083.

92. Farrell, P.J., MacGibbon, B., and Tomberlin, T.J. (1997b), Bootstrap Adjustments for Empirical Bayes Interval Estimates of Small Area Proportions, *Canadian Journal of Statistics*, **25**, 75–89.

93. Farrell, P.J., MacGibbon, B., and Tomberlin, T.J. (1997c), Empirical Bayes Small Area Estimation Using Logistic Regression Models and Summary Statistics, *Journal of Business & Economic Statistics*, **15**, 101–108.

94. Fay, R.E. (1987), Application of Multivariate Regression to Small Domain Estimation, in R. Platek, J.N.K. Rao, C.E. Särndal, and M.P. Singh (Eds.), *Small Area Statistics*, New York: Wiley, pp. 91–102.

95. Fay, R.E., and Herriot, R.A. (1979), Estimation of Income from Small Places: An Application of James–Stein Procedures to Census Data, *Journal of the American Statistical Association*, **74**, 269–277.

96. Fay, R.E. (1992), Inferences for Small Domain Estimates from the 1990 Post Enumeration Survey, Unpublished Manuscript, U.S. Bureau of the Census.

97. Fienberg, S.E. (1992), Bibliography on Capture-Recapture Modelling with Application to Census Undercount Adjustment, *Survey Methodology*, **18**, 143–154.

98. Folsom, R., Shah, B.V., and Vaish, A. (1999), Substance Abuse in States: A Methodological Report on Model Based Estimates from the 1994–1996 National Household Surveys on Drug Abuse, *Proceedings of the Section on Survey Research Methods*, American Statistical Association, Washington, DC, pp. 371–375.

99. Freedman, D.A., and Navidi, W.C. (1986), Regression Methods for Adjusting the 1980 Census (with discussion), *Statistical Science*, **18**, 75–94.

100. Fuller, W.A. (1975), Regression Analysis for Sample Surveys, *Sankhyā, Series C*, **37**, 117–132.

101. Fuller, W.A. (1989), Prediction of True Values for the Measurement Error Model, in *Conference on Statistical Analysis of Measurement Error Models and Applications*, Humboldt State University.

102. Fuller, W.A. (1999), Environmental Surveys Over Time, *Journal of Agricultural, Biological and Environmental Statistics*, **4**, 331–345.

103. Fuller, W.A., and Battese, G.E. (1973), Transformations for Estimation of Linear Models with Nested-Error Structure, *Journal of the American Statistical Association*, **68**, 626–632.

104. Fuller, W.A., and Battese, G. (1981), Regression Estimation for Small Areas, in D.M. Gilford, G.L. Nelson, and L. Ingram (Eds.), *Rural America in Passage: Statistics for Policy*, Washington, DC: National Academy Press, pp. 572–586.

105. Fuller, W.A., and Harter, R.M. (1987), The Multivariate Components of Variance Model for Small Area Estimation, in R. Platek, J.N.K. Rao, C.E. Särndal, and M.P. Singh (Eds.), *Small Area Statistics*, New York: Wiley, pp. 103–123.

106. Gelfand, A.E. (1996), Model Determination Using Sample-Based Methods, in W.R. Gilks, S. Richardson, and D.J. Spiegelhalter (Eds.), *Markov Chain Monte Carlo in Practice*, London: Chapman and Hall, pp. 145–161.

107. Gelfand, A.E., and Smith, A.F.M. (1990), Sample-based Approaches to Calculating Marginal Densities, *Journal of the American Statistical Association*, **85**, 972–985.

108. Gelfand, A.E., and Smith, A.F.M. (1991), Gibbs Sampling for Marginal Posterior Expectations, *Communications in Statistics - Theory and Methods*, **20**, 1747–1766.

109. Gelman, A., and Meng, S-L. (1996), Model Checking and Model Improvement, in W.R. Gilks, S. Richardson, and D.J. Spiegelhalter (Eds.), *Markov Chain Monte Carlo in Practice*, London: Chapman and Hall, pp. 189–201.

110. Gelman, A., and Rubin, D.B. (1992), Inference from Interactive Simulation Using Multiple Sequences, *Statistical Science*, **7**, 457–472.

111. Ghangurde, P.D., and Singh, M.P. (1977), Synthetic Estimators in Periodic Household Surveys, *Survey Methodology*, **3**, 152–181.

112. Ghosh, M. (1992a), Hierarchical and Empirical Bayes Multivariate Estimation, in M. Ghosh, and P.K. Pathak (Eds.), *Current Issues in Statistical Inference: Essays in Honor of D. Basu*, IMS Lecture Notes - Monograph Series, Volume 17.

113. Ghosh, M. (1992b), Constrained Bayes Estimation with Applications, *Journal of the American Statistical Association*, **87**, 533–540.

114. Ghosh, M., and Auer, R. (1983), Simultaneous Estimation of Parameters, *Annals of the Institute of Statistical Mathematics, Part A*, **35**, 379–387.

115. Ghosh, M., and Lahiri, P. (1987), Robust Empirical Bayes Estimation of Means from Stratified Samples, *Journal of the American Statistical Association*, **82**, 1153–1162.

116. Ghosh, M., and Lahiri, P. (1998), Bayes and Empirical Bayes Analysis in Multistage Sampling, in S.S. Gupta, and J.O. Berger (Eds.), *Statistical Decision Theory and Related Topics IV, Vol. 1*, New York: Springer, pp. 195–212.

117. Ghosh, M., and Maiti, T. (1999), Adjusted Bayes Estimators with Applications to Small Area Estimation, *Sankhyā, Series B*, **61**, 71–90.

118. Ghosh, M.,and Nangia, N., (1993), Estimation of Median Income of Four-person Families: A Bayesian Time Series Approach, Technical Report, De-

partment of Statistics, University of Florida, Gainsville.

119. Ghosh, M., Nangia, N., and Kim, D. (1996), Estimation of Median Income of Four-Person Families: A Bayesian Time Series Approach, *Journal of the American Statistical Association*, **91**, 1423–1431.

120. Ghosh, M., Natarajan, K., Stroud, T.W.F., and Carlin, B.P. (1998), Generalized Linear Models for Small Area Estimation, *Journal of American Statistical Association*, **93**, 273–282.

121. Ghosh, M., Natarajan, K., Waller, L.A., and Kim, D.H. (1999), Hierarchical Bayes GLMs for the Analysis of Spatial Data: An Application to Disease Mapping, *Journal of Statistical Planing and Inference*, **75**, 305–318.

122. Gilks, W.R., Richardson, S., and Spiegelhalter, D.J. (Eds.) (1996), *Markov Chain Monte Carlo in Practice*, London: Chapman and Hall.

123. Gilks, W.R., and Wild, P. (1992), Adaptive Rejection Sampling for Gibbs Sampling, *Applied Statistics*, **41**, 337–348.

124. Goldstein, H. (1975), A Note on Some Bayesian Nonparametric Estimates, *Annals of Statistics*, **3**, 736–740.

125. Goldstein, H. (1989), Restricted Unbiased Iterative Generalized Least Squares Estimation, *Biometrika*, **76**, 622–623.

126. Gonzalez, M.E. (1973), Use and Evaluation of Synthetic Estimates, *Proceedings of the Social Statistics Section*, American Statistical Association, pp. 33–36.

127. Gonzalez, M.E., and Hoza, C. (1978), Small–Area Estimation with Application to Unemployment and Housing Estimates, *Journal of the American Statistical Association*, 73, 7–15.

128. Gonzalez, J.F., Placek, P.J., and Scott, C. (1966), Synthetic Estimation of Followback Surveys at the National Center for Health Statistics, in W.L. Schaible (ed.), *Indirect Estimators in U.S. Federal Programs*, Springer–Verlag: New York, pp. 16–27.

129. Gonzalez, M.E., and Wakesberg, J. (1973), Estimation of the Error of Synthetic Estimates, Paper presented at the first meeting of the International Association of Survey Statisticians, Vienna, Austria.

130. Govindarajulu, Z. (1999), *Elements of Sampling Theory and Methods*, Englewood Cliffs, NJ: Prentice-Hall.

131. Griffin, B., and Krutchkoff, R. (1971), Optimal Linear Estimation: An Empirical Bayes Version with Application to the Binomial Distribution, *Biometrika*, **58**, 195–203.

132. Griffiths, R. (1996), Current Population Survey Small Area Estimations for Congressional Districts, *Proceedings of the Section on Survey Research Methods*, American Statistical Association, pp. 314–319.

133. Groves, R.M. (1989), *Survey Errors and Survey Costs*, New York: Wiley.

134. Hansen, M.H., Hurwitz, W.N., and Madow, W.G. (1953), *Sample Survey Methods and Theory I*, New York: Wiley.

135. Hansen, M.H., Madow, W.G., and Tepping, B.J. (1983), An Evaluation of Model-Dependent and Probability Sampling Inferences in Sample Surveys, *Journal of the American Statistical Association*, **78**, 776–793.

136. Hartigan, J. (1969), Linear Bayes Methods, *Journal of the Royal Statistical Society, Series B*, **31**, 446–454.

137. Hartless, G., Booth, J.G., and Littell, R.C. (2000), Local Influence Diagnostics for Prediction of Linear Mixed Models, Technical Report, Department of Statistics, University of Florida, Gainesville.
138. Hartley, H.O. (1959), Analytic Studies of Survey Data, in a *Volume in Honor of Corrado Gini, Instituto di Statistica*, Rome, Italy.
139. Hartley, H.O. (1974), Multiple Frame Methodology and Selected Applications, *Sankhyā, Series C*, **36**, 99–118.
140. Hartley, H.O., and Rao, J.N.K. (1967), Maximum Likelihood Estimation for the Mixed Analysis of Variance Model, *Biometrika*, **54**, 93–108.
141. Harvey, A.C. (1990), *Forecasting, Structural Time Series Models and the Kalman Filter*, Cambridge: Cambridge University Press.
142. Harville, D.A. (1977), Maximum Likelihood Approaches to Variance Component Estimation and to Related Problems, *Journal of the American Statistical Association*, **72**, 322–340.
143. Harville, D.A. (1991), Comment, *Statistical Science*, **6**, 35–39.
144. Harville, D.A., and Jeske, D.R. (1992), Mean Squared Error of Estimation or Prediction Under General Linear Model, *Journal of the American Statistical Association*, **87**, 724–731.
145. Hastings, W.K. (1970), Monte Carlo Sampling Methods Using Markov Chains and Their Applications, *Biometrika*, **57**, 97–109.
146. He, Z., and Sun, H. (1998), Hierarchical Bayes Estimation of Hunting Success Rates, *Environmental and Ecological Statistics*, **5**, 223–236.
147. Hedayat, A.S., and Sinha, B.K. (1991), *Design and Inference in Finite Population Sampling*, New York: Wiley.
148. Henderson, C.R. (1950), Estimation of Genetic Parameters (Abstract), *Annals of Mathematical Statistics*, **21**, 309–310.
149. Henderson, C.R. (1953), Estimation of Variance and Covariance Components, *Biometrics*, **9**, 226–252.
150. Henderson, C.R. (1963), Selection Index and Expected Genetic Advance, in *Statistical Genetics and Plant Breeding*, National Academy of Science, National Research Council, Publication 982, Washington, DC, pp. 141–163.
151. Henderson, C.R. (1973), Maximum Likelihood Estimation of Variance Components, Technical Report, Department of Animal Science, Cornell University, Ithaca, NY.
152. Henderson, C.R. (1975), Best Linear Unbiased Estimation and Prediction Under a Selection Model, *Biometrics*, **31**, 423–447.
153. Henderson, C.R., Kempthorne, O., Searle, S.R., and von Krosigk, C.N. (1959), Estimation of Environmental and Genetic Trends from Records Subject to Culling, *Biometrics*, **13**, 192–218.
154. Hill, B.M. (1965), Inference About Variance Components in the One-Way Model, *Journal of the American Statistical Association*, **60**, 806–825.
155. Hobert, J.P., and Casella, G. (1996), The Effect of Improper Priors on Gibbs Sampling in Hierarchical Linear Mixed Models, *Journal of the American Statistical Association*, **91**, 1461–1473.
156. Hocking, R.R., and Kutner, M.H. (1975), Some Analytical and Numerical Comparisons of Estimators for the Mixed A.O.V. Model, *Biometrics*, **31**, 19–28.

157. Hogan, H. (1992), The 1990 Post-Enumeration Survey, *American Statistician*, **46**, 261–269.

158. Holt, D., and Smith, T.M.F. (1979), Post-Stratification, *Journal of the Royal Statistical Society, Series A*, **142**, 33–46.

159. Holt, D., Smith, T.M.F., and Tomberlin, T.J. (1979), A Model-based Approach to Estimation for Small Subgroups of a Population, *Journal of the American Statistical Association*, **74**, 405–410.

160. Hulting, F.L., and Harville, D.A. (1991), Some Bayesian and Non-Bayesian Procedures for the Analysis of Comparative Experiments and for Small Area Estimation: Computational Aspects, Frequentist Properties and Relationships, *Journal of the American Statistical Association*, **86**, 557–568.

161. IASS Satellite Conference (1999), *Small Area Estimation*, Latvia: Central Statistical Office.

162. Ireland, C.T., and Kullback, S. (1968), Contingency Tables with Given Marginals, *Biometrika*, **55**, 179–188.

163. Isaki, C.T., Huang, E.T., and Tsay, J.H. (1991), Smoothing Adjustment Factors from the 1990 Post Enumeration Survey, *Proceedings of the Social Statistics Section*, American Statistical Association, Washington, DC, pp. 338–343.

164. Isaki, C.T., Tsay, J.H., and Fuller, W.A. (2000), Estimation of Census Adjustment Factors, *Survey Methodology*, **26**, 31–42.

165. James, W., and Stein, C. (1961), Estimation with Quadratic Loss, *Proceedings of the Fourth Berkeley Symposium on Mathematical Statistics and Probability*, University of California Press, Berkeley, pp. 361–379.

166. Jiang, J. (1996), REML Estimation: Asymptotic Behavior and Related Topics, *Annals of Statistics*, **24**, 255–286.

167. Jiang, J. (1997), A Derivation of BLUP – Best Linear Unbiased Predictor, *Statistics & Probability Letters*, **32**, 321–324.

168. Jiang, J. (1998), Consistent Estimators in Generalized Linear Mixed Models, *Journal of the American Statistical Association*, **93**, 720–729.

169. Jiang, J. (2001), Private Communication to J.N.K. Rao.

170. Jiang, J., Lahiri, P., and Wu, C-H. (1991), A Generalization of Pearson's Chi-Square Goodness-of-Fit Test with Estimated Cell Frequencies, *Sankhyā, Series A*, **63**, 260–276.

171. Jiang, J., Lahiri, P., and Wan, S-M. (2002), A Unified Jackknife Theory, *Annals of Statistics*, **30**, in press.

172. Jiang, J., Lahiri, P., Wan, S-M., and Wu, C-H. (2001), Jackknifing the Fay-Herriot Model with an Example, Technical Report, Department of Statistics, University of Nebraska, Lincoln.

173. Jiang, J., and Lahiri, P. (2001), Empirical Best Prediction for Small Area Inference with Binary data, *Annals of the Institute of Statistical Mathematics*, **53**, 217–243.

174. Jiang, J., and Rao, J.S. (2002), Consistent Procedures for Mixed Linear Model Selection, *Sankhyā, Series A.*, **63**, in press.

175. Jiang, J., and Zhang, W. (2001), Robust Estimation in Generalized Linear Mixed Models, *Biometrika*, **88**, 753–765.

176. Kackar, R.N., and Harville, D.A. (1981), Unbiasedness of Two-stage Estimation and Prediction Procedures for Mixed Linear Models, *Communications in Statistics, Series A*, **10**, 1249–1261.

177. Kackar, R.N., and Harville, D.A. (1984), Approximations for Standard Errors of Estimators of Fixed and Random Effects in Mixed Linear Models, *Journal of the American Statistical Association*, **79**, 853–862.

178. Kackar, D.A., and Jeske, D.R. (1992), Mean Squared Error of Estimation or Prediction Under a General Linear Model, *Journal of the American Statistical Association*, **87**, 724–731.

179. Kalton, G., Kordos, J., and Platek, R. (1993), *Small Area Statistics and Survey Designs Vol. I: Invited Papers: Vol. II: Contributed Papers and Panel Discussion*, Warsaw, Poland: Central Statistical Office.

180. Karunamuni, R.J., and Zhang, S. (2003), Optimal Linear Bayes and Empirical Bayes Estimation and Prediction of the Finite Population Mean, *Journal of Statistical Planning and Inference.*

181. Kass, R.E., and Raftery, A. (1995), Bayes Factors, *Journal of the American Statistical Association*, **90**, 773–795.

182. Kass, R.E., and Steffey, D. (1989), Approximate Bayesian Inference in Conditionally Independent Hierarchical Models (Parametric Empirical Bayes Models), *Journal of the American Statistical Association*, **84**, 717–726.

183. Kim, H., Sun, D., and Tsutakawa, R.K. (2001), A Bivariate Bayes Method for Improving the Estimates of Mortality Rates with a Twofold Conditional Autoregressive Model, *Journal of the American Statistical Association*, **96**, 1506–1521.

184. Kish, L. (1999), Cumulating/Combining Population Surveys, *Survey Methodology*, **25**, 129–138.

185. Kleffe, J., and Rao, J.N.K. (1992), Estimation of Mean Square Error of Empirical Best Linear Unbiased Predictors Under a Random Error Variance Linear Model, *Journal of Multivariate Analysis*, **43**, 1–15.

186. Kleinman, J.C. (1973), Proportions with Extraneous Variance: Single and Independent Samples, *Journal of the American Statistical Association*, **68**, 46–54.

187. Kott, P.S. (1990), Robust Small Domain Estimation Using Random Effects Modelling, *Survey Methodology*, **15**, 3–12.

188. Lahiri, P. (1990), "Adjusted" Bayes and Empirical Bayes Estimation in Finite Population Sampling, *Sankhyā, Series B*, **52**, 50–66.

189. Lahiri, P., and Maiti, T. (1999), Empirical Bayes Estimation of Relative Risks in Disease Mapping. Technical Report, Department of Statistics, University of Nebraska, Lincoln.

190. Lahiri, P., and Rao, J.N.K. (1995), Robust Estimation of Mean Squared Error of Small Area Estimators, *Journal of the American Statistical Association*, **82**, 758–766.

191. Laird, N.M. (1978), Nonparametric Maximum Likelihood Estimation of a Mixing Distribution, *Journal of the American Statistical Association*, **73**, 805–811.

192. Laird, N.M., and Louis, T.A. (1987), Empirical Bayes Confidence Intervals Based on Bootstrap Samples, *Journal of the American Statistical Association*,

82, 739–750.

193. Laird, N.M., and Ware, J.H. (1982), Random-Effects Models for Longitudinal Data, *Biometrics*, **38**, 963–974.
194. Lange, N., and Ryan, L. (1989), Asymptotic Normality in Random Effects Models, *Annals of Statistics*, **17**, 624–642.
195. Langford, I.H., Leyland, A.H., Rasbash, J., and Goldstein, H. (1999), Multilevel Modelling of the Geographical Distribution of Diseases, *Applied Statistics*, **48**, 253–268.
196. Laud, P. and Ibrahim, J. (1995), Predicitive Model Selection, *Journal of the Royal Statistical Society, Series B*, **57**, 247–262.
197. Levy, P.S. (1971), The Use of Mortality Data in Evaluating Synthetic Estimates, *Proceedings of the Social Statistics Section*, American Statistical Association, pp. 328–331.
198. Lohr, S.L. (1999), *Sampling: Design and Analysis*, Pacific Grove; CA: Duxbury.
199. Lohr, S.L., and Prasad, N.G.N. (2001), Small Area Estimation with Auxiliary Survey Data, Technical Report, Department of Mathematics, Arizona State University, Tempe.
200. Lohr, S.L., and Rao, J.N.K.. (2000), Inference for Dual Frame Surveys, *Journal of the American Statistical Association*, **95**, 271–280.
201. Louis, T.A. (1984), Estimating a Population of Parameter Values Using Bayes and Empirical Bayes Methods, *Journal of the American Statistical Association*, **79**, 393–398.
202. Lui, K-J., and Cumberland, W.G. (1991), A Model-Based Approach: Composite Estimators for Small Area Estimation, *Journal of Official Statistics*, **7**, 69–76.
203. MacGibbon, B., and Tomberlin, T.J. (1989), Small Area Estimation of Proportions Via Empirical Bayes Techniques, *Survey Methodology*, **15**, 237–252.
204. Maiti, T. (1998), Hierarchical Bayes Estimation of Mortality Rates for Disease Mapping, *Journal of Statistical Planning and Inference*, **69**, 339–348.
205. Malec, D., Davis, W.W., and Cao, X. (1999), Model-Based Small Area Estimates of Overweight Prevalence Using Sample Selection Adjustment, *Statistics in Medicine*, **18**, 3189-3200.
206. Malec, D., Sedransk, J., Moriarity, C.L., and LeClere, F.B. (1997), Small Area Inference for Binary Variables in National Health Interview Survey, *Journal of the American Statistical Association*, **92**, 815–826.
207. Mantel, H.J., Singh, A.C., and Bureau, M. (1993), Benchmarking of Small Area Estimators, in *Proceedings of International Conference on Establishment Surveys*, American Statistical Association, Washington, DC, pp. 920–925.
208. Maritz, J.S., and Lwin, T. (1989), *Empirical Bayes Methods (2nd Ed.)*, London: Chapman and Hall.
209. Marker, D.A. (1995), *Small Area Estimation: A Baysian Perspective*, Unpublished Ph.D. Dissertation, University of Michigan, Ann Arbor.
210. Marker, D.A. (1999), Organization of Small Area Estimators Using a Generalized Linear Regression Framework, *Journal of Official Statistics*, **15**, 1–24.

211. Marker, D.A. (2001), Producing Small Area Estimates From National Surveys: Methods for Minimizing Use of Indirect Estimators, *Survey Methodology*, **27**, 183–188.
212. Marshall, R.J. (1991), Mapping Disease and Mortality Rates Using Empirical Bayes Estimators, *Applied Statistics*, **40**, 283–294.
213. McCulloch, C.E., and Searle, S.R. (2001), *Generalized, Linear, and Mixed Models*, New York: Wiley.
214. Metropolis, N., Rosenbluth, A.W., Rosenbluth, M.N., Teller, A.H., and Teller, E. (1953), Equations of State Calculations by Fast Computing Machine, *Journal of Chemical Physics*, **27**, 1087–1097.
215. Morris, C.A. (1983a), Parametric Empirical Bayes Intervals, in G.E.P. Box, T. Leonard and C.F.J. Wu, (Eds.), *Scientific Inference, Data Analysis, and Robustness*, New York: Academic Press, pp. 25–50.
216. Morris, C.A. (1983b), Parametric Empirical Bayes Inference: Theory and Applications, *Journal of the American Statistical Association*, **78**, 47–54.
217. Moura, F.A.S., and Holt, D. (1999), Small Area Estimation Using Multilevel Models, *Survey Methodology*, **25**, 73–80.
218. Mukhopadhyay, P. (1998), *Small Area Estimation in Survey Sampling*, New Delhi: Narosa Publishing House.
219. Mulry, M.H., and Spencer, B.D. (1991, Total Error in PES Estimates of Population: The Dress Rehearsal Census of 1988 (with discussion), *Journal of the American Statistical Association*, **86**, 839–854.
220. Nandram, B. (1999), An Empirical Bayes Prediction Interval for the Finite Population Mean of Small Area, *Statistica Sinica*, **9**, 325–343.
221. Nandram, B., Sedransk, J., and Pickle, L. (1999), Bayesian Analysis of Mortality Rates for U.S. Health Service Areas, *Sankhyā, Series B*, **61**, 145–165.
222. Nandram, B., Sedransk, J., and Pickle, L. (2000), Bayesian Analysis and Mapping of Mortality Rates for Chronic Obstructive Pulmonary Disease, *Journal of the American Statistical Association*, **95**, 1110–1118.
223. Natarajan, R., and McCulloch, C.E. (1995), A Note on the Existence of the Posterior Distribution for a Class of Mixed Models for Binomial Responses, *Biometrika*, **82**, 639–643.
224. Natarajan, R., and Kass, R.E. (2000), Reference Bayesian Methods for Generalized Linear Mixed Models, *Journal of the American Statistical Association*, **95**, 227–237.
225. National Center for Drug Abuse (1979), *Synthetic Estimates for Small Areas* (Research Monograph 24), Washington, DC: U.S. Government Printing Office.
226. National Center for Health Statistics (1968), *Synthetic State Estimates of disability*, P.H.S. Publications 1759, Washington, DC: U.S. Government Printing Office.
227. National Research Council (1980), *Panel on Small-Area Estimates of Population and Income. Estimating Income and Population of Small Areas*, Washington, DC: National Academy Press.
228. National Research Council (2000), *Small-Area Estimates of School-Age Children in Poverty: Evaluation of Current Methodology*, C.F. Citro and G. Kalton (Eds.), Committee on National Statistics, Washington, DC: National Academy Press.

229. Nichol, S. (1977), A Regression Approach to Small Area Estimation, Unpublished Paper, Australian Bureau of Statistics, Canberra.

230. O'Hare, W. (1976), Report on Multiple Regression Method for Making Population Estimates, *Demography*, **13**, 369–379.

231. Pantula, S.G., and Pollock, K.H. (1985), Nested Analysis of Variance with Autocorrelated Errors, *Biometrics*, **41**, 909–920.

232. Peixoto, J.L., and Harville, D.A. (1986), Comparisons of Alternative Predictors Under the Balanced One-Way Random Model, *Journal of the American Statistical Association*, **81**, 431–436.

233. Petersen, C.J.S. (1986), *The Yearly Immigration of Young Plaice, Into the Limfjord from the German Sea*, Report of the Danish Biological Station to the Ministry of Fisheries, **6**, 1–48.

234. Pfeffermann, D., (2002), Small Area Estimation – New Developments and Directions, *International Statistical Review*, **70**, 125–143.

235. Pfeffermann, D., and Barnard, C. (1991), Some New Estimators for Small Area Means with Applications to the Assessment of Farmland Values, *Journal of Business and Economic Statistics*, **9**, 73–84.

236. Pfeffermann, D., and Burck, L. (1990), Robust Small Area Estimation Combining Time Series and Cross-Sectional Data, *Survey Methodology*, **16**, 217–237.

237. Pfeffermann, D., Feder, M., and Signorelli, D. (1988), Estimation of Autocorrelations of Survey Errors with Application to Trend Estimation in Small Areas, *Journal of Business and Economic Statistics*, **16**, 339–348.

238. Pfeffermann, D., and Tiller, R.B. (2001), Bootstrap Approximation to Prediction MSE for State-space Models with Estimated Parameters, Technical Report, Hebrew University, Jerusalem, Israel.

239. Pinheiro, J.C. and Bates, D.M. (2000), *Mixed-effects Models in S and S-plus*, New York: Springer.

240. Platek, R., Rao, J.N.K., Särndal, C.E., and Singh, M.P. (Eds.) (1987), *Small Area Statistics*, New York: Wiley.

241. Platek, R. and Singh, M.P. (Eds.) (1986), *Small Area Statistics: Contributed Papers*, Laboratory for Research in Statistics and Probability, Carleton University, Ottawa, Canada.

242. Portnoy, S.L. (1982), Maximizing the Probability of Correctly Ordering Random Variables Using Linear Predictors, *Journal of Multivariate Analysis*, **12**, 256–269.

243. Prasad, N.G.N., and Rao, J.N.K. (1990), The Estimation of the Mean Squared Error of Small-Area Estimators, *Journal of the American Statistical Association*, **85**, 163–171.

244. Prasad, N.G.N., and Rao, J.N.K. (1999), On Robust Small Area Estimation Using a Simple Random Effects Model, *Survey Methodology*, **25**, 67–72.

245. Purcell, N.J., and Kish, L. (1979), Estimates for Small Domain, *Biometrics*, **35**, 365–384.

246. Purcell, N.J., and Kish, L. (1980), Postcensal Estimates for Local Areas (or Domains), *International Statistical Review*, **48**, 3–18.

247. Purcell, N.J., and Linacre, S. (1976), Techniques for the Estimation of Small Area Characteristics, Unpublished Paper, Australian Bureau of Statistics, Canberra.

248. Raghunathan, T.E. (1993), A Quasi-Empirical Bayes Method for Small Area Estimation, *Journal of the American Statistical Association*, **88**, 1444–1448.

249. Rao, C.R. (1971), Estimation of Variance and Covariance Components – MINQUE Theory, *Journal of Multivariate Analysis*, **1**, 257–275.

250. Rao, C.R. (1976), Simultaneous Estimation of Parameters – A Compound Decision Problem, in Gupta, S.S. and Moore, D.S. (Eds.), *Statistical Decision Theory and Related Topics II*, New York: Academic Press, pp. 327–350.

251. Rao, J.N.K. (1974), Private Communication to T.W. Anderson.

252. Rao, J.N.K. (1986), Synthetic Estimators, SPREE and Best Model Based Predictors, *Proceedings of the Conference on Survey Research Methods in Agriculture*, U.S. Department of Agriculture, Washington, DC, pp. 1–16.

253. Rao, J.N.K. (1979), On Deriving Mean Square Errors and Their Non-Negative Unbiased Estimators in Finite Population Sampling, *Journal of the Indian Statistical Association*, **17**, 125–136.

254. Rao, J.N.K. (1985), Conditional Inference in Survey Sampling, *Survey Methodology*, **11**, 15–31.

255. Rao, J.N.K. (1994), Estimating totals and Distribution Functions Using Auxiliary Information at the Estimation Stage, *Journal of Official Statistics*, **10**, 153–165.

256. Rao, J.N.K. (2001), EB and EBLUP in Small Area Estimation, in S.E. Ahmed and N. Reid (Eds.), *Empirical Bayes and Likelihood Inference*, Lecture Notes in Statistics 148, New York: Springer, pp. 33–43.

257. Rao, J.N.K. (2001), Small Area Estimation with Applications to Agriculture, in *Proceedings of the Second Conference on Agricultural and Environmental Statistical Applications*, ISTAT, Rome, Italy.

258. Rao, J.N.K., and Choudhry, G.H. (1995), Small Area Estimation: Overview and Empirical Study, in B.G. Cox, D.A. Binder, B.N. Chinnappa, A. Christianson, M.J. Colledge, and P.S. Kott (Eds.), *Business Survey Methods*, New York: Wiley, pp. 527–542.

259. Rao, J.N.K., and Yu, M. (1992), Small Area Estimation by Combining Time Series and Cross-Sectional Data, *Proceedings of the Section on Survey Research Method*, American Statistical Association, pp. 1–9.

260. Rao, J.N.K., and Yu, M. (1994), Small Area Estimation by Combining Time Series and Cross-Sectional Data, *Canadian Journal of Statistics*, **22**, 511–528.

261. Rao, P.S.R.S. (1997), *Variance Component Estimation*, London: Chapman & Hall.

262. Ritter, C., and Tanner, T.A. (1992), The Gibbs Stopper and the Griddy-Gibbs Sampler, *Journal of the American Statistical Association*, **87**, 861–868.

263. Rives, N.W., Serow, W.J., Lee, A.S., and Goldsmith, H.F. (Eds.) (1989) *Small Area Analysis: Estimating Total Population*, Rockville, MD: National Institute of Mental Health.

264. Rivest, L-P. (1995), A Composite Estimator for Provincial Undercoverage in the Canadian Census, *Proceedings of the Survey Methods Section*, Statistical Society of Canada, pp. 33–38.

265. Rivest, L-P., and Belmonte, E. (2000), A Conditional Mean Squared Error of Small Area Estimators, *Survey Methodology*, **26**, 67–78.

266. Roberts, G.O., and Rosenthal, J.S. (1998), Markov-Chain Monte Carlo: Some Practical Implications of Theoretical Results, *Canadian Journal of Statistics*, **26**, 5–31.

267. Robinson, J. (1987), Conditioning Ratio Estimates Under Simple Random Sampling, *Journal of the American Statistical Association*, **82**, 826–831.

268. Robinson, G.K.. (1991), That BLUP Is a Good Thing: The Estimation of Random Effects, *Statistical Science*, **6**, 15–31.

269. Royall, R.M. (1970), On Finite Population Sampling Theory Under Certain Linear Regression, *Biometrika*, **57**, 377–387.

270. Rust, K.F., and Rao, J.N.K. (1996), Variance Estimation for Complex Surveys Using Replication Techniques, *Statistical Methods in Medical Research*, **5**, 283–310.

271. Särdnal, C.E., and Hidiroglou, M.A. (1989), Small Domain Estimation: A Conditional Analysis, *Journal of the American Statistical Association*, **84**, 266–275.

272. Särdnal, C.E., Swensson, B., and Wretman, J.H. (1989), The Weighted Regression Technique for Estimating the Variance of Generalized Regression Estimator, *Biometrika*, **76**, 527–537.

273. Särdnal, C.E., Swensson, B., and Wretman, J.H. (1992), *Model Assisted Survey Sampling*, New York: Springer-Verlag.

274. SAS/STAT User's Guide (1999), Volume 2, Version 8.0, Cary, NC: SAS Institute.

275. Schaible, W.A. (1978), Choosing Weights for Composite Estimators for Small Area Statistics, *Proceedings of the Section on Survey Research Methods*, American Statistical Association, pp. 741–746.

276. Schaible, W.L. (Ed.) (1996), *Indirect Estimation in U.S. Federal Programs*, New York: Springer.

277. Schmitt, R.C., and Crosetti, A.H. (1954), Accuracy of the Ratio–Correlation Method of Estimating Postcentral Population, *Land Economics*, **30**, 279–280.

278. Searle, S.R., Casella, G., and McCulloch, C.E. (1992), *Variance Components*, New York: Wiley.

279. Sekar, C., and Deming, E.W. (1949), On a Method of Estimating Birth and Death Rates and the Extent of Registration, *Journal of the American Statistical Association*, **44**, 101–115.

280. Shao, J., and Tu, D. (1995), *The Jackknife and Bootstrap*, New York: Springer-Verlag.

281. Shapiro, S.S., and Wilk, M.B. (1965), An Analysis of Variance Test for Normality (Complete Samples), *Biometrika*, **52**, 591–611.

282. Shen, W., and Louis, T.A. (1998), Triple-goal Estimates in Two-stage Hierarchical Models, *Journal of the Royal Statistical Society, Series B*, **60**, 455–471.

283. Singh, A.C., and Mian, I.U.H. (1995), Generalized Sample Size Dependent Estimators for Small Areas, *Proceedings of the 1995 Annual Research Conference*, U.S. Bureau of the Census, Washington, DC, pp. 687–701.

284. Singh, A.C., and Folsom, R.E. (2001a), Benchmarking of Small Area Estimators in a Bayesian Framework, Technical Report, Research Triangle Institute.

285. Singh, A.C., and Folsom, Jr., R.E. (2001b), Estimating Function Based Approach to Hierarchical Bayes Small Area Estimation for Survey Data, *Proceedings of the FCSM Conference*, Arlington, VA,

286. Singh, A.C., Mantel, J.H., and Thomas, B.W. (1994), Time Series EBLUPs for Small Areas Using Survey Data, *Survey Methodology*, **20**, 33–43.

287. Singh, A.C., Stukel, D.M., and Pfeffermann, D. (1998), Bayesian Versus Frequentist Measures of Error in Small Area Estimation, *Journal of the Royal Statistical Society, Series B*, **60**, 377–396.

288. Singh, M.P., and Tessier, R. (1976), Some Estimators for Domain Totals, *Journal of the American Statistical Association*, **71**, 322–325.

289. Singh, M.P.., Gambino, J., and Mantel, H.J. (1994), Issues and Strategies for Small Area Data, *Survey Methodology*, **20**, 3–22.

290. Singh, R., and Goel, R.C. (2000), Use of Remote Sensing Satellite Data in Crop Surveys, Technical Report, Indian Agricultural Statistics Research Institute, New Delhi, India.

291. Skinner, C.J. (1994), Sample Models and Weights, *Proceedings of the Section on Survey Research Methods*, American Statistical Association, Washington, DC, pp. 133–142.

292. Skinner, C.J., and Rao, J.N.K. (1996), Estimation in Dual Frame Surveys with Complex Designs, *Journal of the American Statistical Association*, **91**, 349–356.

293. Smith, S.K., and Lewis, B.B. (1980), Some New Techniques for Applying the Housing Unit Method of Local Population Estimations, *Demography*, **17**, 323–340.

294. Smith, T.M.F. (1983), On the Validity of Inferences from Non-Random Samples, *Journal of the Royal Statistical Society, Series A*, **146**, 394–403.

295. Spiegelhalter, D.J., Best, N.G., Gilks, W.R., and Inskip, H (1996), Hepatitis B: A Case Study in MCMC Methods, in W.R. Gilks, S. Richardson, and D. J. Spiegelhalter (Eds.), *Markov Chain Monte Carlo in Practice*, London: Chapman and Hall, pp. 21–43.

296. Spiegelhalter, D.J., Thomas, A., Best, N.G., and Gilks, W.R. (1997), BUGS: Bayesian Inference Using Gibbs Sampling, Version 6.0, Cambridge: Medical Research Council Biostatistics Unit.

297. Spjøtvoll, E., and Thomsen, I. (1987), Application of Some Empirical Bayes Methods to Small Area Statistics, *Bulletin of the International Statistical Institute (Vol. 2)*, pp. 435–449.

298. S-PLUS 6 for Windows Guide to Statistics (2001), *Volume 1*, Seattle: Insightful Corporation.

299. Srivastava, M.S., and Bilodeau, M. (1989), Stein Estimation Under Elliptical Distributions, *Journal of Multivariate Analysis*, **28**, 247–259.

300. Starsinic, D.E. (1974), *Development of Population Estimates for Revenue Sharing Areas*, Census Tract Papers, Ser GE40, No. 10, Washington, DC: U.S. Government Printing Office.

301. Stasny, E., Goel, P.K., and Rumsey, D.J. (1991), County Estimates of Wheat Production, *Survey Methodology*, **17**, 211–225.

302. Statistics Canada (1987), *Population Estimation Methods in Canada*, Catalogue 91–528E, Statistics Canada, Ottawa.

303. Statistics Canada (1993), *1991 Census Technical Report: Coverage*, 1991 Census of Canada, Catalogue No. 92–341E, Ottawa: Supply and Services Canada, 1994.

304. Stein, C. (1981) Estimation of the Mean of a Multivariate Normal Distribution, *Annals of Statistics*, **9**, 1135–1151.

305. Stukel, D.M. (1991), *Small Area Estimation Under One and Two-fold Nested Error Regression Models*, Unpublished Ph.D. Thesis, Carleton University, Ottawa, Canada.

306. Stukel, D.M., and Rao, J.N.K. (1997), Estimation of Regression Models with Nested Error Structure and Unequal Error Variances Under Two and Three Stage Cluster Sampling, *Statistics & Probability Letters*, **35**, 401–407.

307. Stukel, D.M., and Rao, J.N.K. (1999), Small-Area Estimation Under Two-fold Nested Errors Regression Models, *Journal of Statistical Planning and Inference*, **78**, 131–147.

308. Swamy, P.A.V.B. (1971), *Statistical Inference in Random Coefficient Regression Models*, Berlin: Springer-Verlag.

309. Thibault, C., Julien, C., and Dick, P. (1995). The 1996 Census Coverage Error Measurement Program, *Proceedings of the 1995 Annual Research Conference*, U.S. Bureau of the Census, Washington, DC, pp. 17–36.

310. Thompson, M.E. (1997), *Theory of Sample Surveys*, London: Chapman and Hall.

311. Tiller, R.B. (1982), Time Series Modeling of Sample Survey Data from the U.S. Current Population Survey, *Journal of Official Statistics*, **8**, 149–166.

312. Tsutakawa, R.K., Shoop, G.L., and Marienfield, C.J. (1985), Empirical Bayes Estimation of Cancer Mortality Rates, *Statistics in Medicine*, **4**, 201–212.

313. U.S. Bureau of the Census (1966), *Methods of Population Estimation: Part I, Illustrative Procedure of the Bureau's Component Method II*, Current Population Reports, Series P–25, No. 339, Washington, DC: U.S. Government Printing Office.

314. Valliant, R., Dorfman, A.H., and Royall, R.M. (2001), *Finite Population Sampling and Inference: A Prediction Approach*, New York: Wiley.

315. Wang, J. (2000), *Topics in Small Area Estimation with Applications to the National Resources Inventory*, Unpublished Ph.D. Thesis, Iowa State University, Ames.

316. Wolter, K. (1975), *Introduction to Variance Estimation*, New York: Springer-Verlag.

317. Wolter, K. (1985), *Introduction to Variance Estimation*, New York: Springer-Verlag.

318. Wolter, K. (1986), Some Coverage Error Models for Census Data, *Journal of the American Statistical Association*, **81**, 338–346.

319. Woodruff, R.S. (1966), Use of a Regression Technique to Produce Area Breakdowns of the Monthly National Estimates of Retail Trade, *Journal of the American Statistical Association*, **61**, 496–504.

320. Wu, C.F.J. (1986), Jackknife, Bootstrap and Other Resampling Methods in Regression Analysis (with discussion), *Annals of Statistics*, **14**, 1261–1350.

321. You, Y. (1999), *Hierarchical Bayes and Related Methods for Model-Based Small Area Estimation*, Unpublished Ph.D. Thesis, Carleton University, Ottawa, Canada.

322. You, Y., Rao, J.N.K., and Gambino, J. (2001), Model-Based Unemployment Rate Estimation for the Canadian Labour Force Survey: A Hierarchical Bayes Approach, Technical Report, Household Survey Methods Division, Statistics Canada.

323. You, Y., and Rao, J.N.K. (2000), Hierarchical Bayes Estimation of Small Area Means Using Multi-Level Models, *Survey Methodology*, **26**, 173–181.

324. You, Y., and Rao, J.N.K. (2002a), A Pseudo-Empirical Best Linear Unbiased Prediction Approach to Small Area Estimation Using Survey Weights, *Canadian Journal of Statistics*, **30**, 431–439.

325. You, Y., and Rao, J.N.K. (2002b), Small Area Estimation Using Unmatched Sampling and Linking Models, *Canadian Journal of Statistics*, **30**, 3–15.

326. You, Y., and Rao, J.N.K. (2003), Pseudo Hierarchical Bayes Small Area Estimation Combining Unit Level Models and Survey Weights, *Journal of Statistical Planning and Inference*, **111**, 197–208.

327. You, Y, and Reiss, P. (1999), Hierarchical Bayes Estimation of Response Rates for an Expenditure Survey, *Proceedings of the Survey Methods Section*, Statistical Society of Canada, pp. 123–128.

328. Zaslavsky, A.M. (1993), Combining Census, Dual-system, and Evaluation Study Data to Estimate Population Shares, *Journal of the American Statistical Association*, **88**, 1092–1105.

329. Zeger, S.L., and Karim, M.R. (1991), Generalized Linear Models with Random Effects: A Gibbs Sampling Approach, *Journal of the American Statistical Association*, **86**, 79–86.

330. Zhang, D., and Davidian, M. (2001), Linear Mixed Models with Flexible Distributions of Random Effects for Longitudinal Data, *Biometrics*, **57**, 795–802.

331. Zidek, J.V. (1982), *A Review of Methods for Estimating Population of Local Areas*, Technical Report 82–4, University of British Columbia, Vancouver, Canada.

Author Index

Subject Index

WILEY SERIES IN SURVEY METHODOLOGY

Established in Part by WALTER A. SHEWHART AND SAMUEL S. WILKS

Editors: *Robert M. Groves, Graham Kalton, J. N. K. Rao, Norbert Schwarz, Christopher Skinner*

The ***Wiley Series in Survey Methodology*** covers topics of current research and practical interests in survey methodology and sampling. While the emphasis is on application, theoretical discussion is encouraged when it supports a broader understanding of the subject matter.

The authors are leading academics and researchers in survey methodology and sampling. The readership includes professionals in, and students of, the fields of applied statistics, biostatistics, public policy, and government and corporate enterprises.

BIEMER, GROVES, LYBERG, MATHIOWETZ, and SUDMAN · Measurement Errors in Surveys
COCHRAN · Sampling Techniques, *Third Edition*
COUPER, BAKER, BETHLEHEM, CLARK, MARTIN, NICHOLLS, and O'REILLY (editors) · Computer Assisted Survey Information Collection
COX, BINDER, CHINNAPPA, CHRISTIANSON, COLLEDGE, and KOTT (editors) · Business Survey Methods
*DEMING · Sample Design in Business Research
DILLMAN · Mail and Internet Surveys: The Tailored Design Method
GROVES and COUPER · Nonresponse in Household Interview Surveys
GROVES · Survey Errors and Survey Costs
GROVES, DILLMAN, ELTINGE, and LITTLE · Survey Nonresponse
GROVES, BIEMER, LYBERG, MASSEY, NICHOLLS, and WAKSBERG · Telephone Survey Methodology
*HANSEN, HURWITZ, and MADOW · Sample Survey Methods and Theory, Volume I: Methods and Applications
*HANSEN, HURWITZ, and MADOW · Sample Survey Methods and Theory, Volume II: Theory
HARKNESS, VAN DE VIJVER, and MOHLER · Cross-Cultural Survey Methods
KISH · Statistical Design for Research
*KISH · Survey Sampling
KORN and GRAUBARD · Analysis of Health Surveys
LESSLER and KALSBEEK · Nonsampling Error in Surveys
LEVY and LEMESHOW · Sampling of Populations: Methods and Applications, *Third Edition*
LYBERG, BIEMER, COLLINS, de LEEUW, DIPPO, SCHWARZ, TREWIN (editors) · Survey Measurement and Process Quality
MAYNARD, HOUTKOOP-STEENSTRA, SCHAEFFER, VAN DER ZOUWEN · Standardization and Tacit Knowledge: Interaction and Practice in the Survey Interview
RAO · Small Area Estimation
SIRKEN, HERRMANN, SCHECHTER, SCHWARZ, TANUR, and TOURANGEAU (editors) · Cognition and Survey Research
VALLIANT, DORFMAN, and ROYALL · Finite Population Sampling and Inference: A Prediction Approach

*Now available in a lower priced paperback edition in the Wiley Classics Library.